THE SCIENCE OF ELECTRONICS

Analog Devices

THOMAS L. FLOYD

DAVID M. BUCHLA

Upper Saddle River, New Jersey
Columbus, Ohio

Library of Congress Cataloging-in-Publication Data

Floyd, Thomas L.
The science of electronics. Analog devices / Thomas L. Floyd, David M. Buchla.
p. cm.
ISBN 0–13–087540–6 (alk. paper)
1. Electronics–Textbooks. 2. Analog electronic systems–Textbooks. I. Buchla, David M. II. Title.

TK7816.F575 2005
621.381–dc22 2003069031

Editor in Chief: Stephen Helba
Acquisitions Editor: Dennis Williams
Development Editor: Kate Linsner
Production Editor: Rex Davidson
Design Coordinator: Diane Ernsberger
Cover Designer: Linda Sorrells-Smith
Cover art: Getty Images
Production Manager: Pat Tonneman
Marketing Manager: Ben Leonard

Pearson Education Ltd.
Pearson Education Singapore Pte. Ltd.
Pearson Education Canada, Ltd.
Pearson Education—Japan
Pearson Education Australia Pty. Limited
Pearson Education North Asia Ltd.
Pearson Educación de Mexico, S.A. de C.V.
Pearson Education Malaysia Pte. Ltd.

ISBN: 0-13-087540-6

PREFACE

Introduction to *The Science of Electronics* Series

The Science of Electronics: Analog Devices is one of a series that also includes *The Science of Electronics: DC/AC* and *The Science of Electronics: Digital.* This series presents basic electronics theory in a clear and simple, yet complete, format and shows the close relationship of electronics to other sciences. These texts are written at a level that makes them suitable as introductory texts for secondary schools as well as technical and community college programs. Pedagogical features are numerous, and they are designed to enhance the learning process and make the use of each text in the series an enjoyable experience. Inexpensive laboratory exercises for these texts are included in the accompanying lab manuals. The same author team prepared all of the texts and lab manuals in the series, providing consistency of approach and format.

The *DC/AC* text begins the series with some of the basic physics behind electronics, such as fundamental and derived units, work, energy, and energy conservation laws. Important ideas in measurement science, such as accuracy, precision, significant digits, and measurement units, are covered. In addition, the text covers passive dc and ac circuits, magnetic circuits, motors, and generators and instruments.

The *Digital* text introduces traditional topics, including number systems, Boolean algebra, combinational logic, and sequential logic, plus topics not normally found in an introductory text. The trend in industry is toward programmable devices, computers, and digital signal processing. A chapter is devoted to each of these important topics. Despite their complexity, these topics are treated with the same basic approach.

The *Analog Devices* text includes five chapters that cover diodes, transistors, and discrete amplifiers, followed by six chapters of operational amplifier coverage. Measurements are particularly important in all of the sciences, so the final chapter covers measurement and control circuits, including transducers and thyristors.

A "Science Highlight" feature opens every chapter in all textbooks. This highlight looks at scientific advances in an area related to the coverage in that chapter. Science Highlights include important related ideas in the fields of physics, chemistry, biology, computer science, and more. Electronics is a dynamic field of science, and we have tried to bring the excitement of some of the latest discoveries and advancements to the forefront for readers even as they begin their studies.

Key Features of *The Science of Electronics* Series

- A Science Highlight in each chapter looks at scientific advances in an area that is related to the coverage in that chapter.
- Easy to read and well illustrated format
- Full-color format
- "To the Student" provides an overview of the field of electronics, including careers, important safety rules, and workplace information, as well as a brief history of electronics.
- Many types of exercises reinforce knowledge and check progress, including worked-out examples, example questions, section review and chapter questions, chapter checkups, basic and basic-plus problems, and Multisim circuit simulations.
- Two-page chapter openers include a chapter outline, key objectives, list of key terms, a computer simulations directory to the appropriate figure in the chapter, and a lab experiments directory with titles of relevant exercises from the accompanying lab manual.

- Computer simulations throughout all of the books allow the student to see how specific circuits actually work.
- Safety Notes in the margins throughout all of the books continually remind students of the importance of safety.
- Historical Notes appear in margins throughout all of the books and are linked to ideas or persons mentioned in the text.
- An "On the Job" feature appears on selected chapter openers and discusses important aspects of employment.
- Key terms are indicated in red throughout the text, in addition to being defined in a key term glossary at the end of each chapter.
- A comprehensive glossary is included at the end of the book with all key terms and boldface terms from the chapters included.
- Important facts and formulas are summarized at the end of the chapters.

Introduction to *The Science of Electronics: Analog Devices*

This text includes coverage of the most important analog devices, including discrete diodes and transistors, operational amplifiers, transducers, and thyristors. The text starts with five chapters that introduce discrete devices and circuits followed by six chapters about operational amplifiers and other integrated circuits. The final chapter covers measurement and control circuits, including transducers and thyristors.

Another important concept that has been mentioned is the close relationship between science and electronics. All electronics instructors are very aware of this link, but many texts seem to ignore it. We have highlighted this tie throughout the text and augmented the coverage with science highlights that illustrate how electronics is rooted in science.

This text includes a range of topics to accommodate a variety of program requirements and is designed to permit freedom in choosing many of the topics. For example, some instructors may choose to omit some or all of the topics in Chapter 5, which deals with multistage, rf, and power amplifiers. This material can be omitted with no loss in continuity.

Accompanying Student Resources

- *The Science of Electronics: Analog Devices Lab Manual* by David M. Buchla
- *Companion Website, www.prenhall.com/SOE.* This website, created for *The Science of Electronics* series, includes:

  Computer simulation circuits designed to accompany selected examples in the text.

  Chapter-by-chapter quizzes in true/false, fill-in-the-blank, and multiple choice formats that enable a student to check his or her understanding of the material presented in the text.
- *Prentice Hall Electronics Supersite, www.prenhall.com/electronics/.* This website offers math study aids, links to career opportunities in the industry, and other useful information.

Instructor Resources

- *PowerPoint® Slides* on CD-ROM.
- *Companion Website, www.prenhall.com/SOE.* For the instructor, this website offers the ability to post your syllabus online with Syllabus Manager™. This is a great solution for classes taught online, self-paced, or in any computer-assisted manner.
- *Online Course Support.* If your program is offered in a distance-learning format, please contact your local Prentice Hall sales representative for a list of product solutions.
- *Instructor's Edition.* Includes answers to all chapter questions and worked-out solutions to all problems.
- *Lab Manual Solutions.* A separate solutions manual for all experiments in the series is available.
- *Test Item File.* A test bank of multiple choice, true/false, and fill-in-the-blank questions.
- *Prentice Hall TestGen.* This is an electronic version of the Test Item File, allowing an instructor to customize tests.
- *Prentice Hall Electronics Supersite, www.prenhall.com/electronics.* Instructors can access various resources on this site, including password-protected files for the instructor supplements accompanying this text. Contact your local Prentice Hall sales representative for your user name and password.

Illustration of Chapter Features

Chapter Opener

Each chapter begins with a two-page spread, as shown in Figure P–1. The left page includes a list of the sections in the chapter and an introduction to the chapter. The right page has a list of key objectives for each section, a computer simulations directory, a laboratory experiments directory, and a list of key terms to look for in the chapter. Selected chapters contain a special feature called "On the Job."

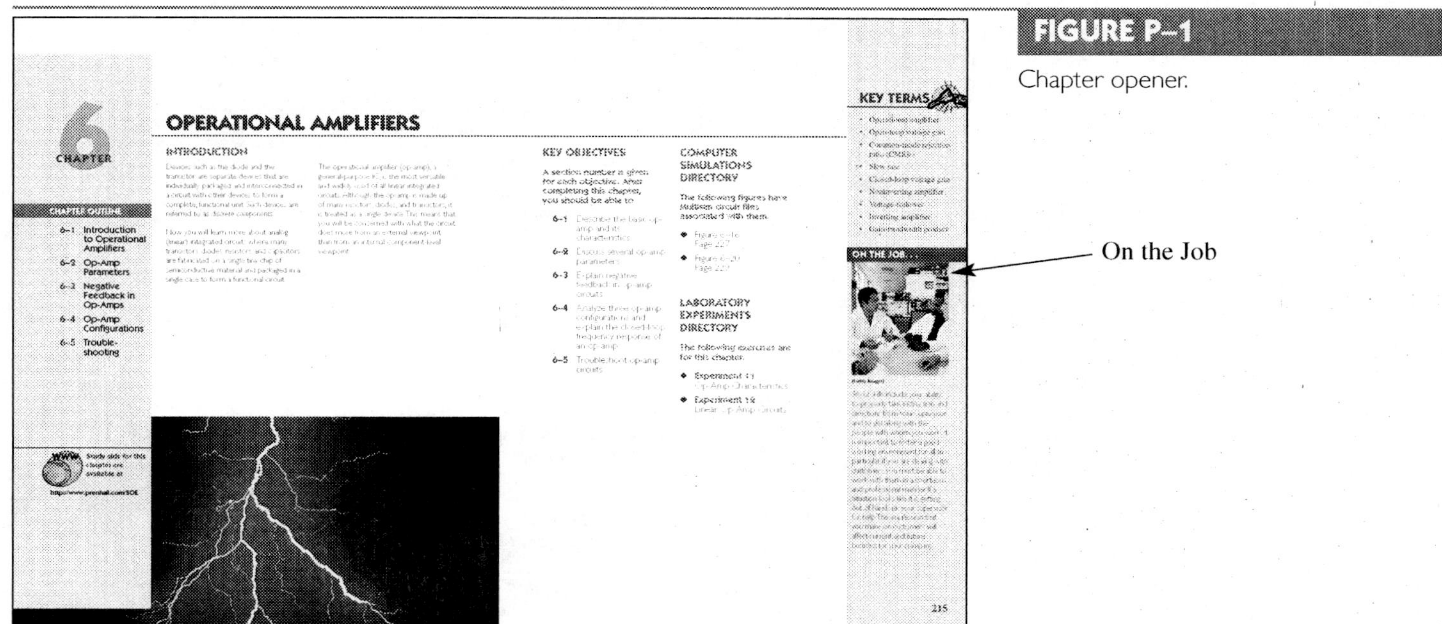

FIGURE P–1

Chapter opener.

Science Highlight

Immediately following the chapter opener is a *Sci Hi* feature that presents advanced concepts and science-related topics that tie in with the coverage of the text. A typical *Sci Hi* is shown in Figure P–2.

FIGURE P–2

Science highlight.

CHAPTER 2

Advances in one branch of science often have an impact on other branches of science. The Periodic Table was developed as a means of classifying the elements in the science of chemistry, but it is applied to many other sciences. Chemistry students are all made aware of the importance of the Periodic Table early in their study of chemistry. Virtually all science labs have the table displayed predominately.

The Periodic Table classifies elements by groups or families. Except for the rare earth elements, each family is in a separate column of the table. For example, the first column is alkali metals. These are all highly reactive silvery metals. The last column is the noble gases, a group of elements that are all unreactive colorless gases.

The family of elements that has carbon on top includes two elements that are widely used as semiconductors — silicon and germanium. These are elements with four electrons in the outer shell, as you will learn in this chapter. Interestingly, carbon, in the form of diamond can also be a semiconductor (but is not used commercially).

Certain materials are used to add impurities to silicon and germanium. The impurities change the conductive properties of pure silicon. If an element from the family to the left of carbon is added to silicon, the outer shell of the impurity will contain one less electron than silicon, and a *p*-type semiconductor will be formed. If an element from the family to the right of carbon is added to silicon, the outer shell of the impurity will contain one more electron than silicon, and an *n*-type semiconductor will be formed. The formation of *p*- and *n*-type materials is part of the story of forming diodes, transistors, and other semiconductor materials.

2–1 THE ATOMIC STRUCTURE OF SEMICONDUCTORS

Electronic devices such as diodes and transistors are constructed from special materials called semiconductors.

In this section, you will learn about the atomic structure of semiconductors.

Electron Shells and Orbits

The electrical properties of materials are explained by their atomic structure. In the early part of the 20th century, Neils Bohr, a Danish physicist, developed a model of the atom that showed electrons orbiting the nucleus. In Bohr's model of the atom, the electrons orbit only in certain discrete (separate and distinct) distances from the nucleus. The nucleus contains positively charged protons and uncharged neutrons. The orbiting electrons are negatively charged. Modem quantum mechanical models of the atom retain much of the ideas of the original Bohr model but have replaced the concept of electron "particles" with mathematical "matter waves"; however, the Bohr model provides a useful mental picture of the structure of an atom.

Energy is the ability to do work and is subdivided into potential (position), kinetic (motion), and rest (mass). Electrons have potential energy because of their position within the atom. Electrons near the nucleus have less energy than those in more distant orbits. These discrete orbits mean that only certain energy levels are permitted within the atom. These energy levels are known as **shells**. Each shell has a certain maximum permissible number of electrons. The differences in energy levels within a shell are much smaller than the difference in energy between shells. The shells are designated 1, 2, 3, 4, and so on, with 1 being closest to the nucleus, as illustrated in Figure 2–1.

30

Section Opener

Each section in a chapter begins with a brief introduction that includes a general overview. An illustration is shown in Figure P–3. This particular page also shows an example of a Safety Note. Safety Notes are appropriately placed throughout the text.

Review Questions

Each section ends with a review consisting of five questions that emphasize the main concepts presented in the section. This feature is also shown in Figure P–3. Answers to the Section Reviews are at the end of the chapter.

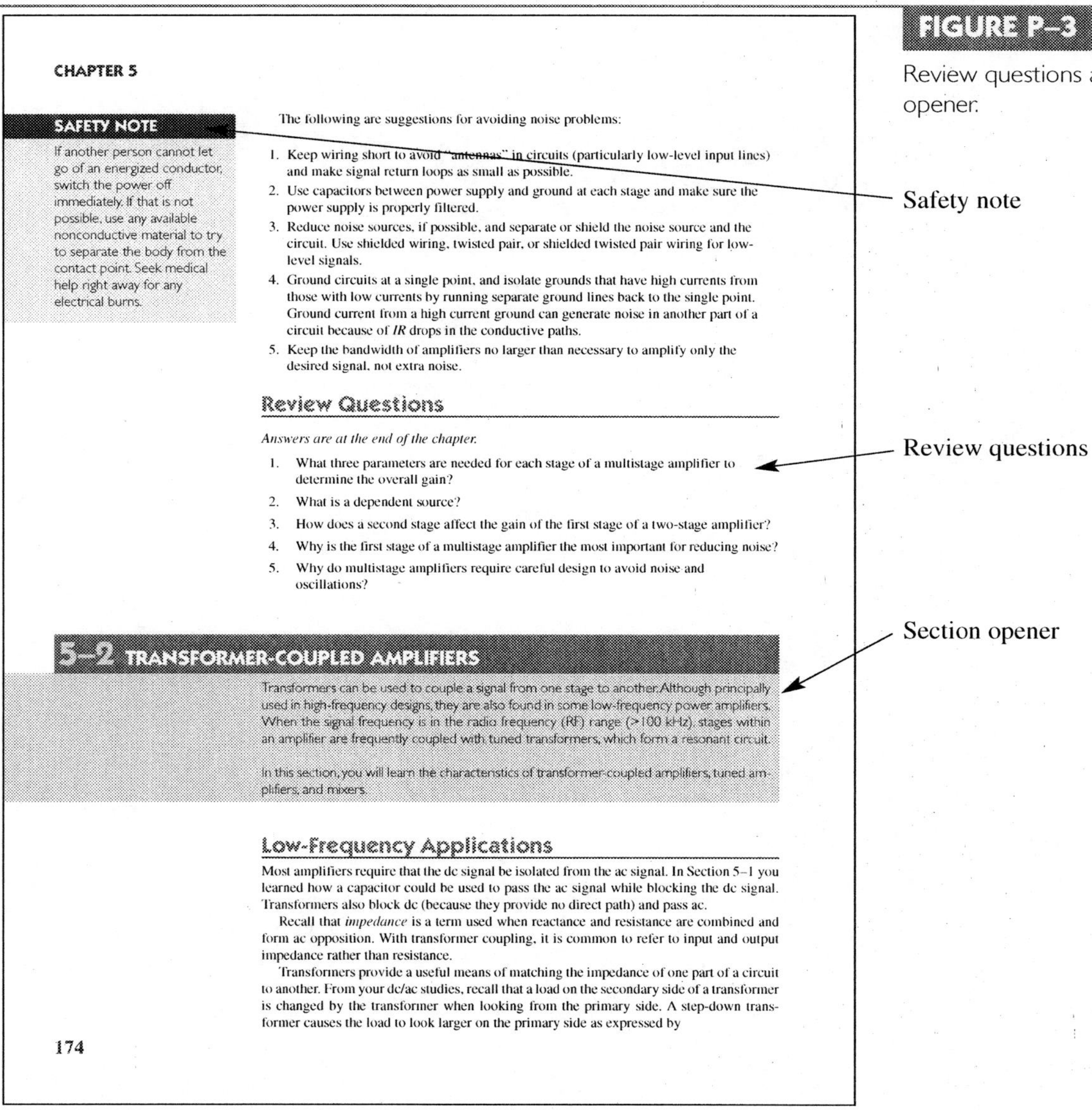

CHAPTER 5

SAFETY NOTE

If another person cannot let go of an energized conductor, switch the power off immediately. If that is not possible, use any available nonconductive material to try to separate the body from the contact point. Seek medical help right away for any electrical burns.

The following are suggestions for avoiding noise problems:

1. Keep wiring short to avoid "antennas" in circuits (particularly low-level input lines) and make signal return loops as small as possible.
2. Use capacitors between power supply and ground at each stage and make sure the power supply is properly filtered.
3. Reduce noise sources, if possible, and separate or shield the noise source and the circuit. Use shielded wiring, twisted pair, or shielded twisted pair wiring for low-level signals.
4. Ground circuits at a single point, and isolate grounds that have high currents from those with low currents by running separate ground lines back to the single point. Ground current from a high current ground can generate noise in another part of a circuit because of *IR* drops in the conductive paths.
5. Keep the bandwidth of amplifiers no larger than necessary to amplify only the desired signal, not extra noise.

Review Questions

Answers are at the end of the chapter.

1. What three parameters are needed for each stage of a multistage amplifier to determine the overall gain?
2. What is a dependent source?
3. How does a second stage affect the gain of the first stage of a two-stage amplifier?
4. Why is the first stage of a multistage amplifier the most important for reducing noise?
5. Why do multistage amplifiers require careful design to avoid noise and oscillations?

5–2 TRANSFORMER-COUPLED AMPLIFIERS

Transformers can be used to couple a signal from one stage to another. Although principally used in high-frequency designs, they are also found in some low-frequency power amplifiers. When the signal frequency is in the radio frequency (RF) range (>100 kHz), stages within an amplifier are frequently coupled with tuned transformers, which form a resonant circuit.

In this section, you will learn the characteristics of transformer-coupled amplifiers, tuned amplifiers, and mixers.

Low-Frequency Applications

Most amplifiers require that the dc signal be isolated from the ac signal. In Section 5–1 you learned how a capacitor could be used to pass the ac signal while blocking the dc signal. Transformers also block dc (because they provide no direct path) and pass ac.

Recall that *impedance* is a term used when reactance and resistance are combined and form ac opposition. With transformer coupling, it is common to refer to input and output impedance rather than resistance.

Transformers provide a useful means of matching the impedance of one part of a circuit to another. From your dc/ac studies, recall that a load on the secondary side of a transformer is changed by the transformer when looking from the primary side. A step-down transformer causes the load to look larger on the primary side as expressed by

174

Safety note

Review questions

Section opener

FIGURE P–3

Review questions and section opener.

FIGURE P–4 Worked example and related question.

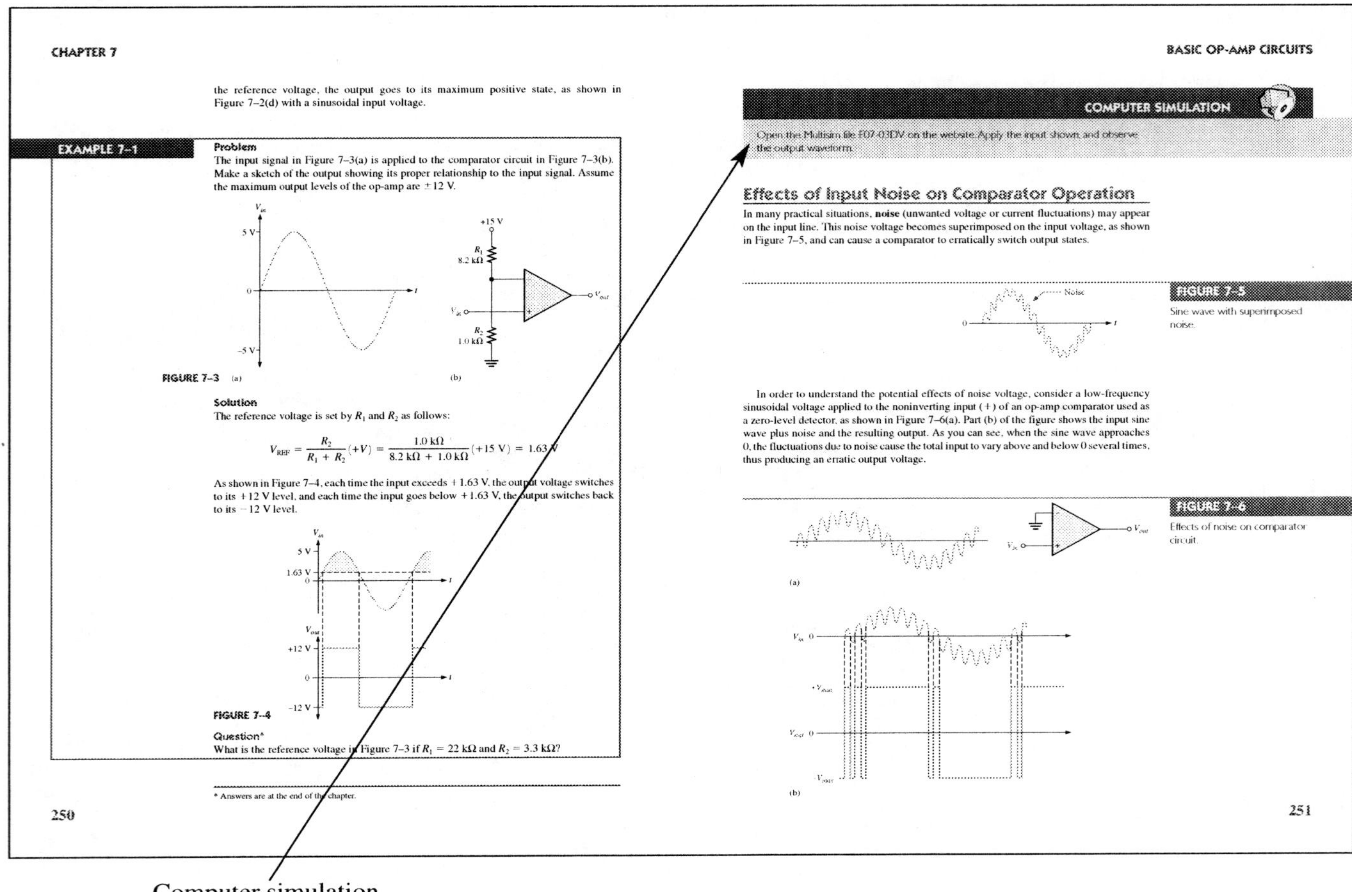

CHAPTER 7

the reference voltage, the output goes to its maximum positive state, as shown in Figure 7–2(d) with a sinusoidal input voltage.

EXAMPLE 7–1

Problem
The input signal in Figure 7–3(a) is applied to the comparator circuit in Figure 7–3(b). Make a sketch of the output showing its proper relationship to the input signal. Assume the maximum output levels of the op-amp are ±12 V.

FIGURE 7–3 (a) (b)

Solution
The reference voltage is set by R_1 and R_2 as follows:

$$V_{REF} = \frac{R_2}{R_1 + R_2}(+V) = \frac{1.0\ k\Omega}{8.2\ k\Omega + 1.0\ k\Omega}(+15\ V) = 1.63\ V$$

As shown in Figure 7–4, each time the input exceeds +1.63 V, the output voltage switches to its +12 V level, and each time the input goes below +1.63 V, the output switches back to its −12 V level.

FIGURE 7–4

Question*
What is the reference voltage in Figure 7–3 if $R_1 = 22\ k\Omega$ and $R_2 = 3.3\ k\Omega$?

* Answers are at the end of the chapter.

250

BASIC OP-AMP CIRCUITS

COMPUTER SIMULATION

Open the Multisim file F07-03DV on the website. Apply the input shown, and observe the output waveform.

Effects of Input Noise on Comparator Operation

In many practical situations, **noise** (unwanted voltage or current fluctuations) may appear on the input line. This noise voltage becomes superimposed on the input voltage, as shown in Figure 7–5, and can cause a comparator to erratically switch output states.

FIGURE 7–5 Sine wave with superimposed noise.

In order to understand the potential effects of noise voltage, consider a low-frequency sinusoidal voltage applied to the noninverting input (+) of an op-amp comparator used as a zero-level detector, as shown in Figure 7–6(a). Part (b) of the figure shows the input sine wave plus noise and the resulting output. As you can see, when the sine wave approaches 0, the fluctuations due to noise cause the total input to vary above and below 0 several times, thus producing an erratic output voltage.

FIGURE 7–6 Effects of noise on comparator circuit.

251

Worked Examples and Questions

There is an abundance of worked-out examples that help to illustrate and clarify basic concepts or specific procedures. Each example ends with a question that is related to the example. A typical example is shown in Figure P–4.

Computer Simulation

Numerous circuits are provided online in Multisim. Filenames are keyed to figures within the text with the format Fxx-yyDV. The xx-yy is the figure number and DV represents a file for this text. These simulations can be used to verify the operation of selected circuits that are studied in the text. An example of a computer simulation feature is shown in Figure P–4. The Multisim circuits can be accessed on the website by going to *www.prenhall.com/SOE* and selecting this text. Choose the chapter you wish to study by clicking on that chapter, then click on the module entitled “Multisim.” There you will see an introductory page with a link to the circuits for the chapter.

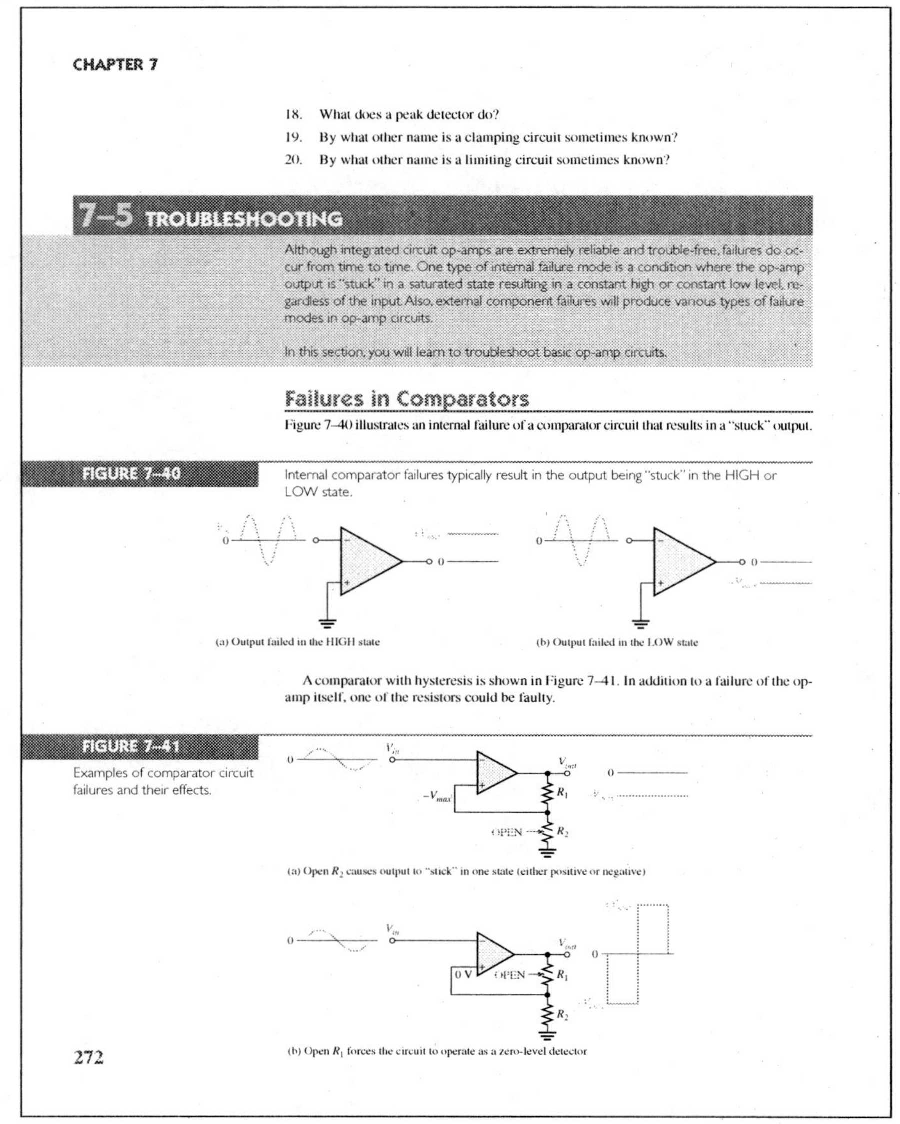
CHAPTER 7

18. What does a peak detector do?
19. By what other name is a clamping circuit sometimes known?
20. By what other name is a limiting circuit sometimes known?

7–5 TROUBLESHOOTING

Although integrated circuit op-amps are extremely reliable and trouble-free, failures do occur from time to time. One type of internal failure mode is a condition where the op-amp output is "stuck" in a saturated state resulting in a constant high or constant low level, regardless of the input. Also, external component failures will produce various types of failure modes in op-amp circuits.

In this section, you will learn to troubleshoot basic op-amp circuits.

Failures in Comparators

Figure 7–40 illustrates an internal failure of a comparator circuit that results in a "stuck" output.

FIGURE 7–40 Internal comparator failures typically result in the output being "stuck" in the HIGH or LOW state.

(a) Output failed in the HIGH state

(b) Output failed in the LOW state

A comparator with hysteresis is shown in Figure 7–41. In addition to a failure of the op-amp itself, one of the resistors could be faulty.

FIGURE 7–41 Examples of comparator circuit failures and their effects.

(a) Open R_2 causes output to "stick" in one state (either positive or negative)

(b) Open R_1 forces the circuit to operate as a zero-level detector

272

FIGURE P–5

Troubleshooting.

Troubleshooting

Many chapters include troubleshooting techniques and the use of test instruments as they relate to the topics covered. Figure P–5 shows typical troubleshooting coverage.

Chapter Review

Each chapter ends in a special color section that is intended to highlight important chapter ideas. Several features are illustrated in Figure P–6. The chapter review includes:

- *Key Terms Glossary.* Terms that are in red within the chapter are defined here and in the glossary at the end of the book.
- *Important facts.* Major points from the chapter are summarized.
- *Formulas.* Numbered formulas in the chapter are summarized.
- *Chapter Checkup.* This is a set of multiple-choice questions with answers at the end of the chapter.
- *Questions.* This is a set of questions pertaining to the chapter. Answers to odd-numbered questions appear at the end of the book.

FIGURE P–6 Chapter review.

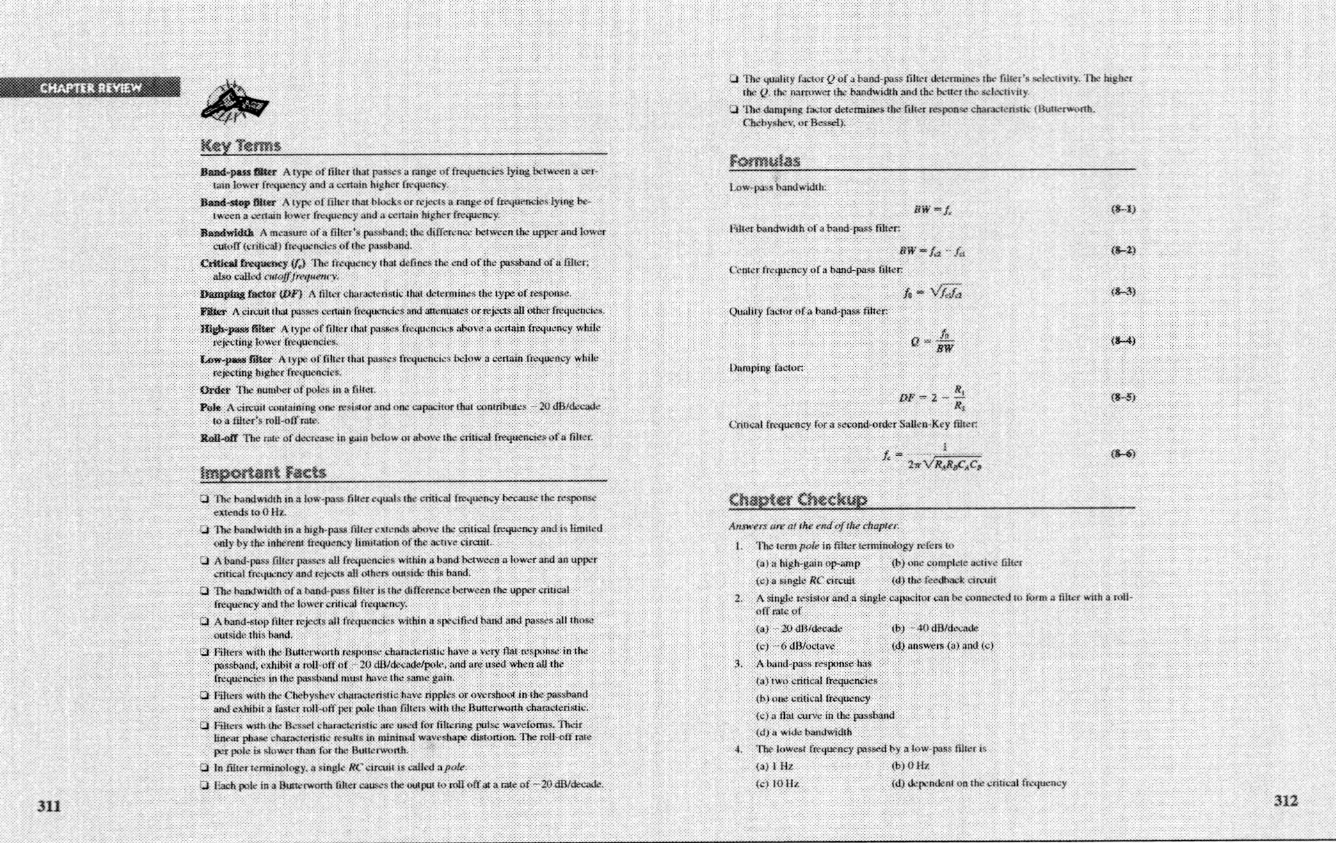

CHAPTER REVIEW

Key Terms

Band-pass filter A type of filter that passes a range of frequencies lying between a certain lower frequency and a certain higher frequency.

Band-stop filter A type of filter that blocks or rejects a range of frequencies lying between a certain lower frequency and a certain higher frequency.

Bandwidth A measure of a filter's passband; the difference between the upper and lower cutoff (critical) frequencies of the passband.

Critical frequency (f_c) The frequency that defines the end of the passband of a filter; also called *cutoff frequency*.

Damping factor (*DF*) A filter characteristic that determines the type of response.

Filter A circuit that passes certain frequencies and attenuates or rejects all other frequencies.

High-pass filter A type of filter that passes frequencies above a certain frequency while rejecting lower frequencies.

Low-pass filter A type of filter that passes frequencies below a certain frequency while rejecting higher frequencies.

Order The number of poles in a filter.

Pole A circuit containing one resistor and one capacitor that contributes −20 dB/decade to a filter's roll-off rate.

Roll-off The rate of decrease in gain below or above the critical frequencies of a filter.

Important Facts

- The bandwidth in a low-pass filter equals the critical frequency because the response extends to 0 Hz.
- The bandwidth in a high-pass filter extends above the critical frequency and is limited only by the inherent frequency limitation of the active circuit.
- A band-pass filter passes all frequencies within a band between a lower and an upper critical frequency and rejects all others outside this band.
- The bandwidth of a band-pass filter is the difference between the upper critical frequency and the lower critical frequency.
- A band-stop filter rejects all frequencies within a specified band and passes all those outside this band.
- Filters with the Butterworth response characteristic have a very flat response in the passband, exhibit a roll-off of −20 dB/decade/pole, and are used when all the frequencies in the passband must have the same gain.
- Filters with the Chebyshev characteristic have ripples or overshoot in the passband and exhibit a faster roll-off per pole than filters with the Butterworth characteristic.
- Filters with the Bessel characteristic are used for filtering pulse waveforms. Their linear phase characteristic results in minimal waveshape distortion. The roll-off rate per pole is slower than for the Butterworth.
- In filter terminology, a single *RC* circuit is called a *pole*.
- Each pole in a Butterworth filter causes the output to roll off at a rate of −20 dB/decade.

311

- The quality factor Q of a band-pass filter determines the filter's selectivity. The higher the Q, the narrower the bandwidth and the better the selectivity.
- The damping factor determines the filter response characteristic (Butterworth, Chebyshev, or Bessel).

Formulas

Low-pass bandwidth:

$$BW = f_c \qquad (8\text{–}1)$$

Filter bandwidth of a band-pass filter:

$$BW = f_{c2} - f_{c1} \qquad (8\text{–}2)$$

Center frequency of a band-pass filter:

$$f_0 = \sqrt{f_{c1} f_{c2}} \qquad (8\text{–}3)$$

Quality factor of a band-pass filter:

$$Q = \frac{f_0}{BW} \qquad (8\text{–}4)$$

Damping factor:

$$DF = 2 - \frac{R_1}{R_2} \qquad (8\text{–}5)$$

Critical frequency for a second-order Sallen-Key filter:

$$f_c = \frac{1}{2\pi\sqrt{R_A R_B C_A C_B}} \qquad (8\text{–}6)$$

Chapter Checkup

Answers are at the end of the chapter.

1. The term *pole* in filter terminology refers to
   (a) a high-gain op-amp (b) one complete active filter
   (c) a single *RC* circuit (d) the feedback circuit
2. A single resistor and a single capacitor can be connected to form a filter with a roll-off rate of
   (a) −20 dB/decade (b) −40 dB/decade
   (c) −6 dB/octave (d) answers (a) and (c)
3. A band-pass response has
   (a) two critical frequencies
   (b) one critical frequency
   (c) a flat curve in the passband
   (d) a wide bandwidth
4. The lowest frequency passed by a low-pass filter is
   (a) 1 Hz (b) 0 Hz
   (c) 10 Hz (d) dependent on the critical frequency

312

Problems

Pedagogical features continue with two levels of problems: Basic and Basic-Plus. In general, the Basic-Plus problems are more difficult than the Basic problems. Answers to all odd-numbered problems are at the end of the book.

Answers

Each chapter concludes with selected answers for questions within that chapter. These include:

- Answers to Example Questions
- Answers to Review Questions
- Answers to Chapter Checkups

End-of-Book Features

- An appendix: Logarithms and Decibels
- Answers to odd-numbered questions
- Answers to odd-numbered problems
- Comprehensive glossary
- Index

CONTENTS

TO THE STUDENT

Introduction to *The Science of Electronics: Analog Devices*

We believe that you will find *The Science of Electronics: Analog Devices* an effective tool in your preparation for a career and should find this text useful in further studies. When you have finished this course, this book should become a valuable reference for more advanced courses or even after you have entered into the job market. We hope it provides a foundation for your continued studies in electronics.

The most complicated system in electronics can be broken down into a collection of simpler circuits. These include passive circuits (resistors, capacitors, inductors) and active circuits (integrated circuits including digital and analog devices). With a solid foundation in these topics, understanding large systems is simplified. Electronics is not an easy subject, but we have endeavored to provide help along the way to make it interesting and informative and to provide you with the preparation you need for a career in this exciting field.

Many examples in the text are worked out in detail. You should follow the steps in the examples and check your understanding with the related question. Check your understanding of each section by answering the review questions and checking your answers. At the end of each chapter are summaries, glossary terms, formulas, questions, and problems as well as many answers. If you can answer all of the questions and work the problems at the end of a chapter, you are well under way toward mastering the material presented.

Careers in Electronics

The field of electronics is diverse, and career opportunities are available in many related areas. Because electronics is currently found in so many different applications and because new technology is being developed at a fast rate, the future appears limitless. There is hardly an area of our lives that is not enhanced to some degree by electronics technology. Those who acquire a sound, basic knowledge of electrical and electronic principles and are willing to continue learning will always be in demand.

The importance of obtaining a thorough understanding of the basic principles contained in this text cannot be overemphasized. Most employers prefer to hire people who have both a thorough grounding in the basics and the ability and eagerness to grasp new concepts and techniques. If you have a good training in the basics, an employer will train you in the specifics of the job to which you are assigned.

(Fluke Corporation. Reproduced with permission.)

There are many types of job classifications for which a person with training in electronics technology may qualify. Common job functions are described in the Bureau of Labor Statistics (BLS) occupational outlook handbook, which can be found on the Internet at *http://www.bls.gov/oco*. Two engineering technician's job descriptions from the BLS are as follows:

- *Electrical and electronics engineering technicians* help design, develop, test, and manufacture electrical and electronic equipment such as communication equipment, radar, industrial and medical measuring or control devices, navigational equipment, and computers. They may work in product evaluation and testing, using measuring and diagnostic devices to adjust, test, and repair equipment.
- *Broadcast and sound engineering technicians* install, test, repair, set up, and operate the electronic equipment used to record and transmit radio and television programs, cable programs, and motion pictures.

Many other technical jobs are available in the electronics field for the properly trained person:

- *Service technicians* are involved in the repair or adjustment of both commercial and consumer electronic equipment that is returned to the dealer or manufacturer for service.
- *Industrial manufacturing technicians* are involved in the testing of electronic products at the assembly-line level or in the maintenance and troubleshooting of electronic and electromechanical systems used in the testing and manufacturing of products.
- *Laboratory technicians* are involved in testing new or modified electronic systems in research and development laboratories.
- *Field-service technicians* repair electronic equipment at the customer's site; these systems include computers, radars, automatic banking equipment, and security systems.
- *User-support technicians* are the first people called when a computer or "high-tech" electronic equipment acts up. User-support technicians must know their product inside and out and be able to troubleshoot a product over the phone. An ability to communicate well is vital.

Related jobs in electronics include technical writers, technical sales people, x-ray technicians, auto mechanics, cable installers, and many others.

Getting a Job

Once you have successfully completed a course of study in the field of electronics, the next step is to find employment. You must consider several things in the process of finding and obtaining a job.

Resources and Considerations for Locating a Job

One consideration is the job location. You must determine if you are willing to move to another town or state or if you prefer to remain near home. Depending on the economy and the current job market, you may not have much choice in location. The important thing is to get a job and gain experience. This will allow you to move up to better or more desirable jobs later. You should also try to find out if a job is one for which you are reasonably suited in terms of personality, skills, and interest.

A good resource for locating a potential employer is the classified ads in your local newspaper or in newspapers from other cities. The Internet is also an important resource for finding employment. Many large employers have a "job-line", which is a phone service dedicated to describing current openings. Another way to find a job is through an employment agency that specializes in technical jobs, but fees may be substantial with private agencies. Often, especially at the college level, employers will come to the campus to interview prospective employees. If you know someone in a technical job, contact that person to find out about job openings at his or her company.

The Resume and Application

A resume is a record of your skills, education, and job experience. Many employers request a resume before you actually apply for a job. This allows the employer to sort through many prospective applicants and narrow the field to a few of the most qualified. For this reason, your resume is very important. Don't wait until you are about to look for a job to start working on it.

The resume is the initial way in which you present yourself to a potential employer, so it is important that you create a well-organized document. There are different types of resumes, but all resumes should have certain specific information. Here are some basic guidelines:

- Your resume should be one page long unless you have significant experience or education. A shorter resume will be more likely to be read, so shorter is generally better.

- Your identification information (name, address, phone number, e-mail address) should come first.
- List your educational achievements such as diplomas, degrees, special certifications, and awards. Include the year for each item.
- List the specific courses that you have completed that relate directly to the type of job for which you are applying. Generally, you should list all the math, science, and electronics courses when applying for a job in electronics technology.
- List all of your prior job experience, especially if it is related to the job you are seeking. Most people prefer to organize it in reverse chronological order (most recent first). Show the employer, dates of employment, and a short description of your duties.
- You may include brief personal data that you feel may be to your advantage such as hobbies and interests (especially if they relate to the job).
- Do not attach any letters of recommendation, certificates, or documents to the resume when you submit it. You may indicate something like, "Letters of recommendation available on request."

If a prospective employer likes your resume, he or she will usually ask you to come in and fill out an application. As with the resume, neatness and completeness are important when completing an employment application. Often, references are required when submitting an application. Make sure you ask the person whom you want to use as reference if it's okay with him or her. Usually a reference must be from outside your family. Previous employers, teachers, school administrators, and friends are excellent choices to use as references.

The Job Interview

The interview is the most critical part of getting a job. The resume and the application are important steps because they get you to the point of having an interview. Although you might look good on paper, it is the personal contact with a prospective employer that usually determines whether or not you get the job. Two main steps in the interview process are preparing for the interview and the interview itself.

(Getty Images)

An interview helps the employer choose the best person for the job. Your goal is to prove to the employer that you are that person. Here are several guidelines for preparing for a job interview:

- Learn as much about the company as you can. Use the Internet or other local resources such as other people who work for the company to obtain information.
- Practice answering some typical questions that an interviewer may ask you.
- Make sure you know how to get to the employer's location and try to get the name of the person who will be interviewing you, if possible.
- Dress appropriately and neatly. Get your clothes and anything you plan to carry ready the night before the interview so that you will not have to rush right before leaving. The employer is looking at you as a potential representative of the company, so make a good impression with your choice of clothes.
- Bring a copy of your resume and diplomas, certificates, or other documents you believe may be of interest to the interviewer.
- Be on time. You should arrive at the interview location at least 15 minutes before the scheduled time.

During the interview itself, there are a few things to keep in mind:

- Greet the interviewer by name and introduce yourself.
- Be polite.
- Maintain good posture.

- Keep eye contact.
- Do not interrupt the interviewer.
- Answer all questions as honestly as you can. If you do not know something, say so.
- Be prepared to tell the interviewer why you believe you are the best person for the job.
- Show interest and enthusiasm. Have some questions in mind and ask the interviewer at appropriate times about the company and the job, but avoid questions about salary, raises, etc.
- When the interview is over, thank the interviewer for his/her time.

After You Get the Job

The job itself is what it's all about. It is the reason that you went through the job search, preparation, and interview. Now you have to come through for the employer because you are being paid to do a job. You must retain interest and enthusiasm for the job and apply your technical skills to the best of your ability. In addition, you will need to work as a "team player," so you must bring basic social and communication skills to the job.

Safety in the Workplace

One of the most important skills a technician brings to the job is knowledge of safe operating practice and recognition of unsafe operation, especially for shock and burn hazards. Employers expect you to work safely. It is important to learn about specific safety issues for an employer when you start a new job, as there may be special requirements that depend on the job. Workplace safety is an important issue to virtually all employers. Most employers will cover issues in an initial job orientation and/or in an employee handbook. Electricity is never to be taken lightly, and safe operation in the laboratory is vital to your own well-being and that of other employees.

(Fluke Corporation. Reproduced with Permission.)

Companies have to comply with regulations prescribed by OSHA to ensure a safe and healthy workplace. OSHA is the Occupational Safety and Health Administration which is part of the U.S. Department of Labor. You can obtain information by accessing the OSHA website at *www.osha.gov*. In addition, you might have to be familiar with certain aspects of the National Electrical Code (NEC) and the standards issued by the American Society for Testing Materials (ASTA). Information on NEC can be obtained from the NFPA website at *www.nfpa.org* and ASTA can be accessed at *www.asta.org*. Of course, all standard electrical and electronic symbols as well as other related areas and standards are issued by IEEE (Institute of Electrical and Electronics Engineers). Their website is *www.ieee.com*.

Product Development, Marketing, and Servicing

The company you work for may be involved in developing, marketing, and servicing of products. Most electronics companies have a role in one or more of these activities. It is useful to understand the various steps in the process of bringing a new product to market.

1. Identifying the need
2. Designing
3. Prototyping and Evaluating
4. Producing
5. Marketing
6. Servicing

Identifying the Need

Before a new product is ever produced, someone (a person or group) must identify a problem or need, and suggest one or more ways a new product can solve the problem or meet the need. A company may be interested in the product but will want to do marketing analysis before expending money and time developing it. If the analysis shows a potential market for the product, design is started.

Designing

Most electronics projects can be implemented with more than one approach, so all ideas for the design of a new product are gathered together and evaluated. Usually this involves meetings with various specialists including electronic designers, test engineers, manufacturing engineers, as well as purchasing and marketing people. Selection considerations for a design must include input from these specialists. The best design is then chosen from the various ideas and a time line for completion is agreed upon. After a preliminary design is accomplished, components are selected based on cost, reliability, and availability.

Prototyping and Evaluating

Once the design of a product is complete, it must be proven that it will work properly by building and testing a prototype. Prototyping is often accomplished in two phases, preproduction and production. Usually technicians will construct a prototype of the design in a preliminary form. Test technicians will measure and evaluate the performance and report results to the design engineer. After approval and any necessary modification, the first prototype is converted to a production prototype by technicians. It is thoroughly tested and evaluated again before final approval for production.

Producing

A manufacturing engineer, familiar with the many processes that must be accomplished to produce a new product, is in charge of the production process. This person must verify the final layout and configuration and determine the sequence of all operations. The manufacturing engineer works with purchasing to determine cost of production and must ensure that the product will meet all necessary safety and reliability standards. Assemblers and technicians will produce the product and test the first production models to assure it works in accordance with the design specifications. Quality assurance technicians will test subsequent models to assure that products are ready to be shipped to customers.

Marketing

Once a product is produced, it is turned over to the marketing organization. The best product in the world is not much good if it can't be sold, so marketing is the key to success for a product. Marketing involves advertising, distribution, and followup. Technical marketing is a specialty requiring persons with both technical and marketing skills.

Servicing

Before a product is marketed, the cost of servicing it must be considered. A product may require servicing in order to assure maximum customer satisfaction and to create repeat business. Most electronics organizations maintain a service center to repair products returned for servicing. The type of servicing will depend on the cost of the product and replacement parts and the level of automation involved in the production process. Many products can be tested with automated test systems that can pinpoint a fault. Test engineers will set up the program to check the product, and service technicians will correct problems discovered as a result of testing.

The Social and Cultural Impact of Electronics Technology

Electronics technology has had a tremendous impact on our lives. The rapid advancement of technology has had both positive and negative influences on our society but by far the positives outweigh the negatives. Three areas of technological advancement that have had the most impact are the computer, communications technology, and medical technology. These areas are all interrelated because computer technology has influenced the other areas.

The Computer and the Internet

In terms of the effect on society, the computer and the Internet have changed in a relatively short time the way we get information, communicate with friends and business associates, learn about new topics, compose letters, and pay bills. In electronics work, instruments can be connected to the computer and data sent anywhere in the world via the Internet. Computers have reduced personal contact to some degree because now we can e-mail someone instead of writing a letter or making a phone call. We tend to spend time on the computer instead of visiting with friends or neighbors, and it has made us, in a sense, more isolated. On the other hand, we can be in touch with just about anyplace in the world via the Internet and we can "talk" to people we have never met through chat rooms.

(Fluke Corporation. Reproduced with permission.)

Communications

Obviously, the computer is closely related to the way we communicate with each other because of the Internet. The world has "shrunk" as a result of modern communications and the instant availability of information. Television is another way in which our cultural and social values have been influenced for both good and bad. We can watch events as they unfold around the world because of television and satellite technology. Political candidates are often elected by how they present themselves on television. We learn about different cultures, people, and topics through the television media in addition to being entertained and enlightened. Unfortunately, we are also exposed to many factors that have a negative influence including significant levels of violence portrayed as "entertainment."

(Seth Joel/Getty Images)

Another recent development that has had a major impact on our ability to communicate is the cellular telephone. Now, it is possible to contact anyone or be contacted no matter where you are. This, of course, facilitates doing business as well as providing personal benefits. Although it has added to our ability to stay in touch, it can be distracting; for example, when one is operating a motor vehicle.

Medical Technology

Great advancements in medical technology have resulted in improvement in the quality and length of life for many people. New medical tools have provided healthcare professionals with the ability to diagnose illnesses, analyze test results, and determine the best course of treatment. Imaging technologies such as MRI, CATSCAN, XRAY, ultrasound, and others make effective diagnosis possible. Electronic monitoring equipment helps to supervise patients, keeping a constant watch on their condition. Operating rooms use electronic tools such as lasers and various video monitors to permit doctors to perform ever more complex surgical procedures and examinations.

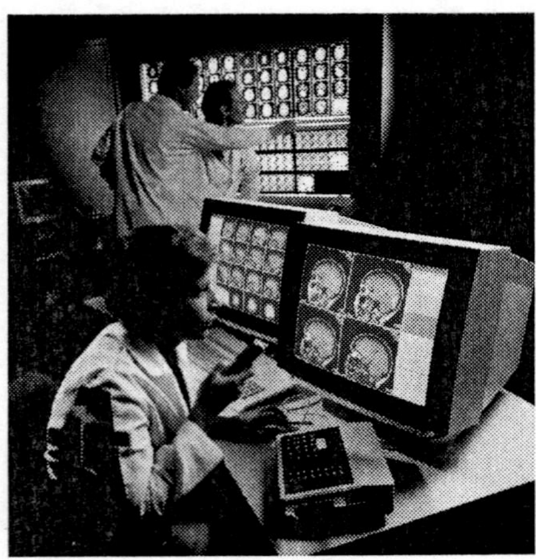

(Allan H. Shoemaker/Getty Images)

Increased lifespan due to modern medical achievements has an impact on our society. People live longer, more productive lives and can contribute more to the improvement of the social and cultural aspects of life. On the other hand, medical advances have extended lives in some cases because of expensive life support, which may put a strain on the social and economic resources.

History of Electronics

Early experiments with electronics involved electric currents in vacuum tubes. Heinrich Geissler (1814–1879) removed most of the air from a glass tube and found that the tube glowed when there was current through it. Later, Sir William Crookes (1832–1919) found the current in vacuum tubes seemed to consist of particles. Thomas Edison (1847–1931) experimented with carbon filament bulbs with plates and discovered that there was a current from the hot filament to a positively charged plate. He patented the idea but never used it.

Other early experimenters measured the properties of the particles that flowed in vacuum tubes. Sir Joseph Thompson (1856–1940) measured properties of these particles, later called *electrons.*

Although wireless telegraphic communication dates back to 1844, electronics is basically a 20th century concept that began with the invention of the vacuum tube amplifier. An early vacuum tube that allowed current in only one direction was constructed by John A. Fleming in 1904. Called the Fleming valve, it was the forerunner of vacuum tube diodes. In 1907, Lee deForest added a grid to the vacuum tube. The new device, called the audiotron, could amplify a weak signal. By adding the control element, deForest ushered in the electronics revolution. It was with an improved version of his device that made transcontinental telephone service and radios possible. In 1912, a radio amateur in San Jose, California, was regularly broadcasting music!

In 1921, the secretary of commerce, Herbert Hoover, issued the first license to a broadcast radio station; within two years over 600 licenses were issued. By the end of the 1920s radios were in many homes. A new type of radio, the superheterodyne radio, invented by Edwin Armstrong, solved problems with high-frequency communication. In 1923, Vladimir Zworykin, an American researcher, invented the first television picture tube, and in 1927 Philo T. Farnsworth applied for a patent for a complete television system.

The 1930s saw many developments in radio, including metal tubes, automatic gain control, "midget" radios, and directional antennas. Also started in this decade was the development of the first electronic computers. Modern computers trace their origins to the work of John Atanasoff at Iowa State University. Beginning in 1937, he envisioned a binary machine that could do complex mathematical work. By 1939, he and graduate student Clifford Berry had constructed a binary machine called ABC, (for Atanasoff-Berry Computer) that used vacuum tubes for logic and condensers (capacitors) for memory. In 1939, the magnetron, a microwave oscillator, was invented in Britain by Henry Boot and John Randall. In the same year, the klystron microwave tube was invented in America by Russell and Sigurd Varian.

The decade of the 1940s opened with World War II. The war spurred rapid advancements in electronics. Radar and very high-frequency communication were made possible by the magnetron and klystron. Cathode ray tubes were improved for use in radar. Computer work continued during the war. By 1946, John von Neumann had developed the first stored program computer, the Eniac, at the University of Pennsylvania. One of the most significant inventions ever occurred in 1947 with the invention of the transistor. The inventors were Walter Brattain, John Bardeen, and William Shockley. All three won Nobel prizes for their invention. PC (printed circuit) boards were also introduced in 1947. Commercial manufacturing of transistors didn't begin until 1951 in Allentown, Pennsylvania.

The most important invention of the 1950s was the integrated circuit. On September 12, 1958, Jack Kilby, at Texas Instruments, made the first integrated circuit, for which he was awarded a Nobel prize in the fall of 2000. This invention literally created the modern computer age and brought about sweeping changes in medicine, communication, manufacturing, and the entertainment industry. Many billions of "chips"—as integrated circuits came to be called—have since been manufactured.

The 1960s saw the space race begin and spurred work on miniaturization and computers. The space race was the driving force behind the rapid changes in electronics that followed. The first successful "op-amp" was designed by Bob Widlar at Fairchild Semiconductor in 1965. Called the μA709, it was very successful but suffered from "latch-up" and other problems. Later, the most popular op-amp ever, the 741, was taking shape at Fairchild. This op-amp became the industry standard and influenced design of op-amps for years to come. Precursors to the Internet began in the 1960s with remote networked computers. Systems were in place within Lawrence Livermore National Laboratory that connected over 100 terminals to a computer system (colorfully called the "Octopus system" and used by one of this text's authors). In an experiment in 1969 with very remote computers, an exchange took place between researchers at UCLA and Stanford. The UCLA group hoped to connect to a Stanford computer and began by typing the word "login" on its terminal. A separate telephone connection was set up and the following conversation occurred.

> The UCLA group asked over the phone, "Do you see the letter L?"
> "Yes, we see the L."
> The UCLA group typed an O. "Do you see the letter O?"
> "Yes, we see the O."

The UCLA group typed a G. At this point the system crashed. Such was technology, but a revolution was in the making.

By 1971, a new company that had been formed by a group from Fairchild introduced the first microprocessor. The company was Intel and the product was the 4004 chip, which had the same processing power as the Eniac computer. Later in that same year, Intel announced the first 8-bit processor, the 8008. In 1975, the first personal computer was introduced by Altair, and *Popular Science* magazine featured it on the cover of the January, 1975, issue. The 1970s also saw the introduction of the pocket calculator and new developments in optical integrated circuits.

By the 1980s, half of all U.S. homes were using cable hookups instead of television antennas. The reliability, speed, and miniaturization of electronics continued throughout the 1980s, including automated testing and calibrating of PC boards. The computer became a part of instrumentation and the virtual instrument was created. Computers became a standard tool on the workbench.

The 1990s saw a widespread application of the Internet. In 1993, there were only 130 websites; by the start of the new century (in 2001) there were over 24 million. In the 1990s, companies scrambled to establish a home page and many of the early developments of radio broadcasting had parallels with the Internet. The exchange of information and e-commerce fueled the tremendous economic growth of the 1990s. The Internet became especially important to scientists and engineers, becoming one of the most important scientific communication tools ever.

In 1995, the FCC allocated spectrum space for a new service called Digital Audio Radio Service. Digital television standards were adopted in 1996 by the FCC for the nation's next generation of broadcast television. As the 20th century drew toward a close, historians could only breathe a sigh of relief. As one person put it, "I'm all for new technologies, but I wish they'd let the old ones wear out first."

The 21st century dawned on January 1, 2001 (although most people celebrated the new century the previous year, known as "Y2K"). The major story was the continuing explosive growth of the Internet; shortly thereafter, scientists were planning a new supercomputer system that would make massive amounts of information accessible in a computer network. The new international data grid will be an even greater resource than the World Wide Web, giving people the capability to access enormous amounts of information and the resources to run simulations on a supercomputer. Research in the 21st century continues along lines of faster and smaller circuits using new technologies. One promising area of research involves carbon nanotubes, which have been found to have properties of semiconductors in certain configurations. See Science Highlight in Chapter 3, page 70.

ACKNOWLEDGMENTS

This text is the result of the work and the skills of many people. We think you will find this book and all the others in *The Science of Electronics* series to be valuable tools in teaching your students the basics of various areas of electronics.

Those at Prentice Hall who have, as always, contributed a great amount of time, talent, and effort to move this project through its many phases in order to produce the book as you see it, include but are not limited to Rex Davidson, Kate Linsner, and Dennis Williams. We are grateful that Lois Porter once again agreed to edit the manuscript on this project. She has done an outstanding job and we appreciate the incredible attention to detail that she has provided. Also, Jane Lopez has done a beautiful job with the graphics. Another individual who contributed significantly to this book is Doug Joksch, of Yuba College, who created all of the Multisim circuit files for the website and helped with checking the problems for accuracy. Our thanks and appreciation go to all of these people and others who were directly involved in this project.

We depend on expert input from many reviewers to create successful textbooks. We wish to express our sincere thanks to the following reviewers who submitted many valuable suggestions and provided lots of constructive criticism: Bruce Bush, Albuquerque Technical-Vocational Institute; Gary DiGiacomo, Broome Community College; Brent Donham, Richland College; J.D. Harrell, South Plains College; Benjamin Jun, Ivy Tech State College; David McKeen, Rogue Community College; Jerry Newman, Southwest Tennessee Community College; Philip W. Pursley, Amarillo College; Robert E. Magoon, Erie Institute of Technology; Dale Schaper, Lane Community College; and Arlyn L. Smith, Alfred State College.

Tom Floyd
David Buchla

CHAPTER

ANALOG CONCEPTS

CHAPTER OUTLINE

INTRODUCTION

The first modern electronic systems trace to the work of Lee DeForest when he discovered in 1907 that a tiny signal on the grid of a vacuum tube could control a larger current. (Marconi and others had earlier experimented with radio but could not amplify the tiny signals received.) William Shockley, John Bardeen, and Walter Brattain started a second revolution of electronic systems with the invention of the point contact transistor at Bell Telephone Laboratories in 1947. This invention was followed in 1959 by the invention of the integrated circuit by Jack Kilby at Texas Instruments and Robert Noyce at Fairchild Semiconductor. If the transistor represented a second revolution in electronics, the integrated circuit was the third.

With the influence of high-speed digital devices, it's easy to overlook the fact that virtually all natural phenomena that we measure (for example, pressure, flow rate, and temperature) originate as analog signals that must be amplified and processed in some way. Depending on the application, either digital or analog techniques may be more efficient for processing these analog signals. Analog circuits are found in nearly all power supplies, in many "real-time" applications (such as motor speed controls), and in high-frequency communication systems. Digital processing is more effective when mathematical operations must be performed and has major advantages in reducing the noise inherent in processing analog signals. In short, the two sides of electronics (analog and digital) complement each other, and a competent technician needs to be knowledgeable of both.

Study aids for this chapter are available at

http://www.prenhall.com/SOE

KEY TERMS

- AC resistance
- Signal
- Analog signal
- Digital signal
- Sampling
- Quantizing
- Domain
- Spectrum

KEY OBJECTIVES

A section number is given for each objective. After completing this chapter, you should be able to

1–1 Explain how the *IV* curve is used as a means of describing components used in analog circuits

1–2 Describe analog signals

1–3 Describe the process for troubleshooting an analog circuit

LABORATORY EXPERIMENTS DIRECTORY

The following exercise is for this chapter. The laboratory book is entitled *The Science of Electronics: Analog Devices Lab Manual,* by David M. Buchla (ISBN 0-13-087559-7). © 2005 Prentice Hall.

◆ **Experiment 1**
Wave Shaping and Measurement

The evacuated glass tube with three electrodes was one of the first truly electronic devices. Lee deForest first inserted a metallic grid in a vacuum tube in the year 1907 and was able to control the current in a circuit. The glass tube has its roots in the experiments of Robert Boyle, an Englishman who first experimented with evacuated glass tubes in the 17th century to study gases. (A famous law in chemistry is named for him.)

A century after Boyle's work, Heinrich Geissler, a glassblower, created a mercury vacuum pump that produced a better vacuum than Boyle was able to produce. He made tubes with electrodes and observed that electricity caused the low pressure gas inside to glow. William Crookes, an Englishman, improved on Geissler's pump, which made it possible for the system to produce vacuums exceeding one millionth of an atmosphere. The Crookes-type tubes showed beautiful colors when a voltage was applied between the electrodes, due to ionized gases inside the tube. Crookes experimented with shadow patterns from a Maltese cross he placed in his tube and concluded that the radiation he observed from unseen particles emitted from the negative terminal travels in a straight line.

A diode is a one-way device for electrical current. The Crookes tube was an early form of a diode, (discussed in chapter 2) although there was no direct application of it for that purpose. Today, cathode ray tubes (CRTs) are commonly used in televisions. These are evacuated glass tubes with a very high vacuum to avoid the ionization of the gas in the tube. One end of the tube has coated phosphors, which glow when struck by electrons to produce a television picture.

1–1 DEVICES AND QUANTITIES

The field of electronics is subdivided into digital electronics and analog electronics, which are classified by the type of signals they use. Digital circuits use discrete voltage levels to represent numbers and quantities, whereas analog circuits use continuously variable quantities. Digital electronics includes all arithmetic and logic operations such as performed in computers and calculators. Analog electronics includes signal-processing functions such as amplification, differentiation, and integration.

In this section, you will learn how the *IV* curve is used as a means of describing components used in analog circuits.

Electronic Devices

During the twentieth century, electronics progressed from spark-gap transmitters to vacuum tubes, transistors, and integrated circuits. Integrated circuits have become smaller, much faster, much more capable and amazingly, less expensive. The transistor and integrated circuit have taken their places as two of the most important inventions ever made.

Integrated circuits (ICs) are complete functioning devices contained on a small piece of silicon substrate and mounted in a package with pins for inputs, outputs, and power. The majority of integrated circuits available can generally be classified as either *digital integrated circuits,* that only use two distinct voltages, or *analog integrated circuits,* that deal with continuously varying inputs and outputs. In this text, our focus is on various analog circuits, including analog integrated circuits. These circuits include transistors, analog ICs, and basic components.

Basic electronic components, such as resistors or diodes and even transistors, can be represented with graphs that show their characteristics in a more intuitive manner than mathematical equations. In this section, you will examine graphs representing resistors and diodes. In Chapter 3, you will see how transistors can also be illustrated with *IV* curves to provide a graphical picture of circuit operation.

Linear Equations

In basic algebra, a linear equation is one that can plot a straight line between the variables and is usually written in the following form (called the slope-intercept form):

$$y = mx + b$$

where y is the dependent variable, x is the independent variable, m is the slope, and b is the y-axis intercept. If the plot of the equation goes through the origin, then the y-axis intercept is zero, and the equation reduces to

$$y = mx$$

which has the same form as Ohm's law.

$$I = \frac{V}{R} \tag{1-1}$$

As written here, the dependent variable in Ohm's law is current (I), the independent variable is voltage (V), and the slope is the reciprocal of resistance ($1/R$). The reciprocal of resistance is simply the conductance, (G). By substitution, the linear form of Ohm's law is more obvious; that is,

$$I = GV$$

A linear component is one in which an increase in current is proportional to the applied voltage as given by Ohm's law. In general, a plot that shows the relationship between two variable properties of a device defines a characteristic curve. For most electronic devices, a characteristic curve refers to a plot of the current, I, plotted as a function of voltage, V. For example, two resistors have *IV* characteristic curves described by the straight lines given in Figure 1–1. Notice that current is plotted on the y-axis because it is the dependent variable.

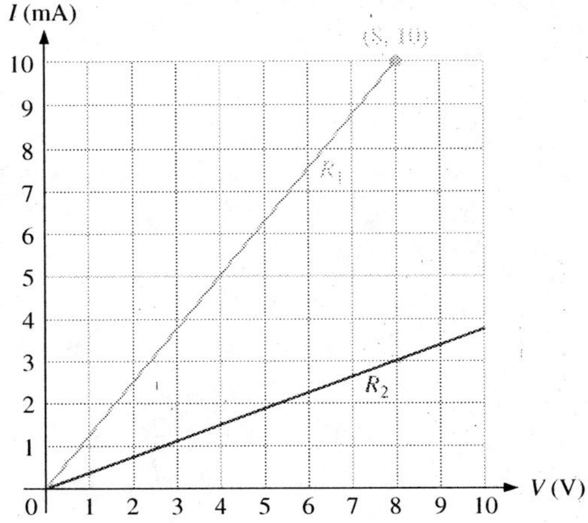

FIGURE 1–1

IV characteristic curves for two resistors.

The circuit for measuring the *IV* characteristic is shown in Figure 1–2. The power supply voltage is set to a value by reading the voltmeter (indicated with a circled V), and the current is measured with the ammeter (indicated with the circled A). Current readings for different voltages produce data points that can be plotted on an *IV* graph. The *IV* characteristic curves in Figure 1–1 describe the current in the resistors for positive voltages, which are plotted in quadrant 1.

FIGURE 1–2

Circuit for measuring the *IV* characteristic for a resistor.

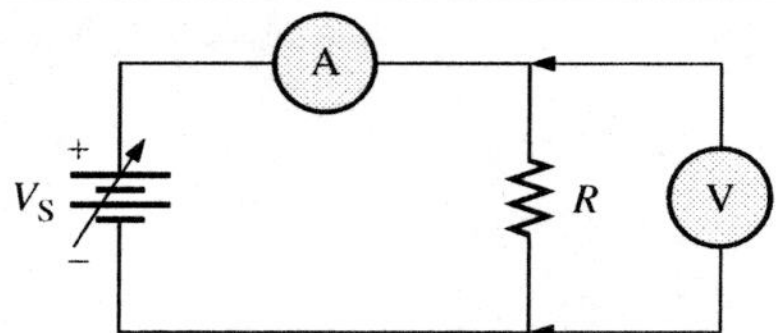

If the power supply is reversed, what happens? In this case, more readings are taken (with the ammeter connected in reverse also). Now the plot for a resistor extends into quadrant 3, which represents negative quantities for both voltage and current. This is illustrated in Figure 1–3.

FIGURE 1–3

An *IV* characteristic curve for a resistor showing the result of both positive and negative voltages.

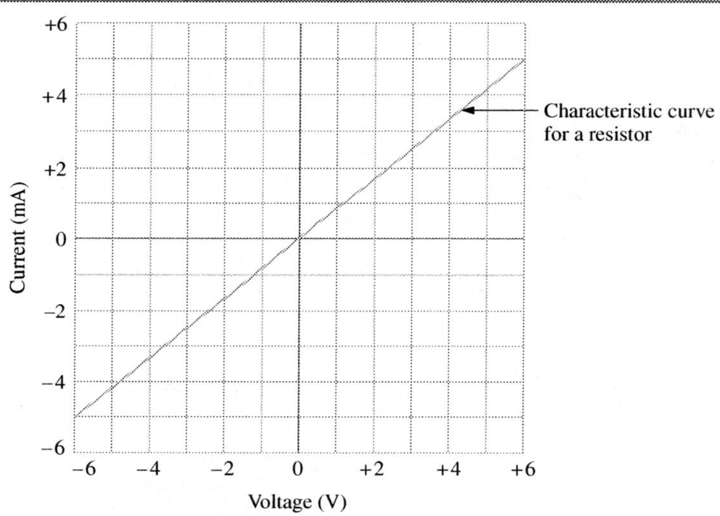

EXAMPLE 1–1

Problem

Figure 1–1 shows the characteristic curves for two resistors. What are the conductance and resistance of R_1?

Solution

Find the conductance, G_1, by measuring the slope of the *IV* characteristic curve for R_1. The slope is the change in the *y* variable (written Δy) divided by the corresponding change in the *x* variable (written Δx).

$$\text{slope} = \frac{\Delta y}{\Delta x}$$

For a straight line, the slope is constant, so you can use any two points to determine the conductance. Choosing the point ($x = 8$ V, $y = 10$ mA) from Figure 1–1 as one point and the origin ($x = 0$ V, $y = 0$ mA) as the other point, you can find the slope and, therefore, the conductance as

$$G_1 = \frac{10\text{ mA} - 0\text{ mA}}{8.0\text{ V} - 0\text{ V}} = \mathbf{1.25\ mS}$$

The resistance is the reciprocal of the conductance.

$$R_1 = \frac{1}{G_1} = \frac{1}{1.25\text{ mS}} = \mathbf{0.8\ k\Omega}$$

Question*

Refer to Figure 1–1. What are the conductance and resistance of R_2?

AC Resistance

As you have seen, the graph of the characteristic curve for a resistor is a straight line that passes through the origin. The slope of the line is constant and represents the conductance of the resistor; the reciprocal of the slope represents resistance. The ratio of voltage at some point to the corresponding current at that point is referred to as *dc resistance.* DC resistance is defined by Ohm's law, $R = V/I$.

Many devices studied in analog electronics have a characteristic curve for which the current is *not* proportional to the voltage. These devices are nonlinear by nature but are included in the study of analog electronics because they take on a continuous range of input signals.

Figure 1–4 shows the *IV* characteristic curve for a diode, a nonlinear analog device. (This curve is discussed in more detail Chapter 2.) Generally, it is more useful to define the resistance of a nonlinear device such as a diode as a small change in voltage divided by the corresponding small change in current, that is, $\Delta V/\Delta I$. The ratio of a small change in voltage divided by the corresponding small change in current is defined as the **ac resistance**** of an analog device.

$$r_{ac} = \frac{\Delta V}{\Delta I}$$

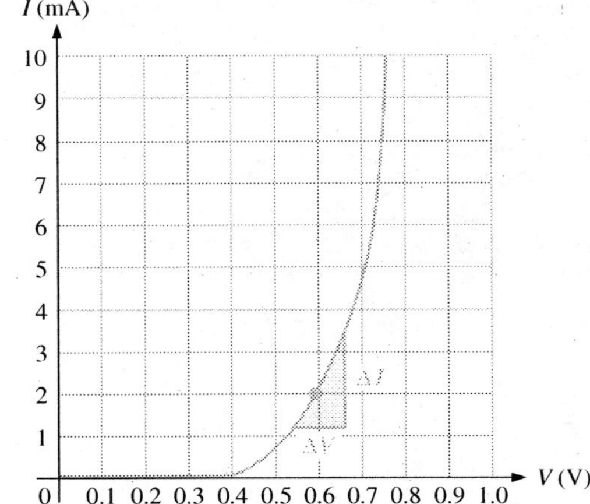

FIGURE 1–4

An *IV* characteristic curve for a diode.

* *Answers are at the end of the chapter.*

** *Key terms are shown in red.*

This resistance (indicated with a lowercase italic r) is also called the *dynamic resistance* of the device. It is an internal resistance, and its value depends on the particular point on the IV characteristic curve where the measurement is made.

For the diode in Figure 1–4, the slope varies dramatically; the point where the ac resistance is measured needs to be specified with any measurement. For example, the slope at the point $x = 0.6$ V, $y = 2$ mA is found by computing the ratio of the change in current to the change in voltage as defined by the small blue triangle shown in the figure. The change in current, ΔI, is 3.4 mA − 1.2 mA = 2.2 mA and the change in voltage, ΔV, is 0.66 V − 0.54 V = 0.12 V. The ratio of $\Delta V/\Delta I$ is 2.2 mA/1.2 V = 18.3 mS. This represents the conductance of the specified point. The internal ac resistance at this point is the reciprocal of this value.

$$r = \frac{1}{g} = \frac{1}{18.3 \text{ mS}} = 54.5\ \Omega$$

Conventional Current Versus Electron Flow

Current is the rate of flow of charge. The original definition of current was based on Benjamin Franklin's belief that electricity was an unseen substance that moved from positive to negative. *Conventional current* assumes for analysis purposes that current is out of the positive terminal of a voltage source, through the circuit, and into the negative terminal of the source. Engineers use this definition, and many textbooks show current with arrows drawn with this viewpoint.

Today, it is known that in solid metallic conductors, the moving charge is actually negatively charged conduction electrons. Electrons move from the negative to the positive point in a circuit, opposite to the defined direction of conventional current. The movement of electrons in a conductor is called *electron flow.* Many schools and textbooks show electron flow with current arrows drawn out of the negative terminal of a voltage source.

Unfortunately, the controversy between whether it is better to show conventional current or electron flow in representing circuit behavior has continued for many years and does not appear to be subsiding. It is not important which direction you use to form a mental picture of current. In practice, there is only one correct direction to connect a dc ammeter to make current measurements. Throughout this text, dc ammeters instead of arrows are shown with the proper polarity to indicate current. Current paths are indicated with special bar meter symbols. In a given circuit, larger or smaller currents are indicated by the relative number of bars shown on a bar graph meter, as illustrated in Figure 1–5.

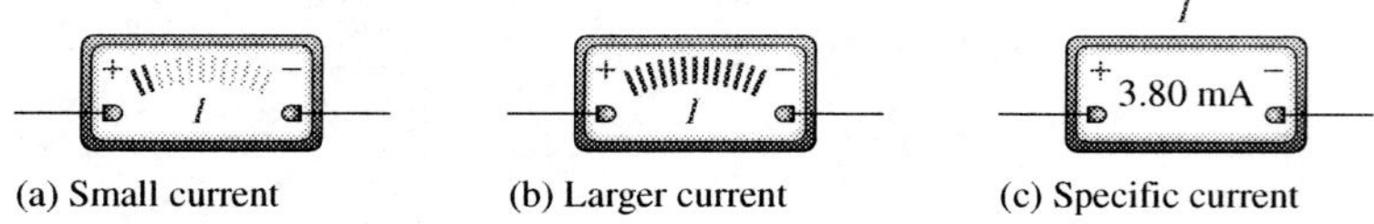

(a) Small current (b) Larger current (c) Specific current

FIGURE 1–5

Special meter symbols for showing current. The number of bars indicates the relative current.

Review Questions

Answers are at the end of the chapter.

1. What are two categories of integrated circuits?
2. What does the slope of an IV characteristic curve represent?
3. How does the IV characteristic curve of a large resistor compare to the IV characteristic curve of a small resistor?
4. What is the difference between dc and ac resistance?
5. What is the difference between conventional current and electron flow?

SIGNALS 1–2

A signal is any physical quantity that carries information. It can be an audible, visual, or other indication of information. In electronics, the term *signal* refers to the information that is carried by electrical waves, either in a conductor or as an electromagnetic field.

In this section, you will learn to describe analog signals.

Analog and Digital Signals

In electronics, a **signal** is an electrical voltage or current that contains information. In electronic systems, the signal is usually processed in some way—either the shape is modified or power level is increased through amplification. The processing can be by analog (continuous) or digital (discrete) methods.

Signals can be classified as either continuous or discrete. A continuous signal changes smoothly, without interruption. A discrete signal can have only certain values. The terms *continuous* and *discrete* can be applied either to the amplitude or to the time characteristic of a signal.

In nature, most signals take on a continuous range of values within limits; any signal that is continuous is referred to as an **analog signal**. For example, consider a potentiometer that is used as a shaft encoder, as shown in Figure 1–6. The output voltage can be continuously varied within the limit of the supply voltage, resulting in an analog signal that is related to the angular position of the shaft. The signal has the position of the potentiometer encoded as a voltage level.

FIGURE 1–6

Analog and digital shaft encoders.

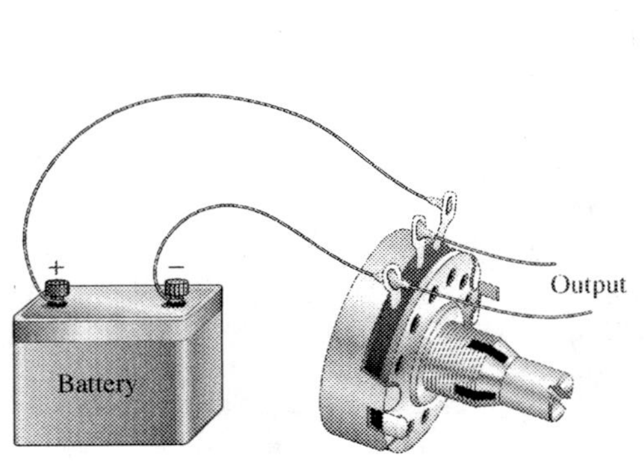

(a) Analog shaft encoder

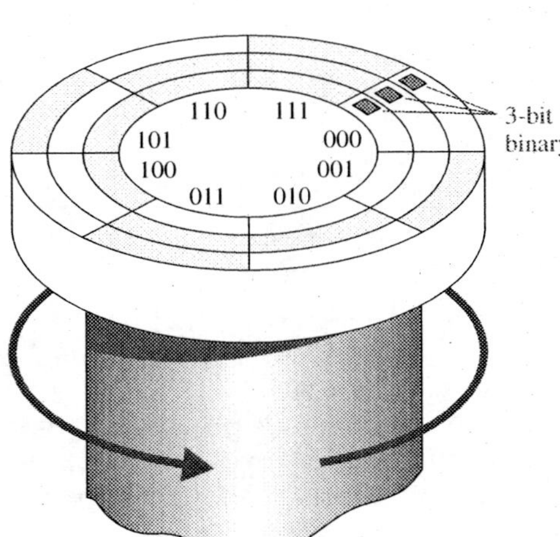

(b) Digital shaft encoder

On the other hand, another type of encoder has a certain number of steps that can be selected, as shown in Figure 1–6(b). When numbers are assigned to these steps, the result is called a digital signal. A **digital signal** is a noncontinuous signal that has discrete numerical values assigned to specific steps.

Analog circuits process analog signals. Analog circuits are generally simple, have high speed and low cost, and can readily simulate natural phenomena. They are often used for operations such as performing linearizing functions, wave shaping, transforming voltage to current or current to voltage, multiplying, and mixing. By contrast, digital circuits, which process digital signals, have high noise immunity, no drift, high speed, and the ability to perform various mathematical calculations. In many electronic systems, a mix of analog and digital signals are required to optimize the overall system's performance or cost.

Many signals have their origin in a natural phenomenon such as a measurement of pressure, temperature, or motion. These signals can be converted to an electrical signal by a **transducer**, a device that converts energy from one form to another. After conversion, the electrical signal is typically analog; a microphone, for example, provides an analog signal to an amplifier. Frequently, the analog signal is then converted to digital form for storing, processing, or transmitting.

A two-step process is required for analog-to-digital conversion: sampling and quantizing. **Sampling** is the process of breaking an analog waveform into time "slices" that approximate the original wave. This process always loses some information; however, the advantages of digital systems (noise reduction, digital storage, and processing) outweigh the disadvantages. After sampling, the time slices are assigned a numeric value. This process, called **quantizing**, produces numbers that can be processed by digital computers or other digital circuits. Figure 1–7 illustrates the sampling and quantizing process.

Frequently, digital signals need to be converted back to their original analog form to be useful in their final application. For instance, the digitized sound on a CD must be converted to an analog signal and eventually back to sound by a loudspeaker.

FIGURE 1–7

Digitizing an analog waveform.

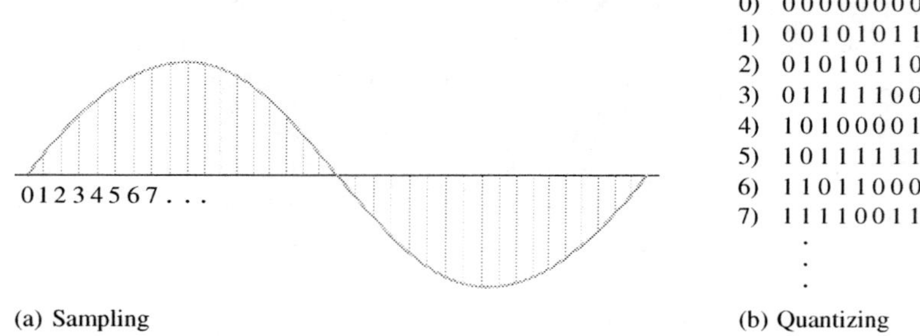

(a) Sampling (b) Quantizing

Periodic Signals

To carry information, some property such as voltage or frequency of an electrical wave needs to vary. Frequently, an electrical signal repeats at a regular interval of time. Repeating waveforms are said to be periodic. The **period (*T*)** is the time for a periodic wave to complete one cycle. A **cycle** is the complete sequence of values that a waveform exhibits before another identical pattern occurs. The period can be measured between any two corresponding points on successive cycles.

Periodic waveshapes are used extensively in electronics. Many practical electronic circuits such as oscillators generate periodic waves. Most oscillators are designed to produce a particular shaped waveform—either a sinusoidal wave or nonsinusoidal waves such as the square, rectangular, triangle, and sawtooth waves.

Sinusoidal Waves

The most basic and important periodic waveform is the sinusoidal wave. In electronics, a **sinusoidal wave** is a voltage or current waveform that has the same shape as the mathematical trig function called the sine wave. The term *sine wave* usually implies the trigonometric function, whereas the term *sinusoidal wave* means a waveform with the shape of a sine wave (although occasionally the terms are interchanged). A sinusoidal waveform is generated as the natural waveform from many ac generators and in radio waves. Sinusoidal waves are also present in physical phenomena from the generation of laser light, the vibration of a tuning fork, or the motion of ocean waves.

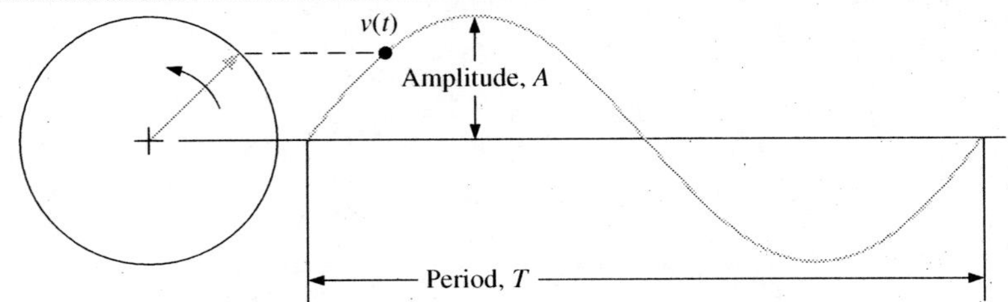

FIGURE 1–8

Generation of a sinusoidal waveform from the projection of a rotating vector.

A **vector** is any quantity that has both magnitude and direction. A sinusoidal curve can be generated by plotting the projection of a rotating vector that is turning with uniform circular motion, as illustrated in Figure 1–8. Successive revolutions of the point generate a periodic curve, which can be expressed mathematically as

$$y(t) = A\sin(\omega t \pm \phi) \tag{1–2}$$

where $y(t)$ = vertical displacement of a point on the curve from the horizontal axis. The quantity (t) is an optional indicator, called *functional notation,* to emphasize that the signals vary with time. Functional notation is frequently omitted when it isn't important to emphasize the time relationship. In the equation, A is the amplitude, the maximum displacement from the horizontal axis. The symbol ω is the angular frequency of the rotating vector in radians per second, t is time in seconds to a point on the wave, and ϕ is the phase angle in radians. The phase angle is simply a fraction of a cycle that a waveform is shifted from a reference waveform of the same frequency. It is positive if the wave begins before $t = 0$ and is negative if the wave starts after $t = 0$.

Equation 1–2 illustrates that the sinusoidal wave can be defined in terms of three basic parameters: the frequency, amplitude, and phase angle.

HISTORICAL NOTE

Heinrich Hertz (1857–1894) was the first to demonstrate the production and detection of electrical waves, which he did in 1888 in his classroom. He generated waves by discharging a "condenser" (now called a capacitor) through a coil. His condenser was made from two metal plates with a small gap for a spark placed between them. When the charge was large enough to cause a spark, the current set up oscillations between the capacitor and coil. He detected the waves with a similar circuit. Hertz's discovery of electronic waves was the beginning of the electronic era. Sadly, he died from food poisoning when he was only 37 years old. The unit of frequency is named in his honor.

Frequency and Period When a rotating vector has made one complete cycle, it has rotated through 2π radians. The number of complete cycles generated per second is called the **frequency**. Dividing the angular frequency (ω in rad/s) of the rotating vector by the number of radians in one cycle (2π rad/cycle) gives the frequency in hertz.

$$f(\text{Hz}) = \frac{\omega\ (\text{rad/s})}{2\pi\ (\text{rad/cycle})} \tag{1–3}$$

One cycle per second is equal to 1 Hz. The frequency (f) of a periodic wave is the number of cycles in one second, and the period (T) is the time for one cycle. Therefore, it is logical that the reciprocal of the frequency is the period and the reciprocal of the period is the frequency.

$$T = \frac{1}{f} \tag{1–4}$$

and

$$f = \frac{1}{T} \tag{1–5}$$

For example, if a signal repeats every 10 ms, then its period is 10 ms and its frequency is

$$f = \frac{1}{T} = \frac{1}{10\text{ ms}} = 0.1\text{ kHz}$$

Instantaneous Value If the sinusoidal waveform shown in Figure 1–8 represents a voltage, Equation 1–2 is written

$$v(t) = V_p \sin(\omega t \pm \phi)$$

In this equation, *v(t)* is a variable that represents the voltage. Since it changes as a function of time, it is often referred to as the *instantaneous voltage.*

Peak Value The amplitude of a sinusoidal wave is the maximum displacement from the horizontal axis, as shown in Figure 1–8. For a voltage waveform, the amplitude is called the peak voltage, V_p. When making voltage measurements with an oscilloscope, it is often easier to measure the peak-to-peak voltage, V_{pp}. The peak-to-peak voltage is twice the peak value.

Average Value During one cycle, a sinusoidal waveform has equal positive and negative excursions. Therefore, the mathematical definition of the average value of a sinusoidal waveform must be zero. However, the term *average value* is generally used to mean the average over a cycle without regard to the sign. That is, the average is usually computed by converting all negative values to positive values, then averaging. The average voltage is defined in terms of the peak voltage by the following equation:

$$V_{avg} = \frac{2V_p}{\pi}$$

Simplifying,

$$V_{avg} = 0.637V_p \tag{1–6}$$

The average value is useful in certain practical problems. For example, if a rectified sinusoidal waveform is used to deposit material in an electroplating operation, the quantity of material deposited is related to the average current.

$$I_{avg} = 0.637I_p$$

Effective Value (RMS Value) If you apply a dc voltage to a resistor, a steady amount of power is dissipated in the resistor and can be calculated using the following power law:

$$P = IV \tag{1–7}$$

where *V* is dc voltage across the resistor (volts), *I* is dc current in the resistor (amperes), and *P* is power dissipated (watts).

In a resistive circuit, a sinusoidal waveform transfers maximum power at the peak excursions of the curve and no power at all at the instant the voltage crosses zero. In order to compare ac and dc voltages and currents, ac voltages and currents are defined in terms of the equivalent heating value of dc. This equivalent heating value is computed with calculus, and the result is called the **rms** (for *root-mean-square*) voltage or current. For a sinusoidal wave, the rms voltage is related to the peak voltage by the following equation:

$$V_{rms} = 0.707V_p \tag{1–8}$$

Likewise, the effective or rms current is

$$I_{rms} = 0.707I_p$$

EXAMPLE 1–2

Problem

A certain voltage waveform is described by the following equation:

$$v(t) = 15\text{ V} \sin(600t)$$

(a) From this equation, determine the peak voltage and the average voltage. Give the angular frequency in rad/s.

(b) Find the instantaneous voltage at a time of 10 ms.

Solution

(a) The form of the equation is

$$y(t) = A \sin(\omega t)$$

The peak voltage is the same as the amplitude (A).

$$V_p = \mathbf{15\ V}$$

The average voltage is related to the peak voltage.

$$V_{avg} = 0.637V_p = 0.637(15\text{ V}) = \mathbf{9.56\ V}$$

The angular frequency, ω, is **600 rad/s**.

(b) The instantaneous voltage at a time of 10 ms is

$$v(t) = 15\text{ V} \sin(600t) = 15\text{ V} \sin(600\text{ rad/s})(10\text{ ms}) = \mathbf{-4.19\ V}$$

The negative value indicates that the waveform is below the axis at this time.

Question

What are the rms voltage, the frequency in hertz, and the period of the waveform described in the example?

Time-Domain Signals

Thus far, the signals you have looked at vary with time, and it is natural to associate time as the independent variable. Some instruments, such as the oscilloscope, are designed to record signals as a function of time. Time is therefore the independent variable. The values assigned to the independent variable are called the **domain**. Signals that have voltage, current, resistance, or other quantity vary as a function of time are called *time-domain signals.*

Frequency-Domain Signals

Sometimes it is useful to view a signal where frequency is represented on the horizontal axis and the signal amplitude (usually in logarithmic form) is plotted along the vertical axis. Since frequency is the independent variable, the instrument works in the frequency domain, and the plot of amplitude versus frequency is called a **spectrum**. The spectrum analyzer is an instrument used to view the spectrum of a signal. This instrument is extremely useful in radio frequency (RF) measurements for analyzing the frequency response of a circuit, testing for harmonic distortion, checking the percent modulation from transmitters, and many other applications.

A continuous sinusoidal wave can be shown as a time-varying signal defined by the amplitude, frequency, and phase angle. The same sinusoidal wave can also be shown as a single line on a frequency spectrum. The frequency-domain representation gives information about the amplitude and frequency, but it does not show the phase angle. These two representations of a sinusoidal wave are compared in Figure 1–9. The height of the line on the spectrum is the amplitude of the sinusoidal wave.

FIGURE 1–9

Time-domain and frequency-domain representations of a sinusoidal wave.

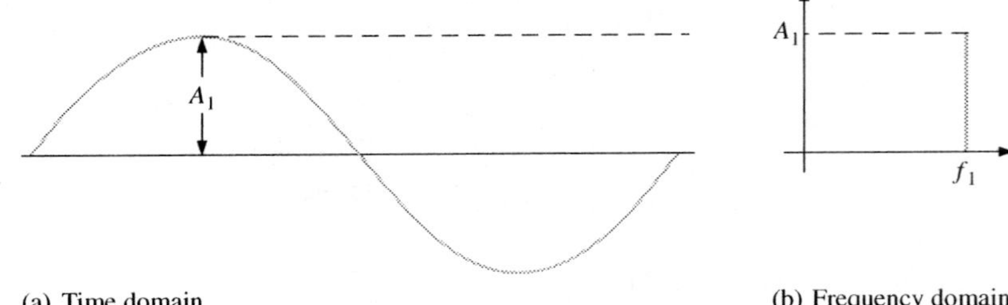

Harmonics

A nonsinusoidal repetitive waveform is composed of a fundamental frequency and harmonic frequencies. The fundamental frequency is the basic repetition rate of the waveform, and the **harmonics** are higher-frequency sinusoidal waves that are integer multiples of the fundamental. Interestingly, these multiples are all related to the fundamental by integers (whole numbers).

Odd harmonics are frequencies that are odd multiples of the fundamental frequency of a waveform. For example, a 1 kHz square wave consists of a fundamental of 1 kHz and odd harmonics of 3 kHz, 5 kHz, and 7 kHz. The 3 kHz frequency in this case is called the third harmonic and the 5 kHz frequency is called the fifth harmonic.

Even harmonics are frequencies that are even multiples of the fundamental frequency. For example, if a certain wave has a fundamental of 200 Hz, the second harmonic is 400 Hz, the fourth harmonic is 800 Hz, and the sixth harmonic is 1200 Hz.

Any variation from a pure sinusoidal wave produces harmonics. A nonsinusoidal wave is a composite of the fundamental and certain harmonics. Some types of waveforms have only odd harmonics, some have only even harmonics, and some contain both. The shape of the wave is determined by its harmonic content. Generally, only the fundamental and the first few harmonics are important in determining the waveshape. For example, a square wave is formed from the fundamental and odd harmonics, as illustrated in Figure 1–10.

FIGURE 1–10 Odd harmonics combine to produce a square wave.

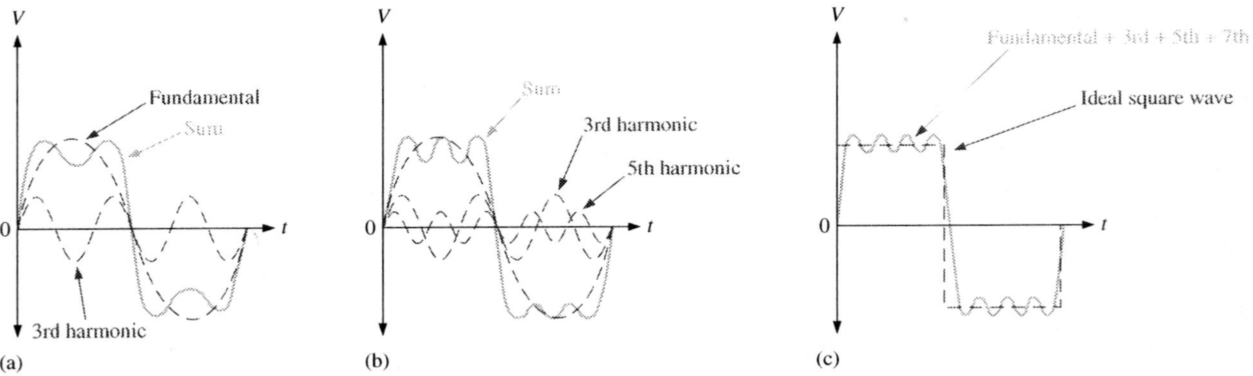

Fourier Series

All periodic waves except the sinusoidal wave itself are complex waveforms composed of a series of sinusoidal waves. Jean Fourier, a French mathematician interested in problems of heat conduction, formed a mathematical series of trigonometry terms to describe periodic waves. This series is appropriately called the Fourier series. With the Fourier series, you can mathematically determine the amplitude of each of the sinusoidal waves that compose a complex waveform.

The frequency spectrum developed by Fourier is often shown as an amplitude spectrum with units of voltage or power on the y-axis plotted against Hz on the x-axis. Figure 1–11(a) illustrates the amplitude spectrum for several different periodic waveforms. Notice that all

FIGURE 1–11 Comparison of the frequency spectrum of repetitive and nonrepetitive waves.

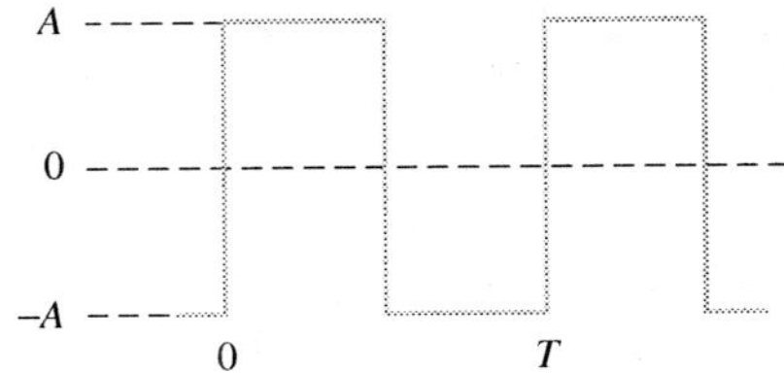

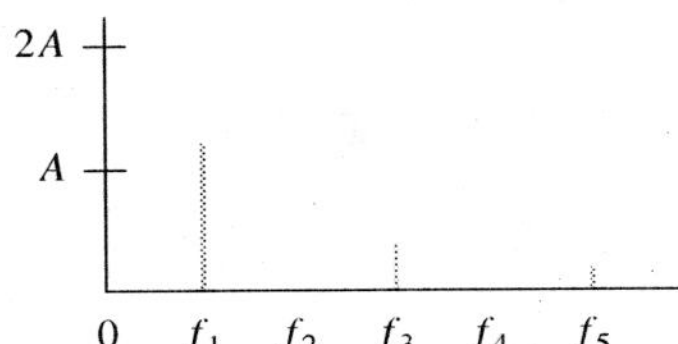

Square

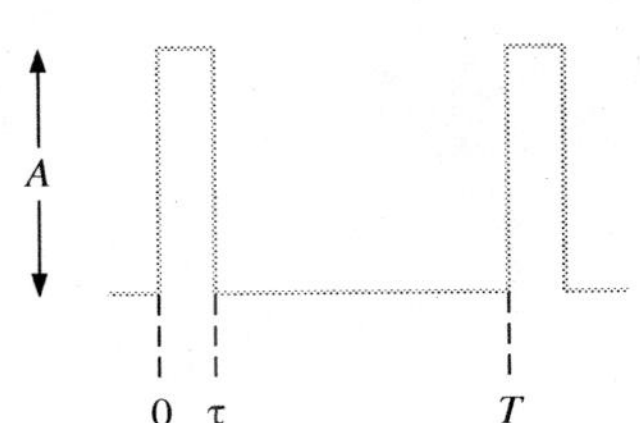

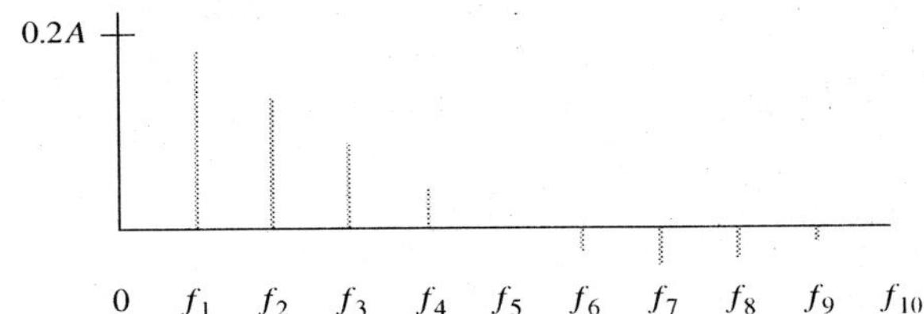

Rectangle

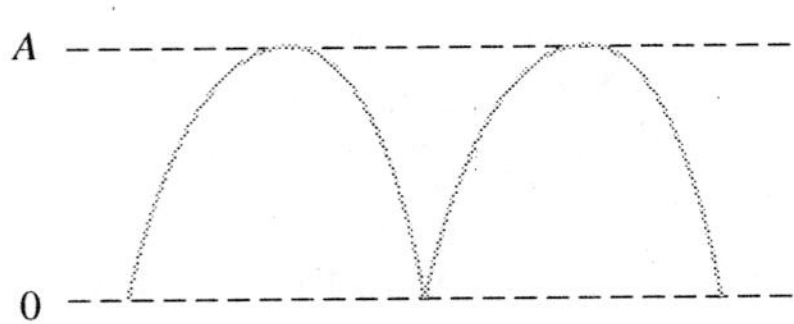

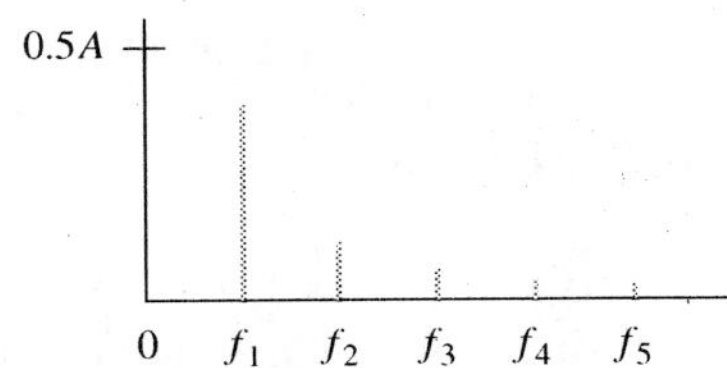

Full-wave rectified sinusoid

(a) Examples of time-domain and frequency-domain representations of repetitive waves

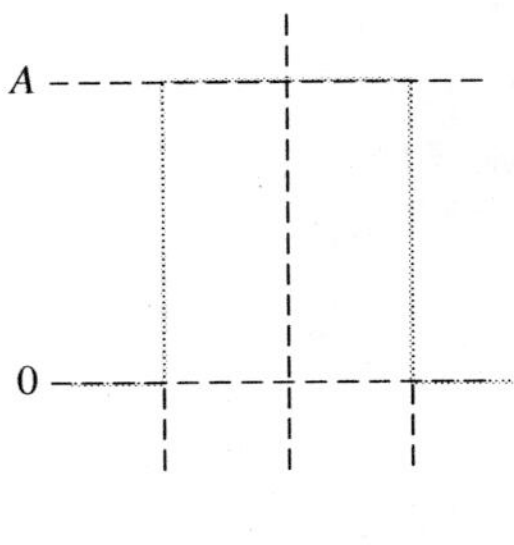

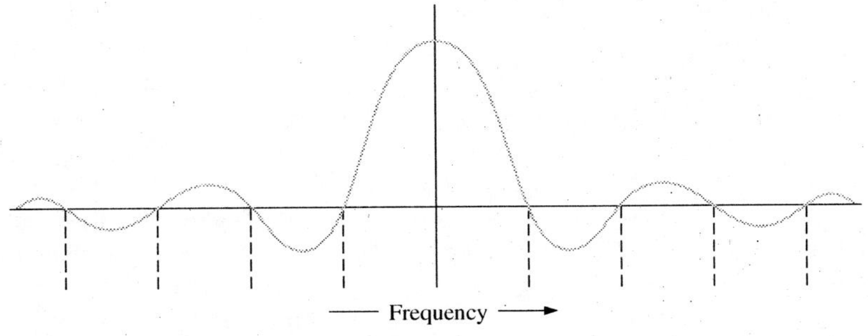

Time-domain representation

Frequency-domain representation

(b) Examples of the frequency spectrum of a nonrepetitive pulse waveform spectrums for periodic waves are depicted as lines located at harmonics of the fundamental frequency. These individual frequencies can be measured with a spectrum analyzer.

Nonperiodic signals such as speech, or other transient waveforms, can also be represented by a spectrum; however, the spectrum is no longer a series of lines as in the case of repetitive waves. Transient waveforms are computed by another method called the *Fourier transform.* The spectrum of a transient waveform contains a continuum of frequencies rather than just harmonically related components. A representative Fourier pair of signals for a nonrepetitive pulse is shown in Figure 1–11(b).

Review Questions

6. What is the difference between an analog signal and a digital signal?
7. What are the two steps required to convert an analog signal into a digital signal?
8. How is the average value of a sinusoidal wave calculated?
9. What is a harmonic?
10. How does the spectrum for a repetitive waveform differ from that of a nonrepetitive waveform?

1–3 TROUBLESHOOTING ANALOG CIRCUITS

Technicians must diagnose and repair malfunctioning circuits or systems. Troubleshooting is the application of logical thinking to correct the malfunctioning circuit or system. Troubleshooting skills will be emphasized throughout the text.

In this section, you will learn to describe the process for troubleshooting an analog circuit.

Analysis, Planning, and Measuring

When you troubleshoot any circuit, the first step is to analyze the clues (symptoms) of a failure. The analysis can begin by determining the answer to several questions: Has the circuit ever worked? If so, under what conditions did it fail? What are the symptoms of a failure? What are possible causes of this failure? The process of asking these questions is part of the analysis of a problem.

After you analyze the clues, the second step in the troubleshooting process is forming a logical plan for troubleshooting. Planning the process can save a lot of time. As part of this plan, you must have a working understanding of the circuit you are troubleshooting. Take the time to review schematics, operating instructions, or other pertinent information if you are not certain how the circuit should operate. It may turn out that the failure was that of the operator, not the circuit! A schematic with proper voltages or waveforms marked at various test points is particularly useful for troubleshooting.

Logical thinking is the most important part of planning to troubleshoot but rarely can solve the problem by itself. The third step is to narrow the number of the possible failures by making carefully thought-out measurements. These measurements usually confirm the direction you are taking in solving the problem or point to a new direction. Occasionally, you may find a totally unexpected result! Measurements in turn give you more information for analysis. As you analyze new information, you will eventually determine the cause of the problem.

The thinking process that is part of analysis, planning, and measuring is best illustrated with an example. Suppose you have a string of 16 decorative lamps connected in series to a 120 V source as shown in Figure 1–12. Assume that this circuit worked at one time and stopped after moving it to a new location. When plugged in, the lamps fail to turn on. How would you go about finding the trouble?

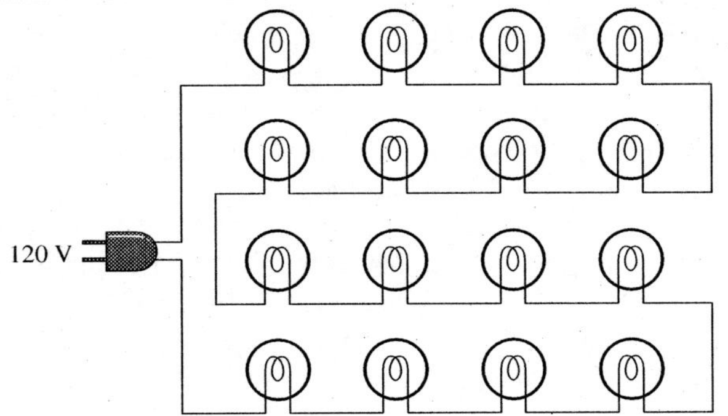

FIGURE 1–12

A series of lights. Is one of them open?

You might think like this: Since the circuit worked before it was moved, the problem could be that there is no voltage at this location. Or perhaps the wiring was loose and pulled apart when moved. It's possible a bulb burned out or became loose. This reasoning has considered possible causes and failures that could have occurred. The fact that the circuit was once working eliminates the possibility that the original circuit may have been incorrectly wired. In a series circuit, the possibility of two open paths occurring together is unlikely. You have analyzed the problem and now you are ready to plan the troubleshooting approach.

The first part of your plan is to measure (or test) for voltage at the new location. If voltage is present, then the problem is in the light string. If voltage is not present, check the circuit breakers at the input panel to the house. Before resetting breakers, you should think about why a breaker may be tripped.

The second part of your plan assumes voltage is present and the string is bad. You can disconnect power from the string and make resistance checks to begin isolating the problem. Alternatively, you could apply power to the string and measure voltage at various points. The decision whether to measure resistance or voltage is a toss-up and can be made based on the ease of making the test. Seldom is a troubleshooting plan developed so completely that all possible contingencies are included. The troubleshooter will frequently need to modify the plan as tests are made. You are ready to make measurements.

Suppose you have a digital multimeter (DMM) handy. You check the voltage at the source and find 120 V present. Now you have eliminated one possibility (no voltage). You know the problem is in the string, so you proceed with the second part of your plan. You might think: Since I have voltage across the entire string, and apparently no current in the circuit (since no bulb is on), there is almost certainly an open in the path either in a bulb or a connection. To eliminate testing each bulb, you decide to break the circuit in the middle and to check the resistance of each half of the circuit.

Now you are using logical thinking to reduce the effort needed. The technique you are using is a common troubleshooting procedure called *half-splitting.* By measuring the resistance of half the bulbs at once, you can reduce the effort required to find the open. Continuing along these lines, by half-splitting again, will lead to the solution in a few tests.

Unfortunately, most troubleshooting is more difficult than this example. However, analysis and planning are important for effective troubleshooting. As measurements are made, the plan is modified; the experienced troubleshooter narrows the search by fitting the symptoms and measurements into a possible cause.

SAFETY NOTE

A sturdy workbench equipped with at least ten grounded outlets is an important aspect of working on and troubleshooting electronic circuits. Metal parts of a workbench should be grounded and covered with insulating material, such as high-pressure laminate, so that metal surfaces are not exposed. Tools used in troubleshooting should have insulated handles, and power tools should be double insulated to protect against shock. Instruments should be periodically checked and inspected. The workbench should be free of clutter. A class-C fire extinguisher rated for electrical fires and an emergency power disconnect should be nearby.

Soldering

When repairing circuit boards, sooner or later the technician will need to replace a soldered part. When you replace any part, it is important to be able to remove the old part without damaging the board by excessive force or heat. Transfer of heat for removal of a part is facilitated with a chisel tip (as opposed to a conical tip) on the soldering iron.

Before installing a new part, the area must be clean. Old solder should be completely removed without exposing adjacent devices to excess heat. A degreasing cleaner or alcohol is suggested for cleaning. Remember, solder won't stick to dirty copper! Solder must be a resin core type; acid solder is never used in electronic circuits and shouldn't even be on your workbench. Solder is applied to the joint (not to the iron). As the solder cools, it must be kept still. A good solder connection is a smooth, shiny one and the solder flows into the printed circuit trace. A poor solder connection looks dull. During repair, it is possible for excessive solder to short together two parts or two pins on an integrated circuit (this rarely happens when boards are machine soldered). This is called a solder bridge, and the technician must be alert for this type of error when repairing boards. After the repair is completed, any flux must be removed from the board with alcohol or other cleaner.

Basic Test Equipment

The ability to troubleshoot effectively requires the technician to have a set of test equipment available and to be familiar with the operation of the instruments. An oscilloscope, DMM, and power supply are basic instruments for troubleshooting. These instruments are shown in Figure 1–13. While no one instrument is best for all situations, it is important to understand the limitations of the test equipment at hand. All electronic measuring instru-

FIGURE 1–13

Test instruments. ((a) and (b) Courtesy of B+K Precision. (c) used with permission from Tekronix, Inc.)

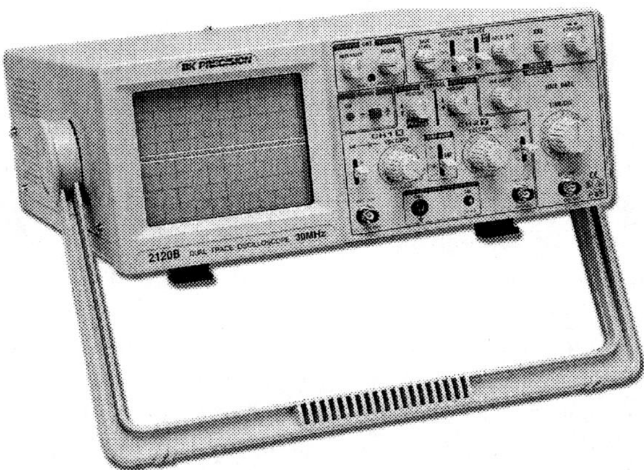

(a) Digital oscilloscope

(b) Digital multimeter

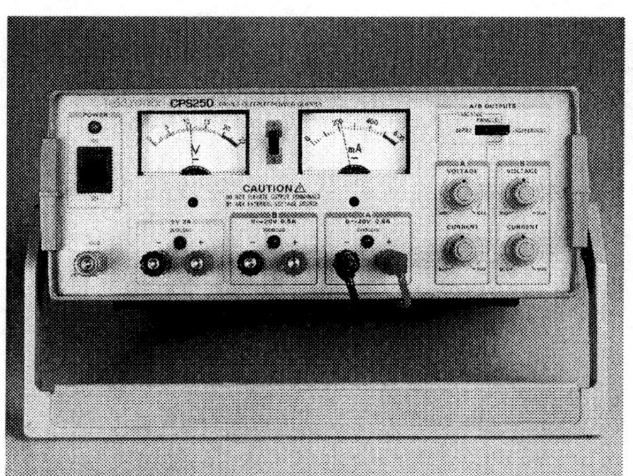

(c) Triple output power supply

ments become part of the circuit they are measuring and thus affect the measurement itself (an effect called *instrument loading*). In addition, instruments are specified for a range of frequencies and must be properly calibrated if readings are to be trusted. An expert troubleshooter must consider these effects when making electronic measurements.

For general-purpose troubleshooting of analog circuits, all technicians need access to an oscilloscope and a DMM. The oscilloscope needs to be a good two-channel scope, fast enough to spot noise or ringing when it occurs. A set of switchable probes with the ability to switch between ×1 (times one attenuation) and ×10 (times ten attenuation) is useful for looking at large or small signals. (Note that in the ×1 position, the scope loses bandwidth.)

The DMM is a general-purpose meter that has the advantage of very high input impedance but may have error if used in circuits with frequencies above a few kilohertz. Many new DMMs offer special features, such as continuity testing and diode checking, and may include capacitance and frequency measurements.

While DMMs are excellent test instruments, the VOM (Volt-Ohm-Milliammeter) has some advantages (for example, spotting trends faster than a digital meter). The VOM is an analog multimeter, with a needle read against a scale. A typical VOM is shown in Figure 1–14. Although generally not as accurate as a DMM, a VOM has very small capacitance to ground, and it is isolated from the line voltage. Also, because a VOM is a passive device, it will not tend to inject noise into a circuit under test.

FIGURE 1–14

A typical VOM. (Photo courtesy of Triplett Corporation)

Many times the circuit under test needs to have a test signal injected to simulate operation in a system. The circuit's response is then observed with a scope or other instrument. This type of testing is called *stimulus-response testing* and is commonly used when a portion of a complete system is tested. For general-purpose troubleshooting, the function generator is used as the stimulus instrument. All function generators have a sine wave, square wave, and triangle wave output; the frequency range varies widely, from a low frequency of 1 Hz to a high of 50 MHz (or more) depending on the generator. Higher-quality function generators offer the user a choice of other waveforms (pulses and ramps, for example) and may have triggered or gated outputs as well as other features.

The basic function generator waveforms (sine, square, and triangle) are used in many tests of electronic circuits and equipment. A common application of a function generator is to inject a sine wave into a circuit to check the circuit's response. The signal is capacitively

FIGURE 1–15

Square wave response of wide-band amplifiers.

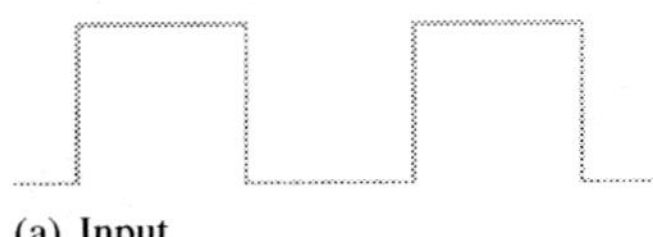

(a) Input

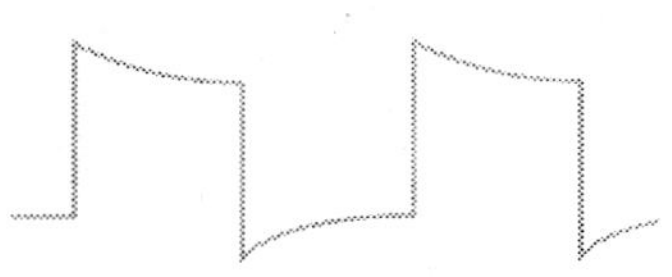

(b) Output: low-frequency attenuation

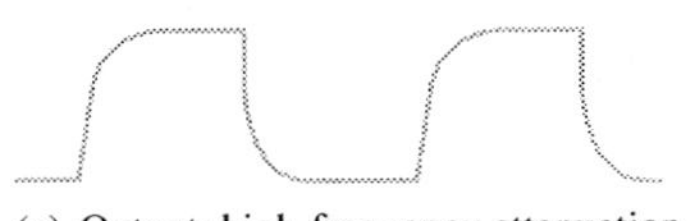

(c) Output: high-frequency attenuation

coupled to the circuit to avoid upsetting the bias network; the response is observed on an oscilloscope. With a sine wave, it is easy to ascertain if the circuit is operating properly by checking the amplitude and shape of the sine wave at various points or to look for possible troubles such as high-frequency oscillation.

A common test for wide-band amplifiers is to inject a square wave into a circuit to test the frequency response. Recall that a square wave consists of the fundamental frequency and an infinite number of odd harmonics. The square wave is applied to the input of the test circuit and the output is monitored. The shape of the output square wave indicates if specific frequencies are selectively attenuated.

Figure 1–15 illustrates square wave distortions due to selective attenuation of low or high frequencies. A good amplifier should show a high-quality replica of the input. If the square wave sags, as in Figure 1–15(b), low frequencies are not being passed properly by the circuit. The rising edge contains mostly higher-frequency harmonics. If the square wave rolls over before reaching the peak, as in Figure 1–15(c), high frequencies are being attenuated. The rise time of the square wave is an indirect measurement of the bandwidth of the circuit.

For testing dc voltages or providing power to a circuit under test, a multiple output power supply with both positive and negative outputs is necessary. The outputs should be variable from 0 to 15 V. A separate low-voltage supply is also handy for powering logic circuits or as a dc source for analog circuits.

Review Questions

11. What is the first step in troubleshooting a circuit?
12. What type of solder is used in electronics work?
13. What is meant by half-splitting?
14. What is meant by instrument loading?
15. How does testing with a square wave indicate an amplifier's response?

CHAPTER REVIEW

Key Terms

AC resistance The ratio of a small change in voltage divided by the corresponding small change in current.

Analog signal A signal that can take on a continuous range of values within certain limits.

Digital signal A noncontinuous signal that has discrete numerical values assigned to specific steps.

Domain The values assigned to the independent variable. Frequency or time is typically used as the independent variable for plotting signals.

Quantizing The process of assigning numbers to sampled data.

Sampling The process of breaking the analog waveform into time "slices" that approximate the original wave.

Signal In electronics, an electrical voltage or current that contains information.

Spectrum A plot of amplitude versus frequency for a signal.

Important Facts

- ❑ A linear component is one in which an increase in current is proportional to the applied voltage.
- ❑ An analog signal takes on a continuous range of values within limits. A digital signal is a discrete signal that can have only certain values.
- ❑ Waveforms that repeat in a certain interval of time are said to be periodic. A cycle is the complete sequence of values that a waveform exhibits before an identical pattern occurs. The period is the time interval for one cycle.
- ❑ A sinusoidal wave is the most basic wave in electronics. Sinusoidal waves are characterized by the frequency and amplitude.
- ❑ Signals that have voltage, current, resistance, or other quantity varies as a function of time are called time-domain signals. When the frequency is made the independent variable, the result is a frequency-domain signal. Any signal can be observed in either the time domain or the frequency domain.
- ❑ A transducer is a device that converts a physical quantity from one form to another for electronic systems. Input transducers convert a physical quantity to an electrical quantity (voltage, current, resistance).
- ❑ Troubleshooting begins with analyzing the symptoms of a failure, then forming a logical plan. Carefully thought-out measurements arc made to narrow the search for the cause of the failure. These measurements may modify or change the plan.
- ❑ For general-purpose troubleshooting, a reasonably fast two-channel oscilloscope and a DMM are the principal measuring instruments. The most common stimulus instruments are a function generator and a regulated power supply.

Formulas

Ohm's law:

$$I = \frac{V}{R} \qquad \textbf{(1–1)}$$

Instantaneous value of a sinusoidal wave:

$$y(t) = A\sin(\omega t \pm \phi) \qquad \textbf{(1–2)}$$

Conversion from radian frequency (rad/s) to hertz (Hz):

$$f(\text{Hz}) = \frac{\omega\ (\text{rad/s})}{2\pi\ (\text{rad/cycle})} \qquad \textbf{(1–3)}$$

Conversion from frequency to period:

$$T = \frac{1}{f} \qquad \textbf{(1–4)}$$

Conversion from period to frequency:

$$f = \frac{1}{T} \qquad \textbf{(1–5)}$$

Conversion from peak voltage to average voltage for a sinusoidal wave:

$$V_{(avg)} = 0.637V_p \qquad \textbf{(1–6)}$$

Power law:

$$P = IV \tag{1–7}$$

Conversion from peak voltage to rms voltage for a sinusoidal wave:

$$V_{rms} = 0.707V_p \tag{1–8}$$

Chapter Checkup

Answers are at the end of the chapter.

1. The graph of a linear equation
   (a) always has a constant slope
   (b) always goes through the origin
   (c) must have a positive slope
   (d) answers (a), (b), and (c)
   (e) none of these answers
2. AC resistance is defined as
   (a) voltage divided by current
   (b) a change in voltage divided by a corresponding change in current
   (c) current divided by voltage
   (d) a change in current divided by a corresponding change in voltage
3. A discrete signal
   (a) changes smoothly
   (b) can take on any value
   (c) is the same thing as an analog signal
   (d) answers (a), (b), and (c)
   (e) none of these answers
4. The process of assigning numeric values to a signal is called
   (a) sampling (b) multiplexing
   (c) quantizing (d) digitizing
5. The reciprocal of the repetition time of a periodic signal is the
   (a) frequency (b) angular frequency
   (c) period (d) amplitude
6. If a sinusoidal wave has a peak amplitude of 10 V, the rms voltage is
   (a) 0.707 V (b) 6.37 V
   (c) 7.07 V (d) 20 V
7. If a sinusoidal wave has a peak-to-peak amplitude of 325 V, the rms voltage is
   (a) 103 V (b) 115 V
   (c) 162.5 V (d) 460 V
8. Assume the equation for a sinusoidal wave is $v(t) = 200 \sin(500t)$. The peak voltage is
   (a) 100 V (b) 200 V
   (c) 400 V (d) 500 V

9. A harmonic is
   (a) an integer multiple of a fundamental frequency
   (b) an unwanted signal that adds noise to a system
   (c) a transient signal
   (d) a pulse
10. An important rule for soldering is
   (a) always use a good acid-based solder
   (b) always apply solder directly to the iron, never to the parts being soldered
   (c) wiggle the solder joint as it cools to strengthen it
   (d) answers (a), (b), and (c)
   (e) none of these answers
11. Assume the circuit in Figure 1–16 is plugged in and operating normally and a bulb is then removed. The voltage across the socket of the removed bulb will
   (a) increase (b) decrease
   (c) not change

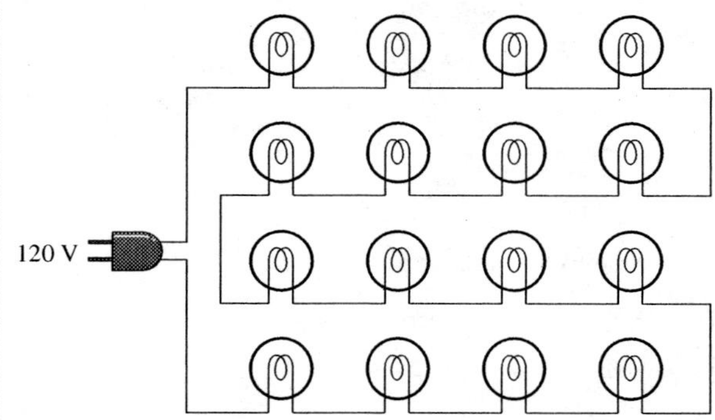

FIGURE 1–16

12. For the circuit described in Question 11, the voltage across all other bulbs will
   (a) increase (b) decrease
   (c) not change
13. For the circuit described in Question 11, the voltage to the circuit will
   (a) increase (b) decrease
   (c) not change
14. Assume one of the sockets in the circuit in Figure 1–16 is shorted with the bulb out, but the other bulbs are on. As a result of the short, the voltage across each of the other bulbs will
   (a) increase (b) decrease
   (c) not change
15. For the circuit described in Question 14, the total voltage applied to the circuit will
   (a) increase (b) decrease
   (c) not change

16. For the circuit described in Question 14, the light output from the other bulbs will
   (a) increase (b) decrease
   (c) not change
17. Assume the circuit in Figure 1–16 is disconnected from the source and the resistance is measured between the prongs. If one of the sockets is shorted, the total resistance will
   (a) increase (b) decrease
   (c) not change
18. For the circuit described in Question 17, if one of the bulbs is open, the total resistance will
   (a) increase (b) decrease
   (c) not change

Questions

Answers to odd-numbered questions are at the end of the book.

1. What are two classifications of integrated circuits?
2. How does the ac resistance of a diode change as the voltage increases?
3. What is the slope-intercept form of the equation for a straight line?
4. What is the difference between rms and average value of a sine wave?
5. What is a transducer?
6. What is a periodic wave?
7. What is the fifth harmonic of a 500 Hz triangular wave?
8. What is the only type of harmonics found in a square wave?
9. What information is obtained when a square wave calibration signal is used as the input to an oscilloscope?
10. How can you protect a static-sensitive circuit from damage when you are working on it?
11. What are two important advantages of a digital storage oscilloscope over an analog oscilloscope?
12. What are basic rules for electronic soldering?
13. What measurements can be made with a DMM?
14. What are three basic waveforms that are found on function generators?

PROBLEMS

Basic Problems

Answers to odd-numbered problems are at the end of the book.

1. What is the conductance of a 22 kΩ resistor?
2. Compute the ac resistance of the diode in Figure 1–4 at the point $V = 0.7$ V, $I = 5.0$ mA.
3. Sketch the shape of an *IV* characteristic curve for a device that has a decreasing ac resistance as voltage increases.
4. Assume a sinusoidal wave is described by the equation $v(t) = 100\text{ V}\sin(200t + 0.52)$.
   (a) From this expression, determine the peak voltage, the average voltage, and the angular frequency in rad/s.

(b) Find the instantaneous voltage at a time of 2.0 ms. (Reminder: the angles are in radians in this equation).

5. Determine the frequency (in Hz) and the period (in s) for the sinusoidal wave described in Problem 4.

6. An oscilloscope shows a wave repeating every 27 μs. What is the frequency of the wave?

7. What is the period of a 10 kHz sine wave?

8. How many cycles of a 1 kHz wave are there in 1 minute?

9. What is the current in a bulb that dissipates 75 W if the supplied voltage is 115 Vac?

10. A DMM indicates the rms value of a sinusoidal wave. If a DMM indicates a sinusoidal wave is 3.5 V, what peak-to-peak voltage would you expect to observe on an oscilloscope?

11. The ratio of the rms voltage to the average voltage for any wave is called the *form factor* (used occasionally to convert meter readings). What is the form factor for a sinusoidal wave?

12. What power is dissipated by a lamp operated from a 12 V supply if the current in the lamp is 0.83 A?

13. What is the resistance of the lamp in Problem 12?

Basic-Plus Problems

14. Two identical bulbs, each rated at 100 W at 110 V, are wired in series and connected to 110 V. Assuming the current is measured and found to be 0.55 A, what power is dissipated in each bulb?

15. A waveform has a radian frequency of 2000 rad/s.

    (a) What is the frequency in Hz?

    (b) What is the period?

16. An oscilloscope shows five complete cycles of a waveform in 10 divisions. The SEC/DIV control is set to 50 μs/div. What is the frequency of the waveform?

17. Figure 1–17 shows a small system consisting of four microphones connected to a two-channel amplifier through a selector switch (SW1). Either the *A* set or the *B* set

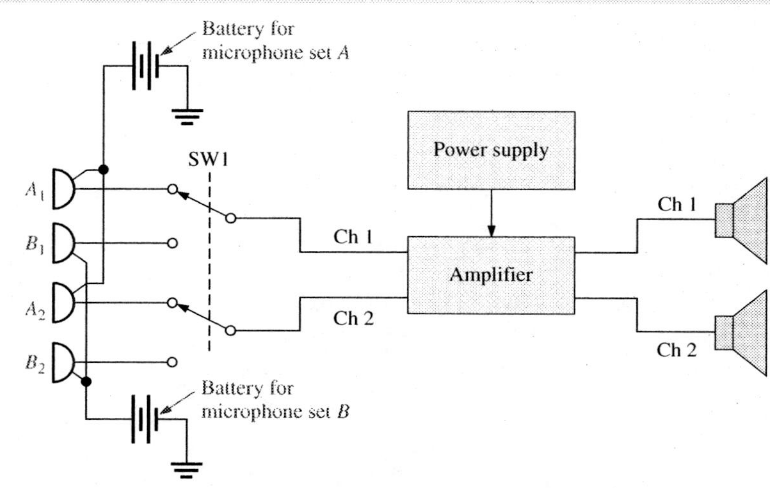

FIGURE 1–17
A small system consisting of a two-channel amplifier and four microphones.

of microphones is selected and amplified. The output of the amplifier is connected to two speakers. Power to the amplifier is supplied by a single power supply that furnishes dc voltages to the amplifier and two batteries that provide power to the microphones as shown.

Assume no sound is heard when the system is plugged in and turned on. Outline a basic troubleshooting plan by indicating the tests you would make to isolate the trouble to either the power supply, amplifier, a microphone, microphone battery, switch, speaker, or other fault.

18. For the system described in Problem 17, outline a basic troubleshooting plan for the case where Channel 1 operates normally but no sound is heard from Channel 2. Indicate the tests you would make to isolate the trouble. (Can you think of a method to do half-splitting?)

ANSWERS

Example Questions

1–1: $G_2 = 375$ mS; $R_2 = 2.67$ kΩ

1–2: $V_{rms} = 10.6$ V; $f = 95$ Hz; $T = 10.5$ mS

Review Questions

1. Digital and analog
2. Conductance
3. The slope of curve is lower for larger resistors.
4. DC resistance is the voltage divided by the current. AC resistance is the *change* in voltage divided by the *change* in current.
5. Conventional current is based on positive-to-negative flow of charge. Electron flow is based on negative-to-positive flow of charge.
6. An analog signal takes on a continuous range of values; a digital signal represents information that has a discrete number of codes.
7. Sampling and quantizing
8. The average is computed based on absolute values. The equation is $V_{avg} = 2V_p/\pi$.
9. A harmonic is a frequency that is an integer multiple of another frequency called the fundamental.
10. The spectrum for the repetitive waveform is a line spectrum; the spectrum for the nonrepetitive waveform is a continuous spectrum.
11. Analyzing the symptoms of a failure by asking questions: Has the circuit ever worked? If so, under what conditions did it fail? What are the symptoms of a failure? What are possible causes of this failure?
12. Rosin core
13. Half-splitting divides a troubleshooting problem into halves and determines which half of the circuit is likely to have the problem.
14. Instrument loading is the effect of changing circuit voltages due to the process of connecting an instrument.
15. If the frequency response is flat, a square wave will be passed unchanged. If it is not flat, the shape reveals if high or low frequencies are selectively attenuated.

Chapter Checkup

1. (a)	2. (b)	3. (e)	4. (c)	5. (a)
6. (c)	7. (b)	8. (b)	9. (a)	10. (e)
11. (a)	12. (b)	13. (c)	14. (a)	15. (c)
16. (a)	17. (b)	18. (a)		

CHAPTER

DIODES AND APPLICATIONS

CHAPTER OUTLINE

Study aids for this chapter are available at

http://www.prenhall.com/SOE

INTRODUCTION

Semiconductors are special crystalline materials that have electrical conductivity between that of a conductor and an insulator. By adding a tiny amount of an impurity to the crystalline material, the conductivity can be controlled. Semiconductors form the backbone of the electronics industry because they are rugged, inexpensive, and can form diodes, transistors, and other important electronic components.

Diodes are an important class of semiconductors that allow current in only one direction, like a one-way valve. In this chapter, you are introduced to the *pn* junction, an important concept essential to the understanding of diodes and transistors. Diodes are semiconductors containing one *pn* junction. After learning about *pn* junctions, you will learn about several important types of diodes and applications.

KEY TERMS

- Energy
- Shell
- Valence electron
- Conduction electron
- Diode
- Barrier potential
- Bias
- Anode
- Cathode
- Rectifier
- Zener diode
- Integrated circuit

KEY OBJECTIVES

A section number is given for each objective, after completing this chapter, you should be able to

2–1 Discuss the basic atomic structure of semiconductors

2–2 Describe the characteristics of a *pn* junction

2–3 Explain how to bias a semiconductor diode

2–4 Describe the diode *IV* characteristic curve

2–5 Analyze the operation of three basic rectifier circuits

2–6 Describe the function of power supply filters and regulators

2–7 Discuss steps in troubleshooting a power supply

COMPUTER SIMULATIONS DIRECTORY

The following figures have Multisim circuit files associated with them. To open a Multisim file, go to the website at http://www.prenhall.com/SOE, click on the cover of this book, choose this chapter, click on "Multisim", and then click on the selected file.

- ◆ Figure 2–16 Page 43
- ◆ Figure 2–20 Page 46
- ◆ Figure 2–28 Page 52

LABORATORY EXPERIMENTS DIRECTORY

The following exercises are for this chapter.

- ◆ **Experiment 2** The Diode Characteristic
- ◆ **Experiment 3** Rectifier Circuits

Advances in one branch of science often have an impact on other branches of science. The Periodic Table was developed as a means of classifying the elements in the science of chemistry, but it is applied to many other sciences. Chemistry students are all made aware of the importance of the Periodic Table early in their study of chemistry. Virtually all science labs have the table displayed predominately.

The Periodic Table classifies elements by groups or families. Except for the rare earth elements, each family is in a separate column of the table. For example, the first column is alkali metals. These are all highly reactive silvery metals. The last column is the noble gases, a group of elements that are all unreactive colorless gases.

The family of elements that has carbon on top includes two elements that are widely used as semiconductors — silicon and germanium. These are elements with four electrons in the outer shell, as you will learn in this chapter. Interestingly, carbon, in the form of diamond can also be a semiconductor (but is not used commercially).

Certain materials are used to add impurities to silicon and germanium. The impurities change the conductive properties of pure silicon. If an element from the family to the left of carbon is added to silicon, the outer shell of the impurity will contain one less electron than silicon, and a *p*-type semiconductor will be formed. If an element from the family to the right of carbon is added to silicon, the outer shell of the impurity will contain one more electron than silicon, and an *n*-type semiconductor will be formed. The formation of *p*- and *n*-type materials is part of the story of forming diodes, transistors, and other semiconductor materials.

2–1 THE ATOMIC STRUCTURE OF SEMICONDUCTORS

Electronic devices such as diodes and transistors are constructed from special materials called semiconductors.

In this section, you will learn about the atomic structure of semiconductors.

Electron Shells and Orbits

The electrical properties of materials are explained by their atomic structure. In the early part of the 20th century, Neils Bohr, a Danish physicist, developed a model of the atom that showed electrons orbiting the nucleus. In Bohr's model of the atom, the electrons orbit only in certain discrete (separate and distinct) distances from the nucleus. The nucleus contains positively charged protons and uncharged neutrons. The orbiting electrons are negatively charged. Modem quantum mechanical models of the atom retain much of the ideas of the original Bohr model but have replaced the concept of electron "particles" with mathematical "matter waves"; however, the Bohr model provides a useful mental picture of the structure of an atom.

Energy is the ability to do work and is subdivided into potential (position), kinetic (motion), and rest (mass). Electrons have potential energy because of their position within the atom. Electrons near the nucleus have less energy than those in more distant orbits. These discrete orbits mean that only certain energy levels are permitted within the atom. These energy levels are known as **shells.** Each shell has a certain maximum permissible number of electrons. The differences in energy levels within a shell are much smaller than the difference in energy between shells. The shells are designated 1, 2, 3, 4, and so on, with 1 being closest to the nucleus, as illustrated in Figure 2–1.

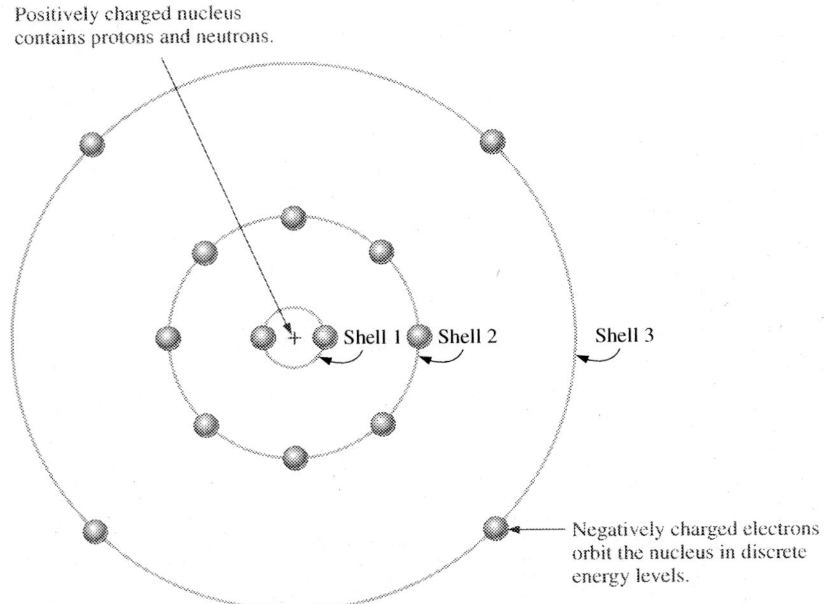

FIGURE 2–1

Energy levels increase as distance from the nucleus increases. In this neutral silicon atom, 14 protons are in the nucleus and 14 electrons orbit the nucleus.

Valence Electrons, Conduction Electrons, and Ions

Electrons in orbits farther from the nucleus are less tightly bound to the atom than those closer to the nucleus. This is because the force of attraction between the positively charged nucleus and the negatively charged electron decreases with increasing distance. Outer-shell electrons are also shielded from the nuclear charge by the inner-shell electrons.

An electron that is in the outermost shell is called a **valence electron**; valence electrons have the highest energy and are relatively loosely bound to their parent atom. A valence electron is identical to any other electron; it is so named because of its location only. For the silicon atom in Figure 2–1, the third-shell electrons are the valence electrons.

Sometimes, a valence electron can acquire enough energy to break free of its parent atom. This free electron is then called a **conduction electron** because it is not bound to any certain atom. When a negatively charged electron is freed from an atom, the rest of the atom is positively charged and is said to be a **positive ion**. In some chemical reactions, the freed electron attaches itself to a neutral atom (or group of atoms), forming a **negative ion**. An ion is always an atom or group of atoms that has acquired a charge due to an imbalance in the number of protons and electrons.

Metallic Bonds

Metals tend to be solids at room temperature. The nucleus and inner-shell electrons of metals occupy fixed lattice positions. The outer valence electrons are held loosely by all of the atoms of the crystal and are free to move about. This "sea" of negatively charged electrons holds the positive ions of the metal together, forming metallic bonding.

With the large number of atoms in the metallic crystal, the discrete energy level for the valence electrons is blurred into a band called the *valence band*. These valence electrons are mobile and account for the thermal and electrical conductivity of metals. In addition to the valence energy band, the next (normally occupied) level from the nucleus in the atom is also blurred into a band of energies called the *conduction band*.

Figure 2–2 compares the energy-level diagrams for three types of solids. Notice that for conductors, shown in Figure 2–2(a), the bands are overlapping. Electrons can easily move between the valence and conduction bands by absorbing light (and radiating light as they

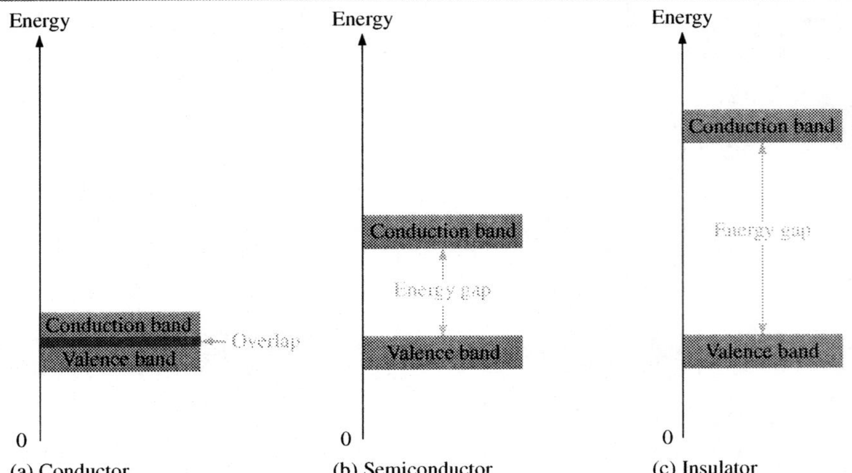

FIGURE 2–2

Energy level diagrams for three types of materials. The upper level is the conduction band; the lower level is the valence band.

move back). This movement of electrons back and forth between the valence band and conduction band accounts for the luster of metals.

Covalent Bonds

Atoms of some solid materials form **crystals**, which are three-dimensional structures held together by strong bonds between the atoms. In diamond, for example, four bonds are formed by the sharing of four valence electrons in each carbon atom with adjacent atoms. This effectively creates eight valence electrons for each atom and produces a state of chemical stability. This sharing of valence electrons produces strong **covalent bonds** that hold the atoms together.

The shared electrons are not mobile; each electron is associated by a covalent bond between the atoms of the crystal. Therefore, there is a large energy gap between the valence band and the conduction band. As a consequence, crystalline materials such as diamond are insulators, or nonconductors, of electricity. Figure 2–2(c) shows the energy bands for a solid insulator.

Electronic devices are constructed from materials called semiconductors. The most common semiconductive material is silicon; however, germanium is sometimes used. At room temperature, silicon forms a covalent crystal. The actual atomic structure is similar to diamond, but the covalent bonds in silicon are not as strong as those in diamond. In silicon, each atom shares a valence electron with each of its four neighbors. As in the case of other crystalline materials, the discrete levels are blurred into a valence band and a conduction band, as shown in Figure 2–2(b).

The important difference between a conductor and a semiconductor is the gap that separates the bands. With semiconductors, the gap is narrow; electrons can easily be promoted to the conduction band with the addition of thermal energy. At absolute zero, the electrons in a silicon crystal are all in the valence band, but at room temperature many electrons have sufficient energy to move to the conduction band. The conduction band electrons are no longer bound to a parent atom within the crystal.

Electrons and Hole Current

When an electron jumps to the conduction band, a vacancy is left in the valence band. This vacancy is called a **hole**. For every electron raised to the conduction band by thermal or light energy, there is one hole left in the valence band, creating what is called an electron-hole pair. *Recombination* occurs when a conduction-band electron loses energy and falls back into a hole in the valence band.

A piece of **intrinsic** (pure) silicon at room temperature has, at any instant, a number of conduction-band (free) electrons that are unattached to any atom and are essentially drifting randomly throughout the material. Also, an equal number of holes are created in the valence band when these electrons jump into the conduction band. Pure silicon has electrons in the conduction band and holes in the valence band.

When a voltage is applied across a piece of intrinsic silicon, as shown in Figure 2–3, the thermally generated free electrons in the conduction band are easily attracted toward the positive end. This movement of free electrons is one type of current in a semiconductor and is called *electron current.*

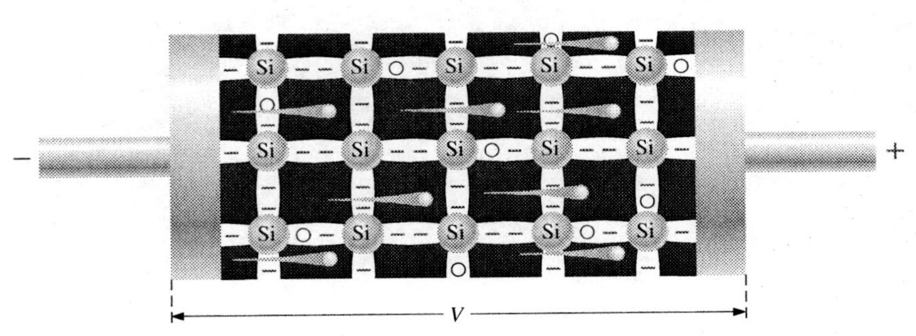

FIGURE 2–3

Electron current in pure silicon. Free electrons are shown with "tails" to indicate mobility.

Another type of current occurs at the valence level, where the holes created by the free electrons exist. Electrons remaining in the valence band are still attached to their atoms and are not free to move randomly in the crystal structure. However, a valence electron can move into a nearby hole, with little change in its energy level, thus leaving another hole where it came from. Effectively, the hole has moved from one place to another in the crystal structure, as illustrated in Figure 2–4. This current is called *hole current.*

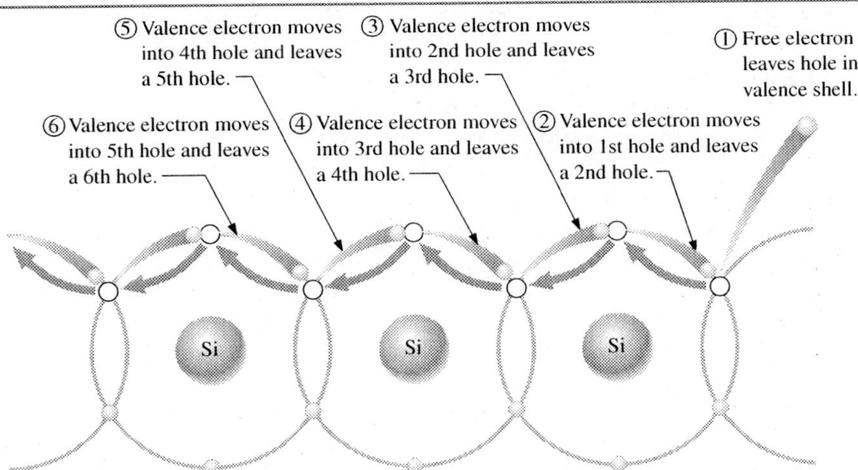

When a valence electron moves left to right to fill a hole while leaving another hole behind, a hole has effectively moved from right to left. Gray arrows indicate effective movement of a hole.

FIGURE 2–4

Hole current in pure silicon.

Review Questions

Answers are at the end of the book.

1. With reference to the atom, what is a shell?
2. In an intrinsic semiconductor, in which energy band do free electrons exist? In which band do holes exist?

3. How are holes created in an intrinsic semiconductor?
4. Why is current established more easily in a semiconductor than in an insulator?
5. How does a metallic bond differ from a covalent bond?

2–2 THE *PN* JUNCTION

Intrinsic silicon is not a good conductor. By adding a small amount of impurity to the silicon crystal, its electrical properties can be changed dramatically. During manufacture, the silicon layers with different impurities are joined and form a boundary called the *pn* junction. Amazingly, it is the characteristics of the *pn* junction that allow diodes and transistors to work.

In this section, you will learn the characteristics of a *pn* junction.

Doping

The conductivity of silicon (or germanium) can be drastically increased by the controlled addition of impurities to the pure (intrinsic) semiconductive material. This process, called **doping**, increases the number of current carriers (electrons or holes), thus increasing the conductivity and decreasing the resistivity. The two categories of impurities are *n*-type and *p*-type.

To increase the number of conduction-band electrons in pure silicon, a controlled number of pentavalent impurity atoms called *donors* are added to the silicon crystal. These are atoms with five valence electrons, such as arsenic, phosphorus, and antimony. Each pentavalent atom forms covalent bonds with four adjacent silicon atoms, leaving one extra electron. This extra electron becomes a conduction (free) electron because it is not bonded to any atom in the crystal. The electrons in these *n* materials are called the *majority carriers;* the holes are called *minority carriers.*

To increase the number of holes in pure silicon, trivalent impurity atoms called *acceptors* are added during manufacture. These are atoms with only three valence electrons, such as aluminum, boron, and gallium. Each trivalent atom forms covalent bonds with four adjacent silicon atoms. All three of the impurity atom's valence electrons are used in the covalent bonds. However, since four electrons are required in the crystal structure, a hole is formed with each trivalent atom added. With *p* materials, the acceptor causes extra holes in the valence band; the majority carrier in *p* materials is holes, and the minority carrier is electrons.

The process of creating *n*-type or *p*-type materials retains the overall electrical neutrality. With *n*-type materials, the extra electron in the crystal is balanced by the additional positive charge of the donor's nucleus.

The *PN* Junction

When a piece of intrinsic silicon is doped so that half is *n* type and the other half is *p* type, a ***pn* junction** is formed between the two regions. The *n* region has many free electrons (majority carriers) and only a few thermally generated holes (minority carriers). The *p* region has many holes (majority carriers) and only a few thermally generated free electrons (minority carriers). The *pn* junction forms a basic diode and is fundamental to the operation of all solid-state devices. A **diode** is a device that allows current in only one direction.

The Depletion Region

When the *pn* junction is formed, some of the conduction electrons near the junction drift across into the *p* region and recombine with holes near the junction, as shown in Figure 2–5(a). For each electron that crosses the junction and recombines with a hole, a pentavalent atom is left with a net positive charge in the *n* region near the junction. Also, when the electron recombines with a hole in the *p* region, a trivalent atom acquires a net negative charge. As a result, positive ions are found on the *n* side of the junction and negative ions are found on the *p* side of the junction. The existence of the positive and negative ions on opposite sides of the junction creates a **barrier potential** (V_B) across the depletion region. The barrier potential depends on temperature, but it is approximately 0.7 V for silicon and 0.3 V for germanium at room temperature. Since germanium diodes are rarely used, 0.7 V is normally the barrier potential found in most diode circuits and will be the value we will assume in this text.

Formation of the *pn* junction. **FIGURE 2–5**

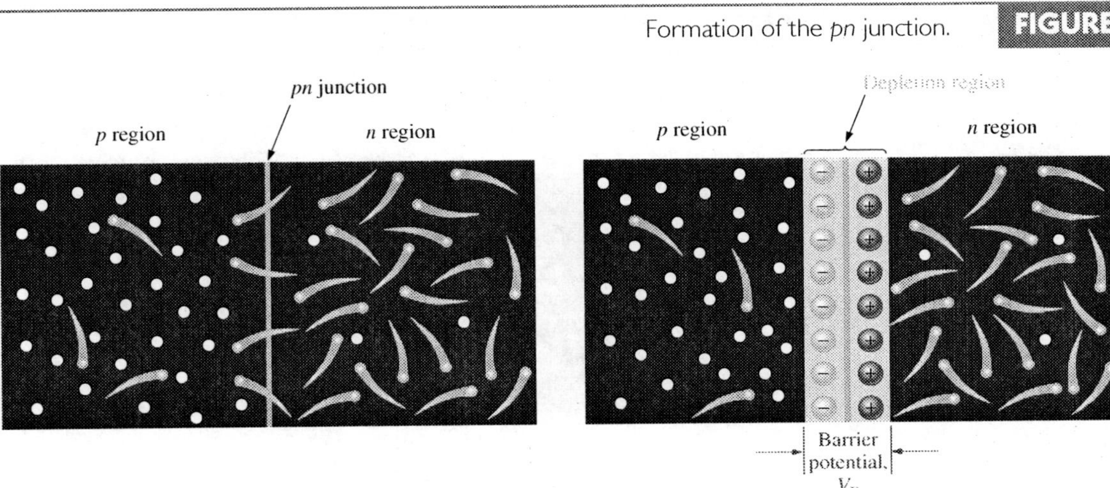

(a) At the instant of junction formation, free electrons in the *n* region near the *pn* junction begin to diffuse across the junction and fall into holes near the junction in the *p* region.

(b) For every electron that diffuses across the junction and combines with a hole, a positive charge is left in the *n* region and a negative charge is created in the *p* region, forming a barrier potential. This action continues until the voltage of the barrier repels further diffusion.

Conduction electrons in the *n* region must overcome both the attraction of the positive ions and the repulsion of the negative ions in order to migrate into the *p* region. After the ion layers build up, the area on both sides of the junction becomes essentially depleted of any conduction electrons or holes and is known as the **depletion region**. This condition is illustrated in Figure 2–5(b). Any further movement of charge across the boundary requires that the barrier potential be overcome.

Review Questions

6. How is an *n*-type semiconductor formed?
7. What are some common impurities used in forming an *n*-type of semiconductor?
8. How is a *p*-type semiconductor formed?
9. What are some common impurities used in forming a *p*-type of semiconductor?
10. What is the value of the barrier potential for silicon?

2–3 BIASING THE SEMICONDUCTOR DIODE

A single *pn* junction forms a semiconductor diode. There is no current across a *pn* junction at equilibrium. The primary usefulness of the semiconductor diode is its ability to allow current in only one direction as determined by the bias.

In this section, you will learn how to bias a semiconductor diode.

Forward Bias

The term **bias** in electronics refers to a fixed dc voltage that sets the operating conditions for a semiconductor device. **Forward bias** is the condition that permits current across a *pn* junction.

Figure 2–6 shows the polarity required from a dc source to forward-bias the semiconductor diode. The negative side of a source is connected to the *n* region (at the cathode **terminal**), and the positive side of a source is connected to the *p* region (at the anode terminal). When the semiconductor diode is forward-biased, the **anode** is the more positive terminal and the **cathode** is the more negative terminal.*

FIGURE 2–6

Electron flow in a forward-biased semiconductor diode.

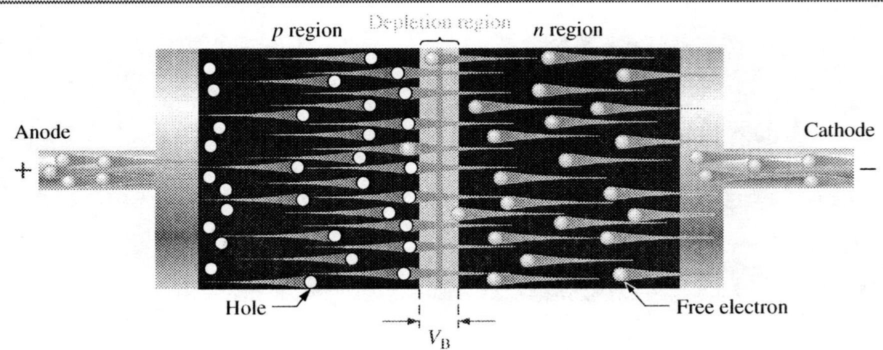

This is how forward bias works: When a dc source is connected to forward-bias the diode, the negative side of the source pushes the conduction electrons in the *n* region toward the junction because of electrostatic repulsion. The positive side pushes the holes in the *p* region also toward the junction. When the external bias voltage is sufficient to overcome the barrier potential, electrons have enough energy to penetrate the depletion region and cross the junction, where they combine with the *p* region holes. As electrons leave the *n* region, more electrons flow in from the negative side of the source. Thus, current through the *n* region occurs by the movement of conduction electrons (majority carriers) toward the junction. When the conduction electrons enter the *p* region and combine with holes, they become valence electrons. Then they move as valence electrons from hole to hole toward the positive anode connection. The movement of these valence electrons essentially creates a movement of holes in the opposite direction. Thus, current in the *p* region occurs by the movement of holes (majority carriers) toward the junction.

Reverse Bias

Reverse bias is the bias condition that prevents current across the *pn* junction. Figure 2–7 shows the polarity required from a dc source to reverse-bias the semiconductor diode. The

* Chemists define *anode* and *cathode* in terms of the type of chemical reaction that occurs in electrochemical cells. For electrochemistry, the anode is the terminal that acts as an electron donor; the cathode is the terminal that acts as an electron acceptor.

FIGURE 2–7

Reverse bias.

(b) There is transient current as depletion region widens.

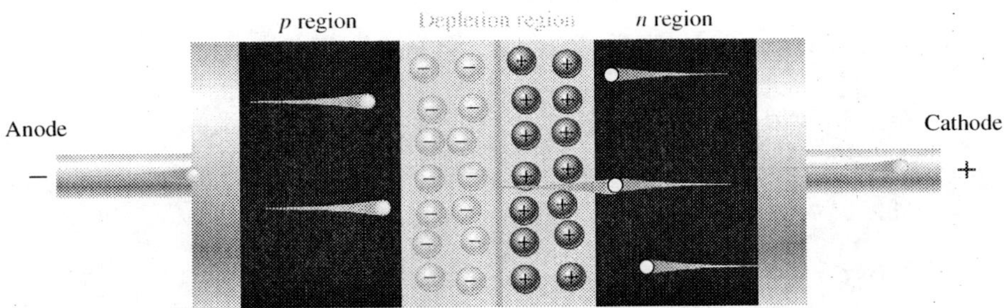

(c) Majority current ceases when barrier potential equals bias voltage. There is an extremely small reverse current due to minority carriers.

negative side of the source is connected to the *p* region, and the positive side to the *n* region. When the semiconductor diode is reverse-biased, the anode is the more negative terminal and the cathode is the more positive terminal.

This is how reverse bias works: The negative side of the source attracts holes in the *p* region away from the *pn* junction, while the positive side of the source attracts electrons away from the junction due to the attraction of opposite charges. As electrons and holes move away from the junction, the depletion region begins to widen; more positive ions are created in the *n* region, and more negative ions are created in the *p* region. The depletion region widens until the potential difference across it is equal to the external bias voltage, as shown in Figure 2–7(b). The depletion region effectively acts as an insulator between the layers of oppositely charged ions when the diode is reverse-biased.

Peak Inverse Voltage (PIV)

When a diode is reverse-biased, it must be able to withstand the maximum value of reverse voltage that is applied or it will break down. The maximum rated voltage for a diode is designated as *peak inverse voltage* (PIV). The required PIV depends on the application; for most cases with ordinary diodes, the PIV rating should be higher than the reverse voltage.

Reverse Breakdown

If the external reverse-bias voltage is increased to a large enough value, *avalanche breakdown* occurs. Here is what happens: Assume that one minority conduction-band electron acquires enough energy from the external source to accelerate it toward the positive end of the diode. During its travel, it collides with an atom and imparts enough energy to knock a valence electron into the conduction band. There are now two conduction-band electrons. Each will collide with an atom, knocking two more valence electrons into the conduction band. There are now four conduction-band electrons, which, in turn, knock four more into the conduction band. This rapid multiplication of conduction-band electrons, known as an *avalanche effect,* results in a rapid buildup of reverse current.

Most diode circuits are not designed to operate in reverse breakdown, and the diode may be destroyed if it is. By itself, reverse breakdown will not harm a diode, but current limiting must be present to prevent excessive heating. One type of diode, the zener diode, is specially designed for reverse-breakdown operation if sufficient current limiting is provided.

Review Questions

11. What is the condition for forward bias?
12. What is the condition for reverse bias?
13. Which type of bias condition produces a majority current carrier toward the junction?
14. Which type of bias condition produces a widening of the depletion region?
15. What is avalanche breakdown?

2–4 DIODE CHARACTERISTICS

A characteristic curve graphically shows the current-voltage relationship for a diode. The simplified model, called the offset model, enables you to apply the *IV* characteristic to nearly all practical diode circuits to calculate basic parameters.

In this section, you will learn to describe the diode *IV* characteristic curve.

Diode Symbol

Figure 2–8(a) shows the standard schematic symbol for a general-purpose diode. The two terminals of the diode are the anode and cathode, labeled A and K on the figure (K is frequently used for the cathode to avoid confusion with capacitors). The arrow on the symbol always points toward the cathode.

FIGURE 2–8

Diode schematic symbol and bias circuits. V_{BIAS} is the bias battery voltage, and V_B is the barrier potential. The resistor limits the forward current to a safe value.

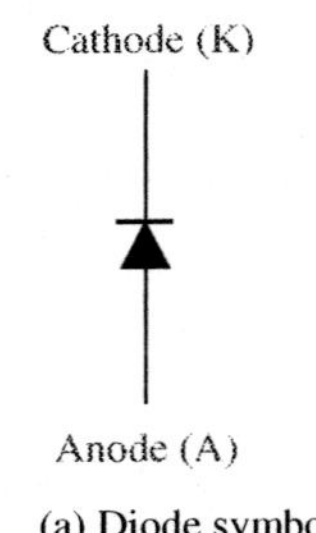

(a) Diode symbol

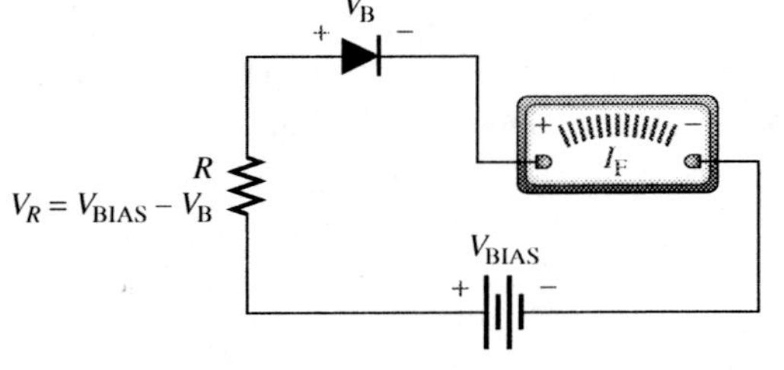

(b) Forward bias

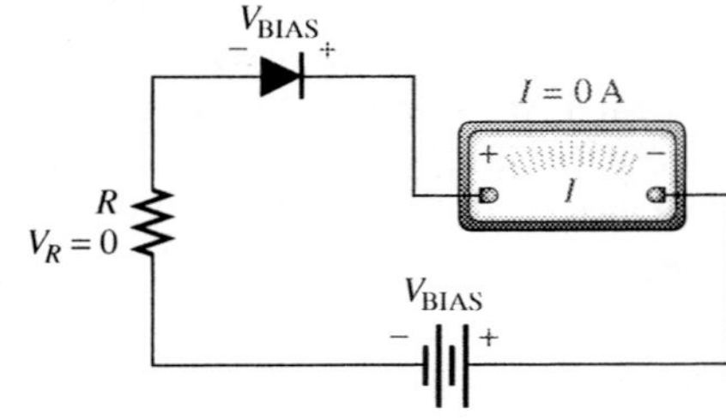

(c) Reverse bias

Figure 2–8(b) shows a forward-biased diode connected to a source through a current-limiting resistor. The anode is positive with respect to the cathode, causing the diode to conduct as indicated by the ammeter symbol. Remember that when the diode is forward-biased, the barrier potential, V_B, always appears between the anode and cathode, as indicated. The voltage across the resistor, V_R, is V_{BIAS} less the barrier potential, V_B.

Figure 2–8(c) shows the diode with reverse bias. The anode is negative with respect to the cathode, and the diode does not conduct as indicated by the ammeter symbol. The entire bias voltage, V_{BIAS}, appears across the diode. There is no voltage across the resistor because there is no current in the circuit.

Some typical diodes are shown in Figure 2–9 to illustrate common packaging. The letter A is used to identify the anode; K is used to identify the cathode.

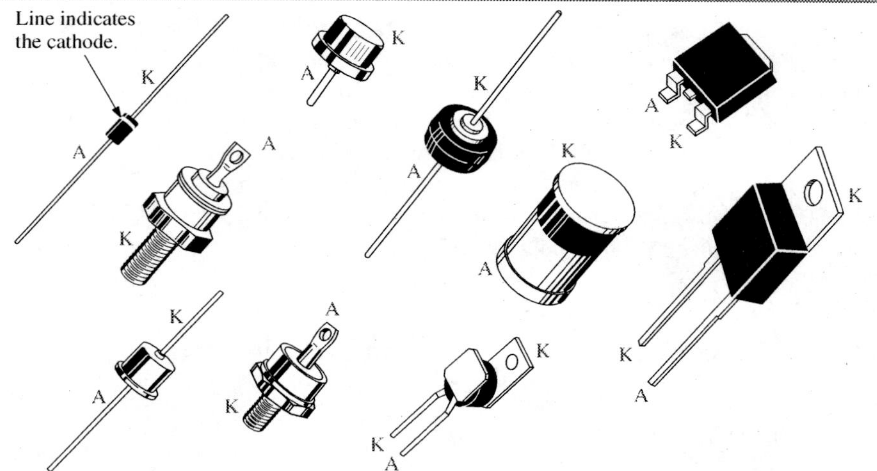

FIGURE 2–9

Typical diode packages and terminal identification. On many diodes a line indicates the cathode side.

Diode Characteristic Curve

Figure 2–10 shows the forward and reverse *IV* curve for a typical diode. The upper right quadrant (quadrant 1) of the graph represents the forward-biased condition. As you can see, there is essentially no forward current (I_F) for forward voltages (V_F) below the barrier potential. As the forward voltage approaches the value of the barrier potential (typically 0.7 V for silicon), the current begins to increase. Once the forward voltage reaches the barrier potential, the current increases drastically and must be limited by a series resistor. The voltage across the forward-biased diode remains approximately equal to the barrier potential but increases slightly with forward current. For a forward-biased diode, this barrier voltage is often referred to as a *diode drop.*

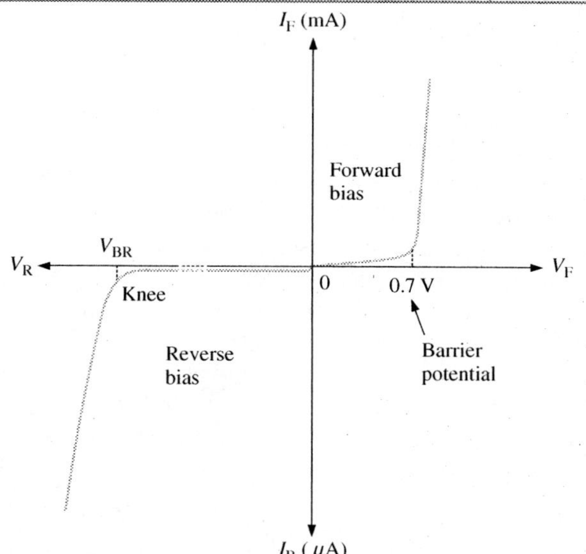

FIGURE 2–10

Diode characteristic curve. The scale for forward and reverse voltages is different.

Quadrant 3 (the lower left) of the graph represents the reverse-biased condition. As the reverse voltage increases to the left, the current remains near zero until the breakdown voltage is reached. When breakdown occurs, there is a large reverse current which, if not limited, can destroy the diode.* Typically, the breakdown voltage is greater than 50 V for most rectifier diodes. Most applications for ordinary diodes do not include operation in the reverse-breakdown region.

Plotting the Characteristic Curve on an Oscilloscope

You can plot the diode's forward characteristic on your oscilloscope by connecting the circuit shown in Figure 2–11. The signal is a 5 V peak-to-peak triangle that is centered about zero volts. This causes the diode to be alternately forward-biased and then reverse-biased. The scope is placed in the X-Y mode. Channel 1 (X channel) senses the voltage drop across the diode; channel 2 (Y channel) shows a signal that is proportional to the current. The common lead on the signal generator must not be the same as the scope ground. Channel 2 must be inverted to display the signal in the proper orientation.

FIGURE 2–11

Plotting the *IV* characteristic curve for a diode on an oscilloscope. The oscilloscope is placed in the X-Y mode and the Y channel is inverted.

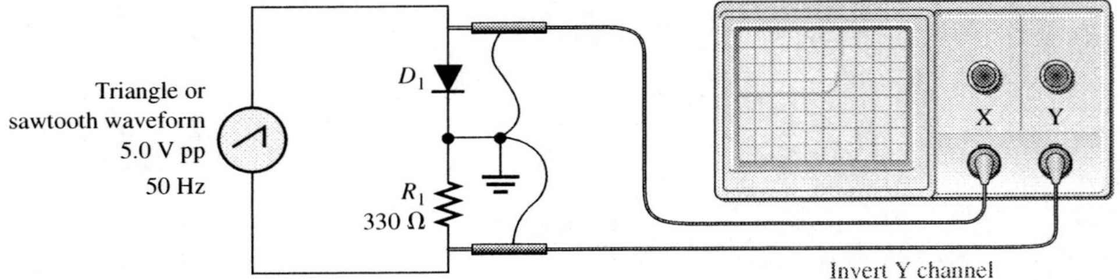

Testing Diodes with an Ohmmeter or a Multimeter

The internal battery in most analog ohmmeters can forward-bias or reverse-bias a diode, permitting a quick relative check of the diode. To check the diode with an analog ohmmeter, select the $R \times 100$ range (to limit current through the diode), connect the meter leads to the diode, then reverse the leads. The meter's internal voltage source will tend to forward-bias the diode in one direction and reverse-bias it in the other. As a result, the resistance will read a lower value in one direction than the other. Look for a high ratio between the forward and reverse readings (typically 1000 or more). The actual reading depends on the internal voltage of the meter, the range selected, and the type of diode, so this is only a relative test.

Many digital multimeters have a diode test position that will indicate the forward diode voltage when a good diode is placed across the test leads. The meter will show an overload when the leads are reversed. This is a better test than a simple resistance test.

The Diode Offset Model

Ideally, a diode acts as a switch — either open or closed. When the diode is forward-biased, it conducts; when it is reversed-biased, it is open. This idea is illustrated in Figure 2–12. This ideal model, of course, neglects the effect of the barrier potential, the internal resistance, and other effects. For almost all practical work, taking into account the diode barrier potential is the only internal effect that you will need to consider. A simplified approximation of the diode *IV* curve, called the *diode offset model,* makes calculations much easier and still enables you to predict circuit behavior with reasonable accuracy.

The diode offset model is shown in Figure 2–13. In this approximation, the forward-biased diode is represented as a closed switch in series with a small "battery" equal to the barrier potential V_B, as shown in Figure 2–13(a). The positive end of the equivalent battery

* With proper current limiting, operation in the reverse-breakdown region does not harm the diode.

Ideal model of a diode as a switch. **FIGURE 2–12**

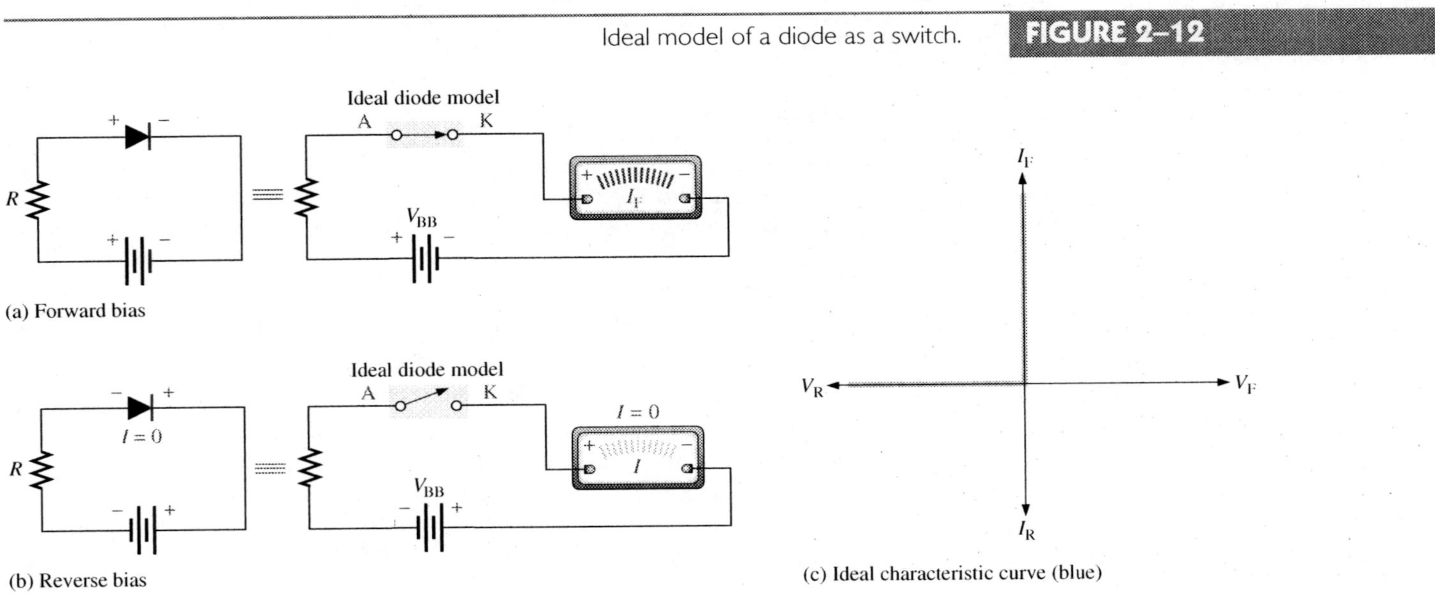

(a) Forward bias

(b) Reverse bias

(c) Ideal characteristic curve (blue)

is toward the anode. Keep in mind that the barrier potential is *not* a voltage source and cannot be measured with a voltmeter; rather it only has the effect of an offsetting battery when forward bias is applied because the forward-bias voltage, V_{BIAS}, must overcome this barrier potential before the diode begins to conduct. The reverse-biased diode is represented by an open switch, as in the ideal case, because the barrier potential does not affect reverse bias,

The offset model for a diode. The barrier potential is included in this model. **FIGURE 2–13**

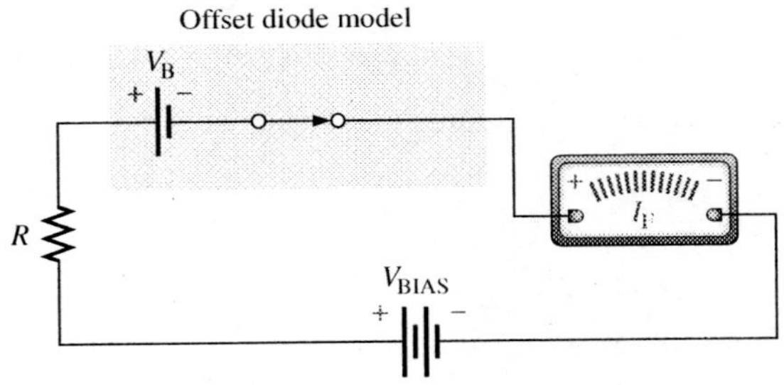

Offset diode model

V_B

R

V_{BIAS}

I_F

$I = 0$

(a) Forward bias

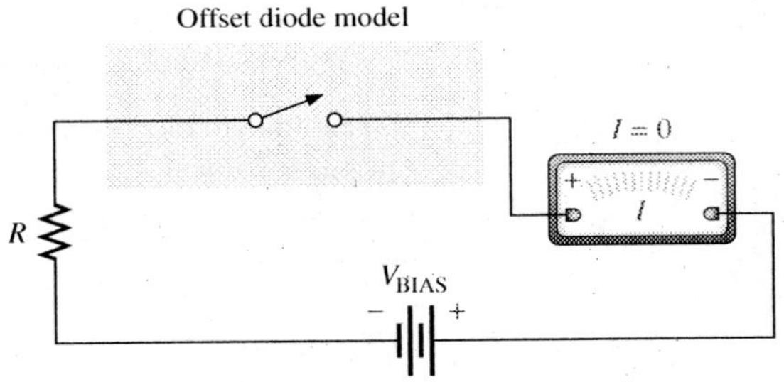

(b) Reverse bias

I_F

V_R 0 V_B V_F

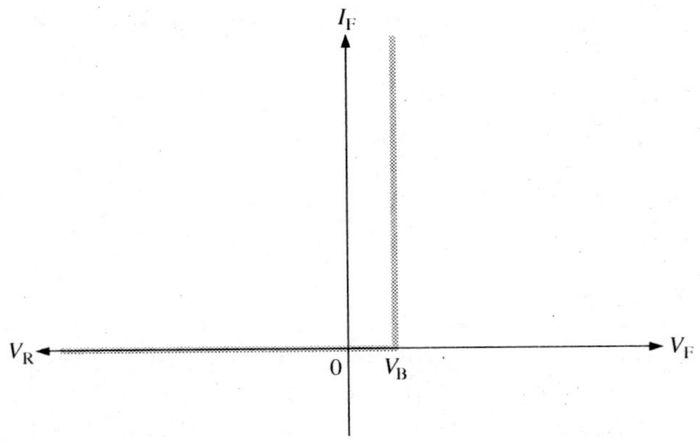

(c) Characteristic curve

as shown in Figure 2–13(b). The characteristic curve is shown in Figure 2–13(c). Unless otherwise stated, this model is used for all analysis of diode circuits in this text.

Types of Diodes

Diodes are used in many circuits. The most important applications for diodes involve their ability to pass current in one direction. Some applications exploit other important characteristics, such as the nonlinear characteristic curve. The following are types of diodes:

- *Rectifier diodes* Designed to carry high current necessary in power supplies.
- *Signal diodes* Optimized for speed, they are used in low current and switching applications.
- *Zener diodes* Designed to conduct in the reverse direction with a precise breakdown voltage, they are used in power supply regulation (discussed in Section 2–6).
- *Light-emitting diodes* Constructed from specialized materials that give off light when conducting, they are useful as displays and indicators.
- *Photodiodes* Operated with reverse bias, reverse current increases with an increase in light.
- *Varactor diodes* Designed to act as a small variable capacitor for tuned circuit applications.

Figure 2–14 shows the symbols for these diodes.

FIGURE 2–14

Diode symbols.

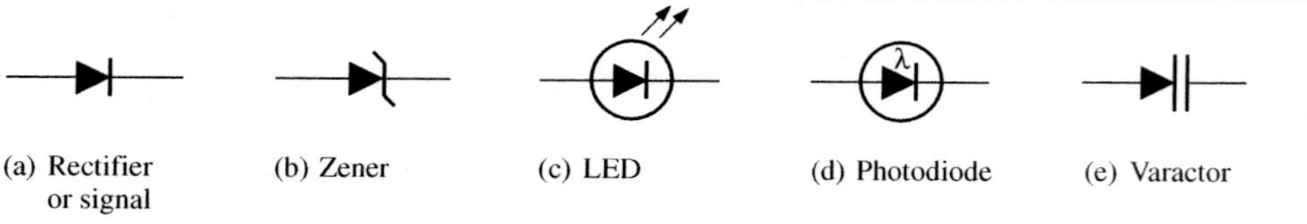

Review Questions

16. What are the two conditions under which a diode is operated?
17. What region of the diode characteristic curve is not part of the normal diode operation?
18. How can you test a diode with an ohmmeter?
19. What is the simplest way to visualize a diode?
20. What approximation is included in the offset model of a diode?

2–5 RECTIFIER CIRCUITS

Power supplies are an essential part of all electronic systems that use ac voltage. After transforming the ac voltage to a usable value, the ac is converted to dc generally using three key elements: a rectifier circuit, a filter, and a regulator.

In this section, you will learn to analyze the half-wave, full-wave, and bridge rectifier circuits.

Half-Wave Rectifiers

A **rectifier** is an electronic circuit that converts ac into pulsating dc. Figure 2–15 illustrates the process called *half-wave rectification.* In a **half-wave rectifier**, an ac source is connected in series with a diode and the load resistor. The diode is a special high current diode called a *rectifier diode.* When the sinusoidal input voltage goes positive, the diode is

FIGURE 2–15

Half-wave rectifier operation.

(a) Operation during positive alternation of the input voltage, diode conducts.

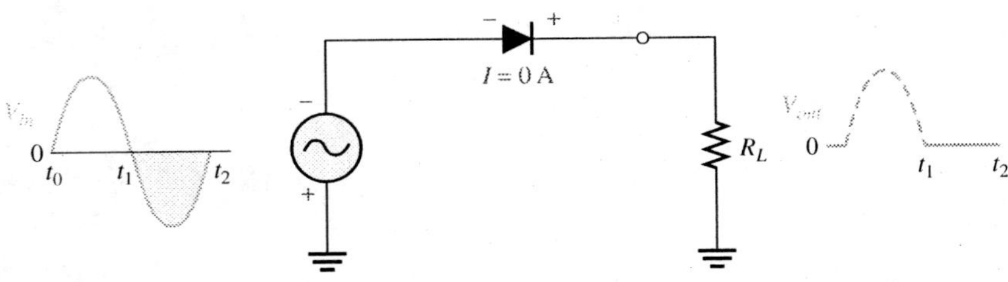

(b) Operation during negative alternation of the input voltage. Diode does not conduct; therefore, the output voltage is zero.

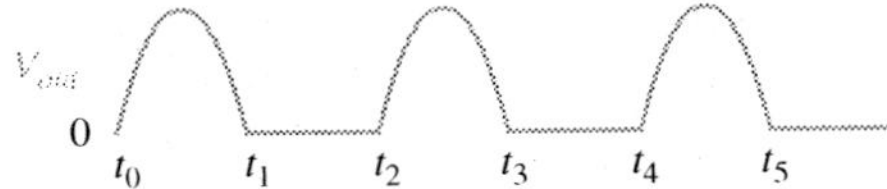

(c) Half-wave output voltage for three input cycles

forward-biased and conducts current to the load resistor, as shown in part (a). The output voltage is equal to the peak voltage less one diode drop.

$$V_{p(out)} = V_{p(in)} - 0.7\text{ V} \tag{2–1}$$

The current produces a voltage across the load, which has the same shape as the positive half-cycle of the input voltage. When the input voltage goes negative during the second half of its cycle, the diode is reverse-biased. There is no current, so the voltage across the load resistor is zero, as shown in part (b). The net result is that only the positive half-cycles of the ac input voltage, less one diode drop, appear across the load, making the output a pulsating dc voltage, as shown in part (c). Notice that during the negative cycle, the diode must withstand the negative peak voltage from the source without breaking down.

EXAMPLE 2–1

Problem

Determine the peak output voltage and the peak inverse voltage (PIV) of the rectifier in Figure 2–16 for the indicated input voltage. Sketch the waveforms you should observe across the diode and the load resistor.

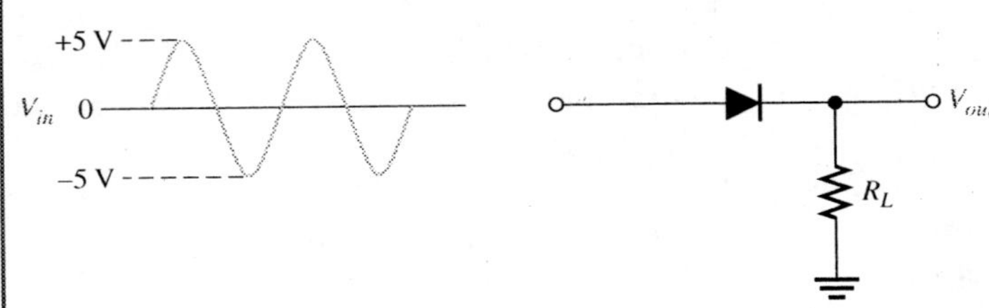

FIGURE 2–16

Solution
The peak half-wave output voltage is

$$V_p = 5\text{ V} - 0.7\text{ V} = \mathbf{4.3\ V}$$

The PIV is the maximum voltage across the diode when it is reverse-biased. The PIV is the maximum voltage during the negative half cycle.

$$\text{PIV} = V_p = \mathbf{5\ V}$$

Waveforms are shown in Figure 2–17. Notice that if you add the load resistor voltage to the diode voltage, you will obtain the input voltage.

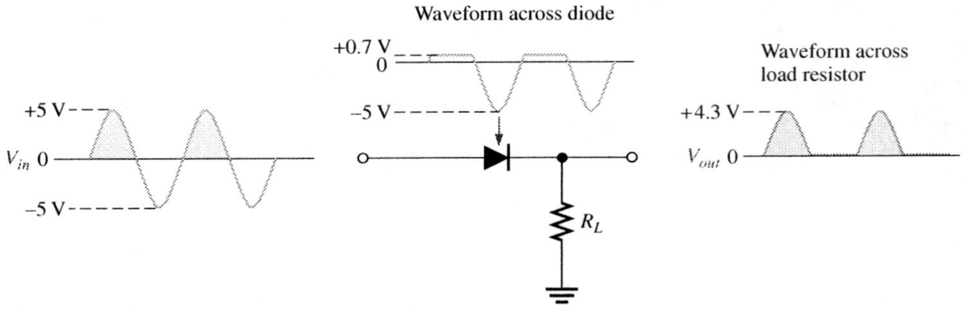

FIGURE 2–17

Question*
What is the peak output voltage and the PIV for the rectifier in Figure 2–16 if the peak input is 3 V?

COMPUTER SIMULATION

Open the Multisim file F02-16DV on the website. Using the oscilloscope, confirm the voltage across R_L.

Full-Wave Rectifiers

The difference between full-wave and half-wave rectification is that a **full-wave rectifier** allows unidirectional current to the load during the entire input cycle, and the half-wave rectifier allows current only during one-half of the cycle. The result of full-wave rectification is a dc output voltage that pulsates every half-cycle of the input, as shown for the center-tapped full-wave rectifier in Figure 2–18.

The **center-tapped** (CT) full-wave rectifier uses two diodes connected to the secondary of a center-tapped transformer, as shown in Figure 2–18. The input signal is coupled through the transformer to the secondary. Half of the total secondary voltage appears between the center tap and each end of the secondary winding as shown.

For a positive half-cycle of the input voltage, the polarities of the secondary voltages are as shown in Figure 2–19(a). This condition forward-biases the upper diode D_1 and reverse-biases the lower diode D_2. The current path is through D_1 and the load resistor, as indicated in red.

---

* *Answers are at the end of the chapter.*

FIGURE 2–18

A center-tapped (CT) full-wave rectifier.

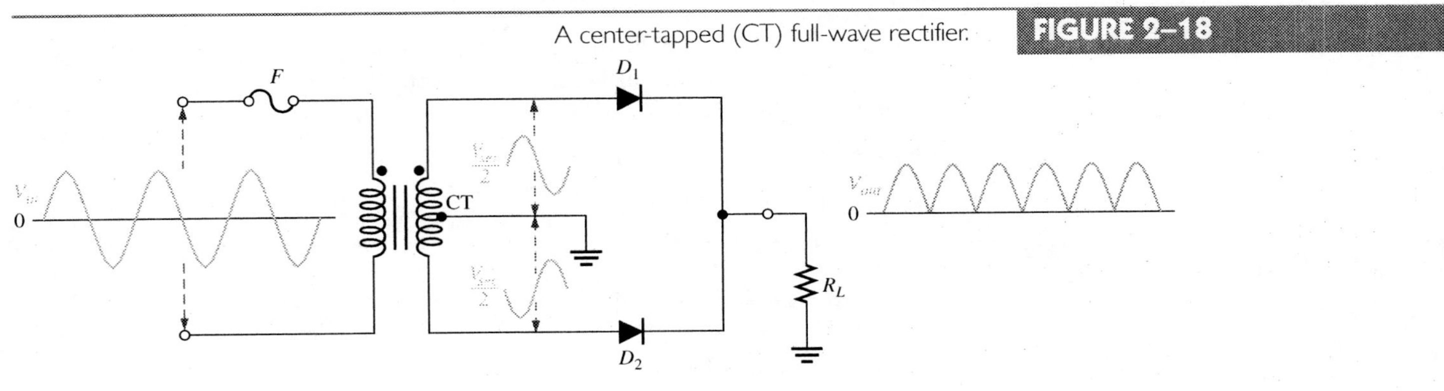

FIGURE 2–19

Conducting paths in the secondary are shown in red.

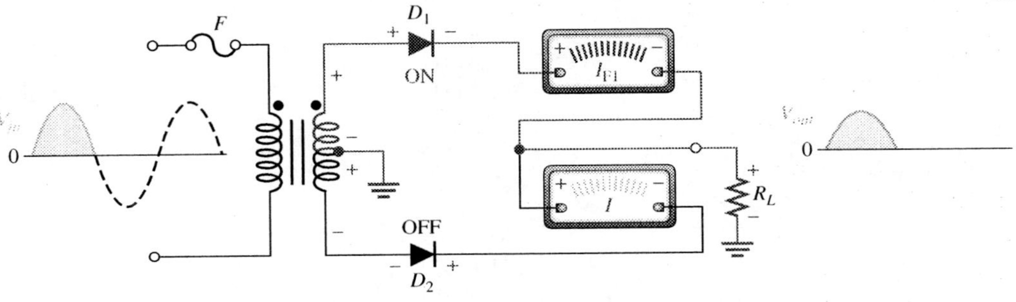

(a) During positive half-cycles, D_1 is forward-biased and D_2 is reverse-biased.

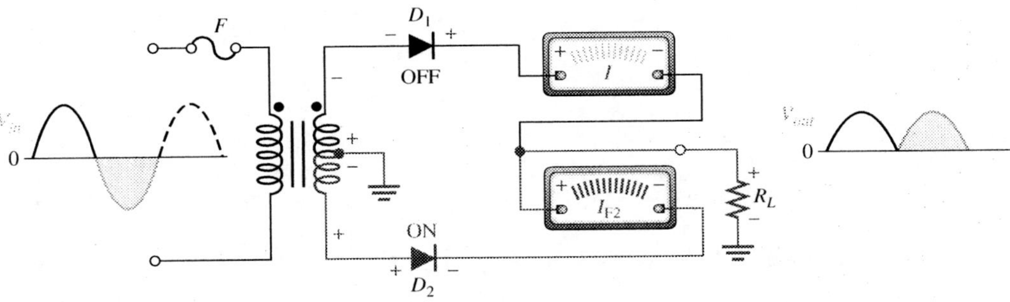

(b) During negative half-cycles, D_2 is forward-biased and D_1 is reverse-biased.

For a negative half-cycle of the input voltage, the voltage polarities on the secondary are as shown in Figure 2–19(b). This condition reverse-biases D_1 and forward-biases D_2. The current path is through D_2 and the load resistor, as indicated in red.

Because the current during both the positive and the negative portions of the input cycle is in the same direction through the load, the output voltage developed across the load is a full-wave rectified dc voltage.

Effect of the Turns Ratio on the Full-Wave Output Voltage

If the turns ratio of a transformer is 1, the peak value of the rectified output voltage equals half the peak value of the primary input voltage less one diode drop. This value occurs because half of the input voltage appears across each half of the secondary winding.

In order to obtain a peak output voltage equal to the peak input voltage (less the barrier potential), you must use a step-up transformer with a turns ratio of 2 (1:2). In this case, the total secondary voltage is twice the primary voltage, so the voltage across each half of the secondary is equal to the input.

Peak Inverse Voltage (PIV)

Each diode in the full-wave rectifier is alternately forward-biased and then reverse-biased. The maximum reverse voltage that each diode must withstand is the peak value of the total secondary voltage (V_{sec}). The peak inverse voltage across either diode in the center-tapped full-wave rectifier is

$$\text{PIV} = V_{p(out)}$$

EXAMPLE 2–2

Problem

(a) Show the voltage waveforms across the secondary winding and across R_L when a 25 V peak sine wave is applied to the primary winding in Figure 2–20.

(b) What minimum PIV rating must the diodes have?

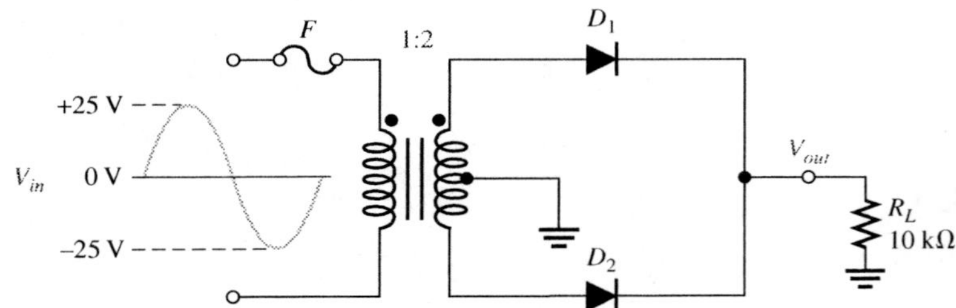

FIGURE 2–20

Solution

(a) The waveforms are shown in Figure 2–21.

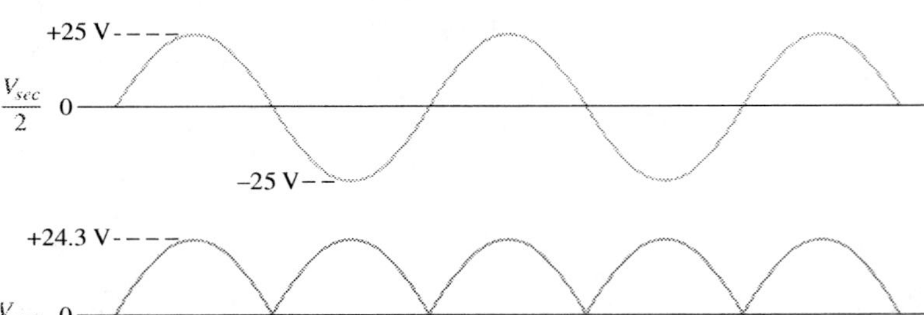

FIGURE 2–21

(b) The total peak secondary voltage is

$$V_{p(sec)} = \left(\frac{N_{sec}}{N_{pri}}\right)V_{p(in)} = 2(25)\text{ V} = 50\text{ V}$$

There is a 25 V peak across each half of the secondary. Using the ideal model, one diode is a short while the other diode has the full secondary voltage across it. Each diode must have a minimum PIV rating of **50 V.**

Question

What diode PIV rating is required to handle a peak input of 160 V in Figure 2–20?

COMPUTER SIMULATION

Open the Multisim file F02-20DV on the website. Observe the output voltage using the oscilloscope.

Bridge Rectifiers

The bridge rectifier uses four diodes, as shown in Figure 2–22 and is the most popular arrangement for power supplies because it does not require a center-tapped transformer. The four diodes are available in a single package, already wired in a bridge configuration. The bridge rectifier is a type of full-wave rectifier because each half of the sine wave contributes to the output.

FIGURE 2–22

Operation of the full-wave rectifier. Conducting paths in the secondary are shown in red.

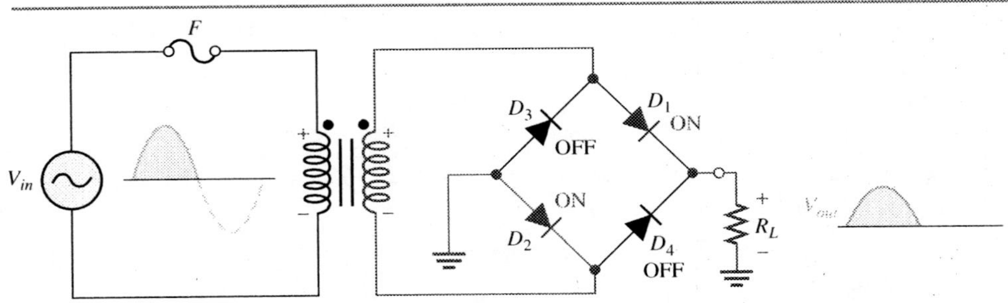

(a) During positive half-cycle of the input, D_1 and D_2 are forward-biased and conduct current. D_3 and D_4 are reverse-biased.

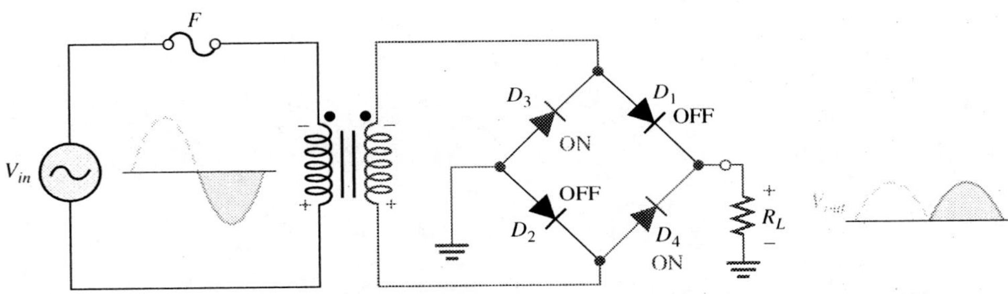

(b) During negative half-cycle of the input, D_3 and D_4 are forward-biased and conduct current. D_1 and D_2 are reverse-biased.

This is how the bridge rectifier works: When the input cycle is positive as in Figure 2–22(a), diodes D_1 and D_2 are forward-biased and conduct current as shown by the red path. A voltage is developed across R_L which looks like the positive half of the input cycle. During this time, diodes D_3 and D_4 are reverse-biased. When the input cycle is negative, as in Figure 2–22(b), diodes D_3 and D_4 are forward-biased and conduct as shown by the red path. A voltage is again developed across R_L in the same direction as during the positive half-cycle. During the negative half-cycle, D_1 and D_2 are reverse-biased. A full-wave rectified output voltage appears across R_L as a result of this action.

Bridge Output Voltage

Neglecting the diode drops, the total secondary voltage, V_{sec}, appears across the load resistor. Thus,

$$V_{out} = V_{sec}$$

As you can see in Figure 2–22, two diodes are always in series with the load resistor during both the positive and the negative half-cycles. If these diode drops are taken into account, the output voltage is

$$V_{out} = V_{sec} - 1.4\text{ V} \tag{2–2}$$

Peak Inverse Voltage

When D_1 and D_2 are forward-biased, the reverse voltage is across D_3 and D_4. Visualizing D_1 and D_2 as shorts (ideally), the peak inverse voltage is equal to the peak secondary voltage.

$$\text{PIV} = V_{p(out)}$$

Review Questions

21. Which type of rectifier (half-wave, full-wave, or bridge) has the greatest output voltage for the same input voltage and transformer turns ratio?
22. For a given output voltage, is the PIV for bridge rectifier diodes less than, the same as, or greater than the PIV for center-tapped rectifier diodes?
23. At what point on the input cycle does the PIV occur in a half-wave rectifier that has a positive output?
24. For a half-wave rectifier, there is current through the load for approximately what percentage of the input cycle?
25. What type of rectifier requires a center-tapped transformer?

2–6 POWER SUPPLY FILTERS AND REGULATORS

After ac is rectified in a power supply, it needs to be filtered or smoothed out by a power supply filter, which is usually just a large-value capacitor. Following the filter, a regulator keeps the voltage constant even with load changes. In small applications, a zener diode or an integrated circuit (IC) regulator can be used for this. Regulators are introduced in this section and covered in more detail in Chapter 11.

In this section, you will learn how power supply filters and regulators function.

Most electronic circuits require a dc source to function properly. The function of a power supply is to convert the standard 60 Hz ac power line voltage to nearly constant dc. After the rectifier has converted the ac to pulsating dc, the filter smoothes out variations and sends the dc on to the regulator. The filter cannot remove all traces of the variations due to the ac, nor can it keep the voltage steady if the load changes. The regulator is the final part of this operation.

A capacitor, an inductor, or a combination of these can accomplish filtering. The capacitor-input filter is the least expensive and most widely used type, by far.

Capacitor-Input Filter

A half-wave rectifier with a capacitor-input filter is shown in Figure 2–23. During the positive first quarter-cycle of the input, the diode is forward-biased, allowing the capacitor to charge to within a diode drop of the input peak, as illustrated in Figure 2–23(a). When the input begins to decrease below its peak, as shown in Figure 2–23(b), the capacitor retains its charge and the diode becomes reverse-biased. During the remaining part of the cycle and the beginning of the next cycle, the capacitor can discharge only through the load resistance at a rate determined by the *RC* time constant. The larger the time constant, the less the capacitor will discharge.

FIGURE 2–23

Operation of a half-wave rectifier with a capacitor-input filter.

(a) Initial charging of capacitor (diode is forward-biased) happens only once when power is turned on.

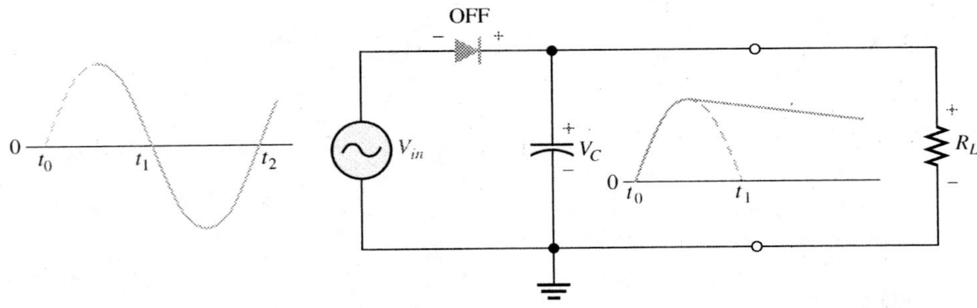

(b) The capacitor discharges through R_L after peak of positive alternation when the diode is reverse-biased. This discharging occurs during the portion of the input voltage indicated by the solid blue curve.

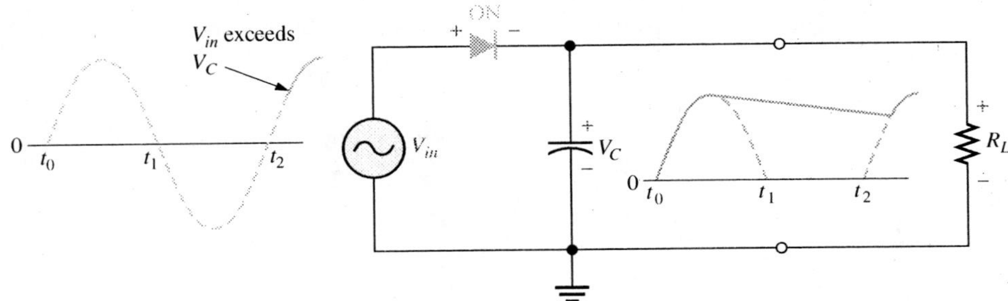

(c) The capacitor charges back to peak of input when the diode becomes forward-biased. This charging occurs during the portion of the input voltage indicated by the solid blue curve. Notice that the diode is not forward biased on the second cycle until the capacitor voltage is overcome.

During the peak of the next cycle, as illustrated in Figure 2–23(c), the diode again will become forward-biased when the input voltage exceeds the capacitor voltage by approximately a diode drop.

Ripple Voltage

As you have seen, the capacitor quickly charges at the beginning of a cycle and slowly discharges through R_L after the positive peak (when the diode is reverse-biased). The variation in the capacitor voltage due to the charging and discharging is called the **ripple voltage**. The smaller the ripple voltage, the better the filtering action.

For a given input frequency, the output frequency of a full-wave rectifier is twice that of a half-wave rectifier. As a result, a full-wave rectifier is easier to filter because of the shorter time between peaks. When filtered, the full-wave rectified voltage has less ripple voltage than does a half-wave voltage for the same load resistance and filter capacitor values. Less ripple voltage occurs because the capacitor discharges less during the shorter interval between full-wave pulses, as shown in Figure 2–24.

FIGURE 2–24

Comparison of ripple voltages for half-wave and full-wave rectifier outputs with the same filter capacitor and derived from the same sinusoidal input.

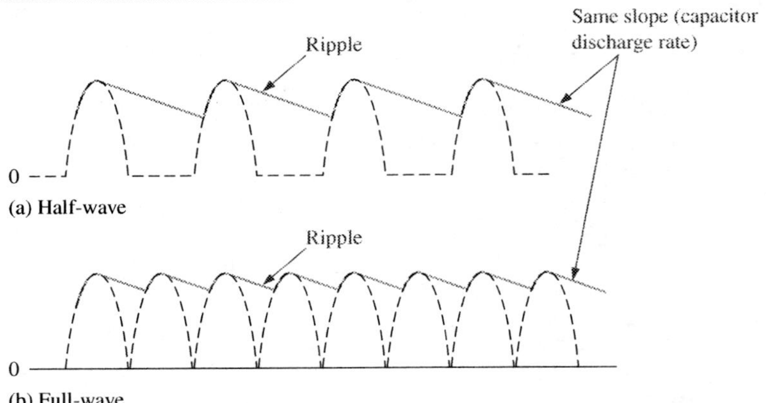

Zener Regulators

A **zener diode** is a special-purpose diode designed to conduct in the reverse direction. Zeners are useful in regulators because of their precise breakdown voltage, which can be controlled at the time of manufacture.

Figure 2–25 shows a zener characteristic in the reverse portion of the characteristic curve. (The forward characteristic is the same as an ordinary diode). Notice that as the reverse voltage (V_R) is increased, the reverse current (I_R) remains extremely small up to the knee of the curve. At this point, the breakdown effect begins. From the bottom of the knee, the zener breakdown voltage (V_Z) remains essentially constant although it increases slightly as I_Z increases. This constant voltage region of the characteristic curve accounts for the zener's ability to regulate.

FIGURE 2–25

Reverse characteristic of a zener diode. V_Z is usually specified at the test current, I_{ZT} and is designated V_{ZT}.

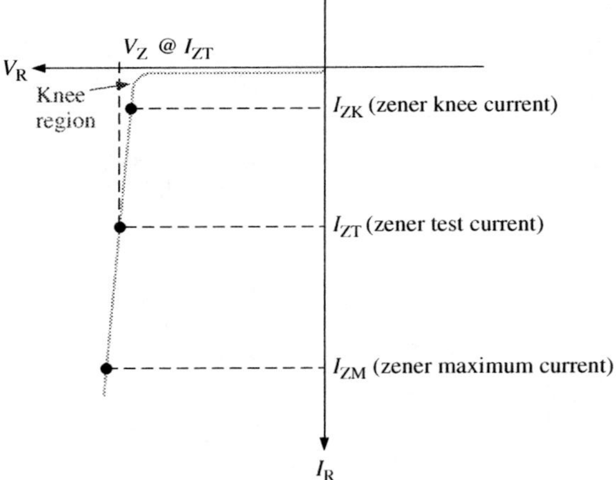

Zener diodes can be used for **voltage regulation** in noncritical applications. Figure 2–26 illustrates how a zener diode can be used to regulate a varying dc input voltage to keep it at a constant level. As the input voltage varies (within limits), the zener diode maintains a nearly constant voltage across the output terminals. However, as V_{IN} changes, I_Z will change proportionally, and therefore the limitations on the input variation are set by the minimum and maximum current values (I_{ZK} and I_{ZM}) with which the zener can operate and on the condition that $V_{IN} > V_Z$. R is the series current-limiting resistor. The bar graph on the DMM symbols indicates the relative values and trends. Many DMMs provide analog bar graph displays in addition to the digital readout.

FIGURE 2–26

Zener regulation of a varying input voltage.

I_Z increasing

$V_{IN} > V_Z$

(a) As the input voltage increases, the output voltage remains constant ($I_{ZK} < I_Z < I_{ZM}$).

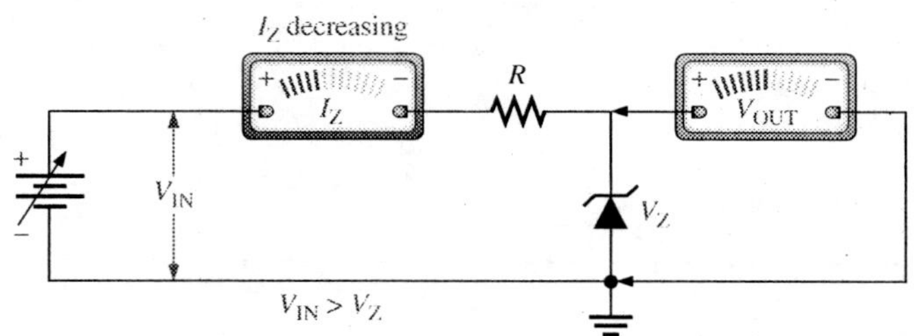

(b) As the input voltage decreases, the output voltage remains constant ($I_{ZK} < I_Z < I_{ZM}$).

EXAMPLE 2–3

Problem

Figure 2-27 shows a zener diode regulator designed to hold 10 V at the output. Assume the zener current ranges from 4 mA minimum (I_{ZK}) to 40 mA maximum (I_{ZM}). What are the minimum and maximum input voltages for these currents?

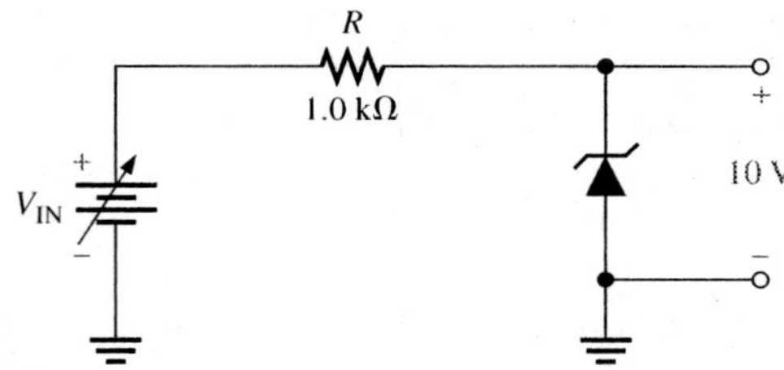

FIGURE 2–27

Solution

For the minimum current, the voltage across the 1.0 kΩ resistor is

$$V_R = I_{ZK}R = (4\text{ mA})(1.0\text{ k}\Omega) = 4\text{ V}$$

Since $V_R = V_{IN} - V_Z$,

$$V_{IN} = V_R + V_Z = 4\text{ V} + 10\text{ V} = \mathbf{14\ V}$$

For the maximum zener current, the voltage across the 1.0 kΩ resistor is

$$V_R = (40\text{ mA})(1.0\text{ k}\Omega) = 40\text{ V}$$

Therefore,

$$V_{IN} = 40\text{ V} + 10\text{ V} = \mathbf{50\ V}$$

As you can see, this zener diode can provide regulation for an input voltage that varies from 14 V to 50 V and maintain approximately a 10 V output. The output will vary slightly from this value because of the internal resistance of the zener diode.

Question

What are the minimum and maximum input voltages that can be regulated by the zener in Figure 2-28 if the minimum current (I_{ZK}) is 2.5 mA and the maximum (I_{ZM}) is 35 mA?

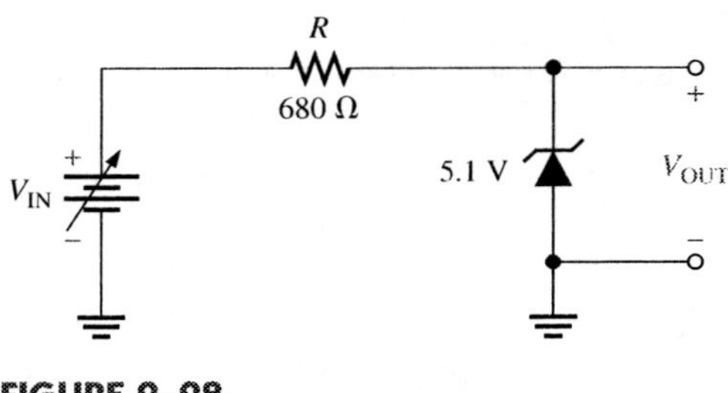

FIGURE 2–28

COMPUTER SIMULATION

Open the Multisim file F02-28DV on the website. Measure V_{OUT} when V_{IN} is 14 V. Change V_{IN} to 50 V and measure V_{OUT}.

IC Regulators

Zener regulators are useful for very low current applications. For most applications, the most effective way to remove ripple is a combination of a capacitor-input filter used with an IC regulator. In general, an IC (**integrated circuit**) is a complete functional circuit constructed on a single, tiny chip of silicon. An integrated circuit regulator is an IC that is connected to the output of a rectifier and maintains a constant output voltage (or current) despite changes in the input, the load current, or the temperature. The capacitor filter reduces the input ripple to the regulator to an acceptable level. The combination of a large capacitor and an IC regulator is inexpensive and helps produce an excellent small power supply.

The most popular IC regulators have three terminals; an input terminal, an output terminal, and a reference (or adjust) terminal. The input to the regulator is first filtered with a capacitor to reduce the ripple to $<10\%$. The regulator reduces the ripple to a negligible amount. In addition, most regulators have an internal voltage reference, consisting of a zener diode, short-circuit protection, and thermal shutdown circuitry. They are available in a variety of voltages, including positive and negative outputs, and can be designed for variable outputs with a minimum of external components. Typically, IC regulators can furnish a constant output of one or more amps of current with high ripple rejection. IC regulators are available that can supply load currents of over 5 A.

Examples of fixed three-terminal regulators are the 78XX and 79XX series* of regulators that are available with various output voltages and can supply up to 1 A of load current (with adequate heat sinking). The last two digits of the number stand for the output voltage; thus, the 7812 has a +12 V output. The negative output versions of the same regulator are numbered as the 79XX series; the 7912 has a −12 V output. The output voltage from these regulators is within 3% of the nominal value but will hold a nearly constant output despite changes in the input voltage or output load. The available voltages and common packaging for the 78XX are shown in Figure 2–29.

* Data sheets for the MC78XX and MC79XX are available at *http://www.onsemi.com*. The XX stands for a specific device number in the series.

FIGURE 2–29

Available voltages and common packages for the 78XX series of regulators. Note that pins for the 79XX series are different than the ones shown.

Type number	Output voltage
7805	+5.0 V
7806	+6.0 V
7808	+8.0 V
7809	+9.0 V
7812	+12.0 V
7815	+15.0 V
7818	+18.0 V
7824	+24.0 V

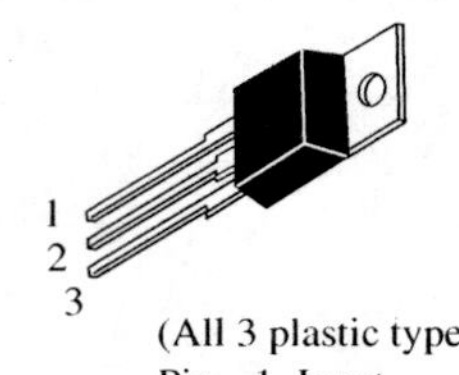
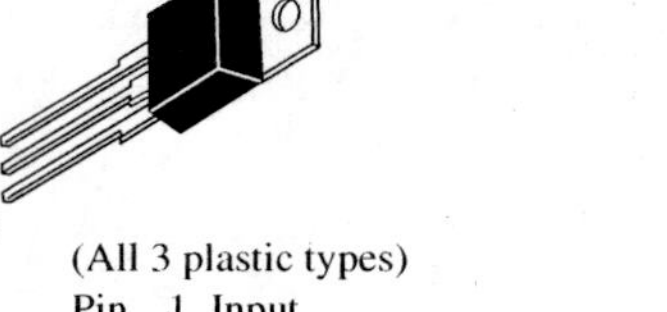

A basic fixed +5 V power supply with a 7805 regulator is shown in Figure 2–30. The three-terminal regulator only requires external capacitors to complete the regulation portion of the power supply. Filtering is accomplished by a large-value capacitor between the input voltage and ground. Sometimes, a second smaller-value input capacitor (not shown) is connected in parallel, especially if the filter capacitor is not close to the IC regulator, to prevent oscillation. This capacitor needs to be located close to the IC regulator. Finally, an output capacitor (typically 0.1 μF to 1.0 μF) is placed in parallel with the output to improve the transient response.

FIGURE 2–30

A basic +5.0 V power supply using the 7805 regulator.

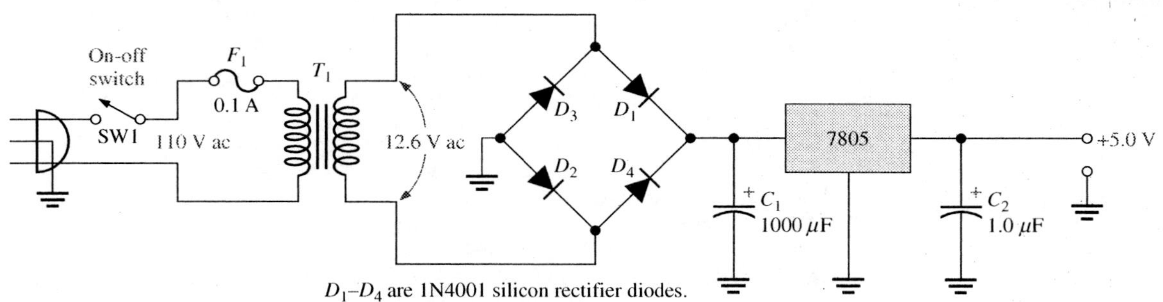

Another type of three-terminal regulator is an adjustable regulator. Figure 2–31 shows a power supply circuit with an adjustable output, controlled by the variable resistor, R_2. Note that R_2 is adjustable from zero to 1.0 kΩ. The LM317 regulator keeps a constant 1.25 V between the output and adjust terminals. This produces a constant current in R_1 of 1.25 V/240 Ω = 52 mA. Neglecting the very small current through the adjust terminal, the current in R_2 is the same as the current in R_1. The output is taken across both R_1 and R_2 and is found from the equation,

$$V_{out} = 1.25\text{ V}\left(\frac{R_1 + R_2}{R_1}\right)$$

FIGURE 2–31 A basic power supply with a variable output voltage (from 1.25 V to 6.5 V).

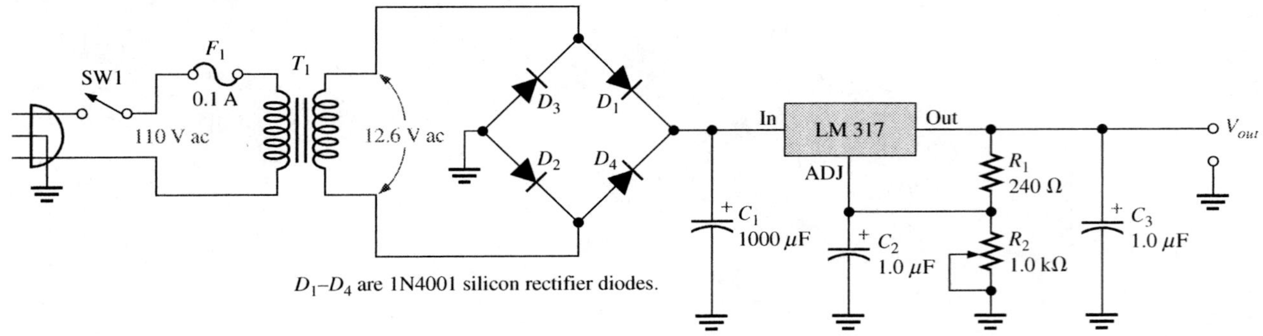

Notice that the output voltage from the power supply is the regulator's 1.25 V multiplied by a ratio of resistances. For the case shown in Figure 2–31; when R_2 is set to the minimum (zero) resistance, the output is 1.25 V. When R_2 is set to the maximum, the output is nearly 6.5 V.

Percent Regulation

The regulation expressed as a percentage is a figure of merit used to specify the performance of any voltage regulator. It can be in terms of input (line) regulation or load regulation. **Line regulation** specifies how much change occurs in the output voltage for a given change in the input voltage. It is typically defined as a ratio of a change in output voltage for a corresponding change in the input voltage expressed as a percentage.

$$\text{Line regulation} = \left(\frac{\Delta V_{\text{OUT}}}{\Delta V_{\text{IN}}}\right)100\% \qquad (2\text{–}3)$$

Load regulation specifies how much change occurs in the output voltage over a certain range of load current values, usually from minimum current (no load, NL) to maximum current (full load, FL). It is normally expressed as a percentage and can be calculated with the following formula:

$$\text{Load regulation} = \left(\frac{V_{\text{NL}} - V_{\text{FL}}}{V_{\text{FL}}}\right)100\% \qquad (2\text{–}4)$$

where V_{NL} is the output voltage with no load and V_{FL} is the output voltage with full (maximum) load. Line and load regulation are discussed further in Section 11–1.

EXAMPLE 2–4

Problem

Assume a certain 7805 regulator has a measured no-load output voltage of 5.185 V and a full-load output of 5.152 V. What is the load regulation expressed as a percentage? Is this within the manufacturer's specification?

Solution

$$\text{Load regulation} = \left(\frac{V_{\text{NL}} - V_{\text{FL}}}{V_{\text{FL}}}\right)100\% = \left(\frac{5.185\text{ V} - 5.152\text{ V}}{5.152\text{ V}}\right)100\% = \mathbf{0.64\%}$$

The data sheet indicates a maximum variation of the output voltage (with a load current from 5 mA to 1.0 A) of 100 mV. This represents a maximum load regulation of 2% (typical is 0.4%). The measured percent regulation is within specifications.

Question

If the no-load output voltage of a regulator is 24.8 V and the full-load output is 23.9 V, what is the load regulation expressed as a percentage?

IC regulators can be adapted to a number of specialized applications or requirements such as current sources or automatic shutdown, current limiting, and the like. For certain other applications (high current, high efficiency, high voltage), more complicated regulators are constructed from integrated circuits and discrete transistors. Chapter 11 discusses some of these applications in more detail.

Review Questions

26. What causes the ripple voltage on the output of a capacitor-input filter?
27. The load resistance of a capacitor-filtered full-wave rectifier is reduced. What effect does this reduction have on the ripple voltage?
28. What advantages are offered by a three-terminal regulator over a zener regulator?
29. What is the difference between input (line) regulation and load regulation?
30. For the circuit in Figure 2–31, what voltage do you expect across R_1?

TROUBLESHOOTING A POWER SUPPLY 2–7

The backbone for nearly all electronic circuits is the power supply. Several types of failures can occur in power supplies.

In this section, you will learn the steps in troubleshooting a power supply.

As discussed in Section 1–3, the first step in troubleshooting any electronic circuit is to analyze the clues (symptoms) of a failure. These clues should lead to a logical plan for troubleshooting. The plan for a circuit that has never worked will be different than one that has been working. The history of past failures, or failures in a similar circuit, can also be a clue to a failure.

It is always useful to have a good understanding of the circuit you are troubleshooting and a schematic. There is no one plan that fits all situations; the one to use depends on the type and complexity of the circuit or system, the nature of the problem, and the preference of the individual technician.

A Troubleshooting Plan

Above all else, efficient troubleshooting requires logical thinking and a plan that will find the simple problems (such as a blown fuse) quickly. As an example, consider the following plan for troubleshooting a power supply that has failed in operation.

Step 1: Ask questions of the person reporting the trouble. When did it fail? (Right after it was plugged in? After running for 2 hours at maximum current out?) How do you know it failed? (Smoke? Low voltage?)

SAFETY NOTE

You need to have eye protection with approved safety glass whenever you are doing electrical or electronic hardware repair or using tools that can produce flying cuttings such as drills, saws, hammers, air hoses, or grinders. Flying cuttings can also be produced whenever small wires are clipped from a circuit board. In addition to wearing eye protection, you should cut in a manner that directs the cuttings downward, not into the lab area.

Step 2: Power check: Make sure the power cord is plugged in and the fuse is not burned out. Check that the controls are set for proper operation. Something this simple is often the cause of the problem. Perhaps the operator did not understand the correct settings required for controls.

Step 3: Sensory check: Beyond the power check, the simplest troubleshooting method relies on your senses of observation to check for obvious defects. Remove power, open the supply, and do a visual check for broken wires, poor solder connections, burned out fuses, and the like. Also, when certain types of components fail, you may be able to detect a smell of smoke if you happen to be there when it fails or shortly afterward. Since some failures are temperature dependent, you can sometimes use your sense of touch (carefully!) to detect an overheated component.

Step 4: Isolate the failure. Apply power to the supply while it is on the bench and trace the voltage. As described in Section 1–3, you can start in the middle of the circuit and do half-splitting or check the voltage at successive test points from the input side until you get an incorrect measurement. Some problems are more difficult than simply finding no voltage, but tracing should isolate the problem to an area or a component.

Power Supply Failures

Having reviewed a plan for troubleshooting, let's turn to fault analysis. When you find a symptom, you next need to ask the question, *If component X fails in the circuit, what are the symptoms?* You will apply fault analysis when you find an incorrect voltage or waveform. For example, assume you observe high ripple at the input to a regulator. From your knowledge of circuit operation, you might reason that a defective or incorrect value capacitor could be the culprit. The following discussion describes four possible power supply failures and gives more examples of fault analysis.

Open Fuse or Circuit Breaker

An overcurrent-protection device is essential to virtually all electronic equipment. These devices prevent damage to the equipment in case of a failure or an overload condition and reduce the probability of a violent failure. Overcurrent-protection devices include fuses, circuit breakers, solid-state current-limiting devices, and thermal overload devices. The circuit breakers for the ac line cannot be counted on to protect electronic devices as they only open when the current is 15 A or more in the ac line, far too high to offer protection to most electronic equipment.

If a single fuse is present, it is usually on the primary side of the transformer and will be rated for 115 or 230 VAC at a current that is consistent with the maximum power rating of the supply. Frequently, protection devices may also be included on the secondary side, especially if a single transformer has multiple outputs. A fuse is designed to carry its rated current indefinitely (and will typically carry 120% of its rated current for several hours). Fuses are available in fast- and slow-blow versions. A fast-blow fuse opens in a few milliseconds when overloaded; a slow-blow fuse can survive transient overloads such as occurs when power is first applied. Most of the time, slow-blow fuses should be used on the ac side of a power supply circuit.

Testing for an open fuse is relatively simple. Glass fuses can be checked by inspection or checked with an ohmmeter (be certain power is disconnected from the circuit). If power is still applied, a blown fuse will have voltage across it (provided there are no other opens in the path, such as a switch). Usually an open fuse is symptomatic of a short circuit or overload; however, fuses can have fatigue failure and may be the only problem in the circuit.

Before replacing a blown fuse, the technician should check for the cause. If the fuse simply opened, it may be a fatigue failure (although this is unusual). If the fuse has blown violently (as evidenced by complete vaporization of the wire inside), it is most likely that it opened due to another problem. Look for a short circuit with an ohmmeter; it can be the load, the filter capacitor, or other component that has shorted. Look for any visible signs of an overheated or damaged component. If the fuse needs to be replaced, it is important to replace it with the identical type and current rating as specified by the manufacturer. The wrong fuse can cause further damage and may be a safety hazard.

Open Diode

Consider the full-wave, center-tapped rectifier in Figure 2–32. Assume that diode D_1 has failed open. This causes diode D_2 to conduct on only the negative cycle. With an oscilloscope connected to the output, as shown in part (a), you would observe a larger-than-normal ripple voltage at a frequency of 60 Hz rather than 120 Hz. Disconnecting the filter capacitor, you would observe a half-wave rectified voltage, as in part (b).

FIGURE 2–32 Symptoms of an open diode in a full-wave, center-tapped rectifier.

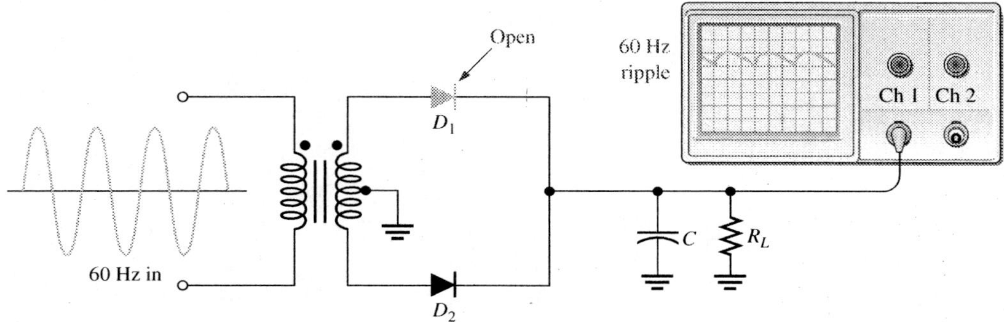

(a) Ripple should be less and have a frequency of 120 Hz. Instead it is greater in amplitude with a frequency of 60 Hz.

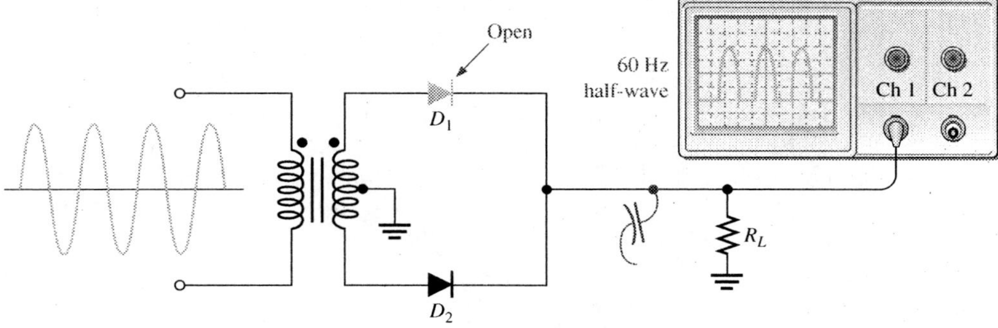

(b) With C removed, output should be a full-wave 120 Hz signal. Instead it is a 60 Hz, half-wave voltage.

With the filter capacitor in the circuit, the half-wave signal will allow it to discharge more than it would with a normal full-wave signal, resulting in a larger ripple voltage. Basically, the same observations would be made for an open failure of diode D_2.

An open diode in a bridge rectifier would create symptoms identical to those just discussed for the center-tapped rectifier. The open diode would prevent current through R_L during half of the input cycle (in this case, the negative half). As a result, there would be a half-wave output and an increased ripple voltage at 60 Hz, as discussed before.

Generally, the easiest test for an open diode in a full-wave power supply is to measure the ripple frequency. If the ripple frequency is the same as the input ac frequency, look for an open diode or a connection problem with a diode (such as a cracked trace).

Shorted Diode

A shorted diode is one that has failed such that it has a very low resistance in both directions. If a diode suddenly became shorted in a bridge rectifier, normally a fuse would blow or other circuit protection would be activated. If the supply was not protected by a fuse, the shorted diode will most likely cause the transformer to be damaged or cause the other diode in series to open, as illustrated in Figure 2–33.

FIGURE 2–33

Effect of shorted diode in a bridge rectifier. Conducting paths are shown in red.

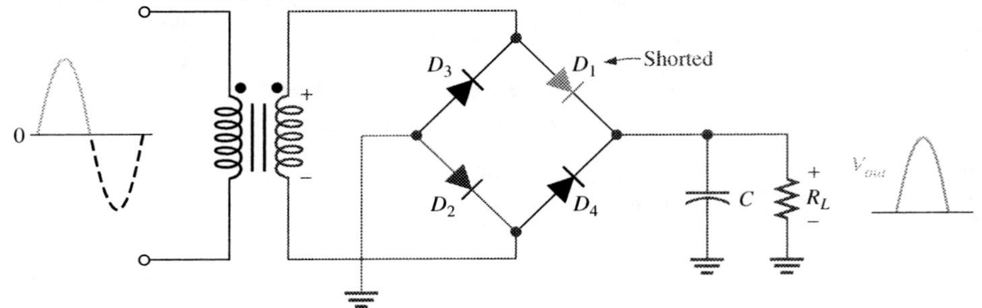

(a) Positive half-cycle: The shorted diode acts as a forward-biased diode, so the load current is normal. D_3 and D_4 are reverse-biased.

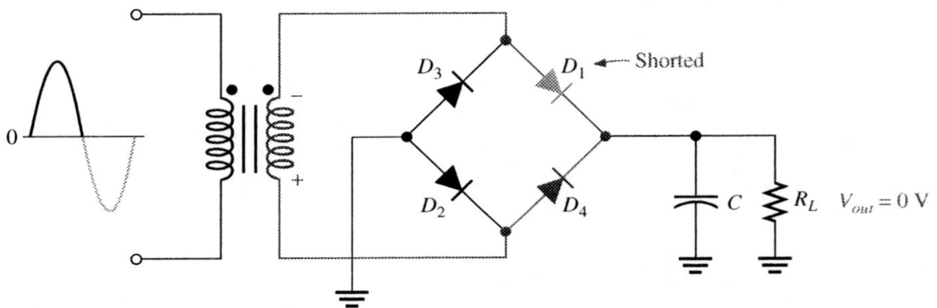

(b) Negative half-cycle: The shorted diode places forward-biased D_4 across the secondary. As a result D_1, D_4, or the transformer secondary will probably burn open, or a fuse (not shown) will open.

In part (a) of Figure 2–33, current is supplied to the load through the shorted diode during the first positive half-cycle, just as though it were forward-biased. During the negative half-cycle, the current is shorted through D_1 and D_4, as shown in part (b). It is this excessive current that leads to the second failure; hence, when a shorted diode is discovered, it is a good idea to check other components for a failure.

Shorted or Leaky Filter Capacitor

Electrolytic capacitors can appear shorted (or have high "leakage" current) when they fail. One cause of failure that produces symptoms of a short occurs when an electrolytic capacitor is put in backwards, an error that can happen with newly manufactured circuit boards. As in the case of a shorted diode, the normal symptom is a blown fuse due to excessive current. A leaky capacitor is a form of partial failure; the effect of the leakage resistance is to reduce the discharging time constant, causing an increase in ripple voltage on the output. A leaky capacitor may simply overheat; a capacitor should never show signs of overheating. For an unfused supply, a shorted capacitor would most likely cause one or more diodes or the transformer to open due to excessive current. In any event, there would be no dc voltage on the output.

When a defective capacitor is replaced, it is important to observe the working voltage specification as well as the size of the capacitor. If the working voltage specification is exceeded, the replacement is likely to fail again. In addition, it is vitally important to observe the polarity of the capacitor. An electrolytic capacitor installed backwards can literally explode.

Troubleshooting a Regulated Power Supply

As indicated at the beginning of this section, the plan for repairing any electronic equipment depends on the observed symptoms. Let's assume you have a supply like the one shown previously in Figure 2–30 that has blown a fuse immediately after it was connected to a printed circuit card. Your thinking might go like this: "The power supply was working fine until I added the card; perhaps I exceeded the current limit of the supply." Here you have considered the conditions and hypothesized a possible cause. The first step then is to remove the load and test the supply to see if this clears the problem. If so, then check the current requirement of the card that was added, or check to see if it is drawing too much current. If not, the problem is in the supply.

What if a power supply is completely dead but has a good fuse? In this case, start tracing the voltage to isolate the problem. For example, you could check for voltage on the primary of the transformer; if there is voltage, then test the secondary voltage. If the primary does not have voltage, check the path for the ac—the switch and connections to the transformer. If a single open is in series with the ac line, the full ac voltage will appear across that open.

Let's assume you have found that the transformer has ac on the primary but no voltage on the secondary. This indicates that the transformer is open, either the primary or the secondary winding. An ohmmeter should confirm this. Before replacing it, you should look why the failure occurred. Transformers seldom fail if they are operated properly. The likelihood is that another component shorted in the circuit. Look for a shorted diode or capacitor as an initial cause.

As previously stated, the exact strategy for troubleshooting depends on what is found at each step, the ease of making a particular test, and the likely cause of a failure. The key is that the technician uses a series of logical tests to reduce the problem to the exact cause.

Review Questions

31. Why is it important to use a fuse on a power supply if the ac line is fused?
32. What would you expect to see if R_1 of the power supply in Figure 2–31 were open?
33. You are checking a 60 Hz full-wave bridge rectifier and observe that the output has a 60 Hz ripple. What failure(s) do you suspect?
34. You observe that the output ripple of a full-wave rectifier is much greater than normal but its frequency is 120 Hz. What component do you suspect?
35. Why is it important to check the polarity before replacing an electrolytic capacitor?

CHAPTER REVIEW

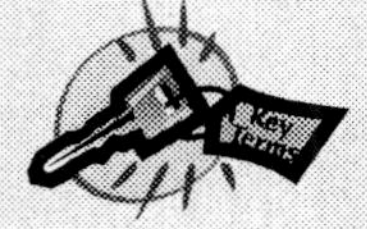

Key Terms

Anode (semiconductor diode definition) The terminal of a semiconductor diode that is more positive with respect to the other terminal when it is biased in the forward direction.

Barrier potential The inherent voltage across the depletion region of a *pn* junction.

Bias The application of dc voltage to a diode or other electronic device to produce a desired mode of operation.

Cathode (semiconductor diode definition) The terminal of a diode that is more negative with respect to the other terminal when it is biased in the forward direction.

Conduction electron An electron that has broken away from the valence band of the parent atom and is free to move from atom to atom within the atomic structure of a material; also called a *free electron.*

Diode An electronic device that permits current in only one direction.

Energy The ability to do work.

Integrated circuit (IC) A type of circuit in which all the components are constructed on a single tiny chip of silicon.

Rectifier An electronic circuit that converts ac into pulsating dc.

Shell An energy level in which electrons orbit the nucleus of an atom.

Valence electron An electron in the outermost shell or orbit of an atom.

Zener diode A type of diode that operates in reverse breakdown (called zener breakdown) to provide voltage regulation.

Important Facts

- ❑ The Bohr model of an atom consists of a nucleus containing positively charged protons and uncharged neutrons orbited by negatively charged electrons.
- ❑ Atomic shells are energy bands. The outermost shell containing electrons is the valence shell.
- ❑ Silicon is the predominant semiconductive material.
- ❑ Atoms within a semiconductor crystal structure are held together with covalent bonds.
- ❑ Electron-hole pairs are thermally produced.
- ❑ A *p*-type semiconductor is made by adding trivalent impurity atoms to an intrinsic (pure) semiconductor.
- ❑ An *n*-type semiconductor is made by adding pentavalent impurity atoms to an intrinsic (pure) semiconductor.
- ❑ The depletion region is a region adjacent to the *pn* junction containing no majority carriers.
- ❑ Forward bias permits majority carrier current through the *pn* junction.
- ❑ Reverse bias prevents majority carrier current.
- ❑ A *pn* junction is called a diode.
- ❑ Reverse breakdown occurs when the reverse-biased voltage exceeds a specified value.
- ❑ Three types of rectifiers are the half-wave, the center-tapped full-wave, and the bridge. The center-tapped and the bridge are both types of full-wave rectifiers.
- ❑ The single diode in a half-wave rectifier conducts for half of the input cycle and has the entire output current in it. The output frequency equals the input frequency.
- ❑ Each diode in the center-tapped full-wave rectifier and the bridge rectifier conduct for one-half of the input cycle but share the total current. The output frequency of a full-wave rectifier is twice the input frequency.
- ❑ The PIV (peak inverse voltage) is the maximum voltage appearing across a reverse-biased diode.
- ❑ A capacitor-input filter provides a dc output approximately equal to the peak of the input.
- ❑ Ripple voltage is caused by the charging and discharging of the filter capacitor.
- ❑ The zener diode operates in reverse breakdown.
- ❑ A zener diode maintains an essentially constant voltage across its terminals over a specified range of zener currents.

- ❑ Zener diodes are used to establish a reference voltage and in basic regulator circuits.
- ❑ Diode symbols are shown in Figure 2–34.

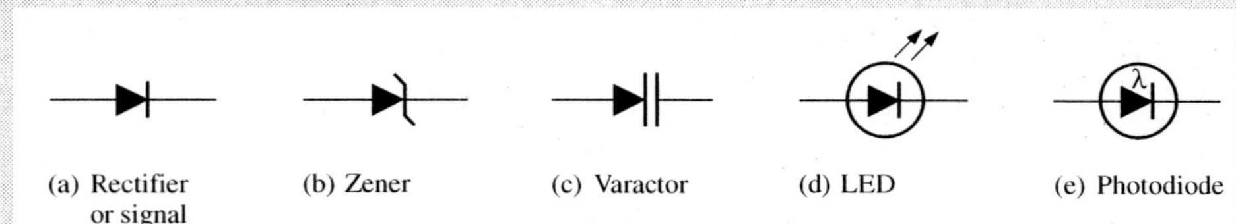

FIGURE 2–34

Diode symbols.

- ❑ Three-terminal integrated circuit regulators provide a nearly constant dc output from an unregulated dc input.
- ❑ Regulation of output voltage over a range of input voltages is called input or line regulation.
- ❑ Regulation of output voltage over a range of load currents is called load regulation.

Formulas

Half-wave and full-wave rectifier peak output voltage:

$$V_{p(out)} = V_{p(in)} - 0.7\text{ V} \tag{2–1}$$

Bridge rectifier peak output voltage:

$$V_{out} = V_{sec} - 1.4\text{ V} \tag{2–2}$$

Line regulation expressed as a percentage:

$$\text{Line regulation} = \left(\frac{\Delta V_{\text{OUT}}}{\Delta V_{\text{IN}}}\right)100\% \tag{2–3}$$

Load regulation expressed as a percentage:

$$\text{Load regulation} = \left(\frac{V_{\text{NL}} - V_{\text{FL}}}{V_{\text{FL}}}\right)100\% \tag{2–4}$$

Chapter Checkup

Answers are at the end of the chapter.

1. When a neutral atom loses or gains a valence electron, the atom becomes
   (a) covalent (b) a metal
   (c) a crystal (d) an ion
2. Atoms within a semiconductor crystal are held together by
   (a) metallic bonds (b) subatomic particles
   (c) covalent bonds (d) the valence band
3. Free electrons exist in the
   (a) valence band (b) conduction band
   (c) lowest band (d) recombination band

4. A hole is a
   (a) vacancy in the valence band
   (b) vacancy in the conduction band
   (c) positive electron
   (d) conduction-band electron
5. The widest energy gap between the valence band and the conduction band occurs in
   (a) semiconductors (b) insulators
   (c) conductors (d) a vacuum
6. The process of adding impurity atoms to a pure semiconductive material is called
   (a) recombination (b) crystallization
   (c) bonding (d) doping
7. In a semiconductor diode, the region near the *pn* junction consisting of positive and negative ions is called the
   (a) neutral zone (b) recombination region
   (c) depletion region (d) diffusion area
8. In a semiconductor diode, the two bias conditions are
   (a) positive and negative (b) blocking and nonblocking
   (c) open and closed (d) forward and reverse
9. The voltage across a forward-biased silicon diode is approximately
   (a) 0.7 V (b) 0.3 V
   (c) 0 V (d) dependent on the bias voltage
10. In Figure 2–35, identify the forward-biased diode(s).
   (a) D_1 (b) D_2
   (c) D_3 (d) D_1 and D_3

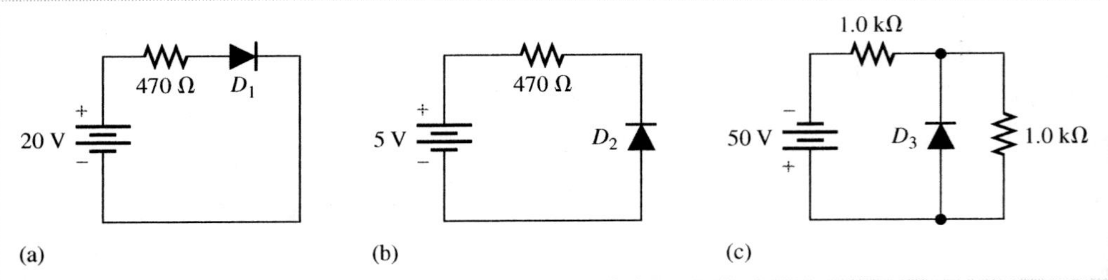

FIGURE 2–35

11. When the positive lead of an analog ohmmeter is connected to the cathode of a diode and the negative lead is connected to the anode, the meter reads
   (a) a very low resistance
   (b) an extremely high resistance or open
   (c) a high resistance initially, decreasing to about 100 Ω
   (d) a gradually increasing resistance
12. The output frequency of a full-wave rectifier with a 60 Hz sinusoidal input is
   (a) 30 Hz (b) 60 Hz
   (c) 120 Hz (d) 0 Hz

13. If a single diode in a center-tapped full-wave rectifier opens, the output is
    (a) 0 V (b) half-wave rectified
    (c) increased in amplitude (d) unaffected
14. During the positive half-cycle of the input voltage in a bridge rectifier,
    (a) one diode is forward-biased
    (b) all diodes are forward-biased
    (c) all diodes are reverse-biased
    (d) two diodes are forward-biased
15. The process of changing a half-wave or a full-wave rectified voltage to a constant dc voltage is called
    (a) filtering (b) ac to dc conversion
    (c) damping (d) ripple suppression
16. The purpose of a small capacitor placed across the output of an IC regulator is to
    (a) improve transient response
    (b) couple the output signal to the load
    (c) filter the ac
    (d) protect the IC regulator
17. The zener diode is designed for operation in
    (a) zener breakdown (b) forward bias
    (c) reverse bias (d) avalanche breakdown
18. Zener diodes are widely used as
    (a) current limiters (b) power distributors
    (c) voltage regulators (d) variable resistors

Questions

Answers to odd-numbered questions are at the end of the book.

1. What is an ion?
2. What type of chemical bonding is in solid copper?
3. What is the difference between a conduction electron and a valence electron?
4. What is meant by *intrinsic* silicon?
5. What is the difference between electron current and hole current?
6. What types of impurities are called donors?
7. How is the depletion region formed in a *pn* junction?
8. What happens to the depletion region when a diode is reverse-biased?
9. What is the maximum voltage that can be placed across a reverse-biased diode called?
10. How can you plot a diode's *IV* curve on an oscilloscope?
11. What are the majority carriers in a *p* material?
12. What does a line indicate on a diode?
13. What is the difference between half-wave and full-wave rectification?
14. What is an advantage of a bridge rectifier over a center-tapped full-wave rectifier?
15. What is meant by *ripple voltage?*

16. What is the output voltage of a 7905 regulator?
17. What type of diode is normally operated in the reverse-breakdown region?
18. What type of diode is designed to act as a small variable capacitor?
19. What type of bias is used on a photodiode?
20. What are some starting ideas for troubleshooting a power supply that has failed?
21. What are symptoms of an open diode in a bridge rectifier?

PROBLEMS

Basic Problems

Answers to odd-numbered problems are at the end of the book.

1. Sketch the waveforms for the load current and voltage for Figure 2–36. Show the peak values.

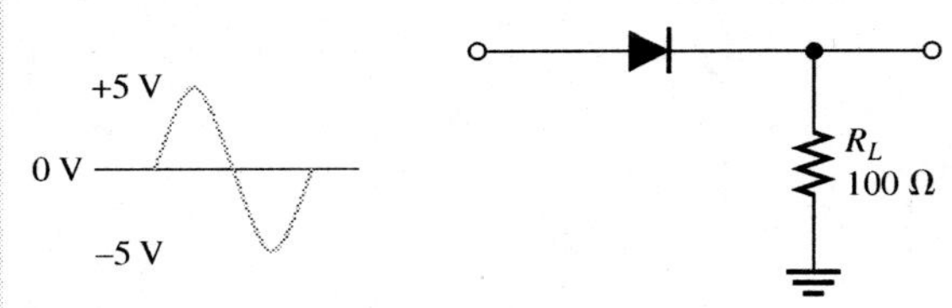

FIGURE 2–36

2. Determine the peak voltage and the peak power delivered to R_L in Figure 2–37.

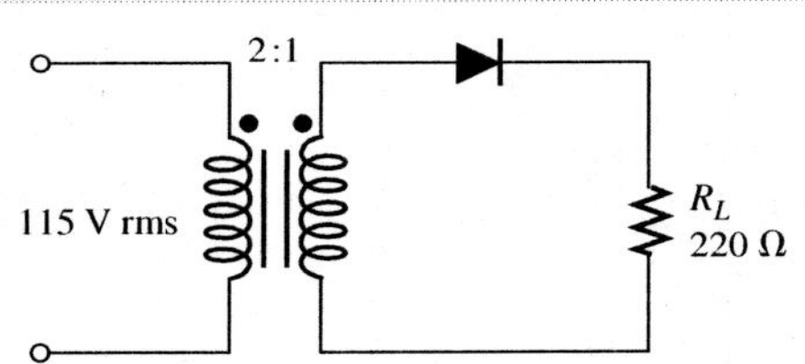

FIGURE 2–37

3. Consider the circuit in Figure 2–38.
   (a) What type of circuit is this?
   (b) What is the total peak secondary voltage?
   (c) Find the peak voltage across each half of the secondary.
   (d) Sketch the voltage waveform across R_L.
   (e) What is the peak current through each diode?
   (f) What is the PIV across each diode?

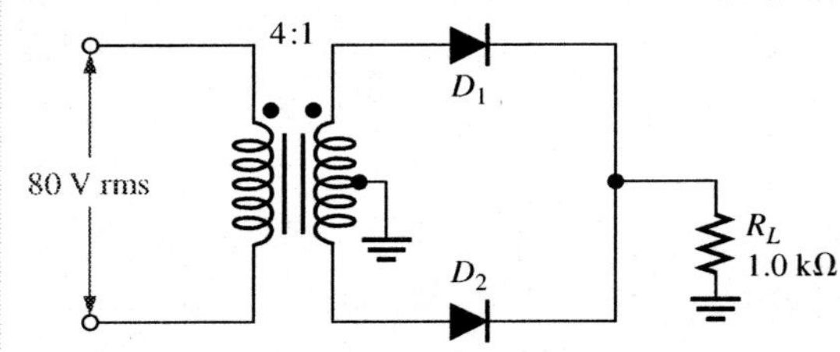

FIGURE 2–38

4. Show how to connect the diodes in a center-tapped rectifier in order to produce a negative-going full-wave voltage across the load resistor.

5. What is the minimum PIV rating required for the diodes in a bridge rectifier that produces an average output voltage of 50 V?

6. The ideal dc output voltage of a capacitor-input filter is the (peak, average) value of the rectified input.

7. If the no-load output voltage of a regulator is 15.5 V and the full-load output is 14.9 V, what is the percent load regulation?

8. Assume a regulator has a percent load regulation of 0.5%. What is output voltage at full-load if the unloaded output is 12.0 V?

9. Figure 2–39 shows a zener diode regulator designed to hold 5.0 V at the output. Assume the zener resistance is zero and the zener current ranges from 2 mA minimum (I_{ZK}) to 30 mA maximum (I_{ZM}). What are the minimum and maximum input source voltages for these currents?

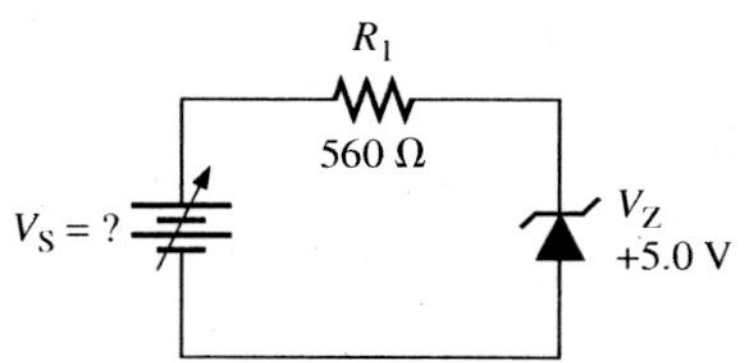

FIGURE 2–39

10. Assume the output of a zener regulator circuit drops from 8.0 V with no load to 7.8 V with a 500 Ω load. What is the percent load regulation?

11. From the meter readings in Figure 2–40, determine the most likely problem. State how you could quickly isolate the exact location of the problem.

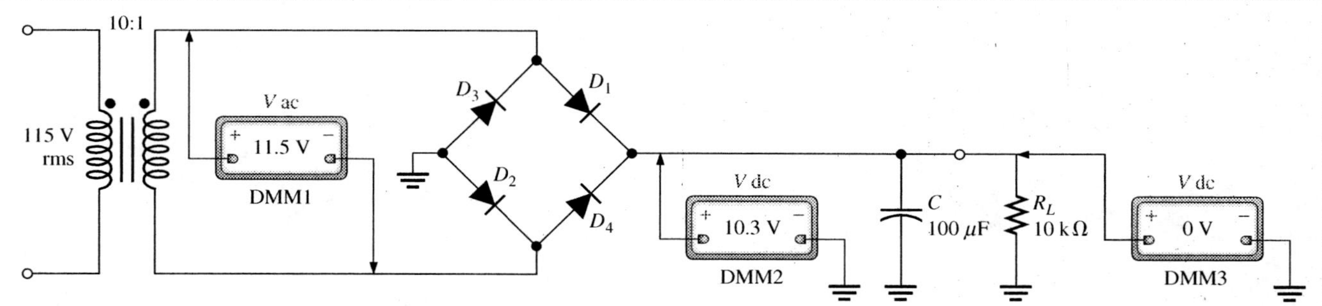

FIGURE 2–40

12. Each part of Figure 2–41 shows oscilloscope displays of rectifier output voltages. In each case, determine whether or not the rectifier is functioning properly and, if it is not, what is (are) the most likely failure(s). Assume all displays are set for 5 ms per division.

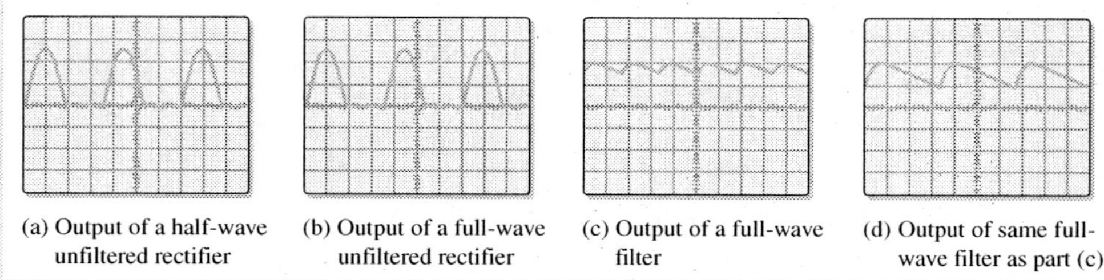

(a) Output of a half-wave unfiltered rectifier (b) Output of a full-wave unfiltered rectifier (c) Output of a full-wave filter (d) Output of same full-wave filter as part (c)

FIGURE 2–41

13. For the variable output supply shown in Figure 2–31, what setting of R_2 would provide an output of 5.0 V?
14. For the variable output supply shown in Figure 2–31, what maximum output voltage could be obtained if R_2 were replaced with a 1.5 kΩ resistor?
15. For each set of measured voltages at the points (1, 2, and 3) indicated in Figure 2–42, determine if they are correct and if not identify the most likely fault(s). State what you would do to correct the problem once it is isolated. The IN963A is a zener diode with a 12 V breakdown voltage.
    (a) $V_1 = 110$ V rms, $V_2 \cong 30$ V dc, $V_3 \cong 12$ V dc
    (b) $V_1 = 110$ V rms, $V_2 \cong 30$ V dc, $V_3 \cong 30$ V dc
    (c) $V_1 = 0$ V, $V_2 = 0$ V, $V_3 = 0$ V
    (d) $V_1 = 110$ V rms, $V_2 \cong 30$ V peak full-wave 120 Hz voltage, $V_3 \cong 12$ V, 120 Hz pulsating voltage
    (e) $V_1 = 110$ V rms, $V_2 = 0$ V, $V_3 = 0$ V

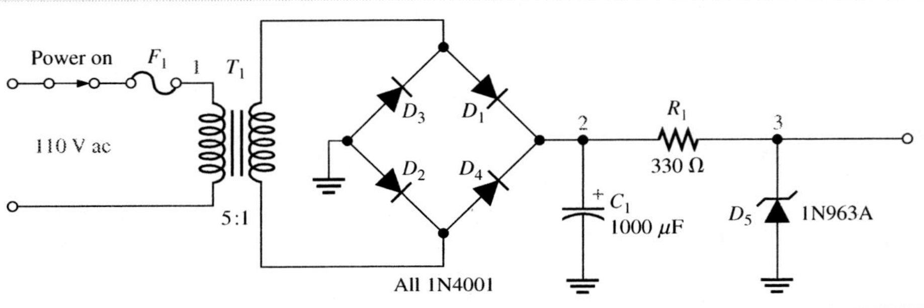

FIGURE 2–42

ANSWERS

Example Questions

2–1: 2.3 V, 3.0 V

2–2: 320 V

2–3: 6.8 V and 28.9 V

2–4: 3.7%

Review Questions

1. An energy level within the atom
2. Conduction band; valence band
3. An electron is thermally raised to the conduction band leaving a vacancy (hole) in the valence band.
4. The gap between the valence band and the conduction band is narrower in a semiconductor than in an insulator.
5. A metallic bond has electrons that are loosely held and move freely; a covalent bond has electrons that are not mobile but rather are shared by atoms in the crystal.
6. By the addition of pentavalent impurities into the semiconductive material
7. Arsenic, phosphorous, and antimony

8. By the addition of trivalent impurities into the semiconductive material
9. Aluminum, boron, and gallium
10. 0.7 V
11. Forward bias permits current by making the anode positive with respect to the cathode; the difference must be sufficient to overcome the barrier potential.
12. Reverse bias blocks current by making the anode negative with respect to the cathode.
13. Forward
14. Reverse
15. A rapid buildup of current when sufficient reverse bias is applied to the diode
16. Forward-bias and reverse-bias
17. The reverse-breakdown region
18. With an analog meter, select $R \times 100$ and check the forward and reverse resistance. Look for a high ratio between the readings.
19. As a switch
20. The barrier potential
21. Bridge
22. Less
23. The peak of the negative alteration
24. 50% (with no filter)
25. Center-tapped full-wave
26. The charging and discharging of the capacitor
27. Increases ripple
28. The IC regulator has better ripple rejection, line and load regulation, thermal protection, and ability to handle a larger load current than a zener regulator.
29. *Line regulation:* Constant output voltage for varying input voltage
    Load regulation: Constant output voltage for varying load current
30. 1.25 V
31. The ac line fuse is generally for a larger current that can damage the power supply.
32. The output would not change as the potentiometer was varied; it would be slightly more than 1.25 V.
33. Open diode
34. The filter capacitor
35. The capacitor will overheat and could explode if connected in reverse.

Chapter Checkup

1. (d)	2. (c)	3. (b)	4. (a)	5. (b)
6. (d)	7. (c)	8. (d)	9. (a)	10. (d)
11. (b)	12. (c)	13. (b)	14. (d)	15. (a)
16. (a)	17. (a)	18. (c)		

BIPOLAR JUNCTION TRANSISTORS

CHAPTER OUTLINE

Study aids for this chapter are available at

http://www.prenhall.com/SOE

INTRODUCTION

Two major categories of transistors are the bipolar junction transistor (BJT) and the field effect transistor (FET). In this chapter, you are introduced to the BJT. Bipolar junction transistors were invented in 1947 at the Bell Telephone laboratories, the forerunner of Lucent Technologies. The transistor was one of the most important inventions of the last century and paved the way for modern electronic systems. It is the fundamental component of virtually all electronic circuits.

The chapter begins with a discussion of the bipolar junction transistor and its dc operation and bias circuits. You will be introduced to the three basic types of BJT amplifiers: the common-emitter, common-collector, and common-base. The chapter concludes with a discussion of basic switching circuits and package identification.

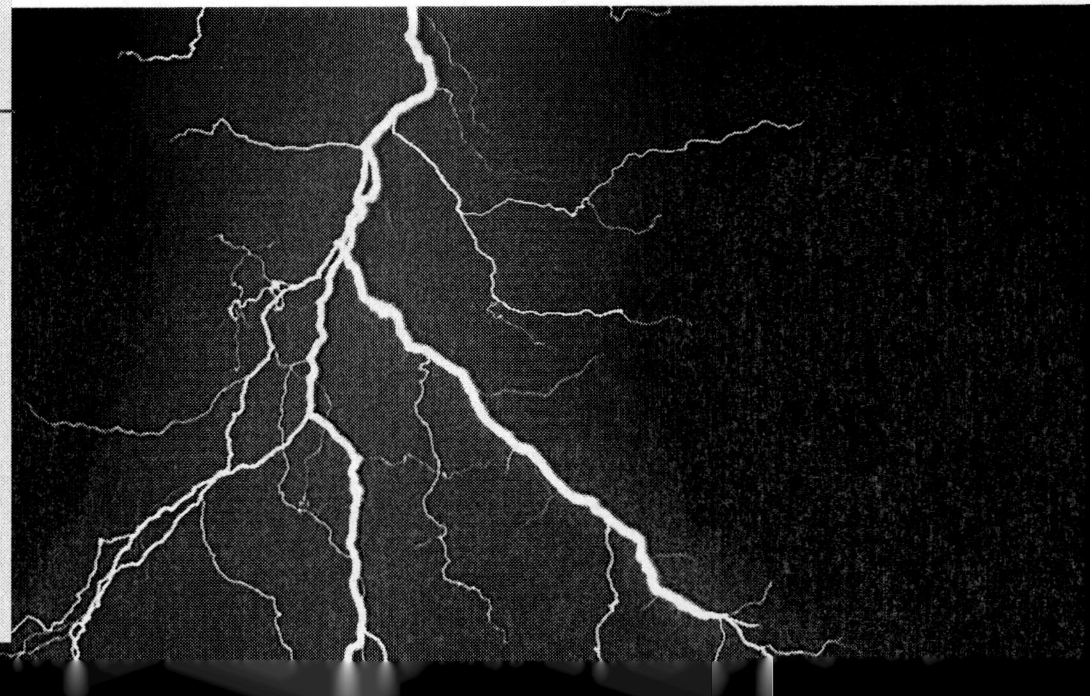

KEY OBJECTIVES

A section number is given for each objective. After completing this chapter, you should be able to

3–1 Describe the structure and operation of bipolar junction transistors (BJTs)

3–2 Explain how to bias a BJT with any of three types of biasing

3–3 Define basic amplifier terms and describe ac considerations for amplifiers

3–4 Analyze the operation of a small-signal amplifier

3–5 Describe how BJTs are used in switching circuits

3–6 Identify various packages used for transistors

COMPUTER SIMULATIONS DIRECTORY

The following figures have Multisim circuit files associated with them.

- Figure 3–7
  Page 74
- Figure 3–19
  Page 85
- Figure 3–23
  Page 87
- Figure 3–28
  Page 93
- Figure 3–37
  Page 102

LABORATORY EXPERIMENTS DIRECTORY

The following exercises are for this chapter.

- **Experiment 4**
  Bipolar Transistor Biasing
- **Experiment 5**
  The Common-Emitter Amplifier
- **Experiment 6**
  Transistor Switches

KEY TERMS

- Bipolar junction transistor
- Emitter
- Base
- Collector
- DC beta (β_{DC})
- Cutoff
- Saturation
- Amplifier
- Gain
- AC beta (β_{ac})
- Common-emitter
- Common-collector
- Common-base

ON THE JOB. . .

Technical skills include competency in the area of electronics in which you are working. You should learn as much as possible about any product on which you work. You must be able to apply your knowledge to solve problems on the job and be able to acquire more advanced knowledge. Also, you must be proficient in the use of tools and instruments that are used to perform the job duties. Look for ways to improve your technical skills by independent study and taking training offered by your employer and area schools.

Carbon nanotubes are tiny structures of carbon that were discovered in 1991 and have remarkable electronic and mechanical properties. This scientific work illustrates the close relationship between chemistry and electronics. The nanotubes are descendants of "buckyballs," a molecule composed of 60 atoms of carbon and shaped like a soccer ball. The nanotubes are shaped similar to tiny chicken-wire cylinders, with groups of six carbon atoms forming hexagons, which are the fundamental building block for the buckyball and the nanotube.

Scientists calculated, and later confirmed, that nanotubes could have very interesting electrical properties, depending on how the hexagons assembled themselves. They calculated that a straight row of hexagons should act as a good conductor of electricity; but that if the line of hexagons were warped into a helix pattern, the properties would be more like a semiconductor.

A natural nanotube with a junction between straight rows of hexagons and helical rows has been shown to have properties of a rectifying diode. Another demonstration showed that a single nanotube could act as a transistor. Much research is needed to learn how to control the specific properties of nanotubes and to find ways to connect them into useful circuits. If these problems can be solved, nanotube electronics may provide an alternative to conventional silicon circuits in the future.

In another related line of research, it was discovered that nanotubes emit electrons under the influence of an electric field (field-emission). This property may prove to be useful in flat panel displays because a source of electrons is necessary for the operation of the display. The nanotubes will take the place of an electron gun to light up the screen. A number of organizations are researching the possibility of using nanotubes in this application.

3–1 STRUCTURE OF BIPOLAR JUNCTION TRANSISTORS

The basic structure of the bipolar junction transistor (BJT) determines its operating characteristics.

In this section, you will learn the basic construction and operation of bipolar junction transistors (BJTs).

The **bipolar junction transistor (BJT)** is constructed with three doped semiconductor regions called **emitter**, **base**, and **collector**. These three regions are separated by two *pn* junctions. The two types of bipolar transistors are shown in Figure 3–1. One type consists

FIGURE 3–1

Construction of bipolar junction transistors.

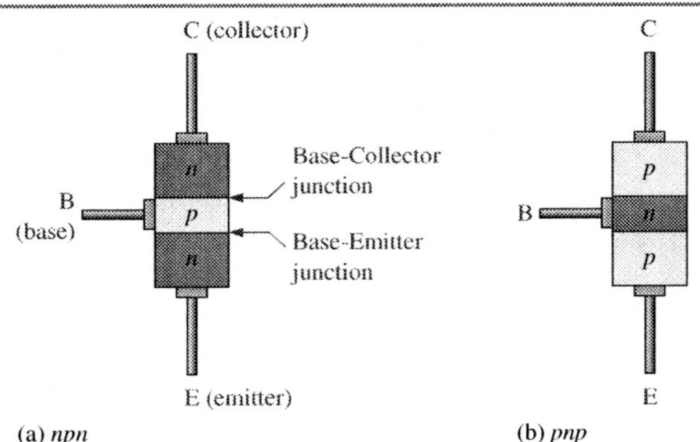

of two *n* regions separated by a thin *p* region (*npn*), and the other type consists of two *p* regions separated by a thin *n* region (*pnp*). Both types are widely used; however, because the *npn* type is more prevalent, it will be used for most of the discussion that follows.

The *pn* junction joining the base region and the emitter region is called the *base-emitter junction.* The *pn* junction joining the base region and the collector region is called the *base-collector junction,* as indicated in Figure 3–1(a). These junctions act just like the diode junctions discussed in Chapter 2 and are frequently referred to as the base-emitter diode and the base-collector diode. Each region is connected to a wire lead, labeled E, B, and C for emitter, base, and collector, respectively. Although the emitter and collector regions are made from the same type of material, the doping level and other characteristics are different.

Figure 3–2 shows the schematic symbols for the *npn* and *pnp* bipolar transistors. (Note that the arrow on an *npn* transistor is *not pointing in.*) The term **bipolar** refers to the use of both holes and electrons as carriers in the transistor structure.

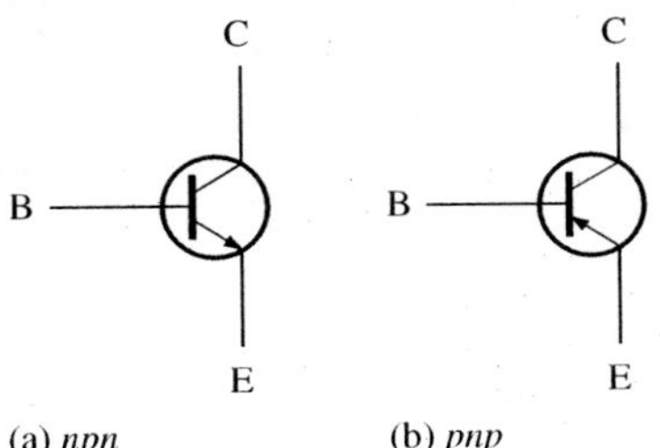

FIGURE 3–2

Standard bipolar junction transistor symbols.

Transistor Operation

In order for the transistor to operate properly, the two *pn* junctions must be supplied with external dc bias voltages to set the proper operating conditions. Figure 3–3 shows the proper bias arrangement for both *npn* and *pnp* transistors. In both cases, the base-emitter (BE) junction is forward-biased and the base-collector (BC) junction is reverse-biased. This is called *forward-reverse bias.* Both *npn* and *pnp* transistors normally use this forward-reverse bias, but the bias voltage polarities and the current directions are reversed between the two types.

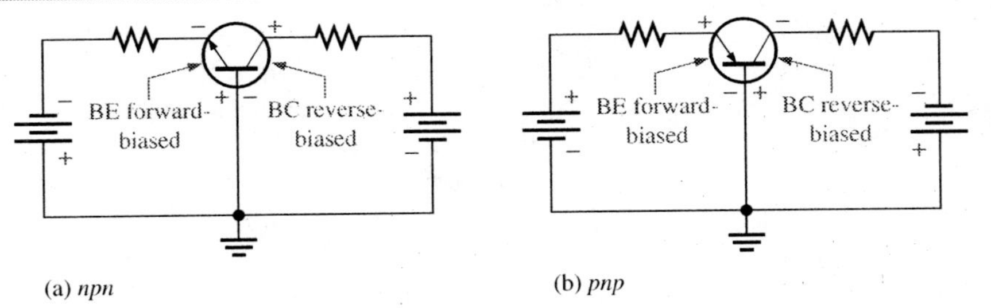

FIGURE 3–3

Forward-reverse bias of a bipolar junction transistor.

To illustrate transistor action, let's examine what happens inside an *npn* transistor when the junctions are forward-reverse biased. (The same concepts can be applied to a *pnp* transistor by reversing the polarities.) The forward bias from base to emitter narrows the BE depletion region, and the reverse bias from base to collector widens the BC depletion region, as depicted in Figure 3–4. The heavily doped *n*-type emitter region is teeming with conduction-band (free) electrons that easily diffuse through the forward-biased BE junction into the *p*-type base region, just as in a forward-biased diode.

The base region is lightly doped and very narrow so that it has a very limited number of holes. Thus, only a small percentage of all the electrons flowing through the BE junction can combine with the available holes in the base. These relatively few recombined electrons

FIGURE 3–4 Illustration of BJT action. The base region is very narrow, but it is shown wider here for clarity.

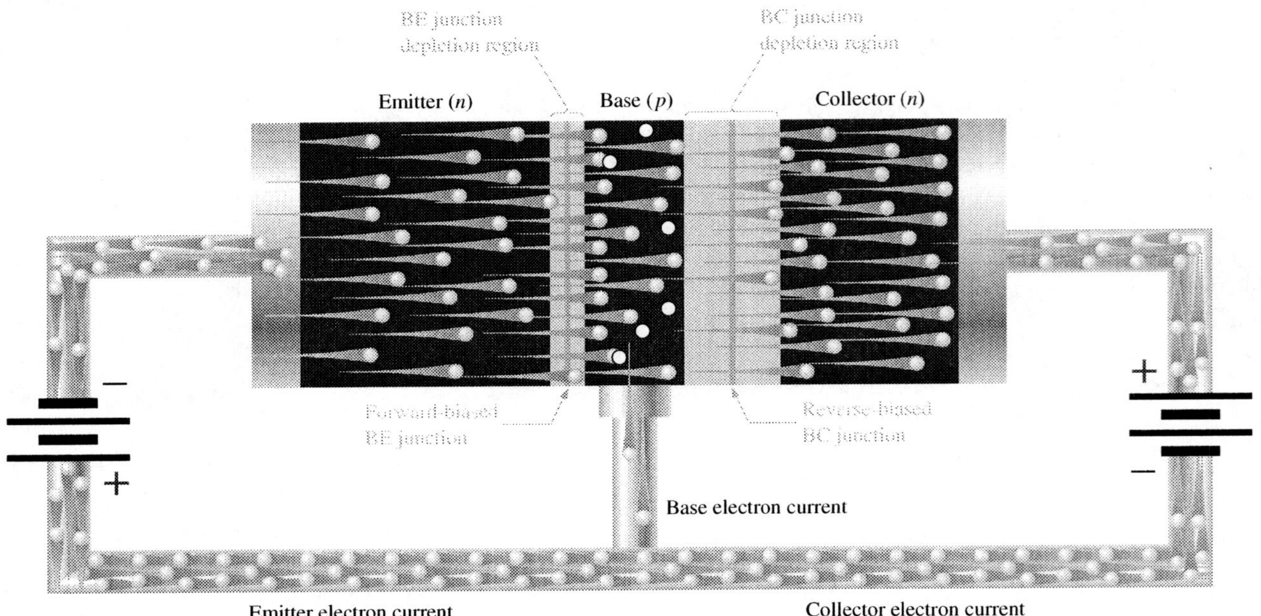

flow out of the base lead as valence electrons, forming the small base current, as shown in Figure 3–4.

Most of the electrons flowing from the emitter into the narrow lightly doped base region do not recombine but diffuse into the BC depletion region. Once in the region, they are pulled through the reverse-biased BC junction by the electric field set up by the force of attraction between the positive and negative ions. Actually, you can think of the electrons as being pulled across the reverse-biased BC junction by the attraction of the collector supply voltage. The electrons now move through the collector region, out through the collector lead, and into the positive terminal of the external dc source, thereby forming the electron flow in the collector circuit, as shown. The amount of collector current depends directly on the amount of base current and is essentially independent of the dc collector voltage.

The bottom line is this: *A small base current can control a larger collector current.* Because the controlling element is base current and it controls a larger collector current, the bipolar junction transistor is essentially a current amplifier. This concept of a small control element for a large current is analogous to deForest's control grid mentioned in the Science Highlight of Chapter 1.

HISTORICAL NOTE

The transistor had its beginnings at Bell Labs in 1947. It was invented by Walter Brattain, John Bardeen, and William Shockley. Their lab notebook entry for December 23, 1947, described measured parameters and included a simple schematic of their one-transistor amplifier. The entry went on to state, "This circuit was actually spoken over, and by switching the device in and out a distinct gain in speech level could be heard and seen on the scope presentation with no noticeable change in quality. . . ." In 1956 Brattain, Bardeen, and Shockley were awarded the Nobel Prize in Physics for their development of the transistor.

Transistor Currents

Kirchhoff's current law (KCL) says the total current entering a junction must be equal to the total current leaving that junction. Applying this law to both the *npn* and the *pnp* transistors shows that the emitter current (I_E) is the sum of the collector current (I_C) and base current (I_B), expressed as follows:

$$I_E = I_C + I_B \tag{3-1}$$

The base current, I_B, is very small compared to I_E or I_C, which leads to the approximation $I_E \cong I_C$, a useful assumption for analyzing transistor circuits. Examples of an *npn* and a *pnp* small-signal transistor with representative currents are shown on the meters in

Figures 3–5(a) and 3–5(b), respectively. Notice the polarity of the ammeters and supply voltages are reversed between the *npn* and the *pnp* transistors. The capital-letter subscripts indicate dc values.

FIGURE 3–5

Currents in small-signal transistors.

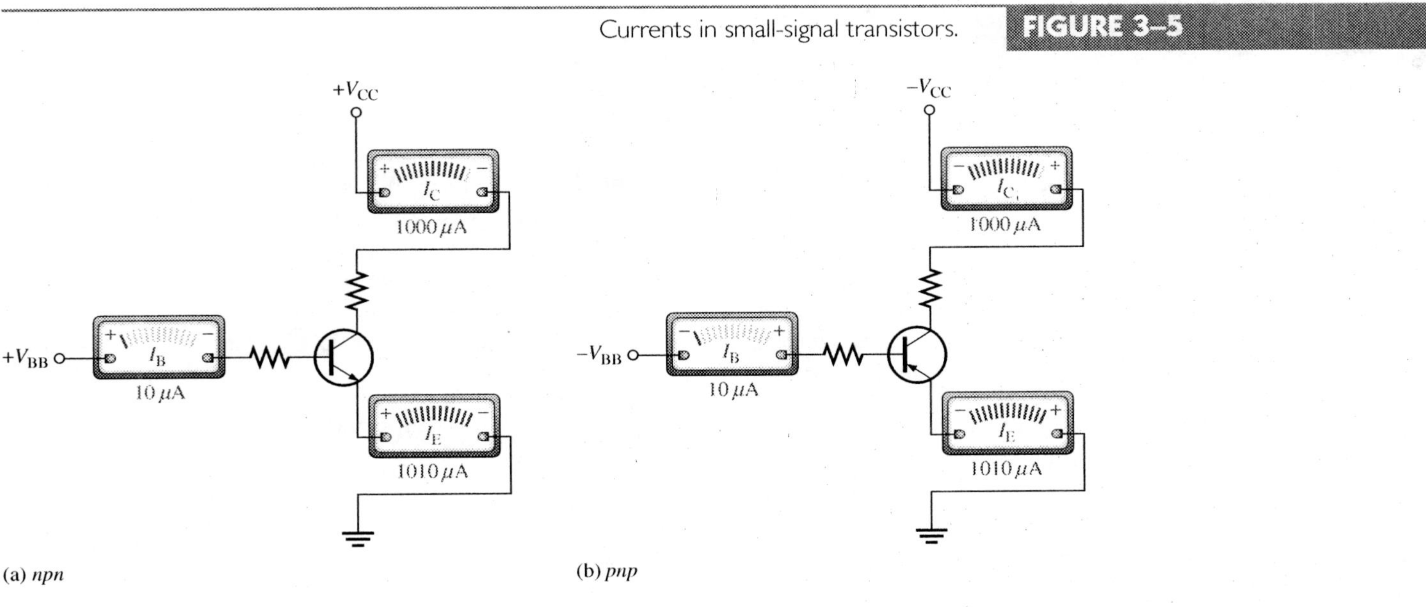

DC Beta (β_{DC})

When a transistor is operated within certain limits, the collector current is proportional to the base current. The **dc beta** (β_{DC}), the current gain of a transistor, is the ratio of the dc collector current to the dc base current.

$$\beta_{DC} = \frac{I_C}{I_B} \qquad (3\text{–}2)$$

The dc beta (β_{DC}) represents a constant of proportionality called the current gain and is usually designated as h_{FE} on transistor data sheets. It is valid as long as the transistor is operated within its linear range. For this case, the collector current is equal to β_{DC} multiplied by the base current. For the examples in Figure 3–5, $\beta_{DC} = 100$ because $I_C = 100I_B$.

The values for β_{DC} vary widely and depend on the type of transistor. They are typically from 20 (power transistors) to 200 (small-signal transistors). Even two transistors of the same type can have current gains that are quite different. Although the current gain is necessary for a transistor to be useful as an amplifier, good designs do not rely on a particular value of β_{DC} to operate properly.

Transistor Voltages

The three dc voltages for the biased transistor in Figure 3–6 are the emitter voltage (V_E), the collector voltage (V_C), and the base voltage (V_B). These single-subscript voltages mean that they are referenced to ground. The collector power supply voltage, V_{CC}, is shown with repeated subscript letters. The collector voltage is equal to the dc supply voltage, V_{CC}, less the drop across R_C.

$$V_C = V_{CC} - I_C R_C$$

FIGURE 3–6

Bias voltages.

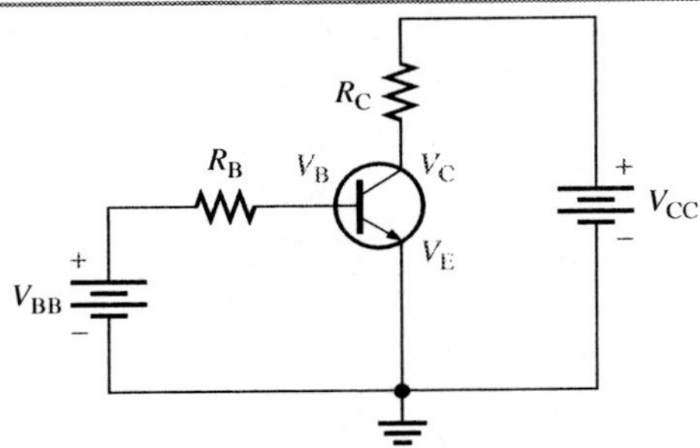

Kirchhoff's voltage law (KVL) says the sum of the voltage drops and rises around a closed path is zero. The previous equation is an application of this law because $V_C = V_{CE}$ in this case.

As mentioned earlier, the base-emitter diode is forward-biased when the transistor is operating normally. The forward-biased base-emitter diode drop, V_{BE}, is approximately 0.7 V. This means that the base voltage is one diode drop larger than the emitter voltage. In equation form,

$$V_B = V_E + V_{BE} = V_E + 0.7 \text{ V}$$

In the configuration of Figure 3–6, the emitter is the reference terminal, so $V_E = 0$ V and $V_B = 0.7$ V.

EXAMPLE 3–1

Problem

Determine I_B, I_C, I_E, V_B, and V_C in Figure 3–7, where β_{DC} is 48.

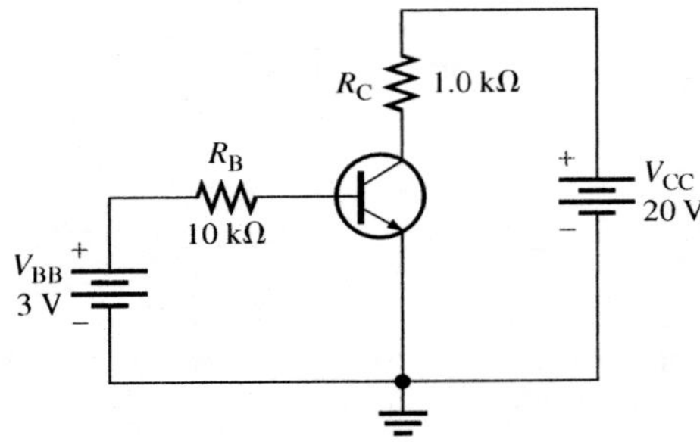

FIGURE 3–7

Solution

Since V_E is ground, V_B = **0.7 V.** The drop across R_B is $V_{BB} - V_B$, so I_B is calculated as follows:

$$I_B = \frac{V_{BB} - V_B}{R_B} = \frac{3 \text{ V} - 0.7 \text{ V}}{10 \text{ k}\Omega} = \mathbf{0.23 \text{ mA}}$$

Now you can find I_C, I_E, and V_C.

$$I_C = \beta_{DC} I_B = 48(0.23 \text{ mA}) = \mathbf{11.0 \text{ mA}}$$
$$I_E = I_C + I_B = 11.0 \text{ mA} + 0.23 \text{ mA} = \mathbf{11.3 \text{ mA}}$$
$$V_C = V_{CC} - I_C R_C = 20 \text{ V} - (11.0 \text{ mA})(1.0 \text{ k}\Omega) = \mathbf{9.0 \text{ V}}$$

Question*

Assume $R_B = 22\ k\Omega$, $R_C = 220\ \Omega$, $V_{BB} = 6\ V$, $V_{CC} = 9\ V$, and $\beta_{DC} = 90$. What is I_B, I_C, I_E, V_{CE}, and V_{CB} in Figure 3–7?

COMPUTER SIMULATION

Open the Multisim file F03-07DV on the website. Using the DMM, confirm V_B and V_C.

Characteristic Curves for a BJT

Base-Emitter Characteristic

The characteristic *IV* curve for the base-emitter junction is shown in Figure 3–8. As you can see, it is identical to that of an ordinary diode. You can model the base-emitter junction with the diode offset model presented in Chapter 2. This means, if you are troubleshooting a bipolar transistor circuit, you can look for 0.7 V across the base-emitter junction (as in a forward-biased diode) to determine if the transistor is conducting. If the voltage is zero, the transistor is not conducting; if it is much larger than 0.7 V, it is likely that the transistor has an open base-emitter junction.

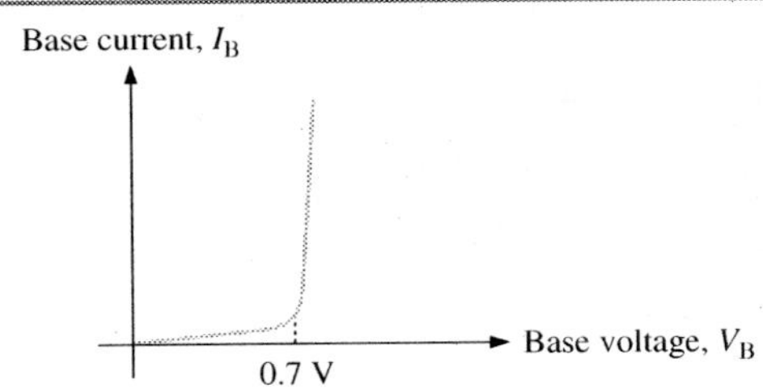

FIGURE 3–8

Base-emitter characteristic.

Collector Characteristic

Recall that the collector current is proportional to the base current ($I_C = \beta_{DC}I_B$). If there is no base current, the collector current is zero. In order to plot the collector characteristic, a base current must be selected and held constant. A circuit such as that in Figure 3–9(a) can be used to generate a set of collector *IV* curves to show how I_C varies with V_{CE} for a given base current. These curves are called the **collector characteristic curves**.

Notice in the circuit diagram that both dc supply voltages, V_{BB} and V_{CC}, are adjustable. If V_{BB} is set to produce a specific value of I_B and V_{CC} is zero, then $I_C = 0$ and $V_{CE} = 0$. Now, as V_{CC} is gradually increased, V_{CE} will increase and so will I_C, as indicated between points *A* and *B* on the curve in Figure 3–9(b).

When V_{CE} reaches $\approx$0.7 V, the base-collector junction becomes reverse-biased and I_C reaches its full value determined by the relationship $I_C = \beta_{DC}I_B$. Ideally, I_C levels off to an almost constant value as V_{CE} continues to increase. This action appears between points *B* and *C* on the curve. In practice, I_C increases slightly as V_{CE} increases due to the widening of the base-collector depletion region, which results in fewer holes for recombination in the base region. The steepness of this rise is determined by a parameter called the *forward Early voltage,* named after J. M. Early.

By setting I_B to other constant values, you can produce additional I_C versus V_{CE} curves, as shown in Figure 3–9(c). These curves constitute a "family" of collector curves for a given transistor. A family of curves allows you to visualize the complex situation when the three variables interact. By holding one of them (I_B) constant, you can see the relation between the other two (I_C versus V_{CE}).

* *Answers are at the end of the chapter.*

FIGURE 3–9

Collector characteristic curves.

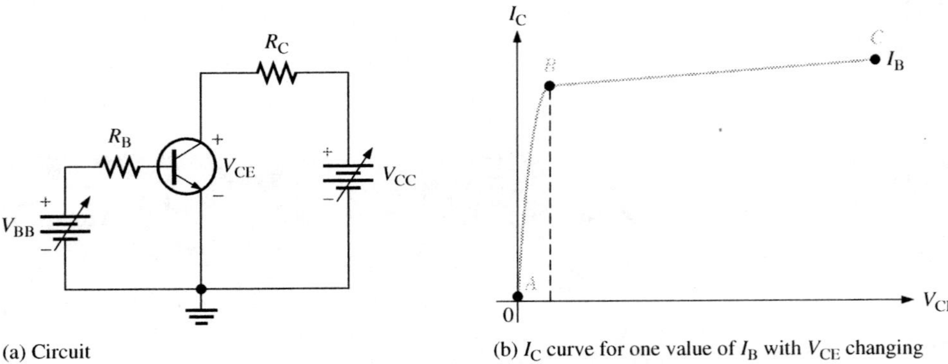

(a) Circuit

(b) I_C curve for one value of I_B with V_{CE} changing

(c) Family of collector curves ($I_{B1} < I_{B2} < I_{B3}$, etc.)

EXAMPLE 3–2

Problem

Sketch the family of collector curves for the circuit in Figure 3–10 for $I_B = 5\ \mu A$ to $25\ \mu A$ in $5\ \mu A$ increments. Assume that $\beta_{DC} = 100$.

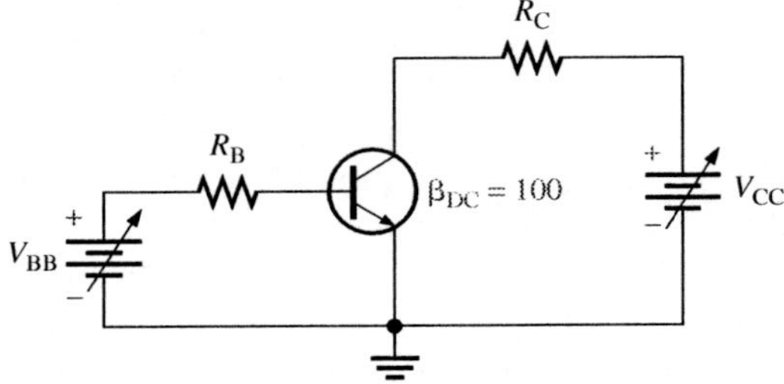

FIGURE 3–10

Table 3–1

I_B	I_C
5 μA	0.5 mA
10 μA	1.0 mA
15 μA	1.5 mA
20 μA	2.0 mA
25 μA	2.5 mA

Solution

Table 3–1 shows the calculations of I_C using the relationship $I_C = \beta_{DC}I_B$. The resulting curves are plotted in Figure 3–11. To account for the forward Early voltage, the resulting curves are shown with a slight upward slope as previously discussed.

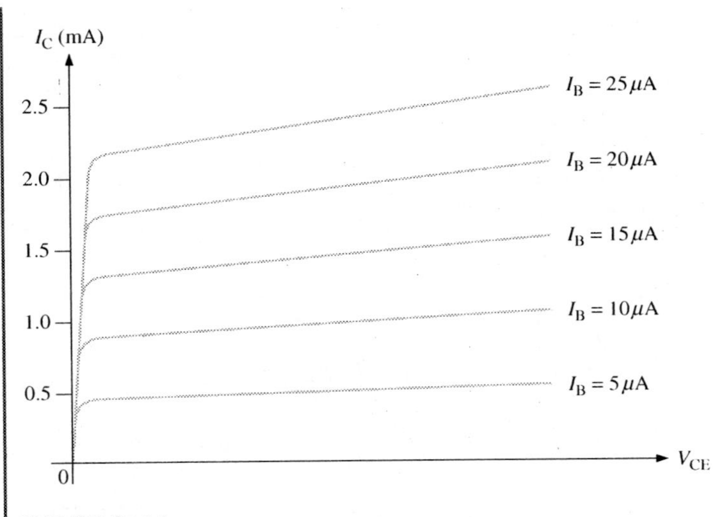

FIGURE 3–11

Question
Ideally, where would the curve for $I_B = 0$ appear on the graph?

Cutoff and Saturation

When $I_B = 0$, the transistor is in **cutoff** and there is essentially no collector current except for a very tiny amount of collector leakage current, I_{CEO}, which can usually be neglected. In cutoff, both the base-emitter and the base-collector junctions are reverse-biased. When you are troubleshooting a transistor that is in cutoff, you can assume the collector current is zero; therefore, there is no voltage drop across the collector resistor. As a result, the collector-to-emitter voltage will be very nearly equal to the supply voltage.

Now let's consider the opposite situation. When the base-emitter junction in Figure 3–9 becomes forward-biased and the base current is increased, the collector current also increases and V_{CE} decreases as a result of more drop across R_C. According to Kirchhoff's voltage law, if the voltage across R_C increases, the drop across the transistor must decrease. Ideally, when the base current is high enough, the entire V_{CC} is dropped across R_C with no voltage between the collector and emitter. This condition is known as **saturation**. Saturation occurs when the supply voltage, V_{CC}, is across the total resistance of the collector circuit, R_C. The saturation current for this particular configuration is given by Ohm's law.

$$I_{C(sat)} = \frac{V_{CC}}{R_C}$$

Once the base current is high enough to produce saturation, further increases in base current have no effect on the collector current, and the relationship $I_C = \beta_{DC} I_B$ is no longer valid. When V_{CE} reaches its saturation value, $V_{CE(sat)}$, which is ideally zero, the base-collector junction becomes forward-biased.

When you are troubleshooting transistor circuits, a quick check for cutoff or saturation provides useful information. Remember a transistor in cutoff has nearly the entire supply voltage between collector and emitter; a saturated transistor, in practice, has a very small voltage drop between collector and emitter (typically 0.1 V).

DC Load Line

Recall from your dc/ac studies that a Thevenin circuit is a voltage source in series with a resistor, as illustrated in Figure 3–12(a). Any linear, two-terminal network can be replaced with this simple circuit to simplify the operation of the original circuit. Just as an *IV* characteristic curve can be drawn to show the relationship between the current and voltage for a component, an *IV* characteristic curve can be plotted for the Thevenin circuit as well. The characteristic curve is a straight line for the dc load line.

FIGURE 3–12

Thevenin circuit and *IV* characteristic curve for the circuit.

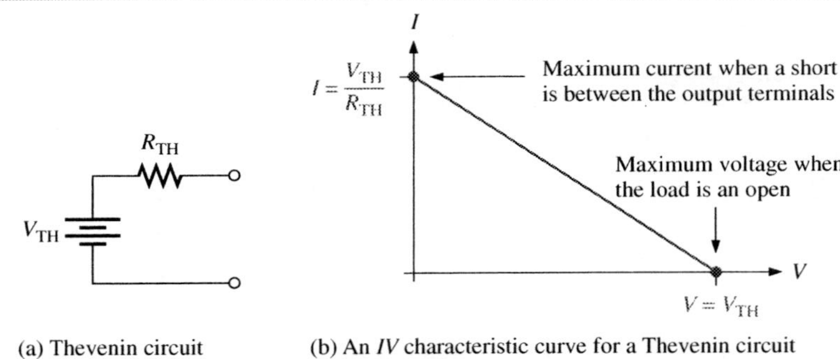

(a) Thevenin circuit

(b) An *IV* characteristic curve for a Thevenin circuit that forms the dc load line

Because it is a linear circuit, two points are sufficient to establish the straight line that represents all possible voltage and current points for the circuit, as viewed from the output terminals. The simplest two points that will establish the relationship are (1) when a short is placed across the output terminals and (2) when the output terminals are left open (no load). For the case of a short, the voltage will be zero and the current will be at its maximum value; from Ohm's law it is V_{TH}/R_{TH}. When the output terminals are left open, the current is zero and the voltage is simply the Thevenin voltage, V_{TH}. A line drawn between these two points is called the *load line* for the circuit. For the Thevenin circuit in Figure 3–12(a), the load line is shown in Figure 3–12(b). *Any load placed across the output terminals will have a current and voltage point located on this line.*

Load lines are a useful visual tool for transistor circuits. Most of the time, you won't draw a load line for a circuit, but it is useful to keep in mind when you are thinking about circuit operation, particularly how the endpoints are found. When a transistor is placed across the output terminals, it may appear to be a short or an open or some value between these extremes, but it will always have an *IV* characteristic somewhere on the load line.

The circuit shown in Figure 3–6 is redrawn in Figure 3–13(a) to emphasize that there is a Thevenin circuit in the collector circuit. The Thevenin collector circuit is shown in the blue box. Notice that the base circuit, shown in the beige box, has another Thevenin circuit as well as the transistor. The transistor and the base circuit act as a load for the collector Thevenin circuit. The minimum current and the maximum current that can be supplied by the collector Thevenin circuit can be calculated by assuming a short or an open for the load (between the transistor's collector and emitter). The result is that the minimum current is zero (open load) and the maximum current is V_{CC}/R_C (shorted load). These are, of course, the cutoff and saturation values as previously defined. Note that the saturation and cutoff points depend only on the Thevenin circuit; the transistor does not affect these points. A straight line drawn between cutoff and saturation defines the dc load line for this circuit, as shown in Figure 3–13(b). This line represents all possible dc operating points for the circuit.

The *IV* characteristic curve for any type of load can be added to the same plot as the dc load line to obtain a graphical picture of the circuit operation. Figure 3–13(c) shows a dc load line superimposed on a set of ideal collector characteristic curves. Any value of I_C and the corresponding V_{CE} will fall on this line, as long as dc operation is maintained.

FIGURE 3–13

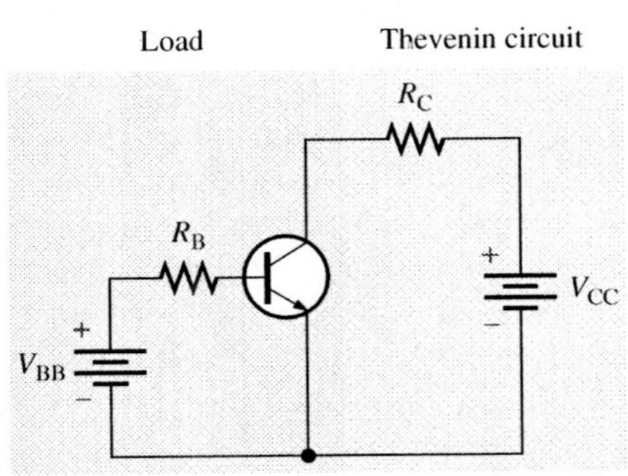

(a) The collector circuit, shown in the blue box, is a Thevenin circuit. The base circuit and the transistor act as a load for this Thevenin circuit.

I_C

$I_{C(sat)} = \frac{V_{CC}}{R_C}$ — Saturation

Cutoff

V_{CE}

$V_{CE(cutoff)} = V_{CC}$

(b) DC load line for the circuit in (a)

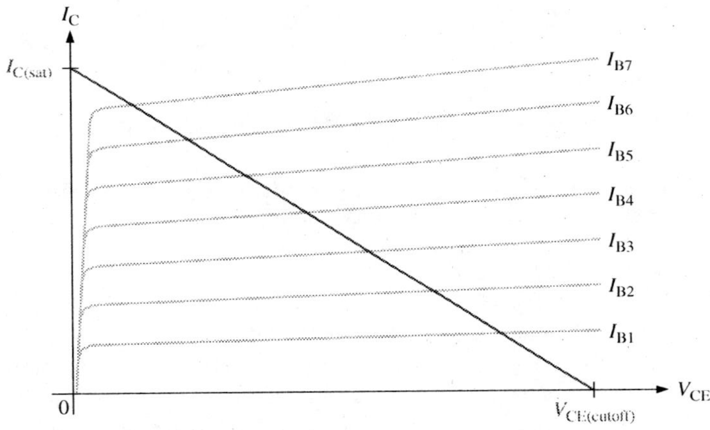

(c) Collector characteristic curves superimposed on the dc load line

Now let's see how the dc load line and the characteristic curves for a transistor can be combined to illustrate transistor operation. Assume you have a transistor test circuit as shown in Figure 3–14(a). A graphical solution can be used to find currents and voltages by drawing a dc load line on the characteristic curves as shown in Figure 3–14(b). The procedure for finding the load line is described next. First, the cutoff point on the load line is determined. When the transistor is cut off, there is essentially no collector current. Thus, the collector-emitter voltage and current are

$$V_{CE(cutoff)} = V_{CC} = 12\text{ V}$$

and

$$I_{C(cutoff)} = 0\text{ mA}$$

Next, the saturation point on the load line is determined. When the transistor is saturated, V_{CE} is nearly zero. Therefore, V_{CC} is dropped across R_C. The saturation value of the collector current, $I_{C(sat)}$, is found by applying Ohm's law to the collector resistor.

FIGURE 3–14

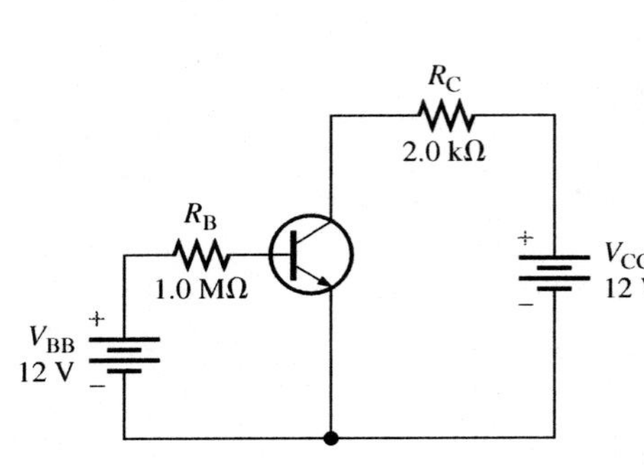

(a) DC test circuit

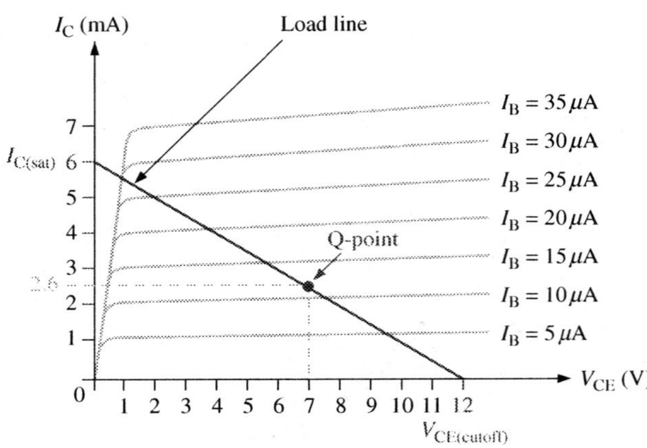

(b) Locating the Q-point

$$I_{C(sat)} = \frac{V_{CC}}{R_C} = \frac{12\text{ V}}{2.0\text{ k}\Omega} = 6.0\text{ mA}$$

This value is the maximum value for I_C. It cannot possibly be increased without changing V_{CC} or R_C.

The cutoff and saturation points establish the dc load line. A line drawn between them represents all possible operating points for the circuit.

Q-Point

Before the actual collector current can be found, the value of the base current, I_B, needs to be established. Referring to the original circuit, it is apparent that the base power supply, V_{BB}, is across the series combination of the base resistor, R_B, and the forward-biased base-emitter junction. This means that the voltage across the base resistor is

$$V_{R_B} = V_{BB} - V_{BE} = 12\text{ V} - 0.7\text{ V} = 11.3\text{ V}$$

By applying Ohm's law, you can find the base current.

$$I_B = \frac{V_{R_B}}{R_B} = \frac{11.3\text{ V}}{1.0\text{ M}\Omega} = 11.3\,\mu\text{A}$$

The point at which the actual base current line intersects the load line is the quiescent or Q-point for the circuit. The Q-point is found on the graph in Figure 3–14(b) by interpolating between the 10 μA and 15 μA base current lines. The coordinates of the Q-point are the values for I_C and V_{CE} at that point, as illustrated. Reading these values from the plot, you find the value of I_C to be approximately 2.6 mA and V_{CE} to be approximately 7.0 V.

The plot in Figure 3–14(b) completely describes the dc operating conditions for the circuit. Keep in mind that the load line is a graphical picture of circuit operation; however, most of the time you will apply the basic math for circuits to obtain an idea of what is occurring with a given circuit. The load line provides a useful mental picture for describing the dc conditions for the transistor.

Review Questions

Answers are at the end of the chapter.

1. What are the three BJT currents called?
2. Explain the difference between *saturation* and *cutoff*.
3. What is the definition of β_{DC}?
4. What is a load line?
5. What is the Q-point?

BJT BIAS CIRCUITS 3–2

Biasing is the application of the appropriate dc voltages to cause a transistor to operate properly. It can be accomplished with any of several basic circuits. The choice of biasing circuit depends largely on the application.

In this section, you will learn about three methods to bias a BJT and see the advantages and disadvantages of each method.

For linear amplifiers, the signal must swing in both the positive and negative directions. Transistors operate with current in one direction only. In order for a transistor to amplify an ac signal, the ac signal needs to be superimposed on a dc level that sets the operating point. Bias circuits set the dc level at a point that allows the ac signal to vary in both the positive and negative directions without driving the transistor into saturation or cutoff.

Base Bias

The simplest biasing circuit is **base bias**. For the *npn* transistor, shown in Figure 3–15(a), a resistor (R_B) is connected between the base and supply voltage. Note that this is essentially the same circuit that was introduced in Figure 3–9(a) and used to plot the characteristic curve. The only difference is that the base and the collector power supplies have been combined into a single supply (referred to as V_{CC}). Although this bias method is simple, it is not a good choice for linear amplifiers as you will see.

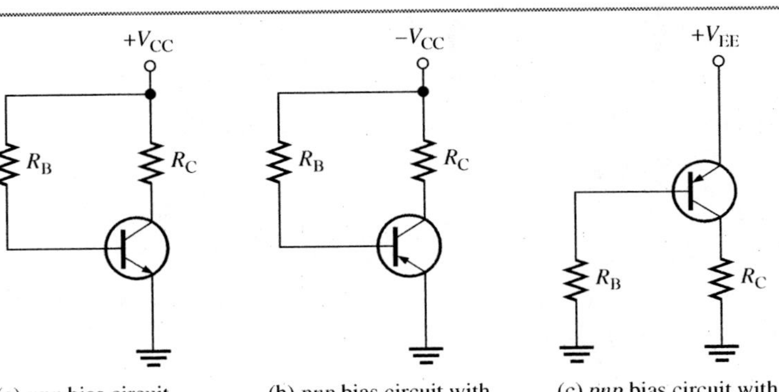

FIGURE 3–15

Base bias circuits.

The *pnp* transistor can be set up using a negative supply as shown in Figure 3–15(b) or it can be run with a positive supply by applying the positive supply voltage to the emitter as shown in Figure 3–15(c). Either of these arrangements provide a path for base current through the base-emitter junction. In turn, this base current causes a collector current that is β_{DC} times larger than the base current (assuming linear operation). Thus, the collector current, for linear operation, is

$$I_C = \beta_{DC} I_B$$

The base resistor, R_B, has the base current, I_B, through it. From Ohm's law, you can substitute for I_B and obtain

$$I_C = \beta_{DC}\left(\frac{V_{R_B}}{R_B}\right)$$

$$I_C = \beta_{DC}\left(\frac{V_{CC} - V_{BE}}{R_B}\right) \qquad (3\text{–}3)$$

This formula gives the collector current for base bias as long as the transistor is not in saturation. It is derived for the case with no emitter resistor, so this formula can only be applied to this configuration.

As mentioned previously, transistors can have very different current gains. Typical transistors of the same type can have β_{DC} values that vary by a factor of 3! In addition, current gain is a function of the temperature; as temperature increases, the base-emitter voltage decreases and β_{DC} increases. As a result, the collector current can vary widely between similar circuits with base bias. Circuits that depend on a particular β_{DC} cannot be expected to operate in a consistent manner. For this reason, base bias is rarely used for linear circuits.

Because it uses only a single resistor for bias, base bias is a good choice in switching applications, where the transistor is always operated in either saturation or cutoff. For switching amplifiers, Equation 3–3 does not apply.

EXAMPLE 3–3

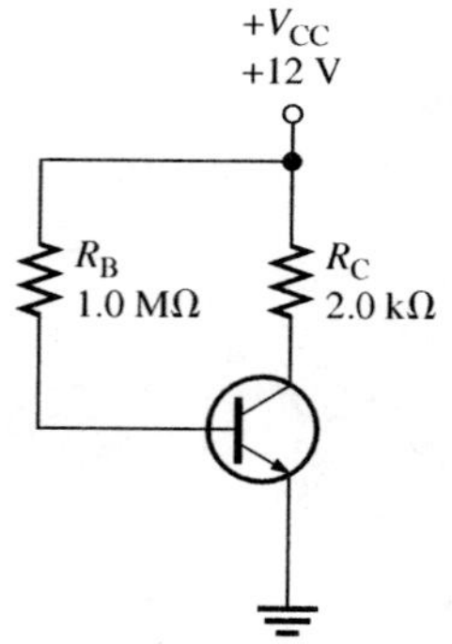

+V_{CC}
+12 V

R_B 1.0 MΩ

R_C 2.0 kΩ

FIGURE 3–16

Problem

The manufacturer's specification for a 2N3904 transistor shows that β_{DC} has a range from 100 to 300. Assume a 2N3904 is used in the base-biased circuit shown in Figure 3–16. Compute the minimum and maximum collector current based on this specification. (Note that this is effectively the same circuit that was solved with load line analysis in Figure 3–14 except it now is shown with a single power supply.)

Solution

The base-emitter junction is forward-biased, causing a 0.7 V drop. The voltage across R_B is

$$V_{R_B} = V_{CC} - V_{BE} = 12\text{ V} - 0.7\text{ V} = 11.3\text{ V}$$

You can find the base current by applying Ohm's law to the base resistor.

$$I_B = \frac{V_{R_B}}{R_B} = \frac{11.3\text{ V}}{1.0\text{ M}\Omega} = 11.3\ \mu\text{A}$$

With linear operation, the collector current is β_{DC} times larger than the base current. Therefore, the minimum collector current is

$$I_{C(min)} = \beta_{DC} I_B = (100)(11.3\ \mu\text{A}) = \mathbf{1.13\ mA}$$

The maximum collector current is

$$I_{C(max)} = \beta_{DC} I_B = (300)(11.3\ \mu A) = \mathbf{3.39\ mA}$$

Notice that a 300% change in β_{DC} caused a 300% change in collector current.

Question

If you measured 2.5 mA of collector current in the circuit of Figure 3–16, what is the β_{DC} for the transistor?

Voltage-Divider Bias

As you have seen, the principal disadvantage to base bias is its dependency on β_{DC}. A high degree of stability can be obtained with **voltage-divider bias**. Voltage-divider bias is the most widely used form of biasing because it uses only one supply voltage and produces bias that is essentially independent of β_{DC}. In fact, looking at the equations for voltage-divider bias reveals that neither β_{DC} nor any other transistor parameter is included. Essentially, good voltage-divider designs are independent of the transistor that is used. Figure 3–17(a) illustrates a basic voltage divider.

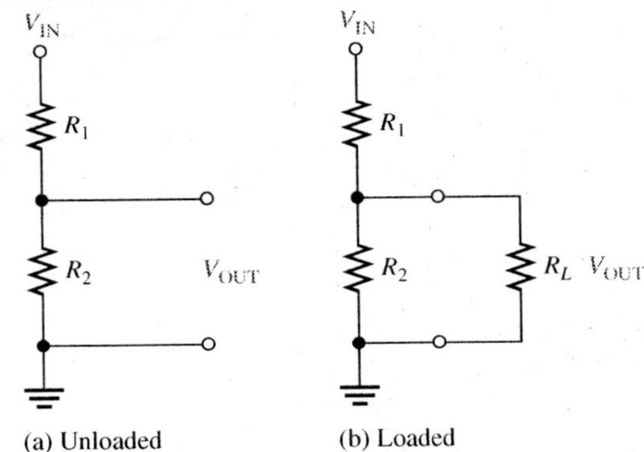

FIGURE 3–17
Voltage dividers.

The voltage-divider rule, one of the most useful rules from your basic dc/ac studies, allows you to compute the voltage across a series resistive branch in a circuit. To find the output voltage, a ratio of the resistance to total resistance is set up and multiplied by the input voltage.

$$V_{OUT} = \left(\frac{R_2}{R_1 + R_2}\right) V_{IN}$$

When setting up the ratio for the voltage-divider rule, the resistor (in this case, R_2) that the output is taken across is the numerator and the sum of the resistances is the denominator.

When a load resistor is placed across the output of a voltage divider, as in Figure 3–17(b), the output voltage decreases because of the loading effect. **Loading effect** is a change in circuit parameters when a load is connected. As long as the load resistor is large compared to the divider resistors, the loading effect is small and can be ignored.

Voltage-divider bias is shown in Figure 3–18. In this configuration, two resistors, R_1 and R_2, form a voltage divider that keeps the base voltage nearly the same for any load that requires a small current. This voltage forward-biases the base-emitter junction, resulting in a small base current. With voltage-divider bias, the transistor acts as a high resistance load on the divider. This will tend to make the base voltage slightly less than the unloaded value. In well-designed voltage-divider bias circuits, the effect is generally small, so the loading effects can be ignored. In any case, this loading effect can be minimized by the choice of R_1 and R_2. As a rule of thumb,

FIGURE 3–18
Voltage-divider bias.

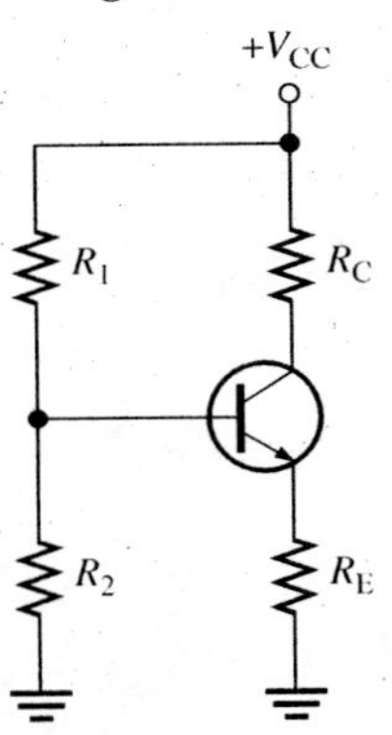

these resistors should have a current that is at least ten times larger than the base current to avoid variations in the base voltage when a transistor with a different β_{DC} is used. This is called *stiff bias* because the base voltage is relatively independent of the base current.

The steps in computing the parameters for a voltage-divider bias circuit are straightforward applications of the voltage-divider rule and Ohm's law. Based on the assumption of no significant loading effect, you can use the voltage-divider rule to compute the base voltage. The voltage-divider rule applied to Figure 3–18 is

$$V_B = \left(\frac{R_2}{R_1 + R_2}\right)V_{CC} \tag{3–4}$$

The emitter voltage is one diode drop less than the base voltage. (For *pnp* transistors, it is one diode drop higher.)

$$V_E = V_B - V_{BE}$$

$$V_E = V_B - 0.7\text{ V} \tag{3–5}$$

With the emitter voltage known, the emitter current is found by Ohm's law.

$$I_E = \frac{V_E}{R_E}$$

The collector current is approximately the same as the emitter current.

$$I_C \cong I_E$$

The collector voltage can now be calculated. It is V_{CC} less the drop across the collector resistor found by Ohm's law. The equation is

$$V_C = V_{CC} - I_C R_C \tag{3–6}$$

To find the collector-emitter voltage, V_{CE}, subtract the emitter voltage, V_E, from the collector voltage, V_C.

$$V_{CE} = V_C - V_E$$

Example 3–4 illustrates this procedure for finding the dc parameters for a circuit.

EXAMPLE 3–4

Problem

Find V_B, V_E, I_E, I_C, and V_{CE} for the circuit in Figure 3–19.

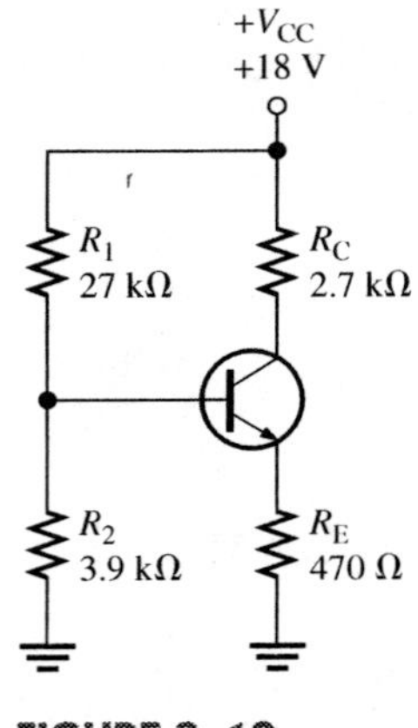

FIGURE 3–19

Solution

Begin by finding the base voltage using the voltage-divider rule.

$$V_B = \left(\frac{R_2}{R_1 + R_2}\right)V_{CC} = \left(\frac{3.9\text{ k}\Omega}{27\text{ k}\Omega + 3.9\text{ k}\Omega}\right)18\text{ V} = \mathbf{2.27\text{ V}}$$

The emitter voltage is one diode drop less than the base voltage.

$$V_E = V_B - V_{BE} = 2.27\text{ V} - 0.7\text{ V} = \mathbf{1.57\text{ V}}$$

Next, find the emitter current from Ohm's law.

$$I_E = \frac{V_E}{R_E} = \frac{1.57\text{ V}}{470\ \Omega} = \mathbf{3.34\text{ mA}}$$

Using the approximation $I_C \cong I_E$,

$$I_C = \mathbf{3.34\ mA}$$

Now find the collector voltage.

$$V_C = V_{CC} - I_C R_C = 18\ \text{V} - (3.34\ \text{mA})(2.7\ \text{k}\Omega) = 8.98\ \text{V}$$

The collector-emitter voltage is

$$V_{CE} = V_C - V_E = 8.98\ \text{V} - 1.57\ \text{V} = \mathbf{7.41\ V}$$

Question

Assume the power supply voltage was incorrectly set to +12 V. What are V_B, V_E, I_E, I_C, and V_{CE} for the circuit in Figure 3–19 for this case?

COMPUTER SIMULATION

Open the Multisim file F03-19DV on the website. Verify the base, emitter, and collector voltages.

Figure 3–20 shows two configurations for voltage-divider bias with a *pnp* transistor. As in the case of base bias, either negative or positive supply voltages can be used for bias. With a negative supply, shown in Figure 3–20(a), the voltage is applied to the collector. With a positive supply, shown in Figure 3–20(b), the voltage is applied to the emitter. The transistor is frequently drawn upside down to place the supply voltage on top; this means that the emitter resistor is also on top. The equations for *npn* transistors can be applied to *pnp* transistors, but you need to be careful of algebraic signs.

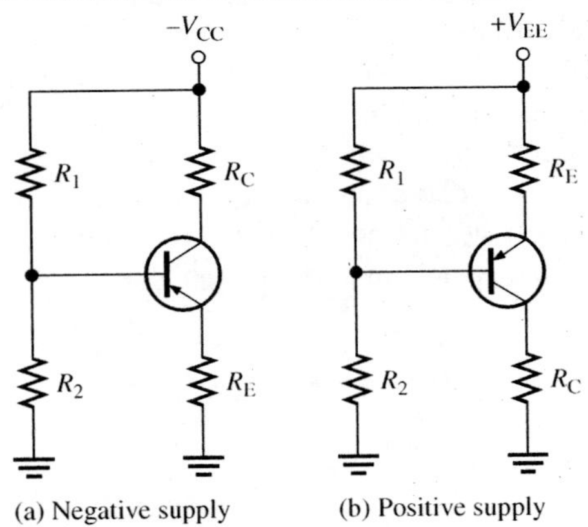

FIGURE 3–20

Voltage-divider bias for *pnp* transistors.

EXAMPLE 3–5

Problem

Find V_B, V_E, I_E, I_C, and V_{CE} for the *pnp* circuit in Figure 3–21.

Solution

Begin by finding the base voltage using the voltage-divider rule.

$$V_B = \left(\frac{R_2}{R_1 + R_2}\right)V_{CC} = \left(\frac{4.7\ \text{k}\Omega}{27\ \text{k}\Omega + 4.7\ \text{k}\Omega}\right)(-12\ \text{V}) = \mathbf{-1.78\ V}$$

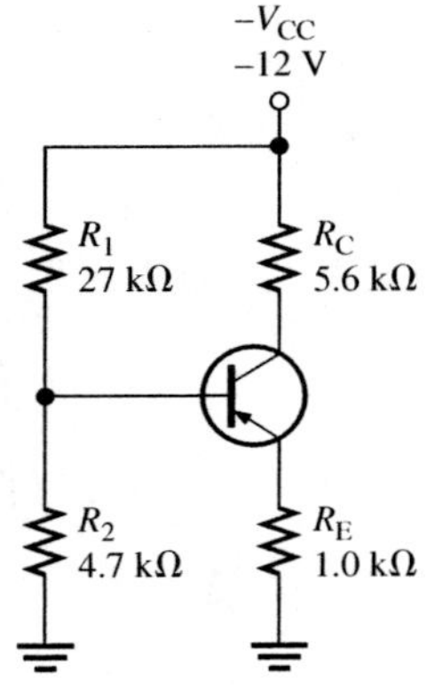

FIGURE 3–21

The equation for V_E is the same one used for the *npn* transistor but note the signs. The emitter voltage is one diode drop *greater* than the base voltage for a forward-biased *pnp* transistor.

$$V_E = V_B - V_{BE} = -1.78 - (-0.7\text{ V}) = \mathbf{-1.08\text{ V}}$$

Now find the emitter current using Ohm's law.

$$I_E = \frac{V_E}{R_E} = \frac{-1.08\text{ V}}{1.0\text{ k}\Omega} = \mathbf{-1.08\text{ mA}}$$

Using the approximation $I_C \cong I_E$,

$$I_C = \mathbf{-1.08\text{ mA}}$$

Now find the collector voltage.

$$V_C = V_{CC} - I_C R_C = -12\text{ V} - (-1.08\text{ mA})(5.6\text{ k}\Omega) = -5.96\text{ V}$$

The collector-emitter voltage is

$$V_{CE} = V_C - V_E = -5.96\text{ V} - (-1.08\text{ V}) = \mathbf{-4.88\text{ V}}$$

Notice that V_{CE} is negative for a *pnp* circuit.

Question

Assume R_E is changed to 1.2 kΩ. What is V_B, V_E, I_E, I_C, and V_{CE} for the circuit in Figure 3–21 for this case?

Emitter Bias

Emitter bias is another very stable form of bias that uses both positive and negative power supplies and a single bias resistor that, in the usual configuration, puts the base voltage near ground potential. It is the type of bias used in most integrated circuit amplifiers. Although very stable, it has the disadvantage of requiring two power supplies.

Emitter bias circuits for *npn* and *pnp* configurations are shown in Figure 3–22. As in the other bias circuits, the key difference between the *npn* and *pnp* versions is that the polarity of the power supplies are opposite to each other.

FIGURE 3–22

Emitter bias circuits.

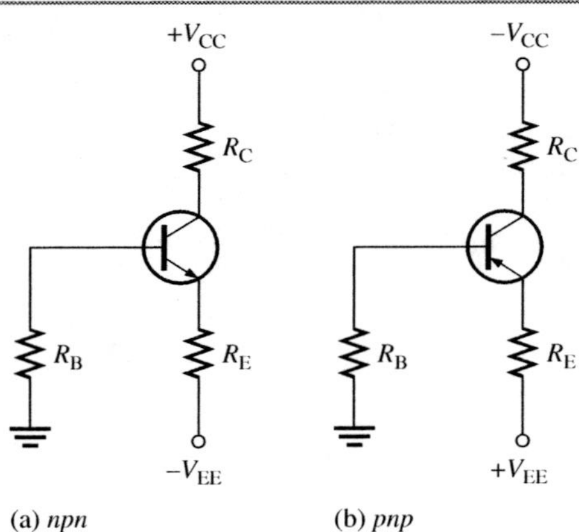

For stable bias, the base resistor is selected to drop only a few tenths of a volt. For the *npn* case, the emitter voltage is approximately −1 V due to the small drop across R_B and the forward-bias drop of the base-emitter junction of 0.7 V. For the *pnp* version, the emitter voltage is approximately +1 V. When troubleshooting, a quick check of the emitter voltage will reveal if the transistor is conducting and if the bias voltage is correct.

The emitter current is computed by applying Ohm's law to the emitter resistor. Thus,

$$I_C = \frac{V_{R_E}}{R_E}$$

The approximation that $I_C \cong I_E$ is then used to calculate the collector current. The collector voltage is again found by applying the following equation:

$$V_C = V_{CC} - I_C R_C$$

These ideas are illustrated in the following example.

EXAMPLE 3–6

Problem

Find V_E, I_E, I_C, and V_{CE} for the emitter bias circuit in Figure 3–23.

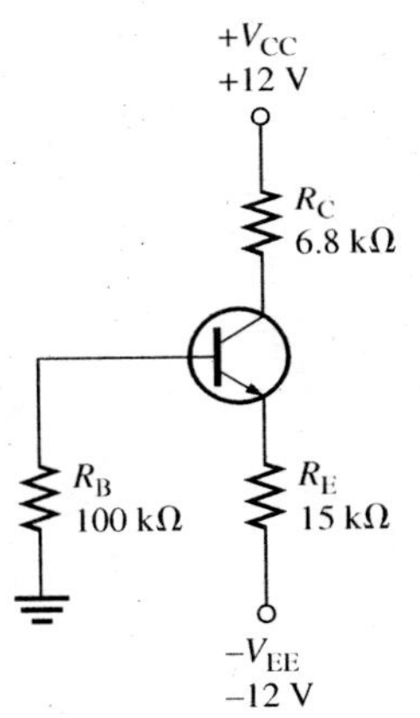

FIGURE 3–23

Solution

Start with the approximation $V_E \cong \mathbf{-1\ V}$. This implies the voltage across R_E is 11 V. Applying Ohm's law to the emitter resistor,

$$I_E = \frac{V_{R_E}}{R_E} = \frac{11\text{ V}}{15\text{ k}\Omega} = \mathbf{0.73\ mA}$$

The collector current is approximately equal to the emitter current.

$$I_C \cong \mathbf{0.73\ mA}$$

Now find the collector voltage.

$$V_C = V_{CC} - I_C R_C = 12\text{ V} - (0.73\text{ mA})(6.8\text{ k}\Omega) = 7.0\text{ V}$$

Find V_{CE} by subtracting V_E from V_C.

$$V_{CE} = V_C - V_E = 7.0\text{ V} - (-1\text{ V}) = \mathbf{8.0\ V}$$

Question

What is V_E if the base of the transistor in Figure 3–23 were shorted to ground?

COMPUTER SIMULATION

Open the Multisim file F03-23DV on the website. Check the base, emitter, and collector voltages.

Troubleshooting A Biased Transistor

Troubleshooting a bias circuit is generally straightforward. Many faults with transistor bias circuits lead to one of two conditions—either the transistor is saturated (maximum conduction) or cut off (no conduction). It is useful to start troubleshooting by checking the base, collector, and emitter voltages to see if they are approximately correct. Keep in mind that these voltages depend to some extent on the specific resistors and the β of the transistor, but they normally should be within 10% of the expected values. If the base, emitter, and collector voltages are normal, the circuit is probably functioning correctly.

Some examples of the voltages you would see for open resistors in a voltage-divider bias are shown in Figure 3–24. (This is the circuit from Example 3–4.) The first two examples (Figure 3–24(a) and 3–24(b) point quickly to a problem in the bias circuit because the base voltage is either at 0 V or V_{CC}.

FIGURE 3–24 Typical open circuit problems in a voltage-divider biased transistor.

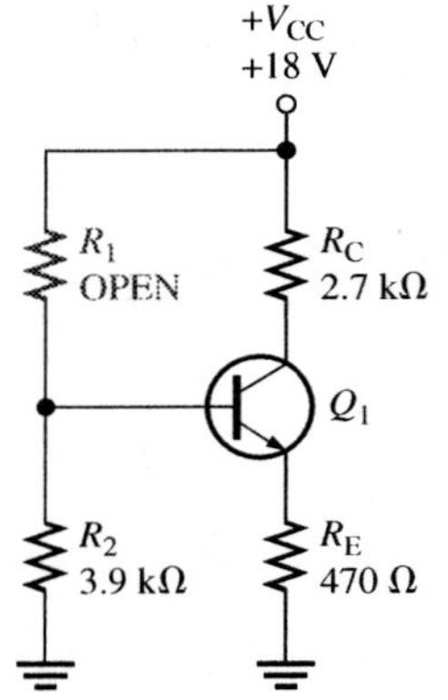

(a) *Fault:* open R_1
Symptoms: transistor cut off
$V_B = 0$ V
$V_E = 0$ V
$V_C = 18$ V

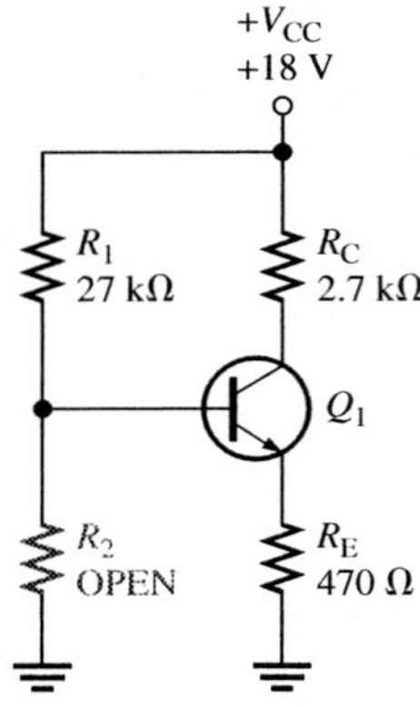

(b) *Fault:* open R_2
Symptoms: transistor saturated
$V_B = +18$ V
$V_E = 17.3$ V
$V_C = 17.3$ V

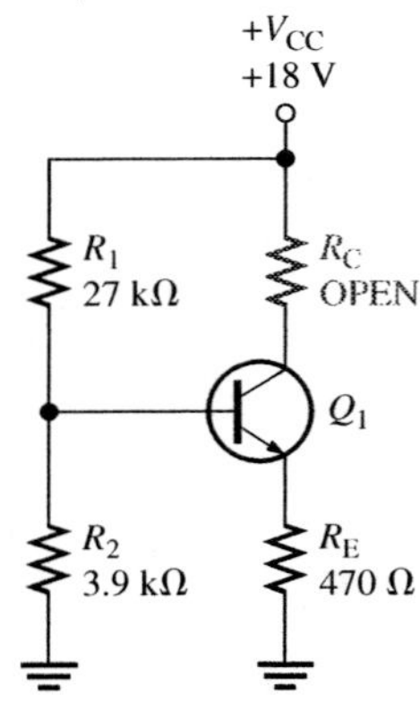

(c) *Fault:* open R_C
Symptoms: transistor cut off
$V_B = 0.9$ V
$V_E = 0.2$ V (due to less current)
$V_C = 0.2$ V

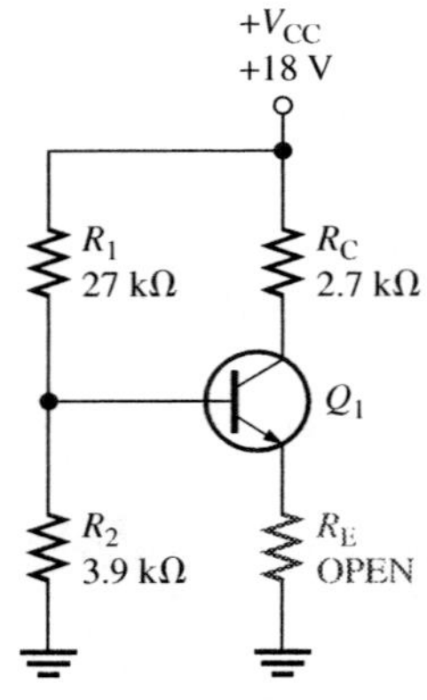

(d) *Fault:* open R_E
Symptoms: transistor cut off
$V_B = 2.3$ V
$V_E = 1.9$ V
$V_C = 18$ V

When the collector resistor is open, as in Figure 3–24(c), the emitter voltage drops because the collector current is no longer in the emitter resistor. The change in the base voltage will depend on the "stiffness" of the bias; a stiff bias will have a smaller effect but in either case the base voltage will drop. The open collector resistor causes the collector terminal to be near ground potential (assuming you measure this on the transistor).

Notice that when the emitter resistor (R_E) is open, as in Figure 3–24(d), the voltage on the base is normal but the emitter voltage, measured on the transistor, is higher than normal because you can read through the transistor. The collector voltage (+18 V) definitely shows that the transistor is cut off. Of course, there are other possible problems, but as in all troubleshooting, the measured parameters lead to further analysis and eventually point to the problem.

+V_{CC} +15 V; R_1 33 kΩ; R_C 2.7 kΩ; Q_1; R_2 5.6 kΩ; R_E 560 Ω

FIGURE 3–25

Review Questions

6. Name three types of bias circuits for BJTs.
7. What are the steps for finding V_{CE} with stiff voltage-divider bias?
8. What dc emitter voltage do you expect to find with emitter bias on a *pnp* transistor?
9. What is the advantage of voltage-divider bias over base bias?
10. What base, emitter, and collector voltage do you expect for the circuit in Figure 3–25 if R_2 is open?

AMPLIFIER CONCEPTS AND AC CONSIDERATIONS 3–3

The backbone of analog electronics is the linear amplifier, a circuit that produces a larger signal that is a replica of a smaller one. In the last section, you saw how bias is used to provide the necessary dc conditions for the transistor to operate.

In this section, you will learn definitions used for amplifiers and ac considerations.

Linear Amplifiers

Linear **amplifiers** produce a magnified replica (**amplification**) of the input signal in order to produce a useful outcome (such as driving a speaker). The concept of an *ideal linear amplifier* means that the amplifier introduces no noise or distortion to the signal; the output varies in time and replicates the input exactly. The amount of amplification is call the **gain**.

Most amplifiers are designed to amplify either an ac voltage or power. For a voltage amplifier, the output signal, V_{out}, is proportional to the input signal, V_{in}, and the ratio of output voltage to input voltage is **voltage gain**.

$$A_v = \frac{V_{out}}{V_{in}} \tag{3–7}$$

where A_v is voltage gain, V_{out} is output signal voltage, and V_{in} is input signal voltage.

Nonlinear Amplifiers

Amplifiers are also used in situations where the output is not intended to be a replica of the input. These amplifiers form an important part of the field of analog electronics. They include two main categories: waveshaping and switching. A *waveshaping amplifier* is used to change the shape of a waveform. A *switching amplifier* produces a rectangular output from some other waveform. The input can be any waveform, for example, sinusoidal, triangle, or sawtooth. The rectangular output wave is often used as a control signal for some digital applications.

EXAMPLE 3–7

Problem

The input and output signals for a linear amplifier are shown in Figure 3–26 and represent an oscilloscope display. What is the voltage gain of the amplifier?

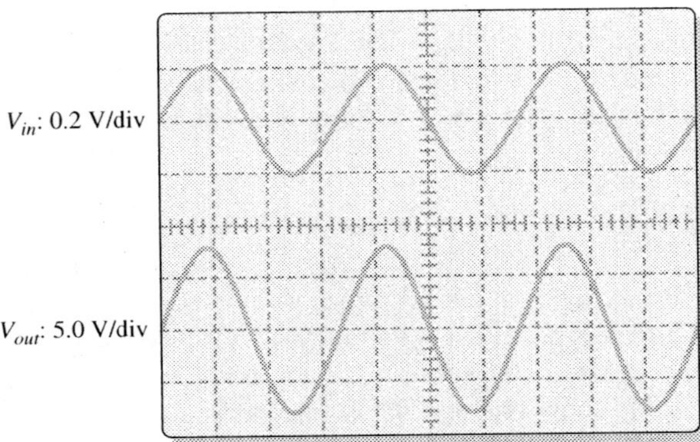

FIGURE 3–26
Oscilloscope display.

Solution

The input signal is 2.0 divisions from peak-to peak.

$$V_{in} = 2.0 \text{ div} \times 0.2 \text{ V/div} = 0.4 \text{ V}$$

The output signal is 3.2 divisions from peak-to peak.

$$V_{out} = 3.2 \text{ div} \times 5.0 \text{ V/div} = 16 \text{ V}$$

$$A_v = \frac{V_{out}}{V_{in}} = \frac{16 \text{ V}}{0.4 \text{ V}} = \mathbf{40}$$

Note that voltage gain is a ratio of voltages and therefore has no units. The answer is the same if rms or peak values had been used for both the input and output voltages.

Question

The input to an amplifier is 20 mV. If the voltage gain is 300, what is the output signal?

Another gain parameter is **power gain**, A_p, defined as the ratio of the signal power out to the signal power in. Power is generally calculated using rms values of voltage or current; however, power gain is a ratio so you can use any consistent units. Power gain, shown as a function of time, is given by the following equation:

$$A_p = \frac{P_{out}}{P_{in}} \quad (3\text{–}8)$$

where A_p is power gain, P_{out} is power out, and P_{in} is power in.

Power can be expressed by any of the standard power relationships studied in basic electronics. For instance, given the voltage and current of the input and output signals, the power gain can be written

$$A_p = \frac{I_{out}V_{out}}{I_{in}V_{in}}$$

where I_{out} is output signal current to the load and I_{in} is input signal current. Power gain can also be expressed by substituting $P = V^2/R$ for the input and output power.

$$A_p = \left(\frac{V_{out}^2/R_L}{V_{in}^2/R_{in}}\right)$$

where R_L is load resistor and R_{in} is input resistance of the amplifier. The particular equation you choose depends on what information is given. Keep in mind that power gain refers to signals. No amplifier creates power; it can only convert some of the power from a dc source into signal power.

DC and AC Quantities

In the first part of this chapter, dc values were used to set up the operating conditions for transistors. These dc quantities of voltage and current used the standard italic capital letters with nonitalic capital-letter subscripts such as V_{E}, I_{E}, I_{C}, and V_{CE}. Lowercase italic subscripts are used to show ac quantities of rms, peak, and peak-to-peak voltages and currents such as V_e, I_e, I_c, and V_{ce}. Instantaneous quantities are indicated with both lowercase italic letters and subscripts such as v_e, i_e, i_c, and v_{ce}.

In addition to currents and voltages, resistances often have different values from an ac viewpoint compared to a dc viewpoint (see Section 1–1 for a review of dc versus ac resistance). Lowercase italic subscripts are used to identify ac resistance values. For example, R_C represents a dc collector resistance and R_c represents an ac collector resistance. You will see the need for this distinction as we discuss amplifiers. Internal resistances that are part of the transistor's equivalent circuit are written as lowercase italic letters (sometimes with a prime) and subscripts. For example, r'_e represents the internal ac emitter resistance and $R_{in(tot)}$ represents the total ac resistance that an amplifier presents to a signal source.

One parameter that is different for dc and ac circuits is β. The dc beta (β_{DC}) for a circuit was previously defined as the ratio of the collector current, I_C, to the base current, I_B. The **ac beta** (β_{ac}) is defined as a small *change* in collector current divided by a corresponding *change* in base current. A changing quantity is written using ac notation and is a ratio of the collector current, I_c, to the base current, I_b (note the lowercase italic subscripts). On manufacturer's data sheets, β_{ac} is usually shown as h_{fe}. In equation form,

$$\beta_{ac} = \frac{I_c}{I_b} \quad (3\text{–}9)$$

The difference between β_{ac} and β_{DC} for a given transistor is normally quite small and is due to small nonlinearities in the characteristic curves. For almost all designs, these differences are not important but should be understood when reading data sheets.

AC Equivalent Circuits

In Section 3-2, you solved the dc bias conditions necessary to set the operating conditions for the transistor. The first step in analyzing or troubleshooting any transistor amplifier is to find the dc conditions. After checking that the dc voltages are correct, the next step is to check ac signals. The equivalent ac circuit is quite different from the dc circuit. For example, a capacitor prevents dc from passing; thus, it appears as an open circuit to dc but looks like a short circuit to most ac signals. For this reason, you need to be able to look at the dc and ac equivalent circuits quite differently.

Recall from your dc/ac circuits studies that the superposition principle allows you to find the voltage or current anywhere in a linear circuit due to a single voltage or current source acting alone. This is done by reducing all other sources to zero. To compute ac parameters, reduce the dc power supply to zero by mentally replacing it with a short and computing the ac parameters as if they were acting alone. Replacing the power supply with a short means that V_{CC} is actually at ground potential for the ac signal. This is called an *ac ground.* The concept of a ground point that is an ac signal ground but not a dc ground may be new to you. Just remember that an ac ground is a common reference point for the ac signal.

Coupling and Bypass Capacitors

A basic BJT amplifier is shown in Figure 3–27. The difference between this circuit and the one in Figure 3–18 is the addition of an ac signal source, three capacitors, and a load resistor. In addition, the emitter resistor is divided into two resistors.

The ac signal is brought into and out of the amplifier through series capacitors (C_1 and C_3) called **coupling capacitors**. As mentioned previously, a capacitor normally appears as a short to the ac signal and an open to dc. This means that coupling capacitors can pass the ac signal while simultaneously blocking the dc voltage. The input coupling capacitor, C_1, passes the ac signal from the source to the base while isolating the source from the dc bias voltage. The output coupling capacitor, C_3, passes the signal on to the load while isolating the load from the power supply voltage. Notice that the coupling capacitors are in series with the signal path.

Capacitor C_2 is different; it is connected in parallel with one of the emitter resistors. This causes the ac signal to *bypass* the second emitter resistor (R_{E2}), thus, it is called a **bypass**

SAFETY NOTE

Be aware of fire safety in the lab. Know the location of fire extinguishers. Electrical fires are never fought with water, which can conduct electricity. Use only a class-C extinguisher, which uses a nonconductive agent—typically CO_2. It is important to know what type of extinguisher you use for any fire. The wrong choice can make the fire worse.

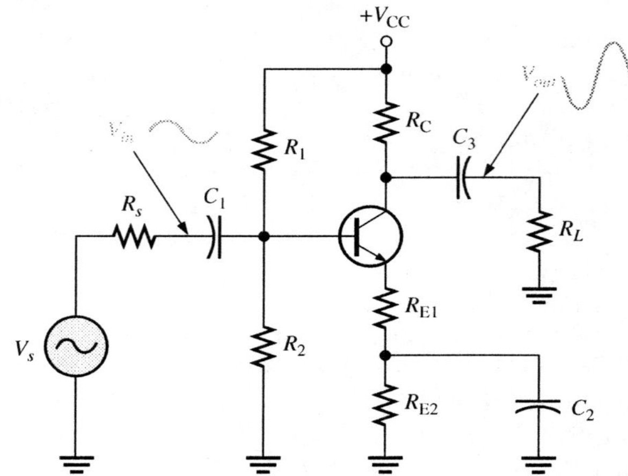

FIGURE 3–27

A basic transistor amplifier.

capacitor. The purpose of the bypass capacitor is to increase the gain of the amplifier. Since the bypass capacitor is an ac short, both ends of the capacitor are at ac ground. Whenever one side of a capacitor is connected to ground, the other side is also a ground to the ac signal. Remember this if you are troubleshooting; you shouldn't find an ac signal on either side of a bypass capacitor. If you do, the capacitor may be open.

Amplification

The signal source, V_s, shown in Figure 3–27 causes variations in base current which, in turn, cause the much larger emitter and collector currents to vary about the Q-point in phase with the base current. However, when the collector current increases, the collector voltage decreases and vice-versa. Thus, the sinusoidal collector-to-emitter voltage varies above and below its Q-point value 180° out of phase with the base voltage. A transistor always inverts the signal between the base and the collector. Amplification occurs because a small swing in base current produces a large variation in collector voltage.

Input and Output Resistance

Two important considerations for any amplifier are the input and output resistance (called *impedance* when inductive and capacitive reactance effects are included). Amplifiers are always used between a signal source and a load. The source can be a previous stage of amplification or come from a transducer such as a microphone or antenna. The load can be another stage of amplification or a speaker or even a picture tube.

Whenever an amplifier is hooked to a signal source or previous stage, it has a loading effect on the previous stage. In general, it is better that an amplifier has a minimal loading effect on the stage (although there are exceptions to this rule, particularly at radio frequencies, where it is common to match impedances). Recall that the loading effect is a change in circuit parameters due to connecting a load. As a rule of thumb, if the loading effect is less than 10%, it is not a problem.

To minimize loading effect, the input resistance of the amplifier should be large compared to the output resistance of the source. This idea often requires compromises in the design of an amplifier. For example, with voltage-divider bias, a stiff bias string tends to produce stable bias that is independent of the transistor. Unfortunately, stiff bias has a low resistance and will tend to load the source more.

The output resistance is another consideration. An amplifier with a low output resistance will tend to be loaded less than one with high output resistance. Bottom line: the ideal amplifier does not exist. If it did, it would have infinite input resistance and zero output resistance. When you study op-amps, in Chapter 6, you will see that certain configurations can approach this ideal amplifier.

Review Questions

11. What is the difference between β_{DC} and β_{ac}?
12. What are two characteristics of an ideal linear amplifier?
13. What are two important applications for nonlinear amplifiers?
14. What is the measurement unit for voltage gain?
15. What is the difference between coupling and bypass capacitors?

SMALL-SIGNAL AMPLIFIERS 3–4

The common-emitter amplifier is the most widely used type of small-signal amplifier. Two other types of amplifiers are the common-collector amplifier and the common-base amplifier.

In this section, you will analyze the operation of a small-signal amplifier.

The Common-Emitter Amplifier

The **common-emitter (CE)** amplifier, the most widely used type of BJT amplifier, has the emitter as the reference terminal for the input and output signal. Figure 3–28(a) shows a CE amplifier that produces an amplified and inverted output signal at the load resistor. The input signal, V_{in}, is capacitively coupled to the base through coupling capacitor C_1, causing the base current to vary above and below its dc bias value. This variation in base current produces a corresponding variation in collector current. The variation in collector current is much larger than the variation in base current because of the current gain through the transistor. This produces a larger variation in the collector voltage which is out of phase with the base signal voltage. This variation in collector voltage is then capacitively coupled through capacitor C_3 to the load and appears as the output voltage, V_{out}.

A basic common-emitter amplifier. **FIGURE 3–28**

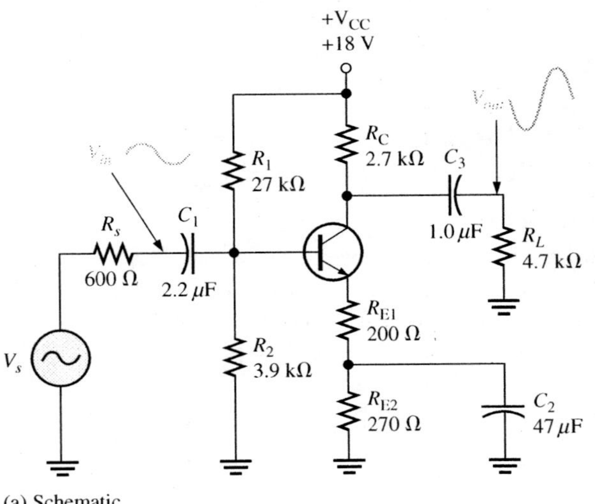

(a) Schematic

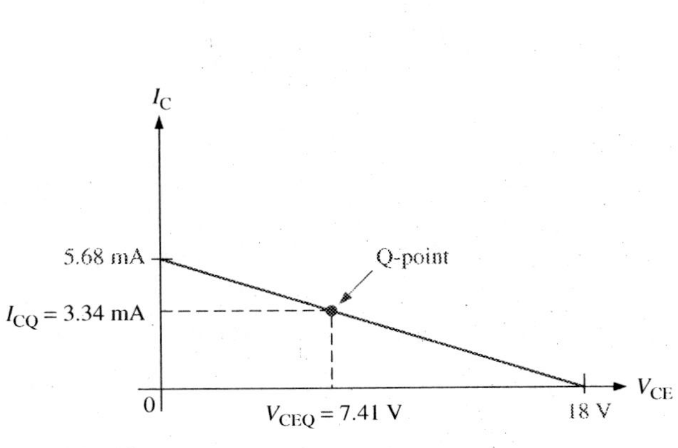

(b) DC load line

Let's look at the amplifier in Figure 3–28(a) in more detail. Whenever you calculate parameters for a transistor amplifier, always start with the dc parameters. The dc parameters for this circuit were calculated in Example 3–4. The original 470 Ω emitter resistor is now composed of two resistors that add up to 470 Ω, so this change has no effect on the dc parameters (but will affect the ac parameters). The dc parameters found in Example 3–4 are $I_C = 3.34$ mA, $V_C = 8.98$ V and $V_{CE} = 7.41$ V. Recall that I_C and V_{CE} define the points for the Q-point, so are given the special labels I_{CQ} and V_{CEQ}, respectively.

A graphical picture of the dc parameters just reviewed may help you visualize them. You can determine the load line by finding the saturation current and the collector-emitter cutoff voltage for the circuit. Recall that the saturation current is the current when the collector-to-emitter voltage is approximately zero. Thus,

$$I_{C(sat)} = \frac{V_{CC}}{R_C + R_{E1} + R_{E2}} = \frac{18\text{ V}}{2.7\text{ k}\Omega + 200\ \Omega + 270\ \Omega} = 5.68\text{ mA}$$

At cutoff, there is no current, so the entire supply voltage, V_{CC}, is across the collector to emitter. These two points, saturation and cutoff, allow you to construct the dc load line, as shown in Figure 3–28(b). The Q-point is located on the load line using the earlier calculation.

The AC Equivalent Circuit

Recall that the ac signal "sees" a different circuit than does the dc source for several reasons. If you apply the superposition theorem to the circuit in Figure 3–28(a) and show the capacitors as shorts, you can redraw the CE amplifier from the perspective of the ac signal. This is shown in Figure 3–29. The power supply has been replaced with an ac ground, shown in red. The capacitors have been replaced with shorts, and R_{E2}, is eliminated because it is bypassed with C_2.

FIGURE 3–29

AC equivalent circuit for Figure 3–28(a).

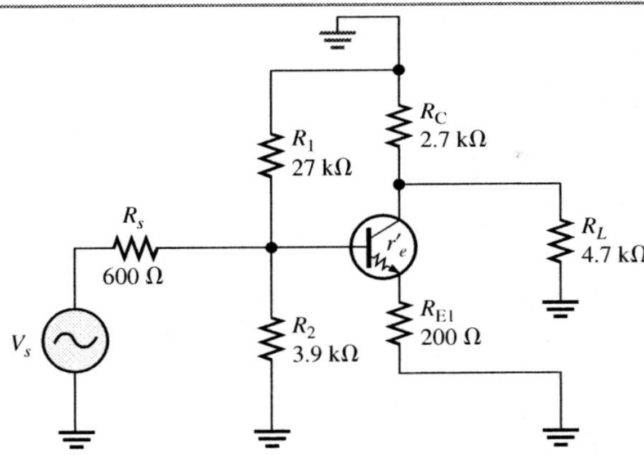

The ac equivalent circuit also shows an internal resistance in the base-emitter diode. This internal resistance, called r'_e, plays a role in the gain and input impedance of the amplifier so is generally included in the ac equivalent circuit. Because it is an ac resistance, it is sometimes called the **dynamic emitter resistance**. The value of this ac resistance is related to the dc emitter current as follows:

$$r'_e = \frac{25\text{ mV}}{I_E} \qquad (3\text{–}10)$$

EXAMPLE 3–8

Problem

Find the dynamic emitter resistance, r'_e, for the circuit in Figure 3–28(a).

Solution

The emitter current was found to be 3.34 mA (see Example 3–4). Substituting into Equation 3–10,

$$r'_e = \frac{25\text{ mV}}{I_E} = \frac{25\text{ mV}}{3.34\text{ mA}} = \mathbf{7.5\ \Omega}$$

Question

What is r'_e for a transistor with an emitter current of 100 μA?

Voltage Gain

The voltage gain, A_v, of the CE amplifier is the ratio of the output signal voltage to the input signal voltage, V_{out}/V_{in}. The output voltage, V_{out}, is measured at the collector and the input voltage, V_{in}, is measured at the base. Because the base-emitter junction is forward-biased, the signal voltage at the emitter is approximately equal to the signal voltage at the base. Thus, since $V_b = V_e$, the voltage gain is

$$A_v = -\frac{V_c}{V_e} = -\frac{I_c R_c}{I_e R_e}$$

Since $I_c \cong I_e$, the voltage gain reduces to the ratio of ac collector resistance to ac emitter resistance.

$$A_v = -\frac{R_c}{R_e} \qquad (3\text{–}11)$$

The negative sign in the gain formula is added to indicate inversion, meaning the output signal is out of phase with the input signal. Note that the gain is written as a ratio of two ac resistances. In the collector circuit, the collector resistor appears to be in parallel with the load resistor. Thus, the ac resistance, R_c, of the collector is simply $R_C \parallel R_L$. In the emitter circuit, you need to include the internal base-emitter diode resistance (r'_e) and any resistor that is not bypassed with a capacitor. The internal r'_e appears to be in series with the unbypassed emitter resistance in the ac emitter circuit.

The gain formula given as Equation 3–11 is a useful and simple way to quickly determine what the voltage gain of a CE amplifier should be. When you're troubleshooting, you need to know what signal to expect; remember that the collector and emitter resistances used in calculating gain are the *total ac* resistances in their respective circuits. Incidentally, this unbypassed resistor, shown as R_{E1} in Figure 3–28(a), reduces the gain but serves an important role in keeping the gain stable. It also increases the input resistance of the amplifier. Sometimes it is called a *swamping resistor* because it tends to "swamp" out the uncertain value of r'_e.

EXAMPLE 3–9

Problem
Find A_v for the circuit in Figure 3–28(a).

Solution
The ac resistance in the emitter circuit, R_e, is composed of r'_e in series with the unbypassed R_{E1}. From Example 3–8, $r'_e = 7.5\ \Omega$. Therefore,

$$R_e = r'_e + R_{E1} = 7.5\ \Omega + 200\ \Omega = 207.5\ \Omega$$

Next, find the ac resistance as viewed from the transistor's collector.

$$R_c = R_C \parallel R_L = 2.7\ \text{k}\Omega \parallel 4.7\ \text{k}\Omega = 1.71\ \text{k}\Omega$$

Substituting into Equation 3–11,

$$A_v = -\frac{R_c}{R_e} = -\frac{1.71\ \text{k}\Omega}{207.5\ \Omega} = \mathbf{-8.3}$$

Again, the negative sign is used to show that the amplifier inverts the signal.

Question
Assume the bypass capacitor in Figure 3–28(a) were open. What effect would this have on the gain?

COMPUTER SIMULATION

Open the Multisim file F03-28DV on the website. Confirm that the gain is −8.3.

Input Resistance

The input resistance for an amplifier, $R_{in(tot)}$, was introduced in the last section. It is an ac parameter that acts like a load in series with the source resistance. As long as the input resistance is high compared to the source resistance, most of the voltage will appear at the input and the loading effect is small. If the input resistance is small compared to the source resistance, most of the source voltage will drop across its own series resistance, leaving little for the amplifier to amplify.

One of the problems with the CE amplifier is that the input resistance is dependent on β_{ac}. As you have seen, this parameter is highly variable, so you can't calculate input resistance exactly for a given amplifier without knowing the β_{ac}. Despite this, you can minimize the effect of β_{ac} and increase the total input resistance by adding a swamping resistor to the emitter circuit. You can then obtain a reasonable estimate of the input resistance, which for most purposes will enable you to determine if a given amplifier is appropriate for the job at hand.

The input circuit for the CE amplifier in Figure 3–28(a) has been redrawn in Figure 3–30 to eliminate the output circuit. R_C is not part of the input circuit because of the reverse-biased base-collector junction. For the input ac signal, there are three parallel paths to

FIGURE 3–30

Equivalent ac input circuit for the CE amplifier in Figure 3–28(a) as viewed from the source.

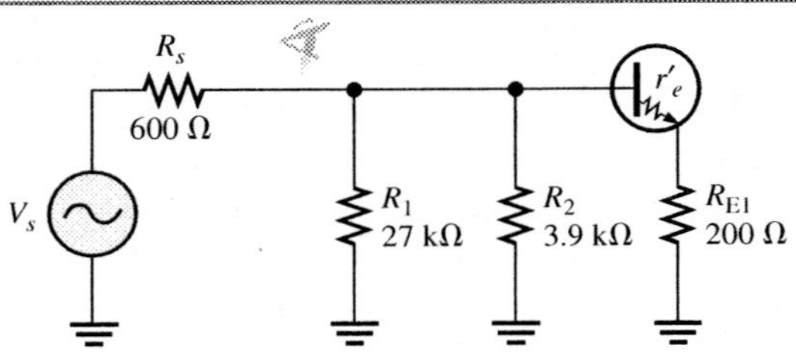

ground. Looking in from the source, the three paths consist of R_1, R_2, and a path through the transistor's base-emitter circuit. It is these three parallel paths that comprise the input resistance of the circuit. We define this resistance as $R_{in(tot)}$ because it represents the total input resistance including the bias resistors. The base-emitter path, however, has β_{ac} dependency because of the transistor's current gain. The equivalent resistances, R_{E1} and r'_e, appear to be larger in the base circuit than in the emitter circuit because of this current gain. The resistors in the emitter circuit must be multiplied by β_{ac} to obtain their equivalent resistance in the base circuit. Therefore, the formula for calculating total input resistance is

$$R_{in(tot)} = R_1 \parallel R_2 \parallel [\beta_{ac}(r'_e + R_{E1})] \quad (3\text{–}12)$$

EXAMPLE 3–10

Problem

Find $R_{in(tot)}$ for the circuit in Figure 3–28(a). Assume the β_{ac} is 120.

Solution

The internal ac emitter resistance, r'_e, was found to be 7.5 Ω in Example 3–8. Substituting into Equation 3–12,

$$R_{in(tot)} = R_1 \parallel R_2 \parallel [\beta_{ac}(r'_e + R_{E1})]$$
$$= 27\text{ k}\Omega \parallel 3.9\text{ k}\Omega \parallel [120(7.5\ \Omega + 200\ \Omega)] = \mathbf{3.0\ k\Omega}$$

Question

What is $R_{in(tot)}$ for the circuit in Figure 3–28(a) if β_{ac} is 200?

Output Resistance

To find the output resistance of any CE amplifier, look back from the output coupling capacitor as illustrated in Figure 3–31. The transistor appears as a current source in parallel with the collector resistor. Recall that the internal resistance of an ideal current source is infinite. With this in mind, you can see that the output resistance for the CE amplifier is simply the collector resistance, R_C.

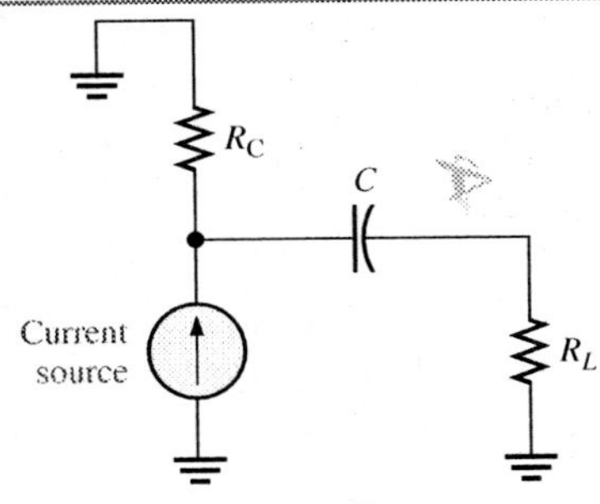

FIGURE 3–31
Equivalent ac output circuit for the CE amplifier as viewed from the load.

The AC Load Line

For most troubleshooting work, it is useful to be able to quickly estimate a circuit's voltage and current values. Although technicians seldom use them in their normal work, load lines are a useful tool for understanding a transistor's operation and may offer insight into a circuit's limitations, such as clipping levels.

As discussed in Section 3–1, a dc load line can be drawn for a basic transistor circuit that consists of a series collector resistor, R_C, and a voltage source, V_{CC}. As shown in Figure 3–13(a), this series combination formed a Thevenin circuit that was represented graphically by a dc load line that crossed the y-axis at saturation. Recall that the load line in Figure 3–13(b) was independent of the transistor, which served as the load.

For ac, the Thevenin resistance is more complicated because of the presence of capacitors and an internal emitter resistance, r_e'. In high-frequency circuits, inductors may also play a role. Even though r_e' is internal to the transistor, it is considered part of the Thevenin resistance. Capacitive coupling and bypass capacitors are also present in most practical ac circuits. Capacitors are normally treated as shorts for the ac signal meaning that the ac resistance (R_{ac}) of the collector-emitter circuit is reduced.

A dc and an ac load line are shown together in Figure 3–32 for a capacitively coupled amplifier. The Q-point is the same for both load lines because when the ac signal is reduced to zero, operation must still occur at the Q-point. Notice that the ac saturation current is greater than the dc saturation current (because the ac resistance is smaller). In addition, the ac collector-emitter cutoff voltage is less than the dc collector-emitter cutoff voltage. The ac load line locates all possible operation points (collector current versus collector-emitter voltage) for the ac signal.

FIGURE 3–32

The dc and ac load lines.

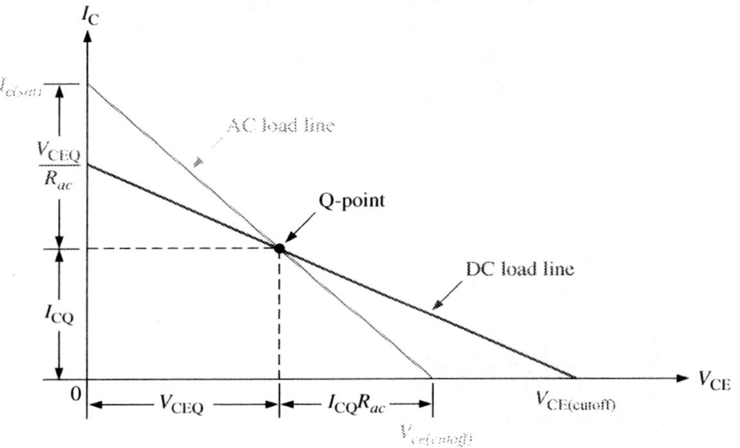

The ac saturation and ac cutoff points can be computed for the ac load line. The ac load line crosses the y axis at $I_{c(sat)}$. This point is found by starting at the dc Q-point (I_{CQ}) and adding a term that includes the ac resistance of the collector-emitter circuit, R_{ac}, as shown in Figure 3–32. The equation for ac saturation is

$$I_{c(sat)} = I_{CQ} + \frac{V_{CEQ}}{R_{ac}}$$

The ac load line touches the x axis at $V_{ce(cutoff)}$. This point is also found by starting at the dc Q-point (V_{CEQ}) and adding a term that includes the ac resistance, R_{ac}. The equation for ac cutoff is

$$V_{ce(cutoff)} = V_{CEQ} + I_{CQ}R_{ac}$$

One interesting way of viewing the operation of an amplifier is to superimpose a set of characteristic curves for the transistor on the ac load line. This is shown in Figure 3–33 for a typical transistor that could be used with the CE amplifier from Figure 3–28(a). Lines projected from the peaks of the base current across to the I_C axis and lines from the ac load line down to the V_{CE} axis indicate the peak-to-peak variations of the collector current and collector-to-emitter voltage, as shown. For the transistor in this example, if an input signal

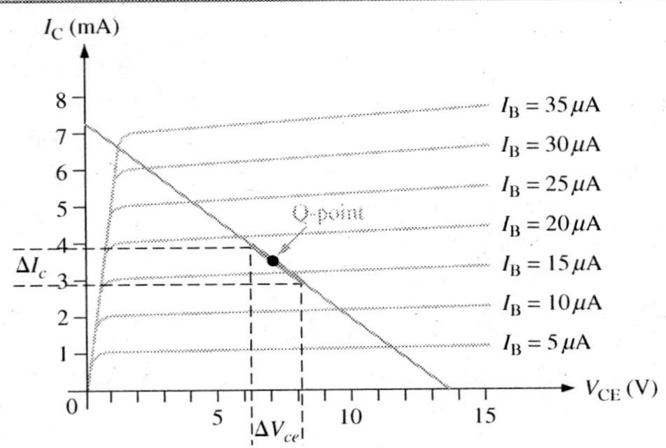

FIGURE 3–33

AC load line superimposed on a transistor characteristic. The line shown is for the amplifier in Figure 3–28(a).

causes the base current to vary from approximately 13 μA to 18 μA, the output collector current will vary from approximately 2.9 mA to 3.9 mA. In addition, V_{CE} varies from approximately 6.3 V to 8.1 V for this same signal. The ac load line also gives a quick visual indication when the signal exceeds the linear range of the amplifier and shows the current and voltage range that a particular signal will encompass.

Common-Collector and Common-Base Amplifiers

Although the CE amplifier is the most widely used, two other configurations have application in certain situations. The first configuration is the **common-collector (CC)** amplifier, named because its collector is at ac ground. The input signal is applied to the base and taken from the emitter. Figure 3–34 shows a typical CC amplifier and its equivalent ac circuit. Notice that there is no collector resistor in a CC amplifier. Not only is it unnecessary, it would waste power and use two extra components (the resistor and a bypass capacitor)!

FIGURE 3–34

Common-collector (CC) amplifier.

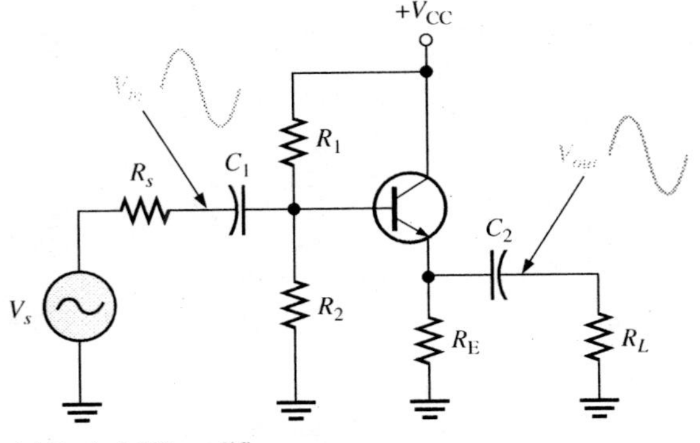

(a) Typical CC amplifier

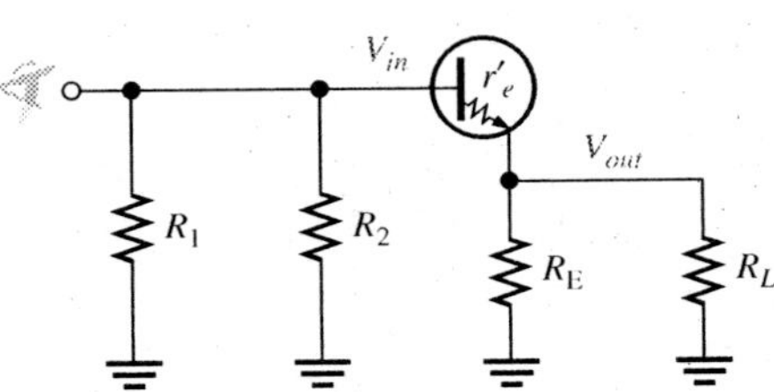

(b) Equivalent ac circuit as viewed from the source

The output of a CC amplifier is taken from the emitter and will appear to be identical to the input (less a very small drop in the transistor that can be ignored). Because the output at the emitter replicates the input, the CC amplifier is also called an **emitter-follower**. The voltage gain of the CC amplifier is therefore

$$A_v \cong 1 \tag{3–13}$$

You might wonder if the output voltage is identical to the input voltage, what good is the amplifier? The answer is that it has current gain but not voltage gain. It can be shown that the current gain is the ratio of the ac input resistance to the load resistance. Thus, it is useful in small power amplifiers when it is necessary to drive a low-impedance load, such as a small speaker. The CC amplifier is discussed again in Chapter 5 as a power amplifier.

The second configuration that is occasionally used in circuits is the **common-base (CB)** amplifier. As the name implies, the base is at ac ground. The common-base amplifier has voltage gain but no current gain. It also has low input resistance, a disadvantage except for some specialized applications, mostly at high frequencies. A CB amplifier works well only if it is driven from a low-resistance source; otherwise, loading effects prevent it from amplifying the input signal. Generally, the source resistance, R_s, should be no more than 50 Ω.

A typical CB amplifier and its equivalent ac circuit are shown in Figure 3–35. The equation for the voltage gain of a CB amplifier is essentially the same as that of a CE amplifier; namely, it is the ac collector resistance divided by the ac emitter resistance. The difference is that the CE amplifier inverts the output and the CB amplifier does not; hence, there is no minus sign in the gain equation.

$$A_v = \frac{R_c}{R_e} \tag{3–14}$$

FIGURE 3–35

Common-base (CB) amplifier.

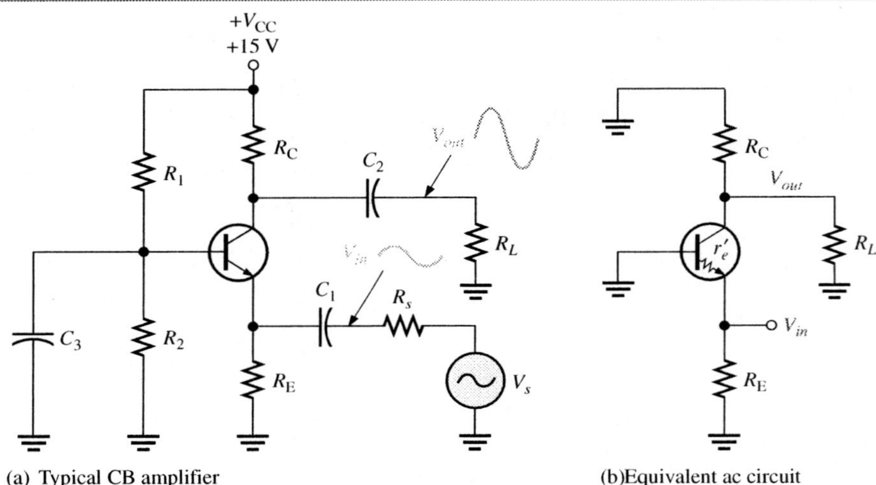

(a) Typical CB amplifier

(b)Equivalent ac circuit

One advantage for CB amplifiers is they have a higher frequency response than CE amplifiers. For this reason, they can be used at radio frequencies as tuned amplifiers, designed to amplify for a specific radio frequency.

Table 3–2 summarizes some of the important parameters of the three basic voltage amplifier configurations. Relative values are indicated for input and output resistance.

Review Questions

16. Which terminal of a CE amplifier is the input terminal? Which is the output terminal?
17. Why does a CE amplifier invert the signal?
18. What is the advantage of a high input resistance in an amplifier?
19. How is the gain determined in a CE amplifier?
20. What is the name of the point where the dc and ac load lines cross?

TABLE 3–2

Comparison of amplifier ac parameters. Voltage-divider bias is assumed for all amplifiers with an unbypassed emitter resistor in the CE and CB configurations.

	CE	CC	CB
Voltage gain	$A_v = -\frac{R_c}{R_e}$ High	$A_v \cong 1$ Low	$A_v = \frac{R_c}{R_e}$ High
Input resistance	$R_{in(tot)} = R_1 \parallel R_2 \parallel [\beta_{ac}(r'_e + R_{E1})]$ Medium	$R_{in(tot)} = R_1 \parallel R_2[\beta_{ac}(r'_e + R_E \parallel R_L)]$ High	$R_{in(tot)} = r'_e + R_{E1}$ Very low
Output resistance	R_C High	$\cong r'_e$ Low	R_C High
Phase relationship between V_{in} and V_{out}	180° inverted	0° not inverted	0° not inverted

THE BIPOLAR TRANSISTOR AS A SWITCH 3–5

The first large-scale use of digital circuits was in telephone systems. Today, computers form the most important application of switching circuits using integrated circuits (ICs). Discrete transistor switching circuits are used when it is necessary to supply higher currents or operate at a different voltage than can be obtained from ICs.

In this section, you will learn how BJTs are used in switching circuits.

Figure 3–36 illustrates the basic operation of a transistor as a switch. A **switch** is a two-state device that is either open or closed. In part (a), the transistor is in cutoff because the base-emitter *pn* junction is not forward-biased. In this condition, there is, ideally, an open between collector and emitter, as indicated by the open switch equivalent. In part (b), the transistor is in saturation because the base-emitter *pn* junction is forward-biased and the base current is large enough to cause the collector current to reach its saturated value. In this condition, there is, ideally, a short between collector and emitter, as indicated by the closed-switch equivalent. Actually, a voltage drop of about 0.1 V normally occurs across the transistor when it is saturated.

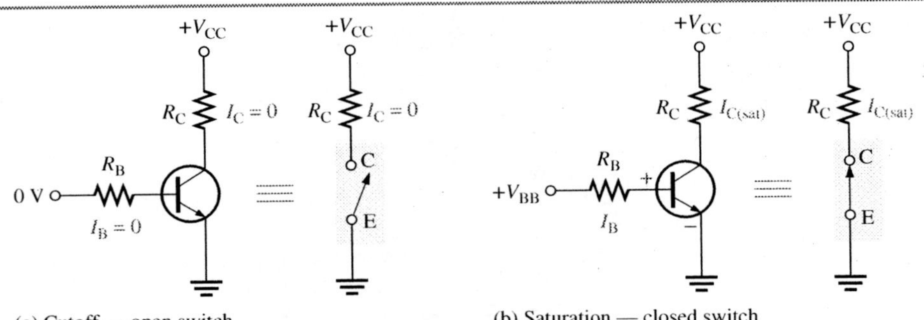

FIGURE 3–36

Ideal switching action of a transistor.

As mentioned before, a transistor is in cutoff when the base-emitter *pn* junction is not forward-biased. Neglecting leakage current, all of the currents are zero, and V_{CE} is equal to V_{CC}.

$$V_{CE(cutoff)} = V_{CC}$$

When the emitter junction is forward-biased and there is enough base current to produce a maximum collector current, the transistor is saturated. Since V_{CE} is very small at saturation, the entire power supply voltage drops across the collector resistor. An approximation for the collector current is

$$I_{C(sat)} \cong \frac{V_{CC}}{R_C}$$

The minimum value of base current needed to produce saturation is

$$I_{B(min)} \cong \frac{I_{C(sat)}}{\beta_{DC}}$$

I_B should be significantly greater than $I_{B(min)}$ to keep the transistor well into saturation and to account for beta variations in different transistors.

EXAMPLE 3–11

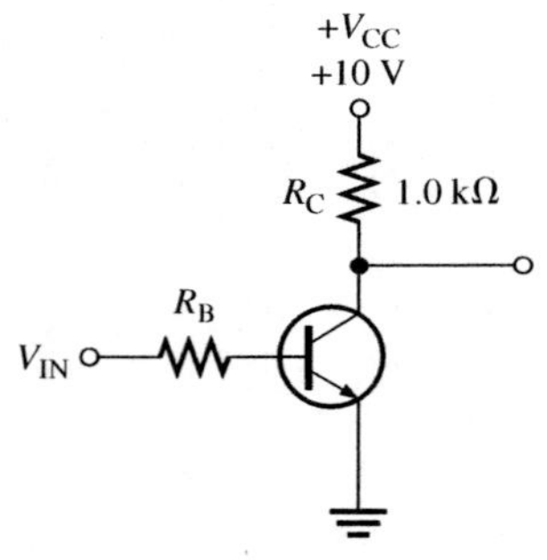

+V_{CC}
+10 V

R_C 1.0 kΩ

R_B

V_{IN}

FIGURE 3–37

Problem

(a) For the transistor switching circuit in Figure 3–37, what is V_{CE} when $V_{IN} = 0$ V?

(b) What minimum value of I_B is required to saturate this transistor if β_{DC} is 200? Assume $V_{CE(sat)} = 0$ V.

(c) Calculate the maximum value of R_B when $V_{IN} = 5$ V.

Solution

(a) When $V_{IN} = 0$ V, the transistor is in cutoff (acts like an open switch) and $V_{CE} = V_{CC} = \mathbf{10\ V}$.

(b) Since $V_{CE(sat)} = 0$ V,

$$I_{C(sat)} \cong \frac{V_{CC}}{R_C} = \frac{10\text{ V}}{1.0\text{ k}\Omega} = 10\text{ mA}$$

$$I_{B(min)} = \frac{I_{C(sat)}}{\beta_{DC}} = \frac{10\text{ mA}}{200} = \mathbf{0.05\ mA}$$

This is the value of I_B necessary to drive the transistor to the point of saturation. Any further increase in I_B will drive the transistor deeper into saturation but will not increase I_C.

(c) When the transistor is saturated, $V_{BE} = 0.7$ V. The voltage across R_B is

$$V_{R_B} = V_{IN} - V_{BE} = 5\text{ V} - 0.7\text{ V} = 4.3\text{ V}$$

The maximum value of R_B needed to allow a minimum I_B of 0.05 mA is calculated by Ohm's law as follows:

$$R_B = \frac{V_{R_B}}{I_B} = \frac{4.3\text{ V}}{0.05\text{ mA}} = \mathbf{86\ k\Omega}$$

Question

What is the minimum value of I_B required to saturate the transistor in Figure 3–37 if β_{DC} is 125 and $V_{CE(sat)}$ is 0.2 V?

COMPUTER SIMULATION

Open the Multisim file F03-37DV on the website. Determine the value of V_{IN} that just saturates the transistor.

Improvements to the One-Transistor Switching Circuit

The basic switching circuit shown in Figure 3–36 has a threshold voltage at which it changes from *off* to *on* or *on* to *off*. Unfortunately, the threshold is not an absolute point because a transistor can operate between cutoff and saturation, a condition not desirable in a switching circuit. A second transistor can dramatically improve the switching action, providing a sharp threshold. The circuit shown in Figure 3–38 is designed with a light-emitting diode (LED) output so that you can construct it if you choose and observe the switching action. The circuit works as follows. When V_{IN} is very low, Q_1 is off since it does not have sufficient base current. Q_2 will be in saturation because it can obtain ample base current through R_2 and the LED is on. As the base voltage for Q_1 is increased, Q_1 begins to conduct. As Q_1 approaches saturation, the base voltage of Q_2 suddenly drops, causing it to go from a saturated to cutoff condition rapidly. The output voltage of Q_2 drops and the LED goes out.

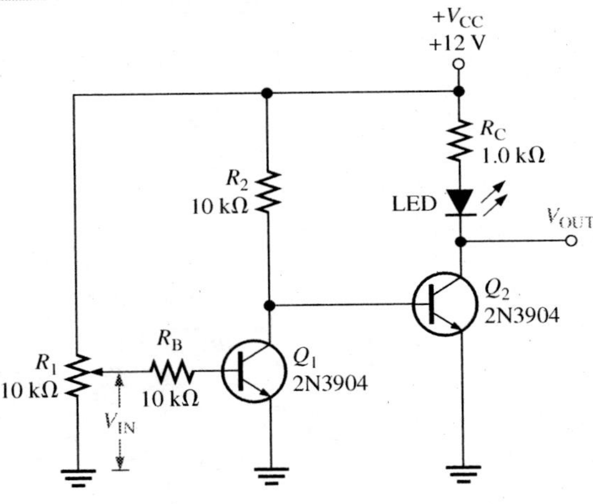

FIGURE 3–38

A two-transistor switching circuit with a sharp threshold.

Another improvement for basic switching circuits is the addition of hysteresis. For switching circuits, **hysteresis** means that there are two threshold voltages depending on whether the output is already high or already low. Figure 3–39 illustrates the situation. When the input voltage rises, it must cross the upper threshold before switching takes place. It does not switch at *A* or *B* because the lower threshold is inactive. When the signal crosses the upper threshold at

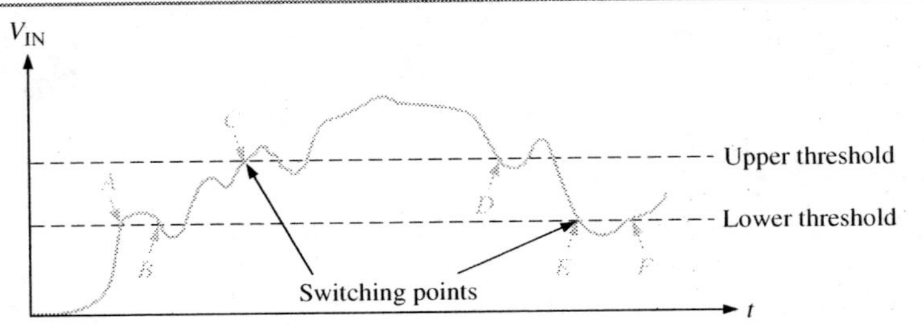

FIGURE 3–39

Hysteresis causes the circuit to switch at points *C* and *E* but not at the other points.

point *C*, the output switches, and immediately the threshold changes to a lower value. The output does not switch back at *D* but instead must cross the lower threshold at *E* before returning to the original state. Again, the threshold changes to the upper level, so switching does not take place at point *F*. The major advantage of hysteresis in a switching circuit is noise immunity. As you can see from this example, the output only changed twice despite a very noisy input.

The schematic for a transistor circuit with hysteresis is shown in Figure 3–40. As the potentiometer is turned in one direction, the output will switch once, even if the potentiometer is "noisy." When the output switches, the common-emitter resistor, R_E, causes the threshold voltage to change. This is caused by the different saturation currents for the two transistors; the threshold is different when the output is in cutoff than when the output is saturated.

FIGURE 3–40

A discrete transistor switching circuit with hysteresis.

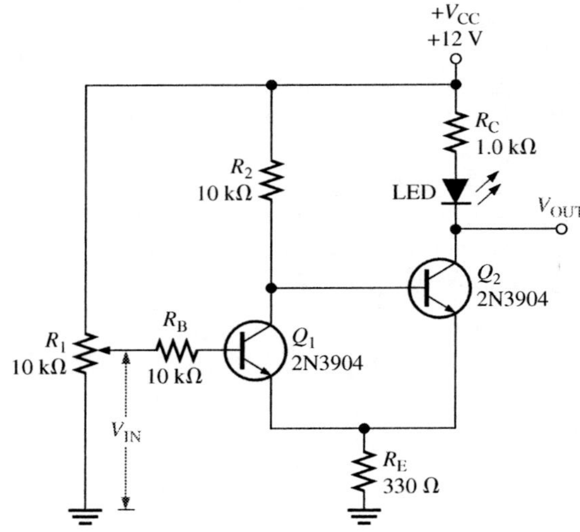

Review Questions

21. When a transistor is used as a switching device, in what two states is it operated?
22. When does the collector current reach its maximum value?
23. When is the collector current approximately zero?
24. When is V_{CE} equal to V_{CC}?
25. What is meant by hysteresis in a switching circuit?

3–6 TRANSISTOR PACKAGES AND TERMINAL IDENTIFICATION

Transistors are available in a wide range of package types for various applications. Those with mounting studs or heat sinks are usually power transistors. Low-power and medium-power transistors are usually found in smaller metal or plastic cases. Still another package classification is for high-frequency devices.

In this section, you will learn to identify various types of transistor packages.

Transistor Categories

Manufacturers generally classify their bipolar junction transistors into three broad categories: general-purpose/small-signal devices, power devices, and RF (radio frequency/microwave) devices. Although each of these categories, to a large degree, has its own unique package types, you will find certain types of packages used in more than one device category. While keeping in mind there is some overlap, we will look at transistor packages for each of the three categories, so that you will be able to recognize a transistor when you see one on a circuit board and have a good idea of what general category it is in.

General-Purpose/Small-Signal Transistors

General-purpose/small-signal transistors are generally used for low- or medium-power amplifiers or switching circuits. The packages are either plastic or metal cases. Certain types of packages contain multiple transistors. Figure 3–41 illustrates common plastic cases, Figure 3–42 shows a metal can, and Figure 3–43 shows multiple-transistor packages. Some of the multiple-transistor packages such as the dual-in-line (DIP) and the small outline (SO) are the same as those used for many integrated circuits. Typical pin connections are shown so you can identify the emitter, base, and collector.

FIGURE 3–41 Plastic cases for general-purpose/small-signal transistors. Both old and new JEDEC TO numbers are given. Pin configurations may vary. Always check the data sheet.

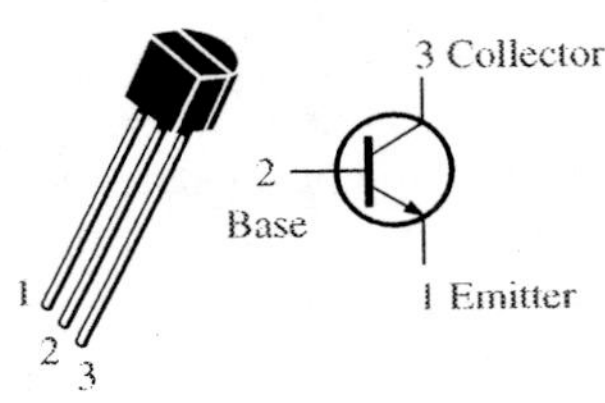

(a) TO-92 or TO-226AA

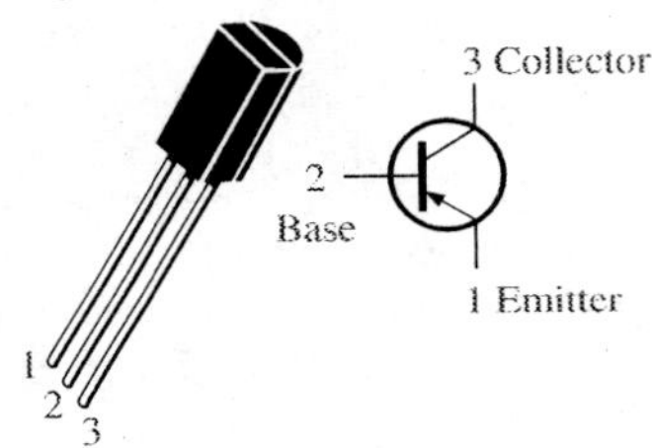

(b) TO-92 or TO-226AE

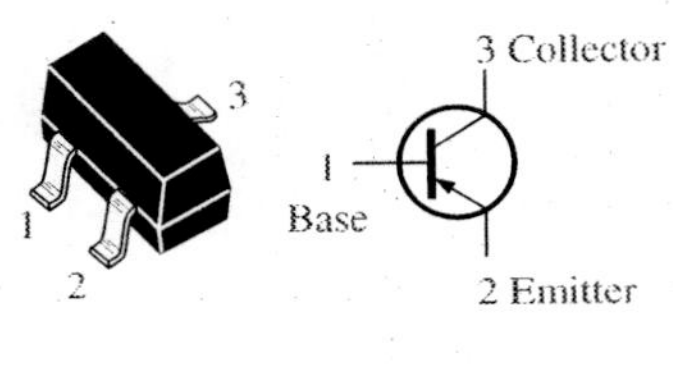

(c) SOT-23 or TO-236AB

FIGURE 3–42 Metal can for general-purpose/small-signal transistors.

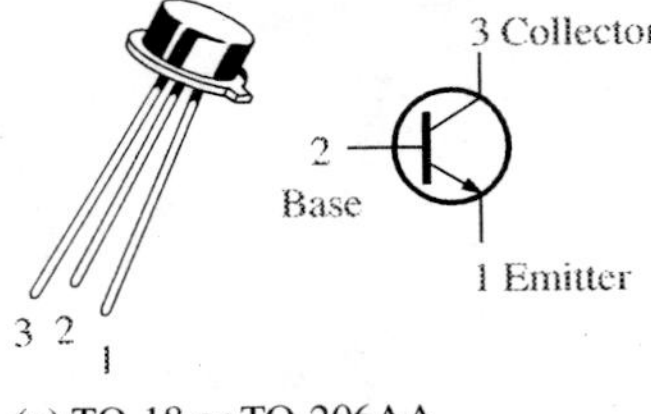

(a) TO-18 or TO-206AA

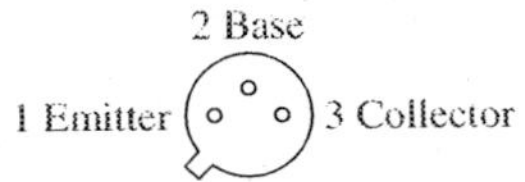

(b) Pin configuration (bottom view). Emitter is closest to tab.

FIGURE 3–43 Typical multiple-transistor packages.

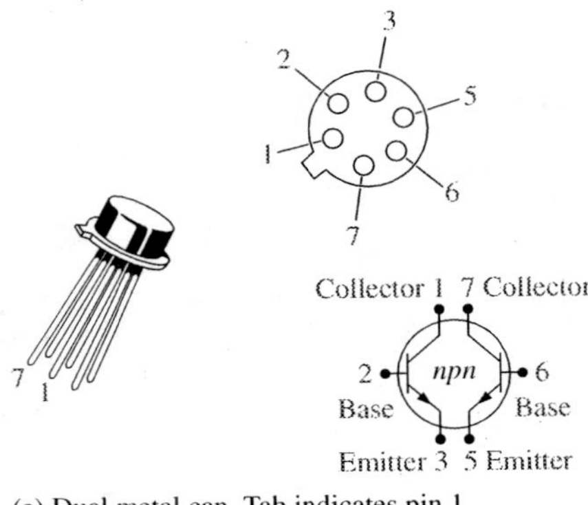

(a) Dual metal can. Tab indicates pin 1.

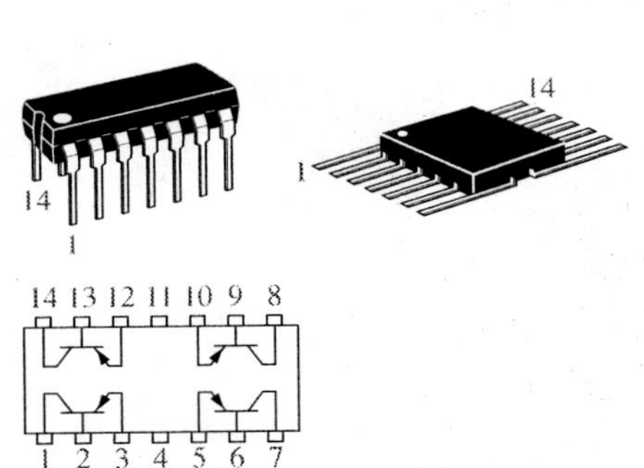

(b) Quad dual-in-line (DIP) and quad flat-pack. Dot indicates pin 1.

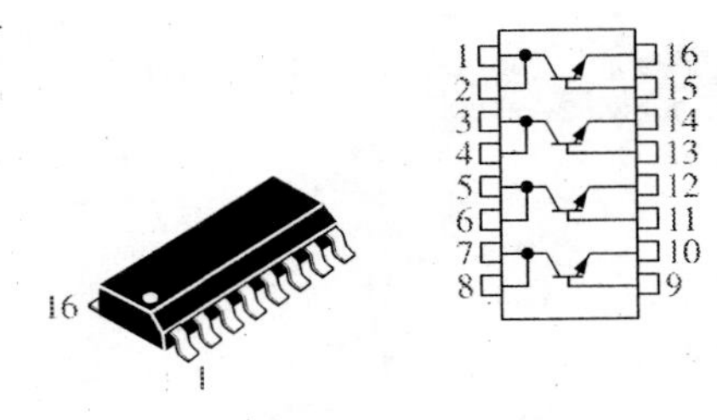

(c) Quad small outline (SO) package for surface-mount technology

Power Transistors

Power transistors are used to handle large currents (typically more than 1 A) and/or large voltages. For example, the final audio stage in a stereo system uses a power transistor amplifier to drive the speakers. Figure 3–44 shows some common package configurations. In most applications, the metal tab or the metal case is common to the collector and is thermally connected to a heat sink for heat dissipation.

FIGURE 3–44 Typical power transistors.

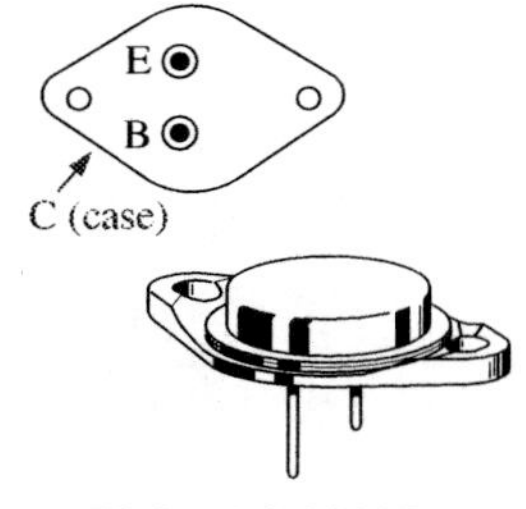

(a) TO-3 or TO-204AE

(b) TO-218

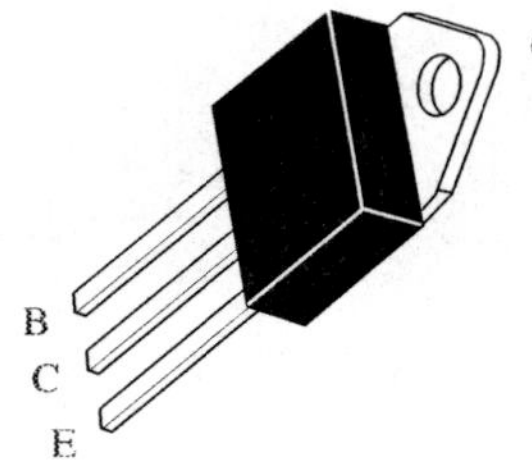

(c) TO-218AC

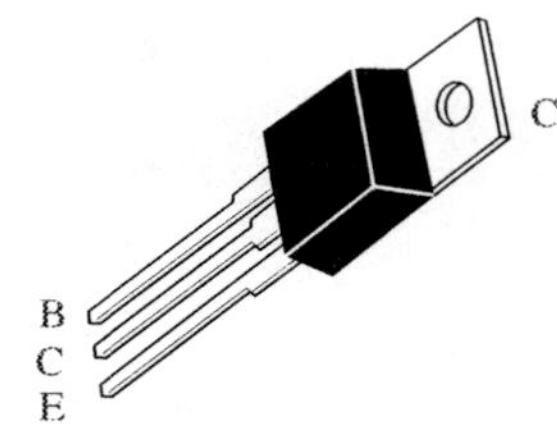

(d) TO-220AB

RF Transistors

RF (radio frequency) transistors are designed to operate at extremely high frequencies and are commonly used for various purposes in communications systems and other high-frequency applications. Their unusual shapes and lead configurations are designed to optimize certain high-frequency parameters. Figure 3–45 shows some examples.

FIGURE 3–45 Examples of RF transistors.

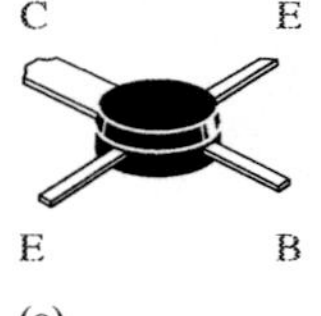

(a)

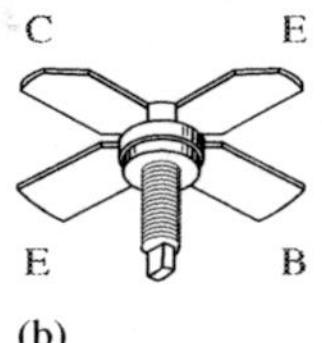

(b)

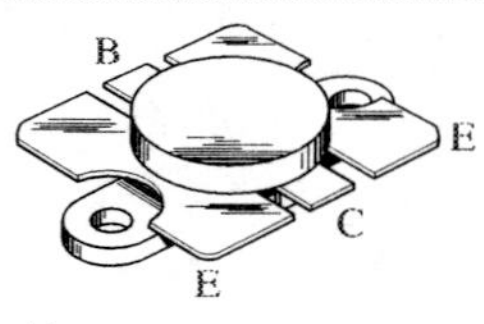

(c)

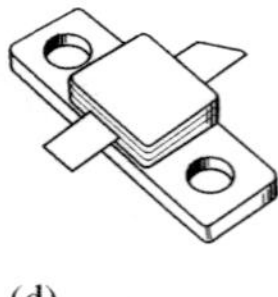

(d)

Review Questions

26. What are the three broad categories of bipolar junction transistors?
27. In a single-transistor metal case, how do you identify the leads?
28. In a quad dual in-line package, how is pin 1 identified?
29. In power transistors, the metal mounting tab or case is connected to which transistor region?
30. What is the purpose of an RF transistor?

CHAPTER REVIEW

Key Terms

ac beta (β_{ac}) The ratio of a change in collector current to a corresponding change in base current in a bipolar junction transistor.

Amplifier An electronic circuit having the capability of amplification and designed specifically for that purpose.

Base One of the semiconductor regions in a BJT.

Bipolar junction transistor (BJT) A transistor constructed with three doped semiconductor regions separated by two *pn* junctions.

Collector One of the semiconductor regions in a BJT.

Common-base (CB) A BJT amplifier configuration in which the base is the common terminal to an ac signal.

Common-collector (CC) A BJT amplifier configuration in which the collector is the common terminal to an ac signal.

Common-emitter (CE) A BJT amplifier configuration in which the emitter is the common terminal to an ac signal.

Cutoff The nonconducting state of a transistor.

dc beta (β_{DC}) The ratio of collector current to base current in a bipolar junction transistor.

Emitter One of the three semiconductor regions in a BJT.

Gain The amount of amplification. Gain is a ratio of an output quantity to an input quantity (e.g., voltage gain is the ratio of the output voltage to the input voltage).

Saturation The state of a BJT in which the collector current has reached a maximum and is independent of the base current.

Important Facts

- ❑ A bipolar junction transistor (BJT) consists of three regions: emitter, base, and collector. The term *bipolar* refers to two types of current: electron current and hole current.
- ❑ The three regions of a BJT are separated by two *pn* junctions.
- ❑ The two types of bipolar junction transistor are the *npn* and the *pnp*.
- ❑ In normal operation, the base-emitter (BE) junction is forward-biased and the base-collector (BC) junction is reverse-biased.
- ❑ The three currents in a BJT are base current, emitter current, and collector current. They are related to each other by this formula: $I_E = I_C + I_B$.
- ❑ Characteristic collector curves for a BJT are a family of curves showing I_C versus V_{CE} for a given set of base currents.
- ❑ When a BJT is in cutoff, there is essentially no collector current except for a very tiny amount of collector leakage current, I_{CEO}. V_{CE} is a maximum.
- ❑ When a BJT is saturated, there is maximum collector current as determined by the external circuit.
- ❑ A load line represents all possible operating points for a circuit, including cutoff and saturation. The point at which the actual base current line intersects the load line is the quiescent or Q-point for the circuit.
- ❑ Base bias uses a single resistor between the power supply and the base terminal.
- ❑ Voltage-divider bias is a very stable form of bias that uses two resistors to form a voltage divider in the base circuit.
- ❑ Emitter bias is a very stable form of bias that uses both positive and negative power supplies and a single resistor between the base terminal and ground.
- ❑ DC values are identified with capital-letter nonitalic subscripts; ac values are identified with lowercase italic subscripts.

- ❑ Coupling capacitors are connected in series with the ac signal to bring it into and out of the amplifier.
- ❑ Bypass capacitors are connected in parallel with a resistor to provide an ac path around the resistor.
- ❑ Common-emitter (CE), common-collector (CC), and common-base (CB) designations refer to the common terminal for the ac signal.
- ❑ Voltage gain for CE and CB amplifiers can be found using a ratio of ac resistances.
- ❑ The voltage gain of a CC amplifier is approximately 1.
- ❑ In switching circuits, transistors are designed to operate at either cutoff or saturation, the equivalent of an open or closed switch.

Formulas

Relationship of key transistor currents:

$$I_E = I_C + I_B \tag{3–1}$$

Definition of β_{DC}:

$$\beta_{DC} = \frac{I_C}{I_B} \tag{3–2}$$

Collector current for base bias:

$$I_C = \beta_{DC}\left(\frac{V_{CC} - V_{BE}}{R_B}\right) \tag{3–3}$$

Base voltage for voltage-divider bias:

$$V_B = \left(\frac{R_2}{R_1 + R_2}\right)V_{CC} \tag{3–4}$$

Emitter voltage for voltage-divider bias:

$$V_E = V_B - 0.7\text{ V} \tag{3–5}$$

Collector voltage for CE and CB amplifiers:

$$V_C = V_{CC} - I_C R_C \tag{3–6}$$

Voltage gain of an amplifier:

$$A_v = \frac{V_{out}}{V_{in}} \tag{3–7}$$

Power gain of an amplifier:

$$A_p = \frac{P_{out}}{P_{in}} \tag{3–8}$$

Definition of β_{ac}:

$$\beta_{ac} = \frac{I_c}{I_b} \tag{3–9}$$

AC emitter resistance:

$$r'_e = \frac{25\text{ mV}}{I_E} \tag{3–10}$$

Voltage gain for CE amplifier:

$$A_v = -\frac{R_c}{R_e} \quad \textbf{(3–11)}$$

Input resistance for CE amplifier with voltage-divider bias (R_{E1} is not bypassed):

$$R_{in(tot)} = R_1 \parallel R_2 \parallel [\beta_{ac}(r'_e + R_{E1})] \quad \textbf{(3–12)}$$

Voltage gain for CC amplifier:

$$A_v \cong 1 \quad \textbf{(3–13)}$$

Voltage gain for a CB amplifier:

$$A_v = \frac{R_c}{R_e} \quad \textbf{(3–14)}$$

Chapter Checkup

Answers are at the end of the chapter.

1. The *n*-type regions in an *npn* bipolar junction transistor are
   (a) collector and base (b) collector and emitter
   (c) base and emitter (d) collector, base, and emitter
2. The *n*-region in a *pnp* transistor is the
   (a) base (b) collector
   (c) emitter (d) case
3. For normal operation of an *npn* transistor, the base must be
   (a) disconnected
   (b) negative with respect to the emitter
   (c) positive with respect to the emitter
   (d) positive with respect to the collector
4. Beta (β_{DC}) is the ratio of
   (a) collector current to emitter current
   (b) collector current to base current
   (c) emitter current to base current
   (d) output voltage to input voltage
5. Two currents that are nearly the same in normal operation of a BJT are
   (a) collector and base
   (b) collector and emitter
   (c) base and emitter
   (d) input and output
6. If the base current for a transistor operating below saturation is increased, the collector current
   (a) increases and the emitter current decreases
   (b) decreases and the emitter current decreases
   (c) increases and the emitter current does not change
   (d) increases and the emitter current increases

7. A saturated bipolar transistor can be recognized by
   (a) a very small voltage between the collector and emitter
   (b) V_{CC} between collector and emitter
   (c) a base emitter drop of 0.7 V
   (d) no base current
8. The voltage gain for a common-emitter (CE) amplifier can be expressed as a ratio of
   (a) ac collector resistance to ac input resistance
   (b) ac emitter resistance to ac input resistance
   (c) dc collector resistance to dc emitter resistance
   (d) none of the above
9. The voltage gain for a common-collector (CC) amplifier
   (a) depends on the input signal
   (b) depends on the transistor's β
   (c) is approximately 1
   (d) none of the above
10. In a CE amplifier, the capacitor from emitter to ground is called the
    (a) coupling capacitor (b) decoupling capacitor
    (c) bypass capacitor (d) tuning capacitor
11. If the capacitor from emitter to ground in a CE amplifier is removed, the voltage gain
    (a) increases (b) decreases
    (c) is not affected (d) becomes erratic
12. When the collector resistor in a CE amplifier is increased in value, the voltage gain
    (a) increases (b) decreases
    (c) is not affected (d) becomes erratic
13. The input resistance of a CE amplifier is affected by
    (a) the bias resistors (b) the collector resistor
    (c) answers (a) and (b) (d) neither (a) nor (b)
14. The output signal of a CB amplifier is always
    (a) in phase with the input signal
    (b) out of phase with the input signal
    (c) smaller than the input signal
    (d) exactly equal to the input signal
15. The output signal of a CC amplifier is always
    (a) in phase with the input signal
    (b) out of phase with the input signal
    (c) larger than the input signal
    (d) exactly equal to the input signal
16. Compared to the CE amplifier, the common-base (CB) amplifier has
    (a) a lower input resistance
    (b) a much larger voltage gain
    (c) a larger current gain
    (d) a higher input resistance

17. Compared to a normal transistor switch, a transistor switch with hysteresis has
(a) high input impedance
(b) faster switching time
(c) higher output current
(d) two switching thresholds
18. If R_2 in Figure 3–46 is open, V_B will
(a) increase (b) decrease
(c) not change

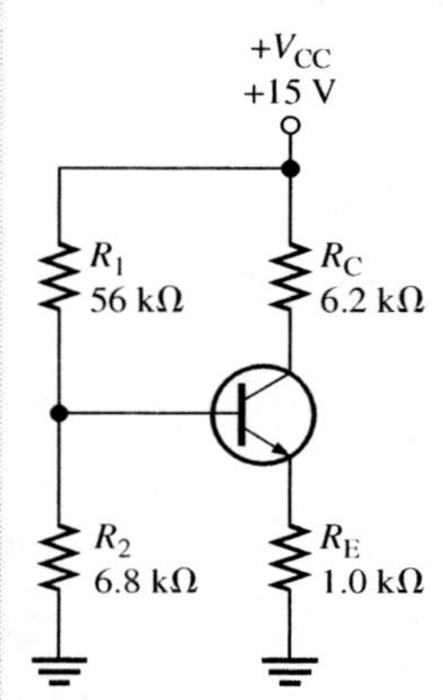

FIGURE 3–46

19. If R_2 in Figure 3–46 is open, V_C will
(a) increase (b) decrease
(c) not change
20. If R_C in Figure 3–46 is open, V_B will
(a) increase (b) decrease
(c) not change
21. If R_C in Figure 3–46 is open, V_C will
(a) increase (b) decrease
(c) not change

Questions

Answers to odd-numbered questions are at the end of the book.

1. Which of the three BJT currents, (base, emitter, or collector) is the largest?
2. How does the schematic of a *pnp* transistor differ from an *npn* transistor?
3. Which BJT junction is normally forward-biased? Which is normally reverse-biased?
4. What is the shape of the base-emitter characteristic curve for a BJT?
5. At what point on the *x*-axis does the dc load line touch?
6. At what point on the *y*-axis does the dc load line touch?
7. Why isn't base bias a good way to bias a transistor used in a linear application?

8. With base bias, how does a higher β transistor affect V_{CE}?
9. What is meant by stiff bias?
10. How is the internal ac resistance of the base-emitter circuit found?
11. What is meant by h_{fe}?
12. What does *partial bypassing* of the emitter resistance mean?
13. Why is the dc power supply said to be an ac ground?
14. What does the term *swamping* mean?
15. How do you find the gain of a CE amplifier?
16. What is the advantage of high input resistance for an amplifier?
17. What is the advantage of low output resistance for an amplifier?
18. Where is the output taken in a CC amplifier?
19. Where is the output taken in a CB amplifier?
20. What V_{CE} do you expect to see across a saturated transistor?

PROBLEMS

Basic Problems

Answers to odd-numbered problems are at the end of the book.

1. What is the exact value of I_C for $I_E = 5.34$ mA and $I_B = 47.5\ \mu$A?
2. A certain transistor has an $I_C = 25$ mA and an $I_B = 200\ \mu$A. Determine the β_{DC}.
3. In a certain transistor circuit, the base current is 2% of the 30 mA emitter current. Determine the approximate collector current.
4. Find V_E and I_C in Figure 3–47.

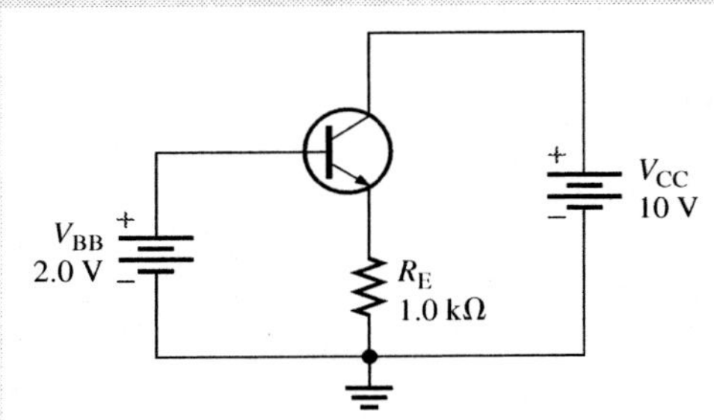

FIGURE 3–47

5. Determine the I_B, I_C, and V_C for the transistor circuit in Figure 3–48. Assume $\beta_{DC} = 75$.

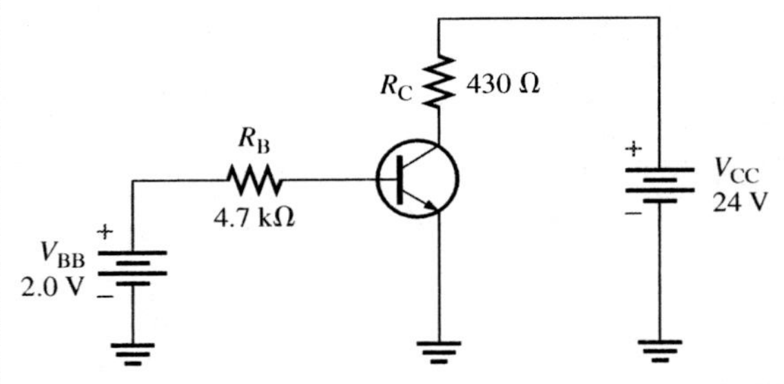

FIGURE 3–48

6. Draw the dc load line for the transistor circuit in Figure 3–49.

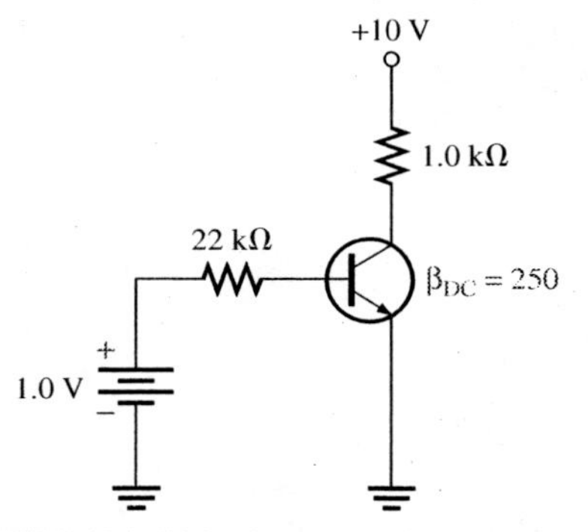

FIGURE 3–49

7. Determine I_B, I_C, and V_C in Figure 3–49.
8. For the base-biased *npn* transistor in Figure 3–50, assume $\beta_{DC} = 100$. Find I_C and V_{CE}.

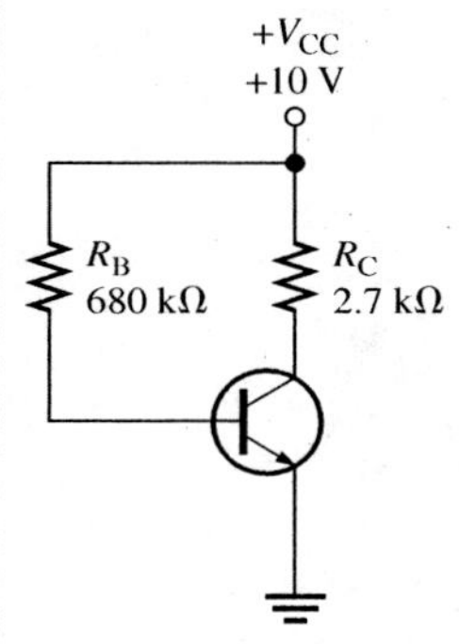

FIGURE 3–50

9. Repeat Problem 8 for $\beta_{DC} = 300$. (*Hint:* The transistor is now saturated!).
10. For the base-biased *pnp* transistor in Figure 3–51, assume $\beta_{DC} = 200$. Find I_C and V_{CE}.

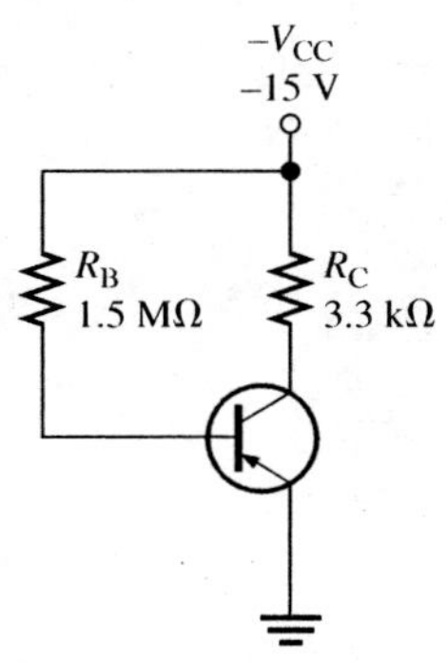

FIGURE 3–51

11. For each of the following conditions in the circuit of Figure 3–51, determine if the collector current will increase, decrease, or remain the same:
    (a) the base is shorted to ground
    (b) R_C is smaller
    (c) the transistor has a higher β
    (d) R_B is smaller

12. For the voltage-divider circuit in Figure 3–52, determine V_B, V_E, and V_C.

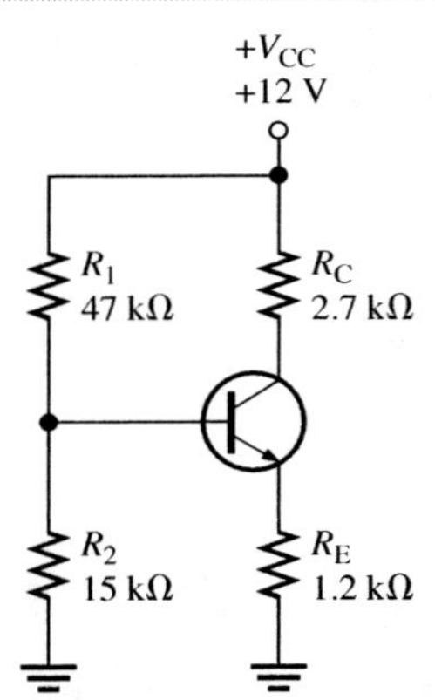

FIGURE 3–52

13. For the voltage-divider biased (*pnp*) circuit in Figure 3–53; determine I_C and V_{CE}.

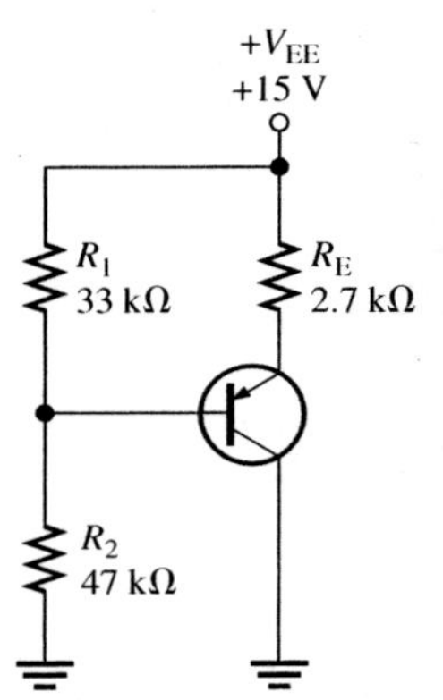

FIGURE 3–53

14. Determine the end points for the dc load line, $I_{C(sat)}$ and $V_{CE(cutoff)}$ for Figure 3–53.
15. For the emitter-bias circuit in Figure 3–54, determine I_C and V_{CE}.

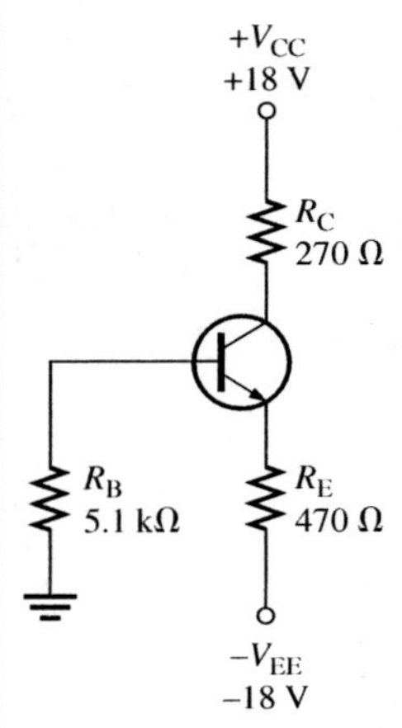

FIGURE 3–54

16. Determine the dc voltages, V_B, V_E, and V_C, with respect to ground in Figure 3–55.

17. Determine the voltage gain for the CE amplifier in Figure 3–55.

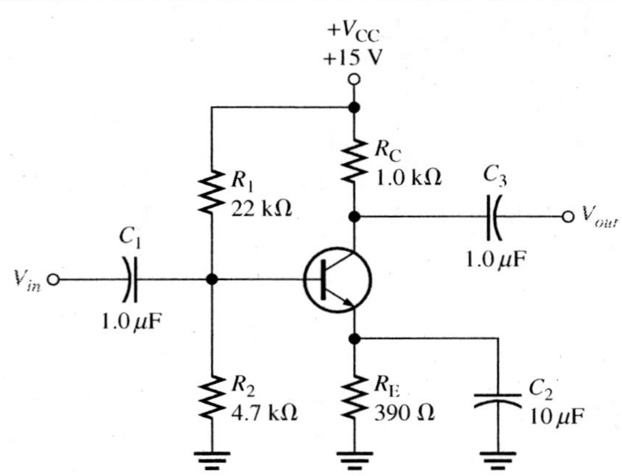

FIGURE 3–55

18. What type of amplifier is shown in Figure 3–56? What is I_C?

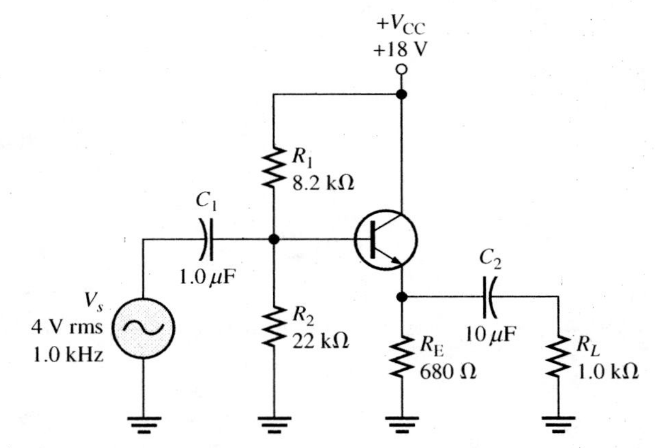

FIGURE 3–56

19. What type of amplifier is shown in Figure 3–57? What is I_C?

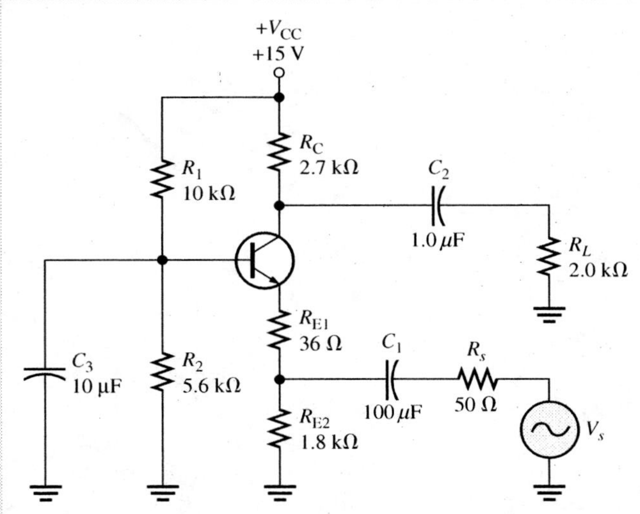

FIGURE 3–57

20. Determine $I_{C(sat)}$ for Q_1 and Q_2 in Figure 3–38.

21. The transistor in Figure 3–58 has $\beta_{DC} = 100$. Determine the maximum value of R_B that will ensure saturation when V_{IN} is 5 V.

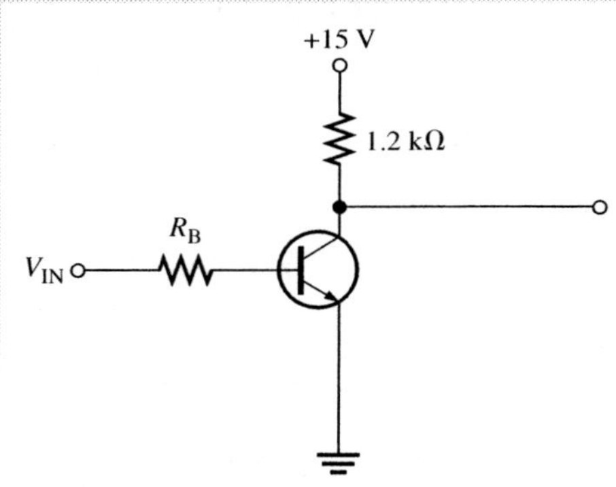

FIGURE 3–58

22. Identify the leads on the transistors in Figure 3–59. Bottom views are shown.

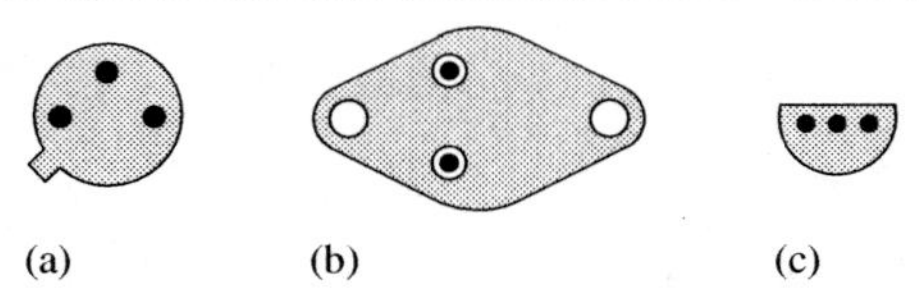

FIGURE 3–59

23. What is the most probable category of each transistor in Figure 3–60?

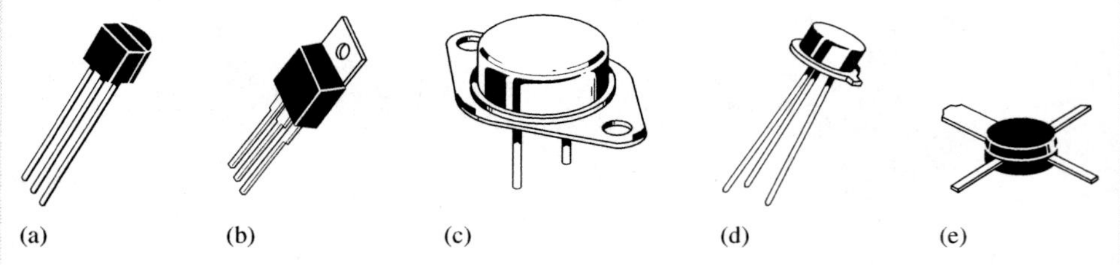

FIGURE 3–60

Basic-Plus Problems

24. Assume that R_C in Figure 3–54 was replaced with a 330 Ω resistor.
    (a) What is the new value of I_C and V_{CE}?
    (b) What is the power dissipated in R_C as a result of this change?

25. The amplifier in Figure 3–61 has a variable gain control, using a 100 Ω potentiometer for R_E with the wiper ac grounded. As the potentiometer is adjusted, more or less of R_E is bypassed to ground, thus varying the gain. The total R_E remains constant to dc, keeping the bias fixed. Determine the maximum and minimum gains for this amplifier.

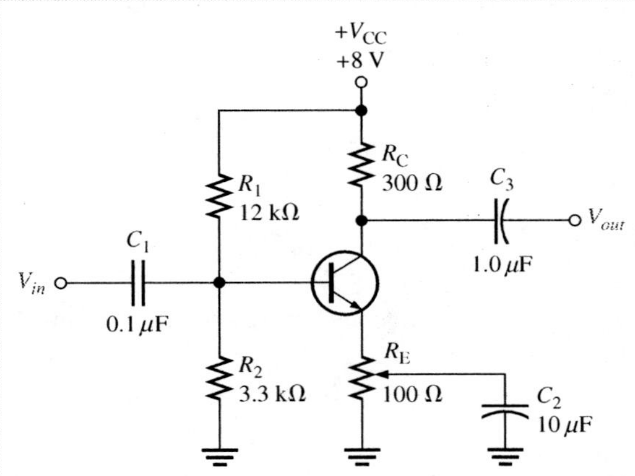

FIGURE 3–61

26. If a load resistance of 600 Ω is placed on the output of the amplifier in Figure 3–61, what is the maximum gain?
27. For the circuit in Figure 3–40, assume Q_1 is saturated and Q_2 is off.
    (a) What is the emitter voltage of Q_1?
    (b) What voltage, V_{IN}, will turn off Q_1?
28. Refer to the ac load line in Figure 3–33. If an input ac signal causes the base current to move between 10 μA and 25 μA, how much does V_{CE} change?

ANSWERS

Example Questions

3–1: $I_B = 0.241$ mA, $I_C = 21.7$ mA, $I_E = 21.9$ mA, $V_{CE} = 4.23$ V, $V_{CB} = 3.53$ V

3–2: Along the *x*-axis

3–3: 221

3–4: $V_B = 1.51$ V, $V_E = 0.81$ V, $I_E = 1.73$ mA, $I_C = 1.73$ mA, $V_{CE} = 6.51$ V

3–5: $V_B = -1.78$ V, $V_E = -1.08$ V, $I_E = 0.90$ mA, $I_C = 0.90$ mA, $V_{CE} = -5.88$ V

3–6: $V_E = -0.7$ V

3–7: 6.0 V

3–8: 250 Ω

3–9: Gain is reduced to −3.65.

3–10: 3.15 kΩ

3–11: 78.4 μA

Review Questions

1. Emitter, base, collector
2. Saturation means there is maximum conduction and the voltage from collector to emitter is close to zero. Cutoff is essentially when there is no collector current and the power supply voltage appears between the collector and the emitter.
3. The ratio of collector current to base current in a bipolar junction transistor
4. A line drawn between saturation on the *y*-axis and cutoff on the *x*-axis that shows all possible dc operating points.

5. The Q-point is the intersection of the base current characteristic and the load line.
6. Base, voltage-divider, and emitter
7. The following steps are for an *npn* transistor with a positive supply voltage:
   (a) Compute the base voltage using the voltage-divider rule.
   (b) Subtract 0.7 V to obtain the emitter voltage.
   (c) Apply Ohm's law to the emitter resistor to obtain the approximate collector current.
   (d) Using the collector current, find the voltage across the collector resistor by Ohm's law.
   (e) Subtract the drop across the collector resistor from the power supply to obtain the collector voltage.
   (f) Subtract the emitter voltage from the collector voltage to obtain V_{CE}.
8. Approximately +1 V. (This result assumes the base resistor is connected to ground and drops a few tenths of a volt.)
9. Voltage-divider bias is much more stable; essentially it is independent of beta.
10. Transistor is saturated. $V_B = 3.3$ V, $V_E = 2.6$ V, $V_C = 2.7$ V.
11. β_{DC} is the ratio of I_C to I_B (for an unsaturated transistor). β_{ac} is the ratio of I_c to I_b (for an unsaturated transistor).
12. An ideal linear amplifier adds no noise or distortion to the signal. It has infinite input resistance, zero output resistance, and replicates the input exactly.
13. Waveshaping and switching
14. Voltage gain is dimensionless.
15. A coupling capacitor is in series with the signal and passes it to or from a transistor. A bypass capacitor is in parallel with the signal and provides an ac path around a resistor.
16. The input terminal is the base; the output terminal is the collector.
17. Current in the collector circuit follows the input signal. When current increases, the collector voltage decreases (and vice versa). As a result, the output is inverted with respect to the input.
18. High input resistance reduces the loading effect on a source.
19. The ac collector resistance is divided by the ac emitter resistance.
20. Q-point
21. Saturation (on) and cutoff (off)
22. At saturation
23. At cutoff
24. At cutoff
25. Two different switching thresholds
26. Three categories of BJT are small signal/general purpose, power, and RF.
27. Going clockwise from tab: emitter, base, and collector (bottom view).
28. A dot
29. The metal mounting tab or case in power transistors is connected to the collector terminal.
30. To amplify radio frequencies and/or microwave frequencies

Chapter Checkup

1. (b)	2. (a)	3. (c)	4. (b)	5. (b)
6. (d)	7. (a)	8. (d)	9. (c)	10. (c)
11. (b)	12. (a)	13. (a)	14. (a)	15. (a)
16. (a)	17. (d)	18. (a)	19. (b)	20. (b)
21. (b)				

CHAPTER

FIELD-EFFECT TRANSISTORS

CHAPTER OUTLINE

4–1 Structure of Field-Effect Transistors
4–2 JFET Characteristics
4–3 JFET Biasing
4–4 MOSFET Characteristics
4–5 MOSFET Biasing
4–6 FET Linear Amplifiers
4–7 FET Switching Circuits

Study aids for this chapter are available at

http://www.prenhall.com/SOE

INTRODUCTION

This chapter introduces the field-effect transistor (FET), a transistor that works on an entirely different principle than the bipolar junction transistor (BJT) introduced in the last chapter. Although the idea for the FET preceded the invention of the BJT by decades, it wasn't until the 1960s that it became commercially available. For certain applications, such as switching, the FET is superior to the BJT. Some applications use a mix of the two types to exploit certain characteristics.

KEY TERMS

- Field-effect transistor
- Source
- Drain
- Gate
- Junction field-effect transistor
- Ohmic region
- Constant-current region
- Pinch-off voltage
- Transconductance
- MOSFET
- Depletion mode
- Enhancement mode
- Common-source
- Common-drain

KEY OBJECTIVES

A section number is given for each objective. After completing this chapter, you should be able to

4–1 Describe the classifications for field-effect transistors (FETs)

4–2 Describe the characteristic curve of a JFET (junction field-effect transistor) and explain common JFET parameters

4–3 Explain how to bias a JFET using self-bias and voltage-divider bias

4–4 Describe the characteristics of a MOSFET (metal-oxide semiconductor field-effect transistor)

4–5 Explain how to bias a D-MOSFET using zero bias, voltage-divider bias, or current-source bias and an E-MOSFET using voltage-divider bias

4–6 Analyze common-source and common-drain FET amplifiers to determine basic operational parameters including gain and input resistance

4–7 Discuss two types of FET switching circuits

COMPUTER SIMULATIONS DIRECTORY

The following figures have Multisim circuit files associated with them.

- Figure 4–11
Page 132
- Figure 4–33
Page 145

LABORATORY EXPERIMENTS DIRECTORY

The following exercises are for this chapter.

- **Experiment 7**
JFET Biasing
- **Experiment 8**
JFET Amplifiers

You have probably seen how iron filings around a magnet line up, enabling you to visualize the magnetic field. An electric field can be visualized in the same manner. In a small-scale experiment, grass seeds, suspended in an insulating liquid, will align along the electric field when opposite charged rods are inserted into the liquid. The electric field between the rods can be drawn by "lines of force" from the positive to the negative rod. The electric potential between the rods has units of joules/coulomb (energy/charge), which is the unit of volts.

On a larger scale, lightning is evidence of the electric field that can build between a cloud and the earth due to friction between rising and falling air particles. Worldwide, there are about 100 lightning bolts each second from the thousands of storm clouds present at any time. These lightning bolts can travel between clouds or from cloud to earth or (rarely) from the earth to the cloud. The potential difference in a lightning strike ranges from 10 to 100 million volts.

Another interesting phenomenon that is due to electric structures even higher in the atmosphere than clouds is the aurora. An aurora, such as the one shown here, may have positive and negative structures found near each other. A positively charged structure has a downward electric field component that accelerates electrons upward. A structure like this may contain a "black aurora," which is embedded in a diffuse aurora that has the oppositely charged structure and accelerates electrons downward.

4–1 STRUCTURE OF FIELD-EFFECT TRANSISTORS (FETs)

Recall that the bipolar junction transistor (BJT) is a current-controlled device; that is, the base current controls the amount of collector current. The field-effect transistor (FET) is a voltage-controlled device in which the voltage at the gate terminal controls the amount of current through the device. Both the BJT and the FET can be used as an amplifier and in switching applications.

In this section, you will learn the basic classifications for field-effect transistors (FETs).

The FET Family

Field-effect transistors (FETs) are a class of semiconductors that operate on an entirely different principle than BJTs. In a FET, a narrow conducting channel is connected to two leads called the **source** and the **drain**. This channel is made from either an *n*-type or *p*-type material. As the name *field-effect* implies, conduction in the channel is controlled by an electric field, established by a voltage applied to a third lead called the **gate**.

FETs are divided into two classes. In a *junction FET* (or JFET), the gate forms a *pn* junction with the channel. The other type of FET, called the *MOSFET* (for *M*etal *O*xide *S*emiconductor *FET*), uses an insulated gate to control conduction in the channel. (The terms *insulated gate* and *MOSFET* refer to the same type of device.) The insulation is an extremely thin layer (<1 μm) of glass (typically SiO_2). Figure 4–1 is an overview of the FET family, showing the various types. There are limited applications for depletion-mode MOSFETs (the exception is some high frequency specialized types). Of the few that are available, all are *n*-channel. The *p*-channel type is shown as generic only because you can experiment with it in simulations like Multisim.

Classification of field-effect transistors. **FIGURE 4–1**

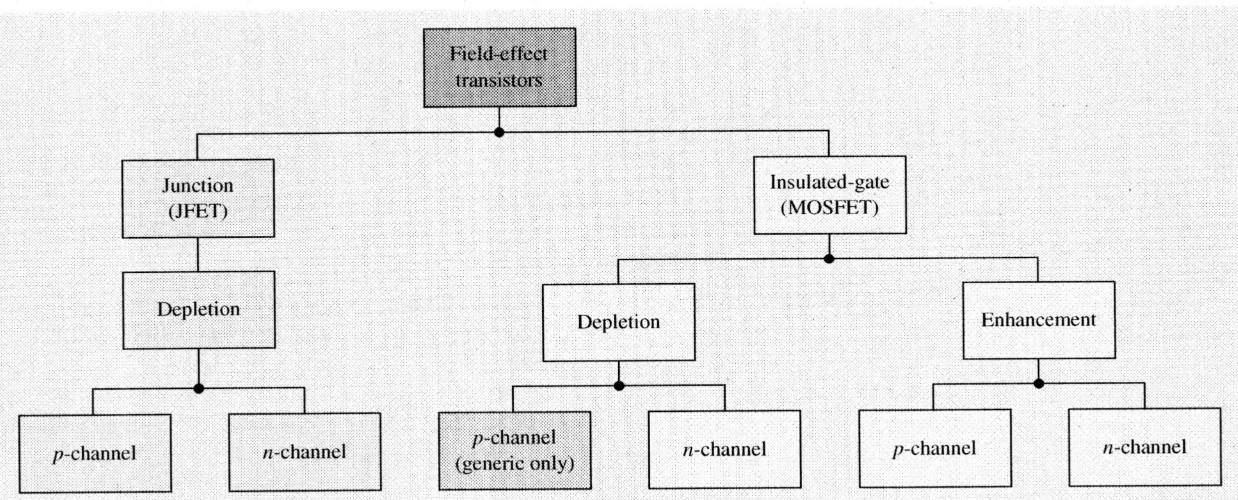

FETs are used in most computer integrated circuits (ICs) because of several important advantages they have over BJTs, particularly with respect to manufacturing of large-scale ICs. For digital circuits, MOSFETs have become the dominant type of transistor for several reasons. They can be fabricated in much smaller areas than BJTs, they are relatively easy to manufacture on ICs, and they produce simpler circuits with no resistors or diodes. Most microprocessors and computer memories use FET technology. A brief look at how FETs are used in certain ICs is included in Section 4–7.

Compared to the BJT, the FET family is more diverse. A characteristic that differs between various types of FETs is their dc behavior. Therefore, JFETs are biased differently than E-MOSFETs. Fortunately, bias circuits are fairly easy to understand. Before proceeding to bias circuits, the characteristics of the transistors that make up the FET family will be discussed.

Common to all FETs is very high input resistance and low electrical noise. In addition, both JFETs and MOSFETs respond the same way to ac signals and have similar ac equivalent circuits. JFETs achieve their high input resistance because the input *pn* junction is always operated with reverse bias; MOSFETs achieve their high input resistance because of the insulated gate. Although all FETs have high input resistance, they do not have the high gain of bipolar junction transistors. BJTs are also inherently more linear than FETs. For certain applications, FETs are superior; for other applications, BJTs are superior. Many designs take advantage of both types and include a mix of FETs and BJTs.

HISTORICAL NOTE

Julius Lilienfeld actually thought of FETs long before the invention of the BJT. He applied for a patent in Canada in 1925 and later in the U.S. for a FET device. He was able to demonstrate signal amplification with his device. His U.S. patent application was entitled "Method and Apparatus for Controlling Electric Currents" and clearly described the field-effect transistor. It wasn't until the 1960s that FETs became commercially available, a decade after BJTs were available.

Review Questions

Answers are at the end of the chapter.

1. What are the three terminals of a FET called?
2. What are two classes of FETs?
3. What is another name for an insulated-gate FET?
4. Why are MOSFETs the dominant type of transistor used in ICs?
5. What are some important differences between BJTs and FETs?

4–2 JFET CHARACTERISTICS

The JFET characteristic curve is divided into the ohmic region, the constant-current region, and the breakdown region. It is normally operated in either the ohmic region or the constant-current region.

In this section, you will learn how to interpret the characteristic curve and explain common JFET parameters.

JFET Operation

Figure 4–2(a) shows the basic structure of an *n*-channel **junction field-effect transistor (JFET)**. Wire leads are connected to each end of the *n* channel; the drain is shown at the upper end, and the source is at the lower end. This channel is a conductor: for an *n*-channel JFET, electrons are the carrier; for a *p*-channel JFET, holes are the carrier. With no external voltages, the channel can conduct current in either direction.

FIGURE 4–2

Basic structure of the two types of JFET.

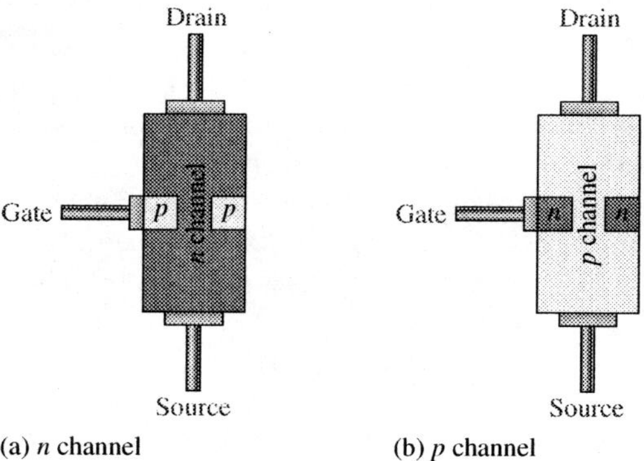

In an *n*-channel device, a *p* material is diffused into the *n*-channel to form a *pn* junction and is connected to the gate lead. The diagram in Figure 4–2(a) shows *p*-material diffused into two regions that are normally connected internally by the manufacturer to form a single gate. (A special-purpose JFET, called a dual-gate JFET, has a separate lead to each of these regions.) In the structure diagrams, the interconnection of both *p* regions is omitted for simplicity, with a connection to only one shown. A *p*-channel JFET is shown in Figure 4–2(b).

To illustrate JFET operation, Figure 4–3(a) shows normal operating voltages applied to an *n*-channel device. V_{DD} provides a positive drain-to-source voltage, causing electrons to flow from the source to the drain. For an *n*-channel JFET, reverse-biasing of the gate-source junction is done with a negative gate voltage. V_{GG} sets the reverse-biased voltage between the gate and the source, as shown. Notice that there should *never* be any forward-biased junctions in a FET; this is one of the principal differences between FETs and BJTs.

FIGURE 4–3

Effects of V_{GG} on channel width, resistance, and drain current ($V_{GG} = V_{GS}$).

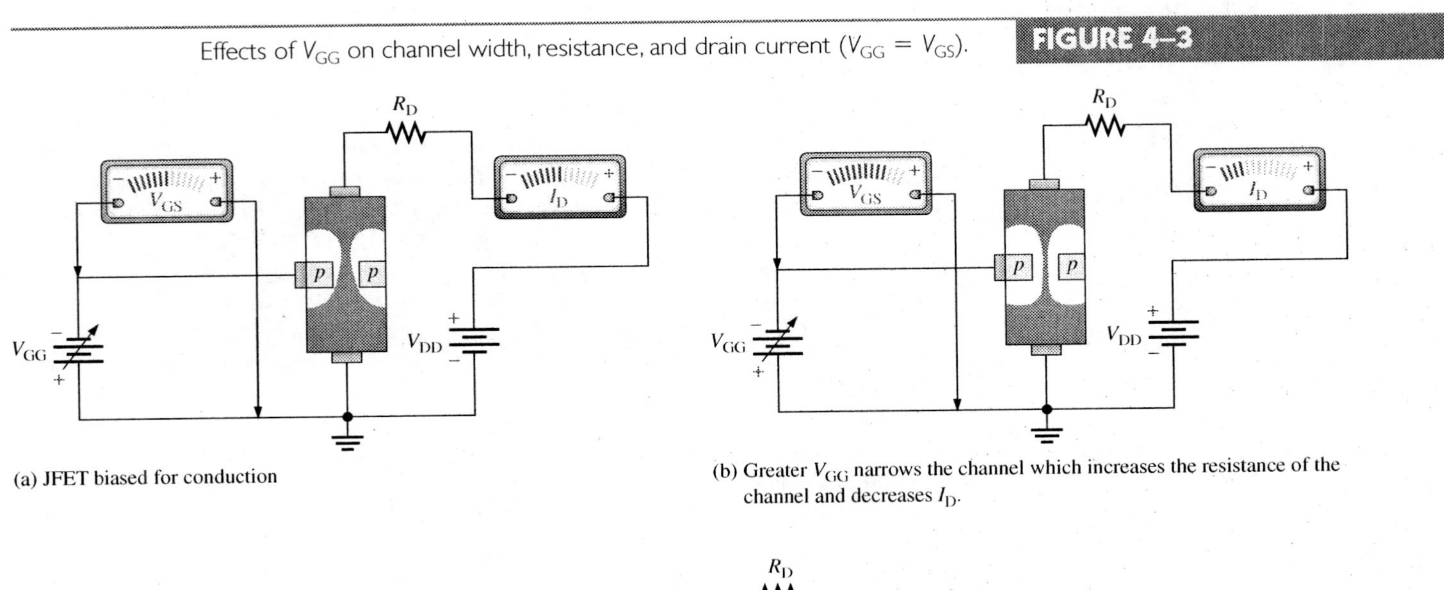

(a) JFET biased for conduction

(b) Greater V_{GG} narrows the channel which increases the resistance of the channel and decreases I_D.

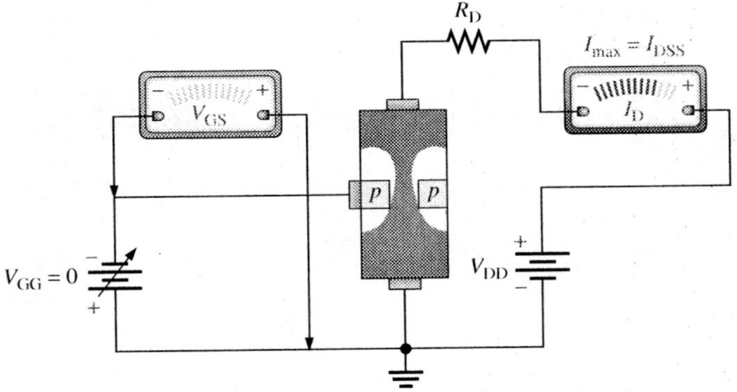

(c) Less V_{GG} widens the channel which decreases the resistance of the channel and increases I_D.

The channel width, and thus the channel resistance, is controlled by varying the gate voltage, thereby controlling the amount of drain current, I_D. This concept is illustrated in Figure 4–3(b) and (c). The white areas represent the depletion region created by the reverse bias. The key idea is that the channel width is controlled by the gate voltage.

JFET Symbols

The schematic symbols for both *n*-channel and *p*-channel JFETs are shown in Figure 4–4. Notice that the arrow on the gate points "in" for *n* channel and "out" for *p* channel.

FIGURE 4–4

JFET schematic symbols.

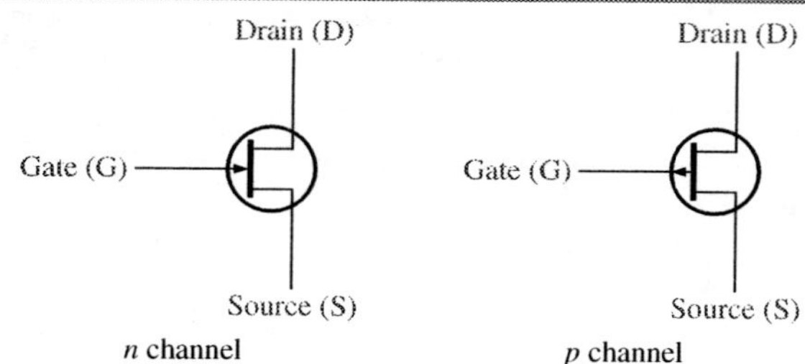

Drain Characteristic Curve

The drain characteristic curve is a plot of the drain current, I_D, versus the drain-to-source voltage, V_{DS}, which corresponds to a BJT's collector current, I_C, versus collector-to-emitter voltage, V_{CE}. There are, however, some significant differences between BJT characteristics and FET characteristics. Since the FET is a voltage-controlled device, the third variable on the FET characteristic (V_{GS}) has units of gate voltage instead of base current (I_B) in the case of the BJT. The characteristics for *n*-channel devices are introduced in this section. *P*-channel devices operate in the same way but with opposite polarities. Generally, *n*-channel JFETs have better specifications than their *p*-channel counterparts, so they are more popular.

Consider an *n*-channel JFET where the gate-to-source voltage is zero ($V_{GS} = 0$ V). This zero voltage is produced by shorting the gate to the source, as in Figure 4–5(a) where both are grounded. As V_{DD} (and thus V_{DS}) is increased from 0 V, I_D will increase proportionally, as shown in the graph of Figure 4–5(b) between points *A* and *B*. In this region, the channel resistance is essentially constant because the depletion region is not large enough to have a significant effect. This region is called the **ohmic region** because V_{DS} and I_D are related by Ohm's law. The value of the resistance can be changed by the gate voltage; thus, it is possible to use a JFET as a voltage-controlled resistor when it is operated in this region. An application will be shown in Figure 10–10 (Wien bridge).

FIGURE 4–5

The drain characteristic curve of a JFET for $V_{GS} = 0$ V showing pinch-off.

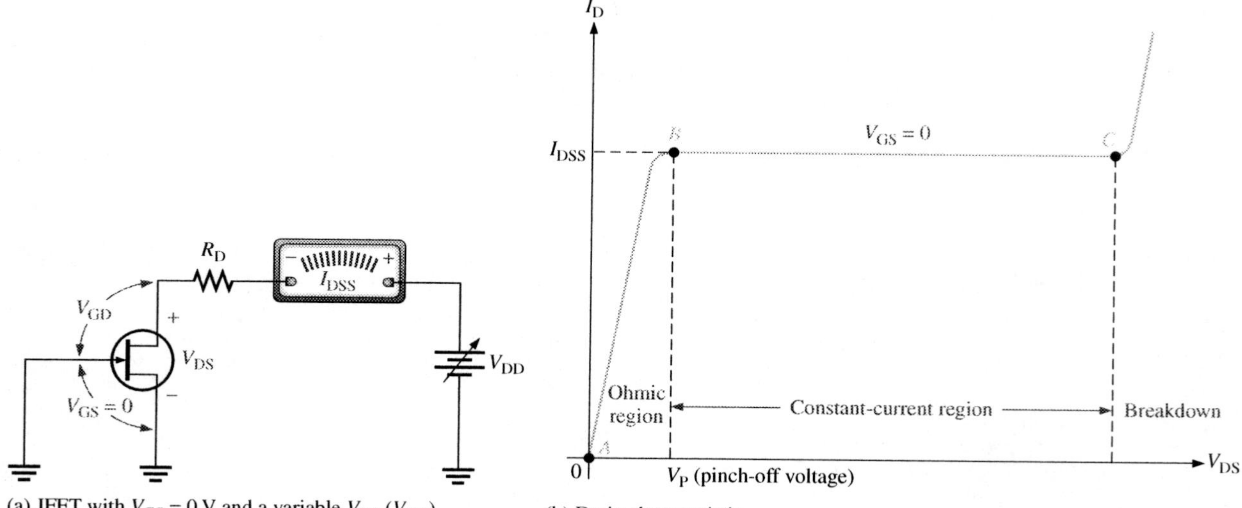

(a) JFET with $V_{GS} = 0$ V and a variable V_{DS} (V_{DD})

(b) Drain characteristic

At point *B* in Figure 4–5(b), the curve levels off and I_D becomes essentially constant. As V_{DS} increases from point *B* to point *C*, the reverse-bias voltage from gate to drain (V_{GD}) produces a depletion region large enough to offset the increase in V_{DS}, thus keeping I_D relatively constant. This region is called the **constant-current region**.

Pinch-Off Voltage

For $V_{GS} = 0$ V, the value of V_{DS} at which I_D becomes essentially constant (point *B* on the curve in Figure 4–5 (b)) is the **pinch-off voltage**, V_P. Notice that the pinch-off voltage is a positive value for an *n*-channel JFET. For a given JFET, V_P has a fixed value. As you can see, a continued increase in V_{DS} above the pinch-off voltage produces an almost constant drain current. This value of drain current is I_{DSS} (*D*rain to *S*ource current with gate *S*horted) and is always specified on JFET data sheets. I_{DSS} is the maximum drain current that a specific JFET can produce regardless of the external circuit, and it is always specified for the condition, $V_{GS} = 0$ V.

Continuing along the graph in Figure 4–5(b), breakdown occurs at point *C* when I_D begins to increase very rapidly with any further increase in V_{DS}. Breakdown can result in irreversible damage to the device, so JFETs are always operated below breakdown and usually within the constant-current region (between points *B* and *C* on the graph).

V_{GS} Controls I_D

Let's connect a bias voltage, V_{GG}, from gate to source as shown in Figure 4–6(a). As V_{GS} is set to increasingly more negative values by adjusting V_{GG}, a family of drain characteristic curves is produced, as shown in Figure 4–6(b). Notice that I_D decreases as the magnitude of V_{GS} is increased to larger negative values because of the narrowing of the channel. Also notice that, for each increase in V_{GS}, the JFET reaches pinch-off (where constant current begins) at values of V_{DS} less than V_P. So, the amount of drain current is controlled by V_{GS}.

FIGURE 4–6

Pinch-off occurs at a lower V_{DS} as V_{GS} is increased to more negative values.

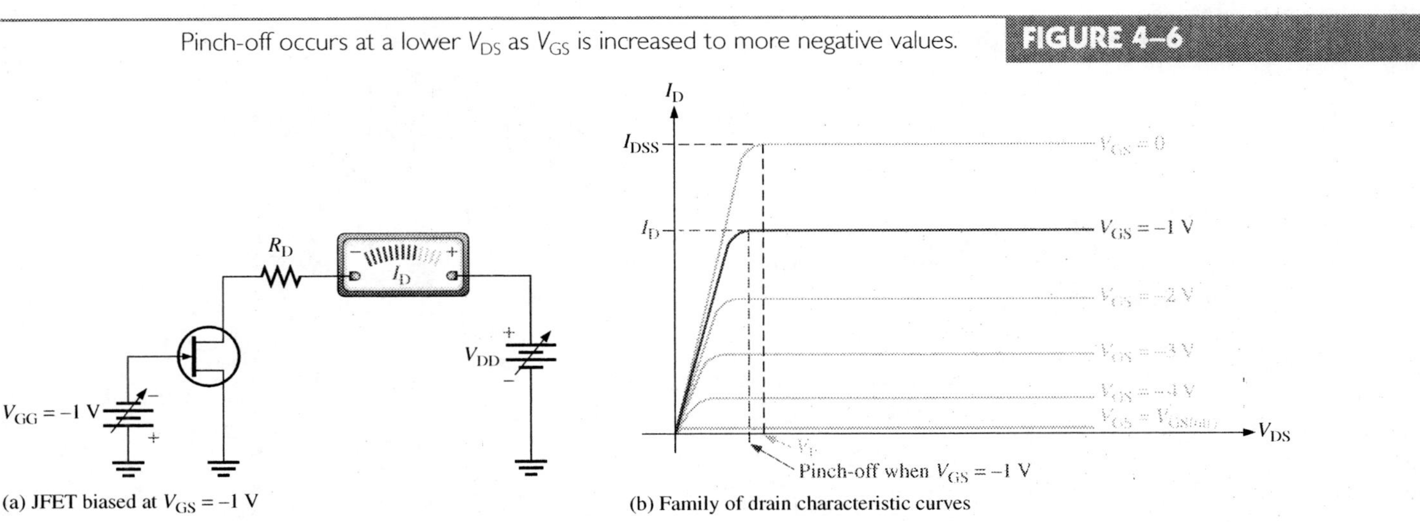

(a) JFET biased at $V_{GS} = -1$ V

(b) Family of drain characteristic curves

Cutoff Voltage

The value of V_{GS} that makes I_D approximately zero is the cutoff voltage, $V_{GS(off)}$. The JFET must be operated between $V_{GS} = 0$ V and $V_{GS(off)}$. For this range of gate-to-source voltages, I_D will vary from a maximum of I_{DSS} to a minimum of almost zero.

As you have seen, for an n-channel JFET, the more negative V_{GS} is, the smaller I_D becomes in the constant-current region. When V_{GS} has a sufficiently large negative value, I_D is reduced to zero. This cutoff effect is caused by the widening of the depletion region to a point where it completely closes the channel. The bottom line on the characteristic curve represents this condition.

Comparison of Pinch-Off and Cutoff

The pinch-off voltage is measured on the drain characteristic. For an n-channel device, it is the positive voltage at which the drain current becomes constant when $V_{GS} = 0$ V. Cutoff can also be measured on the drain characteristic and represents the negative gate-to-source voltage that reduces the drain current to zero.

$V_{GS(off)}$ and V_P are always equal in magnitude but opposite in sign. A data sheet usually will give either $V_{GS(off)}$ or V_P, but not both. However, when you know one, you have the other. For example, if $V_{GS(off)} = -5$ V, then $V_P = +5$ V.

EXAMPLE 4–1

Problem

For the n-channel JFET in Figure 4–7, $V_{GS(off)} = -4$ V and $I_{DSS} = 12$ mA. Determine the *minimum* value of V_{DD} required to put the device in the constant-current region of operation.

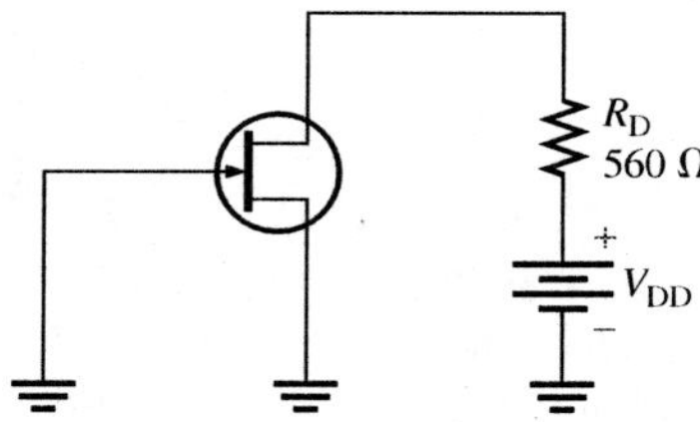

FIGURE 4–7

Solution

Since $V_{GS(off)} = -4$ V, $V_P = 4$ V. The minimum value of V_{DS} for the JFET to be in its constant-current region is

$$V_{DS} = V_P = 4 \text{ V}$$

In the constant-current region with $V_{GS} = 0$ V,

$$I_D = I_{DSS} = 12 \text{ mA}$$

The drop across the drain resistor is

$$V_{R_D} = (12 \text{ mA})(560 \ \Omega) = 6.7 \text{ V}$$

Applying Kirchhoff's law around the drain circuit gives

$$V_{DD} = V_{DS} + V_{R_D} = 4 \text{ V} + 6.7 \text{ V} = \mathbf{10.7 \ V}$$

This is the minimum value of V_{DD} to make $V_{DS} = V_P$ and to put the device in the constant-current region.

Question*

If V_{DD} is increased to 15 V, what is the drain current?

* *Answers are at the end of the chapter.*

JFET Transconductance Curves

A useful way of looking at any circuit is to show the output for a given input. This characteristic is called a *transfer curve.*

Since the JFET is controlled by a negative voltage on the input (gate) and the output is drain current, the transfer curve is a plot of I_D, plotted on the y-axis, as a function of V_{GS}, plotted on the x-axis. When the output unit (mA) is divided by the input unit (V), the result is the unit of conductance (mS). You can think of a voltage at the input being transferred to the output as a current; thus, the prefix "trans" is added to the word *conductance* forming the word **transconductance**. The transconductance curve is a plot of the transfer characteristic (I_D versus V_{GS}) of a FET. Transconductance is listed on data sheets as g_m or y_{fs}.

A representative curve for an n-channel JFET is shown in Figure 4–8(a). Generally, all types of FETs have a transconductance curve with this same basic shape. The curves shown are typical for a general-purpose n-channel JFET.

FIGURE 4–8 Characteristic curves for a typical n-channel JFET.

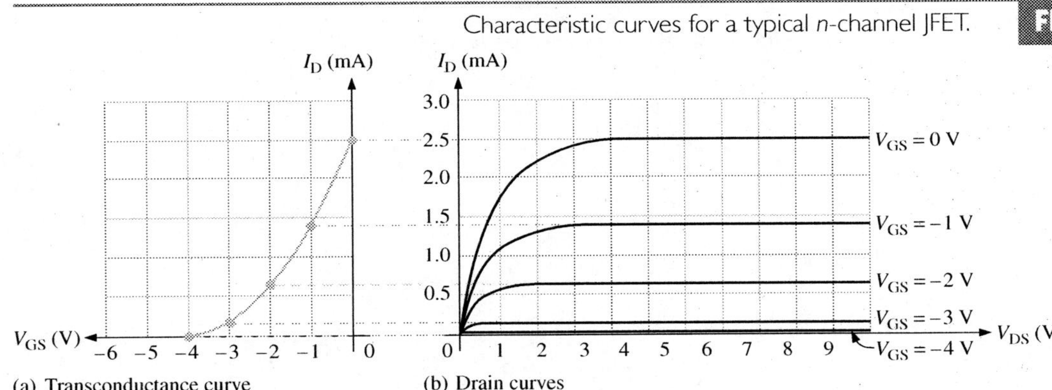

(a) Transconductance curve (b) Drain curves

The transconductance characteristic is directly related to the drain characteristic as shown in Figure 4–8(b). Notice that both plots have the same vertical axis, representing I_D. Transconductance is an ac parameter so its value is found at any point on the curve by dividing a small change in drain current by a small change in gate-to-source voltage.

$$g_m = \frac{\Delta I_D}{\Delta V_{GS}}$$

This equation can be written with ac notation as simply

$$g_m = \frac{I_d}{V_{gs}} \qquad (4\text{–}1)$$

The transconductance curve is not a straight line, implying that the relation between the output current and the input voltage is nonlinear. This is an important point: FETs have a nonlinear transconductance curve. This means that they tend to add distortion to an input signal. Distortion is not always a bad thing; for example, in radio frequency mixers, JFETs have an advantage over BJTs because of this characteristic.

EXAMPLE 4–2

Problem

For the curve in Figure 4–9, determine the transconductance at $I_D = 1.0$ mA.

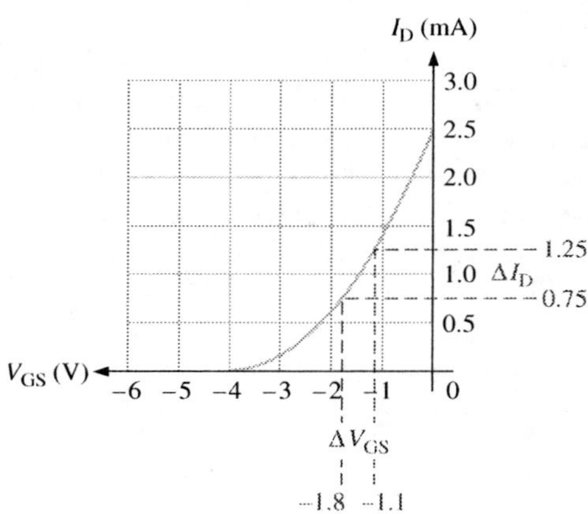

FIGURE 4–9

Solution

Select a small change in I_D and divide it by the corresponding change in V_{GS} at 1.0 mA. The graphical method is shown in Figure 4–9. From the graph, the transconductance is

$$g_m = \frac{\Delta I_D}{\Delta V_{GS}} = \frac{1.25 \text{ mA} - 0.75 \text{ mA}}{-1.1 \text{ V} - (-1.8 \text{ V})} = \mathbf{0.714 \text{ mS}}$$

Question

What is the transconductance at $I_D = 1.5$ mA?

JFET Input Resistance

As you know, a *pn* junction has a very high resistance when it is reverse-biased. A JFET operates with its gate-source junction reverse-biased; therefore, the input resistance at the gate is very high. This very high input resistance is a major advantage of the JFET over the bipolar junction transistor with its forward-biased base-emitter junction.

JFET data sheets often specify the input resistance by giving a value for the gate reverse current, I_{GSS}, at a certain gate-to-source voltage. The input resistance can then be determined using the following equation. The vertical lines indicate an absolute value (an unsigned value).

$$R_{IN} = \left| \frac{V_{GS}}{I_{GSS}} \right| \quad (4\text{–}2)$$

For example, the 2N5457 data sheet lists a maximum I_{GSS} of -1 nA for $V_{GS} = -15$ V. Using these values, you find that the input resistance is

$$R_{IN} = \left| \frac{V_{GS}}{I_{GSS}} \right| = \frac{15 \text{ V}}{1 \text{ nA}} = 15 \text{ G}\Omega$$

As you can see from this result, the input resistance of this JFET is incredibly high. However, in a typical application, the total input resistance will include a resistor connected to the gate. The result is a total input resistance in the 1–10 MΩ range.

Review Questions

6. What is another name for the transfer curve for a JFET?
7. Does a p-channel JFET require a positive or a negative voltage for V_{GS}?
8. How is the drain current controlled in a JFET?
9. The drain-to-source voltage at the pinch-off point of a particular JFET is 7 V. If the gate-to-source voltage is zero, what is V_P?
10. The V_{GS} of a certain n-channel JFET is increased negatively. Does the drain current increase or decrease?

JFET BIASING 4–3

Using some of the JFET characteristics discussed in the previous section, we will now see how to dc bias JFETs. The purpose of biasing is to select a proper dc gate-to-source voltage to establish a desired value of drain current. Because the gate is reverse-biased, the methods for applying bias with a bipolar junction transistor do not work for JFETs.

In this section, you will learn how to bias a JFET using self-bias and voltage-divider bias.

Self-Biasing a JFET

Biasing a FET is relatively easy. An n-channel JFET is shown for the following examples. Keep in mind that a p-channel JFET just reverses the polarities. To set up reverse bias requires a negative V_{GS} for an n-channel JFET. This can be achieved using the self-bias arrangement shown in Figure 4–10. Notice that the gate is biased at 0 V by resistor R_G connected to ground. Although reverse leakage current, I_{GSS}, does produce a very tiny voltage across R_G, it is neglected in most cases; it can be assumed that R_G has no current and no voltage drop across it. The purpose of R_G is to tie the gate to a solid 0 V without affecting any ac signal that will be added later. Since the gate current is negligible, R_G can be large (typically 1.0 MΩ or more), resulting in very high input resistance to low frequency ac signals.

FIGURE 4–10

Self-biased n-channel JFET.

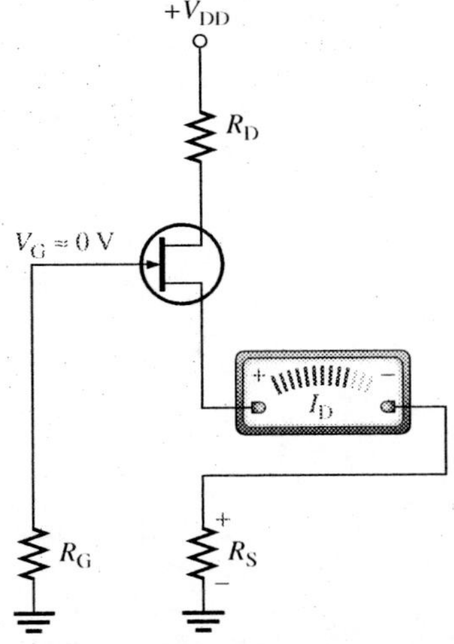

If the gate is at zero volts, how do you obtain the required negative bias on the gate-source junction? The answer is that you make the source *positive* with respect to the gate, producing the required reverse bias. For the n-channel JFET in Figure 4–10, I_D produces a voltage drop across R_S with the polarity shown, making the source terminal positive with respect to ground. Since $V_G = 0$ V, and $V_S = I_DR_S$, the gate-to-source voltage is

$$V_{GS} = V_G - V_S = 0 - I_DR_S$$

Thus,

$$V_{GS} = -I_DR_S$$

This result shows that the gate-to-source voltage is negative, producing the required reverse bias. In this analysis, an n-channel JFET was used for illustration. Again, the p-channel JFET also requires reverse bias, but the polarity of all voltages is opposite those of the n-channel JFET.

The drain voltage with respect to ground is determined as follows:

$$V_D = V_{DD} - I_DR_D \quad (4\text{–}3)$$

Since $V_S = I_D R_S$, the drain-to-source voltage is

$$V_{DS} = V_D - V_S$$

$$V_{DS} = V_{DD} - I_D(R_D + R_S) \tag{4–4}$$

EXAMPLE 4–3

Problem

Find V_{DS} and V_{GS} in Figure 4–11. For the particular JFET in this circuit, assume that the internal parameters are such that a drain current (I_D) of approximately 5.0 mA is produced. Another JFET, even of the same type, may not produce the same results when connected in this circuit due to variations in parameter values.

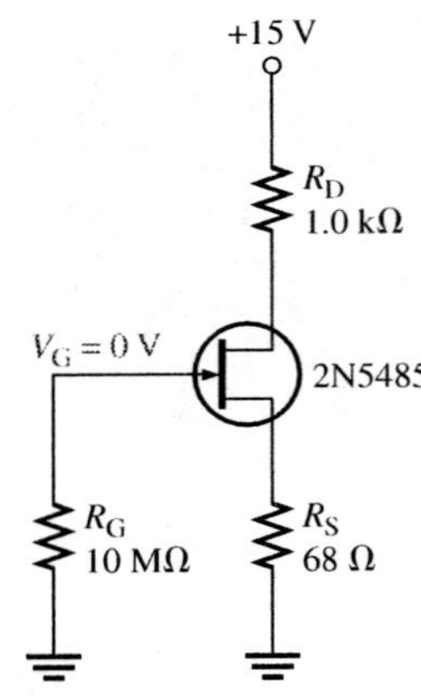

FIGURE 4–11

Solution

$$V_S = I_D R_S = (5.0 \text{ mA})(68\ \Omega) = 0.34 \text{ V}$$
$$V_D = V_{DD} - I_D R_D = 15 \text{ V} - (5.0 \text{ mA})(1.0 \text{ k}\Omega) = 10.0 \text{ V}$$

Therefore,

$$V_{DS} = V_D - V_S = 10.0 \text{ V} - 0.34 \text{ V} = \mathbf{9.66\ V}$$

and

$$V_{GS} = V_G - V_S = 0 \text{ V} - 0.34 \text{ V} = \mathbf{-0.34\ V}$$

Question

What is V_{DS} and V_{GS} in Figure 4–11 if I_D = 3.0 mA?

COMPUTER SIMULATION

Open the Multisim file F04-11DV on the website. Verify the gate-source voltage is negative.

Graphical Methods

Recall that the *IV* characteristic curve for a resistor, *R*, is a straight line with a slope of 1/*R*. To compare the plot of the self-bias resistor with the transconductance curve, both lines are plotted in the second quadrant; the resistance is plotted with a slope of −1/*R*.

The transconductance curve for a typical JFET can be used to illustrate how a reasonable value of a self-bias resistor (R_S) is selected. Assume the transconductance curve is as shown in Figure 4–12. Draw a straight line from the origin to the point where $V_{GS(off)}$ (-4 V) intersects I_{DSS} (2.5 mA). The reciprocal of the slope of this line represents a reasonable choice for R_S.

$$R_S = \frac{|V_{GS\,(off)}|}{I_{DSS}} = \frac{4\text{ V}}{2.5\text{ mA}} = 1.6\text{ k}\Omega$$

The absolute (unsigned) value of $V_{GS(off)}$ is used. The resulting 1.6 kΩ resistor is available as a standard 5% value, or you could select a 1.5 kΩ standard 10% resistor instead. The point where the two lines cross represents the Q-point for determining bias. This Q-point represents V_{GS} and I_D for this particular case; it shows that $V_{GS} = -1.5$ V at $I_D = 0.95$ mA.

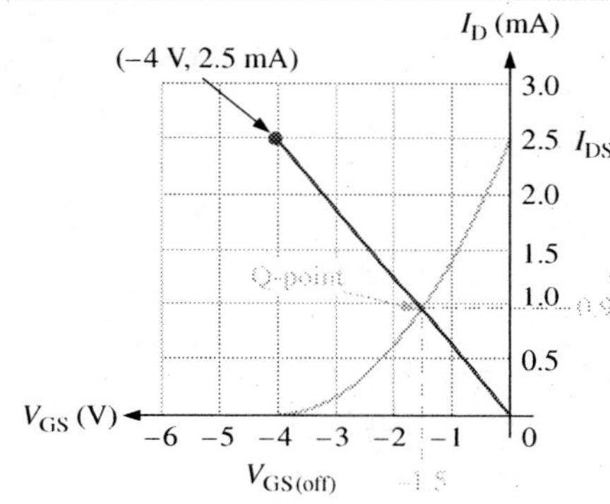

FIGURE 4–12

Graphical analysis of self-bias.

Self-bias helps compensate for different device characteristics between various JFETs. For instance, assume the transistor is replaced with one with a lower transconductance. As a result, the new drain current will be less, causing a smaller voltage drop across R_S. This reduced voltage tends to turn the JFET on more, compensating for the lower transconductance of the new transistor. The effect of a range of transconductance curves is best illustrated by an example.

EXAMPLE 4–4

Problem

A 2N5457 general-purpose JFET has the following specifications: $I_{DSS(min)} = 1$ mA, $I_{DSS(max)} = 5$ mA, $V_{GS(off)(min)} = -0.5$ V, $V_{GS(off)(max)} = -6$ V. Select a self-bias resistor for this JFET.

Solution

Typical of small-signal JFETs, the range of I_{DSS} and $V_{GS(off)}$ is very large. To select the best resistor, check the extremes of the specified values $V_{GS(off)}$ and I_{DSS}.

$$R_S = \frac{|V_{GS\,(off)(min)}|}{I_{DSS\,(min)}} = \frac{0.5\text{ V}}{1.0\text{ mA}} = 500\ \Omega$$

$$R_S = \frac{|V_{GS\,(off)(max)}|}{I_{DSS\,(max)}} = \frac{6\text{ V}}{5.0\text{ mA}} = 1.2\text{ k}\Omega$$

A good choice is **820 Ω,** a standard value between these extremes. To see what this looks like on the transconductance curve, sketch the curves with this resistor and plot the maximum and minimum Q-points. This is done in Figure 4–13. Despite the extreme variation between the minimum and maximum specification, the 820 Ω resistor represents a good choice for either.

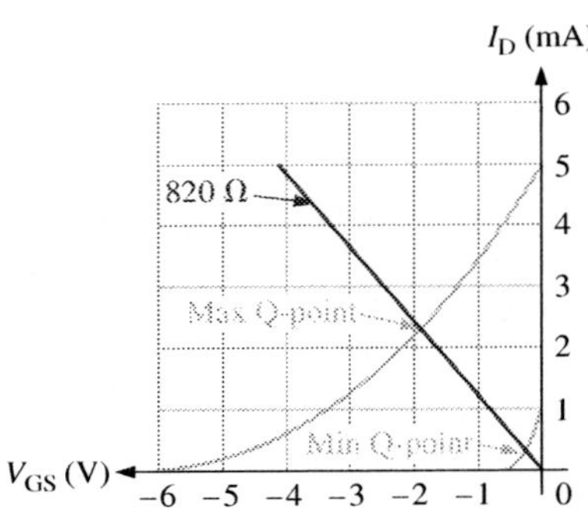

FIGURE 4–13
Effect of a wide range of transconductance curves on the Q-point.

Question
What is the largest and smallest I_D expected for a 2N5457 that is self-biased with an 820 Ω resistor?

Voltage-Divider Bias

Although self-bias is satisfactory for many applications, the operating point is dependent on the transconductance curve as you have seen. The bias can be made more stable with the addition of a voltage divider on the gate circuit, forcing the gate to a positive voltage. Since the JFET must still operate with a negative gate-source bias, a larger source resistor is used than in normal self-bias. The circuit is shown in Figure 4–14. Typical values are shown on the schematic.

The gate voltage is found by applying the voltage-divider rule to R_1 and R_2.

$$V_G = \left(\frac{R_2}{R_1 + R_2}\right)V_{DD} \tag{4–5}$$

To illustrate how the voltage-divider bias works, find the gate voltage from Equation 4–5.

$$V_G = \left(\frac{R_2}{R_1 + R_2}\right)V_{DD} = \left(\frac{2.2\text{ M}\Omega}{10\text{ M}\Omega + 2.2\text{ M}\Omega}\right)12\text{ V} = 2.2\text{ V}$$

The source resistor is generally larger than one used in basic self-bias because it must develop at least 2.2 V to produce the required negative bias between the gate and source. Because of the wide variation of FETs, it is not possible to predict the exact source voltage; about 3 V is expected on the source for the 2N5458.

Remember, if you are troubleshooting any JFET circuit, the source voltage has to be equal or larger than the gate voltage. The drain current is in both R_D and R_S. Since I_D is dependent on the transconductance of the JFET, the precise value of V_D and V_S cannot be determined from the circuit values alone because of the manufacturing spread of FETs. In general, a JFET linear amplifier should be designed such that V_{DS} is in the range from about

FIGURE 4–14
Voltage-divider bias and self-bias.

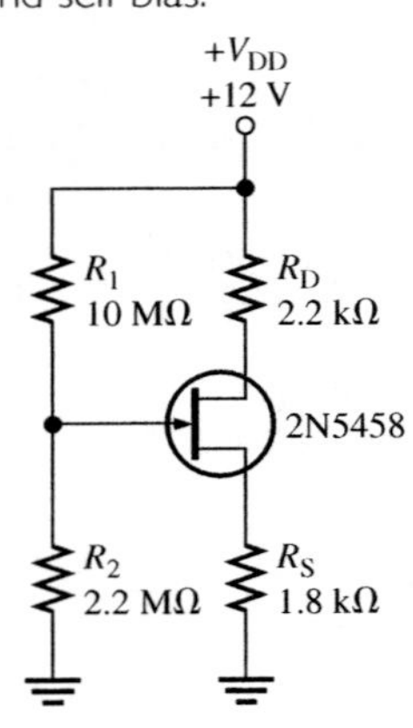

25% to 50% of V_{DD}. Even without knowing the parameters for the transistor, you can verify that the bias is set up correctly by checking V_{DS}.

Review Questions

11. What two parameters for a JFET could you use to choose a reasonable value of self-bias resistor?
12. What is the purpose of the gate resistor in Figure 4–10?
13. Why can't the bias circuits for BJTs be used for JFETs?
14. In a certain self-biased n-channel JFET circuit, $I_D = 8$ mA and $R_S = 1.0$ kΩ. What is V_{GS}?
15. Why is a larger source resistor required for voltage-divider bias than for self-bias in a JFET?

MOSFET CHARACTERISTICS 4–4

The metal-oxide semiconductor field-effect transistor (MOSFET) is the other major category of field-effect transistors. The MOSFET differs from the JFET in that it has no *pn* junction structure; instead, the gate of the MOSFET is insulated from the channel by a very thin silicon dioxide (SiO_2) layer. The two basic types of MOSFETs are depletion (D) and enhancement (E).

In this section, you will learn the characteristics of a MOSFET (metal-oxide semiconductor field-effect transistor).

Depletion MOSFET (D-MOSFET)

One type of **MOSFET** is the depletion MOSFET (D-MOSFET), and Figure 4–15 illustrates its basic structure. The drain and source are diffused into the substrate material and then connected by a narrow channel adjacent to the insulated gate. Although both n-channel and p-channel devices are shown in the figure, only n-channel devices are available from manufacturers. We have retained mention of the p-channel D-MOSFETs because they can be simulated in computer simulation programs like MultiSim. The discussion that follows relates to the n-channel D-MOSFET.

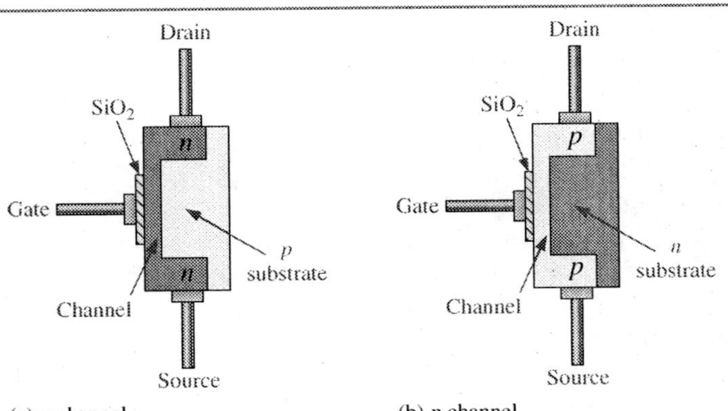

FIGURE 4–15

Structure of D-MOSFET. The SiO_2 layer is exaggerated for clarity.

The D-MOSFET can be operated in either of two modes—the depletion mode or the enhancement mode—and is sometimes called a *depletion-enhancement MOSFET.* Since the gate is insulated from the channel, either a positive or a negative gate voltage can be applied. The n-channel D-MOSFET operates in the **depletion mode** when a negative gate-to-source voltage is applied and in the **enhancement mode** when a positive gate-to-source voltage is applied. These devices are generally operated in the depletion mode.

Depletion Mode

Visualize the gate as one plate of a parallel-plate capacitor and the channel as the other plate. The silicon dioxide insulating layer is the dielectric. With a negative gate voltage, the negative charges on the gate repel conduction electrons from the channel, leaving positive ions in their place. Therefore, the n-channel is depleted of some of its electrons, thus decreasing the channel conductivity. The greater the negative voltage on the gate, the greater the depletion of n-channel electrons. At a sufficiently negative gate-to-source voltage, $V_{GS(off)}$, the channel is totally depleted and the drain current is zero. This depletion mode is illustrated in Figure 4–16(a). Like the n-channel JFET, the n-channel D-MOSFET conducts drain current for gate-to-source voltages between $V_{GS(off)}$ and 0 V. In addition, the D-MOSFET conducts for values of V_{GS} above 0 V.

FIGURE 4–16 Operation of n-channel D-MOSFETs.

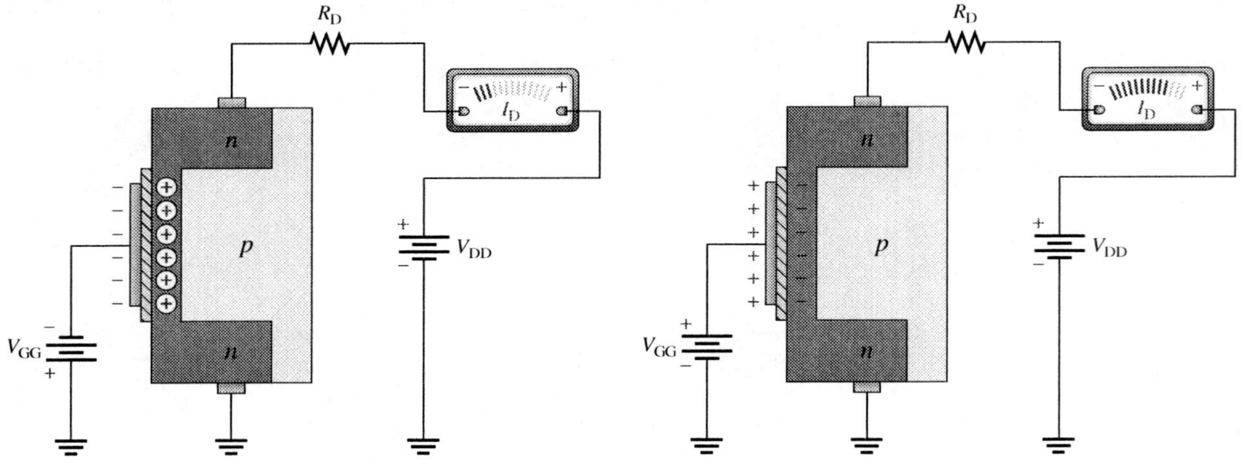

(a) Depletion mode: V_{GS} negative and less than $V_{GS(off)}$

(b) Enhancement mode: V_{GS} positive

Enhancement Mode

With an n-channel device and with a positive gate voltage, more conduction electrons are attracted into the channel, thus increasing (enhancing) the channel conductivity, as illustrated in Figure 4–16(b).

D-MOSFET Symbols

Figure 4–17 shows the schematic symbols for both the n-channel and the p-channel D-MOSFETs. The substrate, indicated by the arrow, is normally (but not always) connected internally to the source. Sometimes the substrate is brought out as another lead. An inward substrate arrow is for n-channel, and an outward arrow is for p-channel.

Because the MOSFET is a field-effect device like the JFET, you might expect it to have similar characteristics as the JFET. This is indeed the case. The transfer characteristic (I_D versus V_{GS}) for a typical n-channel D-MOSFET, is shown in Figure 4–18. It has the same shape as the one given earlier for the n-channel JFET (Figure 4–8(a)), but note

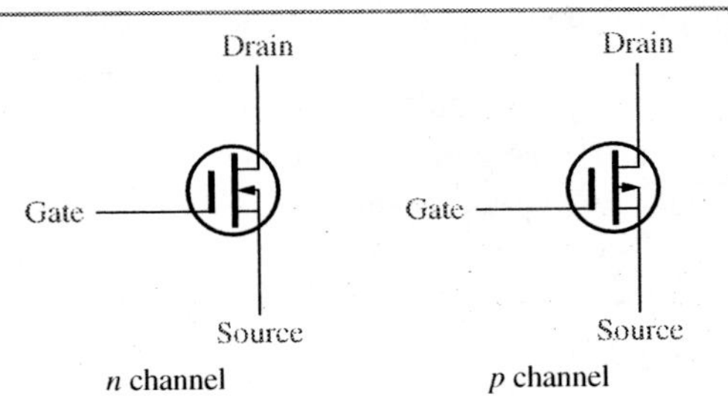

FIGURE 4–17
D-MOSFET schematic symbols.

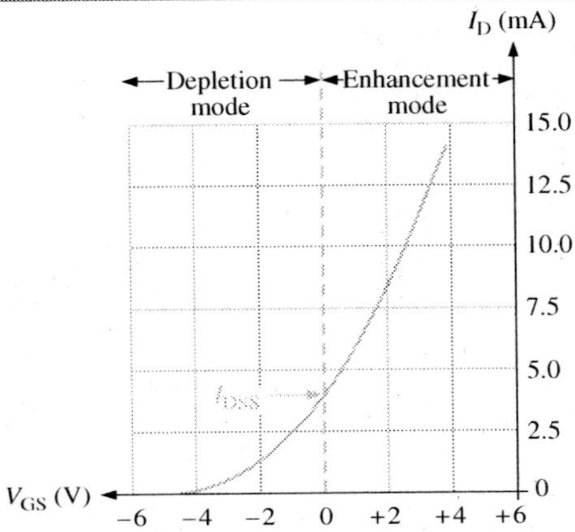

FIGURE 4–18
Transfer characteristic for a typical D-MOSFET.

that both negative and positive values of V_{GS} are shown on the transfer characteristic representing operation in the depletion region and the enhancement region respectively. This particular curve indicates the I_D is approximately 4.0 mA when V_{GS} is 0 V. Since V_{GS} is 0 V, the current is I_{DSS}. Notice that operation with currents higher than I_{DSS} is permissible with a D-MOSFET but not with a JFET.

Enhancement MOSFET (E-MOSFET)

This type of MOSFET operates only in the enhancement mode and has no depletion mode. It differs in construction from the D-MOSFET in that it has no physical channel. Notice in Figure 4–19(a) that the substrate extends completely to the SiO_2 layer.

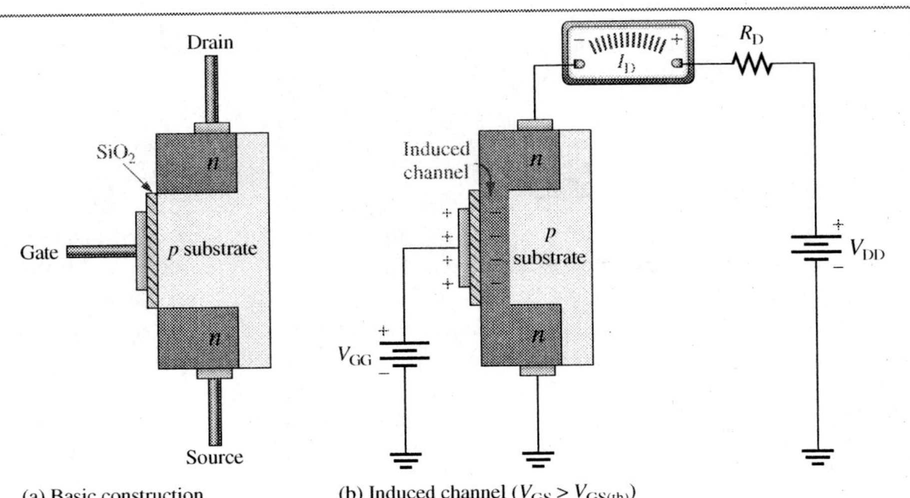

FIGURE 4–19
E-MOSFET construction and operation (n-channel).

For an n-channel device, a positive gate voltage above a threshold value, $V_{GS(th)}$, induces a channel by creating a thin layer of negative charges in the substrate region adjacent to the SiO_2 layer, as shown in Figure 4–19(b). The conductivity of the channel is enhanced by increasing the gate-to-source voltage, thus pulling more electrons into the channel. For any gate voltage below the threshold value, there is no channel.

The schematic symbols for the n-channel and p-channel E-MOSFETs are shown in Figure 4–20. The broken lines symbolize the absence of a physical channel.

FIGURE 4–20

E-MOSFET schematic symbols.

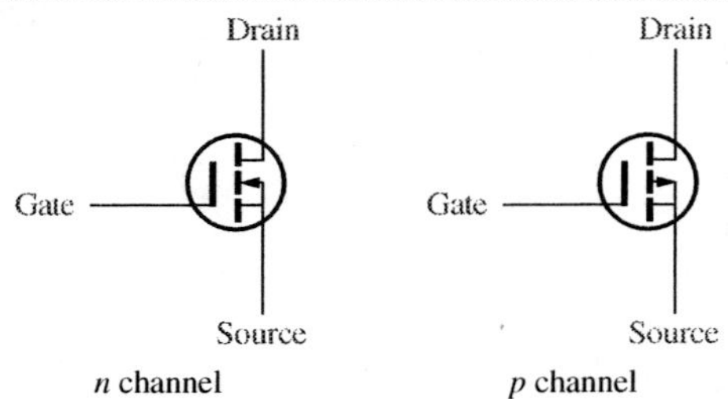

Because the channel is closed unless a voltage is applied to the gate, an E-MOSFET can be thought of as a normally off device. Again the transfer characteristic has the same shape as the JFET and D-MOSFET, but now the gate of an n-channel device must be made positive in order to cause conduction. This means that the $V_{GS(off)}$ specification will be a positive voltage for an n-channel E-MOSFET. A typical characteristic is shown in Figure 4–21. Compare it to the D-MOSFET characteristic in Figure 4–18.

FIGURE 4–21

Transfer characteristic for a typical E-MOSFET.

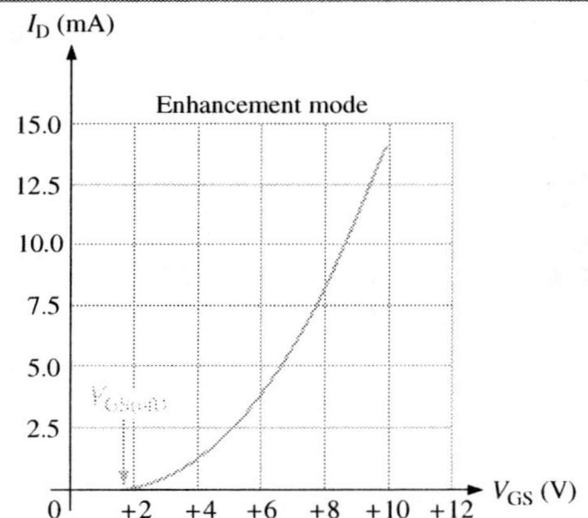

Handling Precautions

Because the gate of a MOSFET is insulated from the channel, the input resistance is extremely high (ideally infinite). The gate leakage current, I_{GSS}, for a typical MOSFET is in the pA range, whereas the gate reverse current for a typical JFET is in the nA range. The input insulated gate structure forms a capacitor with an extremely thin insulating layer that is in series with the input. This capacitance can accumulate a high voltage static charge that can break down the very high input resistance, thus destroying the device. This breakdown is called **electrostatic discharge** or ESD. ESD is the single largest cause of failure with MOSFET devices. To avoid ESD and possible damage to MOSFET devices, the following precautions should be taken:

1. Metal-oxide semiconductor (MOS) devices should be shipped and stored in conductive foam.
2. All instruments and metal benches used in assembly or testing should be connected to earth ground (round prong of wall outlets).
3. The assembler's or handler's wrist should be connected to earth ground with a length of wire and a high-value series resistor.
4. Never remove a MOS device (or any other device, for that matter) from the circuit while the power is on.
5. Do not apply signals to a MOS device while the dc power supply is off.

SAFETY NOTE

A grounding wrist strap for working with static sensitive components *must* have a high value series resistor to protect you. Without the resistor, you could receive a potential lethal shock due to high current if you contacted a voltage source.

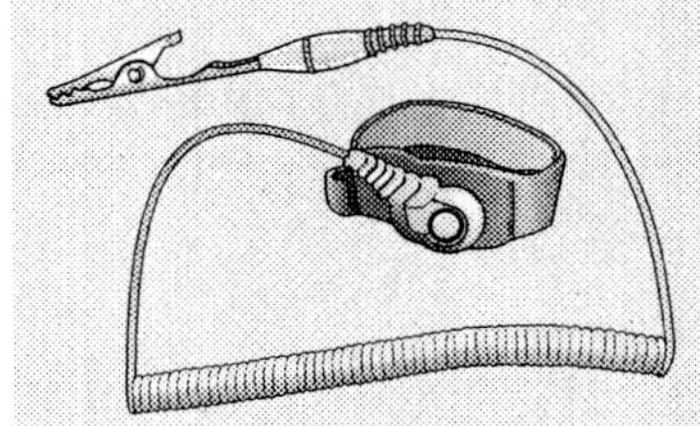

Review Questions

16. Name two types of MOSFETs, and describe the major difference in construction?
17. If the gate-to-source voltage in a D-MOSFET is zero, is there current from drain to source? ($V_D > V_P$)
18. If the gate-to-source voltage in an E-MOSFET is zero, is there current from drain to source? ($V_D > V_P$)
19. Can a D-MOSFET have a higher current than I_{DSS} and remain within the specified drain current?
20. What is ESD?

MOSFET BIASING 4–5

As with BJTs and JFETs, bias establishes the appropriate dc operating conditions that provide a stable operating point for centering the ac signal. MOSFET biasing circuits are similar to those you have already seen for BJTs and JFETs. The particular bias circuit depends on whether one or two supplies are used and the type of MOSFET (depletion or enhancement).

In this section, you will learn how to bias a D-MOSFET using zero bias, voltage-divider bias, or current-source bias and an E-MOSFET using voltage-divider bias.

D-MOSFET Bias

As you know, D-MOSFETs can be operated with either positive or negative values of V_{GS}. When V_{GS} is negative, operation is in the depletion mode; when it is positive, operation is in the enhancement mode. A D-MOSFET has the advantage of being able to operate in both modes; it is the only type of transistor that can do this.

Zero Bias

The most basic bias method is to set $V_{GS} = 0$ V so that an ac signal at the gate varies the gate-to-source voltage above and below this bias point. Figure 4–22 shows the circuit. Because it is effective and simple, it is the preferred method for biasing a D-MOSFET. The operating point is set between depletion and enhancement operation. Since $V_{GS} = 0$ V, $I_D = I_{DSS}$, as indicated. The drain-to-source voltage is expressed as follows:

$$V_{DS} = V_{DD} - I_{DSS}R_D$$

FIGURE 4–22

A zero-biased D-MOSFET.

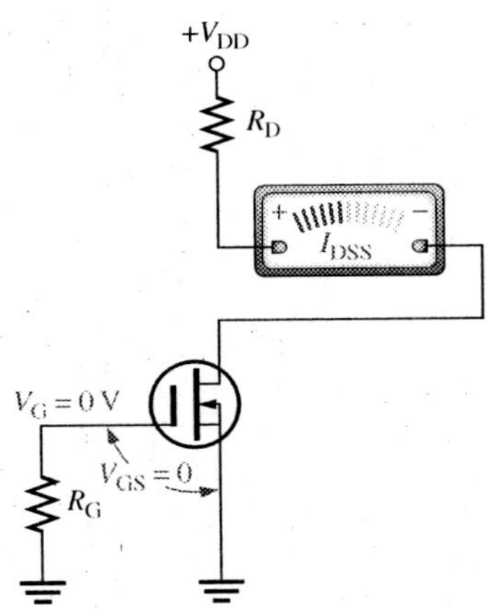

EXAMPLE 4–5

FIGURE 4–23

Problem

Determine the drain-to-source voltage in the circuit of Figure 4–23. The MOSFET data sheet gives $I_{DSS} = 12$ mA.

Solution

Since $I_D = I_{DSS} = 12$ mA, the drain-to-source voltage is

$$V_{DS} = V_{DD} - I_{DSS}R_D = 18\text{ V} - (12\text{ mA})(560\ \Omega) = \mathbf{11.28\ V}$$

Question

What is V_{DS} in Figure 4–23 when $I_{DSS} = 20$ mA?

Other Bias Arrangements

As you know, the D-MOSFET can operate in the depletion or enhancement mode. Because of this versatility, various bias circuits can be applied to D-MOSFETs. Figure 4–24 illustrates two popular methods for biasing, but you may see other methods in practice.

FIGURE 4–24

Other D-MOSFET bias circuits.

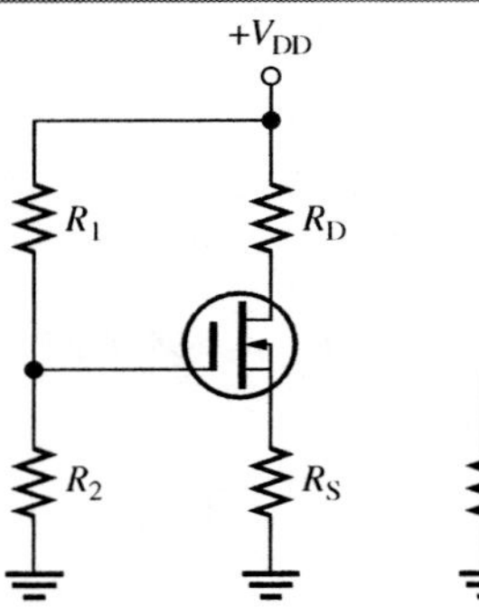

(a) Voltage-divider with self-bias

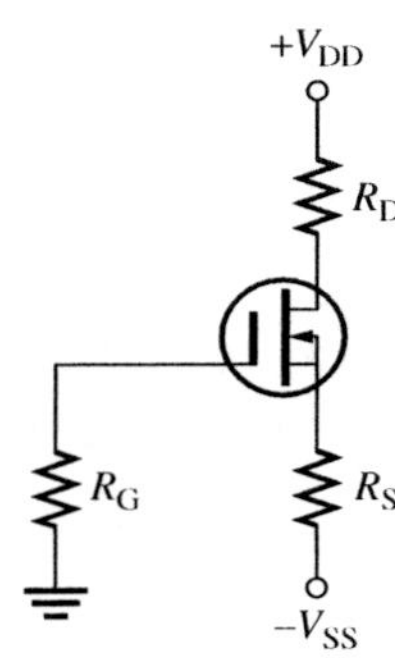

(b) Source bias

The bias circuit in Figure 4–24(a) uses a combination of voltage divider and self-bias as seen earlier with JFETs. The voltage at the gate is computed by the voltage-divider formula, which is quite accurate for any FET device because of the negligible loading effect. The gate voltage is the same as given for JFETs (see Equation 4–5).

$$V_G = \left(\frac{R_2}{R_1 + R_2}\right)V_{DD}$$

The resistors that form the voltage divider are usually quite large (in the megohm range) because of the high input resistance of the gate terminal. The voltages at the other terminals depend on specific device parameters.

When positive and negative supplies are used, the source-bias arrangement in Figure 4–24(b) is frequently used. This is similar to emitter bias seen with BJTs. Ideally, the gate circuit looks like an open circuit so you would expect the gate voltage to be at ground potential.

Another bias technique that is commonly used in operational amplifiers (an analog integrated circuit) is called current-source biasing. This method is not as common in discrete circuits because it requires an extra transistor to act as a current source. Figure 4–25 illus-

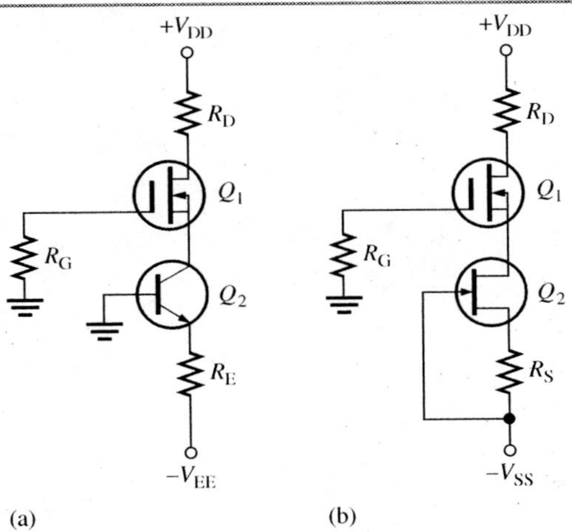

FIGURE 4–25

Current-source biasing for a D-MOSFET. Current-source biasing is common in integrated circuits.

trates the method. In Figure 4–25(a), a BJT acts as a current source for the D-MOSFET. The current is determined by applying Ohm's law to R_E. In Figure 4–25(b), a JFET is used as the current source to the MOSFET. In this case, the current is dependent on the I_{DSS} of the JFET and the value of the source resistor. Example 4–6 illustrates how to find the current for the BJT.

EXAMPLE 4–6

Problem

What is the drain current for the D-MOSFET in Figure 4–26?

Solution

The base of Q_2 is connected to ground; therefore, the emitter is -0.7 V. The voltage across R_E is

$$V_{R_E} = -0.7\text{ V} - (-15\text{ V}) = 14.3\text{ V}$$

The current in the emitter resistor is

$$I_{R_E} = \frac{14.3\text{ V}}{27\text{ k}\Omega} = 0.53\text{ mA}$$

The D-MOSFET acts as a load for the BJT. Consequently, the drain current is approximately equal to the emitter current.

$$I_D = I_{R_E} = \mathbf{0.53\text{ mA}}$$

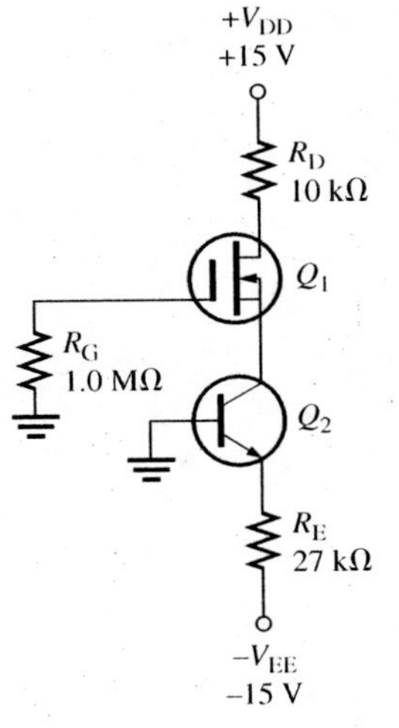

FIGURE 4–26

Question

What is the voltage across R_D?

E-MOSFET Bias

E-MOSFETs must have a V_{GS} greater than the threshold value, $V_{GS(th)}$. Figure 4–27 shows the most common way to bias an n-channel E-MOSFET using a voltage divider. (D-MOSFETs can also be biased this way.) With the voltage-divider bias arrangement, the gate voltage will be more positive than the source by an amount exceeding $V_{GS(th)}$.

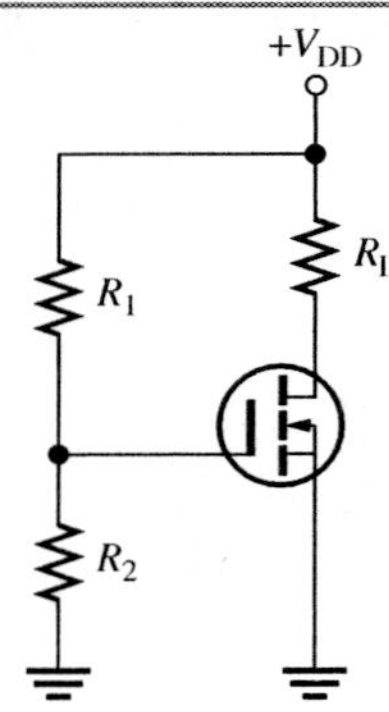

FIGURE 4–27

E-MOSFET biasing using a voltage divider.

Review Questions

21. For a D-MOSFET biased at $V_{GS} = 0$ V, is the drain current equal to 0, I_{GSS}, or I_{DSS}?
22. Why can't an E-MOSFET use zero bias?
23. What type of MOSFET can operate in either depletion mode or enhancement mode?
24. For an *n*-channel E-MOSFET with $V_{GS(th)} = 2$ V, V_{GS} must be in excess of what value in order to conduct?
25. What is current-source biasing?

4–6 FET LINEAR AMPLIFIERS

Field-effect transistors, both JFETs and MOSFETs, can be used as linear amplifiers in any of three circuit configurations similar to the bipolar junction transistor's CE, CC, and CB amplifiers you studied earlier. The FET configurations are common-source (CS), common-drain (CD), and common-gate (CG). The CS and CD amplifiers are excellent choices as the first stage of an amplifier. The CG amplifier has little advantage over other types so it is not considered here, but it may be found in specialized applications.

In this section, you will learn to analyze CS and CD amplifiers to determine basic operational parameters including gain and input resistance.

Transconductance of FETs

The transfer characteristic for a FET is the transconductance curve as was shown in Figure 4–8(a). FETs are fundamentally different than BJTs because they are voltage-controlled devices. *The output drain current is controlled by the input gate voltage.* As an ac parameter, transconductance was earlier defined as

$$g_m = \frac{I_d}{V_{gs}}$$

Considered in terms of output current (I_d) divided by input voltage (V_{gs}), transconductance is essentially the gain of the FET by itself. Unlike β_{ac}, a pure number, transconductance (g_m) has units of the siemens (the reciprocal of resistance). The transconductance of a particular FET can be measured directly as shown in Figure 4–28(a). Notice that the transconductance is the slope of the transfer curve and is *not* a constant, but it depends on the drain current.

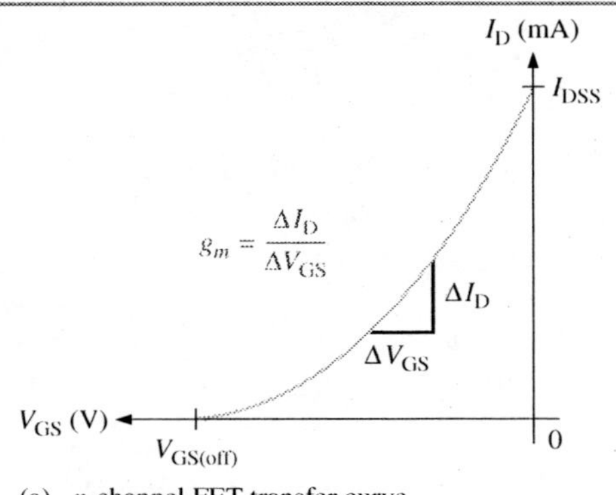

(a) n-channel FET transfer curve

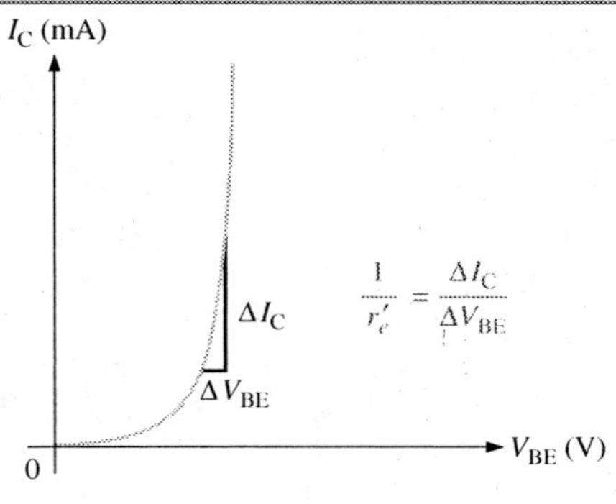

(b) BJT transfer curve

FIGURE 4–28

Comparison of the transfer curve for an n-channel FET with a BJT.

Figure 4–28(b) shows an analogous situation for the input to a BJT. The base voltage, applied across the base-emitter *pn* junction, "sees" an ac resistance that depends on the dc emitter current. This small ac resistance plays an important role in determining the gain of a BJT amplifier as you saw in Section 3–4.

The *reciprocal* of g_m is analogous to r'_e for BJTs. Most ac models for a FET use g_m as one of the key parameters; however, to make the transition from BJT amplifiers to FET amplifiers, it is useful to define a parameter representing the ac source resistance of the FET.

$$r'_s = \frac{1}{g_m} \tag{4–6}$$

The concept of r'_s leads to voltage gain equations that are analogous to those developed in Section 3–4 for BJTs. A mental picture of r'_s for a JFET is shown in Figure 4–29. The gate is shown with a dotted line to remind you that, from the gate's perspective, the input resistance is nearly infinite (because of the input's reverse-biased diode). Although the gate voltage controls the drain current, it does it with negligible current. Unfortunately, r'_s for a FET is not as predictable as r'_e is for a BJT and it is generally larger than r'_e, as illustrated in the plots in Figure 4–28. Data sheets don't show this parameter, but they do show a range of values for g_m (also shown as y_{fs}), so you can obtain an approximate value of r'_s by taking the reciprocal of the typical value of g_m. For example, if y_{gs} is shown as 2000 μS on a data sheet, $r'_s = 500\ \Omega$.

FIGURE 4–29

The internal source resistance r'_s is analogous to r'_e for a BJT. The dotted line is a reminder that gate current is negligible because of extremely high input resistance.

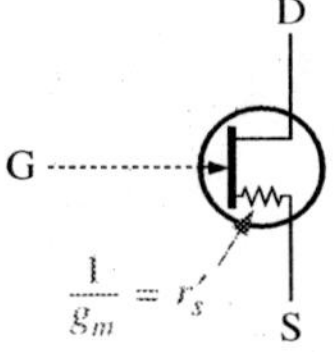

Common-Source Amplifiers

JFET

Figure 4–30 shows a **common-source (CS)** amplifier with a self-biased n-channel JFET. An ac source is capacitively coupled to the gate. The resistor, R_G, serves two purposes: (a) It

FIGURE 4–30

JFET common-source amplifier.

keeps the gate at approximately 0 V dc (Because I_{GSS} is extremely small), and (b) its large value (usually several megohms) prevents loading of the ac signal source. The bias voltage is created by the drop across R_S. The bypass capacitor, C_2, keeps the source of the FET effectively at ac ground.

The signal voltage causes the gate-to-source voltage to swing above and below its Q-point value, causing a swing in drain current. As the drain current increases, the voltage drop across R_D also increases, causing the drain voltage (with respect to ground) to decrease.

The drain current swings above and below its Q-point value in phase with the gate-to-source voltage. The drain-to-source voltage is 180° out of phase with the gate-to-source voltage, as illustrated.

D-MOSFET

Figure 4–31 shows a zero-biased n-channel D-MOSFET with an ac source capacitively coupled to the gate. The gate is at approximately 0 V dc and the source terminal is at ground, thus making $V_{GS} = 0$ V.

FIGURE 4–31

Zero-biased D-MOSFET common-source amplifier.

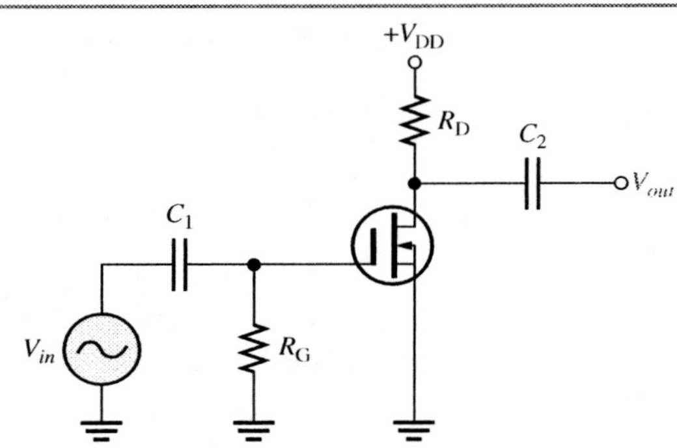

The signal voltage causes V_{gs} to swing above and below its 0 value, producing a swing in I_d. The negative swing in V_{gs} produces the depletion mode, and I_d decreases. The positive swing in V_{gs} produces the enhancement mode, and I_d increases.

E-MOSFET

Figure 4–32 shows a voltage-divider biased, n-channel E-MOSFET with an ac signal source capacitively coupled to the gate. The gate is biased with a positive voltage such that $V_{GS} > V_{GS(th)}$. As with the JFET and D-MOSFET, the signal voltage produces a swing in V_{gs} above and below its Q-point value. This swing, in turn, causes a swing in I_d. Operation is entirely in the enhancement mode.

FIGURE 4–32

Common-source E-MOSFET amplifier with voltage-divider bias.

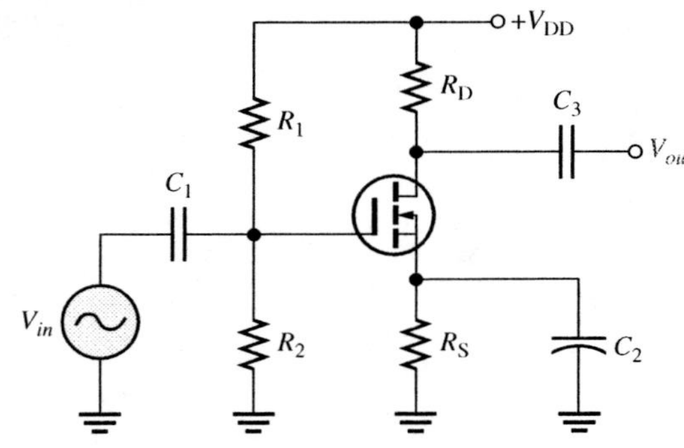

Voltage Gain

Voltage gain, A_v, of any amplifier is V_{out}/V_{in}. In the case of the CS amplifier, V_{in} is equal to V_{gs} (due to the bypass capacitor) and V_{out} is equal to the signal voltage developed across R_d, the ac drain resistance. In a CS amplifier with no load, the ac and dc drain resistances are equal: $R_d = R_D$. Thus, $V_{out} = I_d R_d$.

$$A_v = \frac{V_{out}}{V_{in}} = \frac{I_d R_d}{V_{gs}}$$

Since $g_m = I_d/V_{gs}$, the common-source voltage gain is

$$A_v = -g_m R_d \tag{4–7}$$

Equation 4–7 is the traditional voltage gain equation for the CS amplifier. The negative sign is added to Equation 4–7 to indicate that it is an inverting amplifier. The gain for the CS amplifier can be expressed in a similar form to the common-emitter (CE) amplifier as a ratio of ac resistances. By substituting $1/r'_s$ for g_m, the voltage gain can be written as

$$A_v = -\frac{R_d}{r'_s} \tag{4–8}$$

Compare this equation with Equation 3–11 that gives the voltage gain for a CE amplifier: $A_v = -R_c/R_e$. Both equations show voltage gain as a ratio of ac resistances.

Input Resistance

Because the input to a CS amplifier is at the gate, the input resistance to the transistor is extremely high. As you know, this extremely high resistance is produced by the reverse-biased *pn* junction in a JFET and by the insulated gate structure in a MOSFET. For practical work, the transistor's input circuit looks open.

When the transistor's internal resistance is ignored, the input resistance seen by the signal source is determined only by the bias resistor (or resistors). With self-bias, it is simply the gate resistor, R_G. With voltage-divider bias, the two voltage-divider resistors are seen by the ac source in parallel. The input resistance is the parallel combination of R_1 and R_2.

$$R_{in} \cong R_1 \| R_2$$

EXAMPLE 4–7

Problem

(a) What is the total output voltage (dc + ac) of the amplifier in Figure 4–33? The g_m is 2500 μS and I_D is 1.7 mA.

(b) What is the input resistance seen by the signal source?

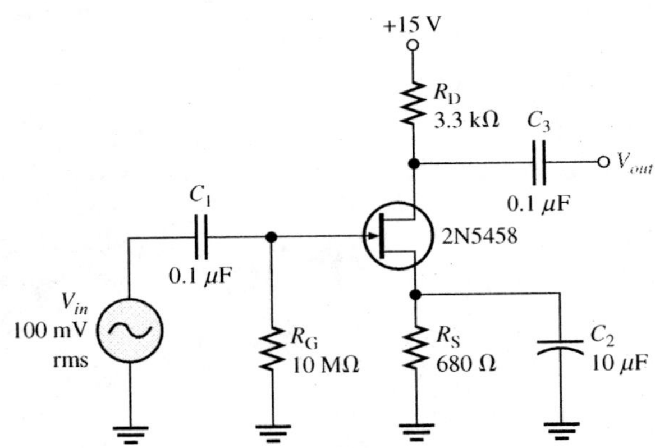

FIGURE 4–33

Solution

(a) First, find the dc output voltage.

$$V_D = V_{DD} - I_D R_D = 15\text{ V} - (1.7\text{ mA})(3.3\text{ k}\Omega) = 5.6\text{ V}$$

Next, find the voltage gain.

$$A_v = -g_m R_d = -(2500\ \mu\text{S})(3.3\text{ k}\Omega) = -8.25$$

Alternatively, the voltage gain could be found by computing r'_s and using the ratio of ac drain resistance to ac source resistance.

$$r'_s = \frac{1}{g_m} = \frac{1}{2500\ \mu\text{S}} = 400\ \Omega$$

$$A_v = -\frac{R_d}{r'_s} = -\frac{3.3\text{ k}\Omega}{400\ \Omega} = -8.25$$

The output voltage is the gain times the input voltage.

$$V_{out} = A_v V_{in} = (-8.25)(100\text{ mV}) = -825\text{ mV rms}$$

The negative sign indicates the output waveform is inverted.

The total output voltage is an ac signal with a peak-to-peak value of 0.825 V × 2.828 = **2.33 V,** riding on a dc level of **5.6 V.**

(b) The input resistance is

$$R_{in} \cong R_G = \mathbf{10\ M\Omega}$$

Question

What happens to the g_m if the source resistor is made larger? Does this affect the gain?

COMPUTER SIMULATION

Open the Multisim file F04-33DV on the website. Measure the ac and dc output voltage.

Common-Drain (CD) Amplifier

A **common-drain (CD)** JFET amplifier is shown in Figure 4–34 with voltages indicated. Self-biasing is used in this circuit. The input signal is applied to the gate through a coupling capacitor, and the output is connected through a capacitor to the load. There is no drain resistor. This circuit, of course, is analogous to the BJT emitter-follower and is sometimes called a *source-follower.* It is a widely used FET circuit because of its very high input resistance.

Voltage Gain

As in all amplifiers, the voltage gain is $A_v = V_{out}/V_{in}$. Like the emitter-follower, the source-follower has an ideal voltage gain of 1, but in practice it is less (typically between 0.5 and

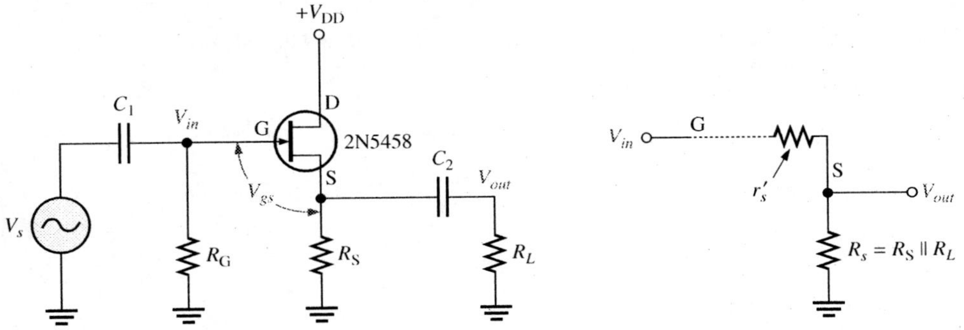

FIGURE 4–34

JFET common-drain amplifier (source-follower).

(a) A basic self-biased CD amplifier

(b) Simplified ac circuit to compute gain

1.0). To compute the voltage gain, the voltage-divider rule can be applied to the ac equivalent circuit shown in Figure 4–34(b). The gate resistor does not affect the ac, so it is not shown. The gate input is shown with a dotted line to remind you that it looks like an open to the ac input signal. The load and source resistors are in parallel and can be combined to form an equivalent ac source resistance, R_s, that is in series with the internal resistance r'_s (or $1/g_m$). The input is across both R_s and r'_s, but the output is taken across R_s only. Therefore, the output voltage is

$$V_{out} = V_{in}\left(\frac{R_s}{r'_s + R_s}\right)$$

Dividing by V_{in} results in the equation for voltage gain.

$$A_v = \frac{R_s}{r'_s + R_s} \quad (4\text{–}9)$$

Again, note that gain can be written as a ratio of ac resistances. This equation is easy to recall if you keep in mind that it is based on the voltage-divider rule.

An alternate voltage-gain equation is as follows:

$$A_v = \frac{g_m R_s}{1 + g_m R_s} \quad (4\text{–}10)$$

This formula yields the same result as Equation 4–9.

Input Resistance

Because the input signal is applied to the gate, the input resistance seen by the input signal source is the same as the CS-amplifier configuration discussed previously. For practical work, you can ignore the very high resistance of the transistor's input. The input resistance is determined by the bias resistor or resistors as done with the CS amplifier. For self-bias, the input resistance is equal to the gate resistor R_G.

$$R_{in} \cong R_G$$

With voltage-divider bias, the voltage-divider resistors are seen by the source as a parallel path to ground. Thus, for voltage-divider bias, the input resistance is

$$R_{in} \cong R_1 \parallel R_2$$

EXAMPLE 4–8

Problem

The p-channel JFET shown in Figure 4–35 has a transconductance between 2000 μS and 6000 μS. Determine the minimum and maximum voltage gain.

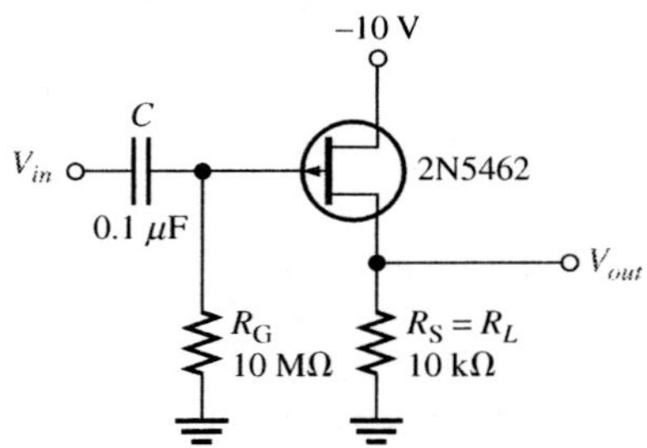

FIGURE 4–35

Solution

The maximum value of r'_s is

$$r'_s = \frac{1}{g_m} = \frac{1}{2000\ \mu\text{S}} = 500\ \Omega$$

The ac source resistance, R_s, is simply the load resistor, R_L, which is R_S. Substituting into Equation 4–9, the minimum voltage gain is

$$A_{v(\text{min})} = \frac{R_s}{r'_s + R_s} = \frac{10\ \text{k}\Omega}{500\ \Omega + 10\ \text{k}\Omega} = \mathbf{0.95}$$

The minimum value of r'_s is

$$r'_s = \frac{1}{g_m} = \frac{1}{6000\ \mu\text{S}} = 167\ \Omega$$

The maximum voltage gain is then

$$A_{v(\text{max})} = \frac{R_s}{r'_s + R_s} = \frac{10\ \text{k}\Omega}{167\ \Omega + 10\ \text{k}\Omega} = \mathbf{0.98}$$

Notice that the gain is slightly less than 1. When r'_s is small compared to the ac source resistance, then a good approximation is $A_v = 1$. Since the output voltage is at the source, it is in phase with the gate (input) voltage.

Question

What is the approximate input resistance seen by the source for the amplifier in Figure 4–35?

Review Questions

26. How is the gain computed for a CS amplifier?
27. How is the gain computed for a CD amplifier?
28. Which configuration (CS or CD) does not invert the input signal?
29. How do you find the input resistance for a voltage-divider-biased CS or CD amplifier?
30. Which characteristics of FETs make them excellent choices for the first stage of an amplifier?

FET SWITCHING CIRCUITS 4–7

There are two basic switching circuits that use FETs: analog switches and digital switches. Switching circuits are often found as the interface between analog and digital circuits. For most (but not all) switching applications, FETs are superior to BJTs because they require almost no drive current. Switching applications are frequently found in industry where the need for automatic control of high currents is common.

In this section, you will learn about two types of FET switching circuits.

Types of Solid-State Switches

There are two distinctly different switching applications for FETs. The first is the *analog switch* (sometimes called an ac switch) which takes advantage of the FET's low internal drain-source resistance when it is on and its high drain-source resistance when it is off to either pass or block an ac signal. Analog switches are connected in series with a signal; and when the switch is closed, the signal should be passed to the load without regard to the waveform polarity and with no significant change. Ideally, when the switch is open, the signal should be completely blocked and the switch should appear as an open circuit. While FETs cannot quite match ideal behavior, they offer the major advantage of very fast electronic control. Usually, analog switches are used in low-power low-voltage applications; for example, to turn on only one of several inputs to an analog-to-digital converter.

The second type of switching circuit is a *digital switch* (sometimes called a dc or logic switch) which turns on or turns off current to a device (such as a motor). This type of switch was introduced in Section 3–5 with BJTs. FET digital switches are frequently designed for power switching where current and voltages may be large. A control voltage causes the switch to appear open or closed. The control signal is frequently generated by a computer or a logic circuit that does not have sufficient drive capability by itself to perform the needed function. FET digital switches are controlled by a voltage applied to the gate, allowing drain current to turn on or turn off the device.

The Analog Switch

JFET Analog Switch

An n-channel JFET analog switch (transmission gate) is shown in Figure 4–36(a). The equivalent circuit is shown in Figure 4–36(b). The JFET is switched between the on and off state with a control voltage applied to the gate. To turn the JFET on, V_{GS} is made equal to zero volts. This can create a potential problem because the input (source) voltage varies and could cause the gate-source pn junction to be forward-biased. The solution is

FIGURE 4–36

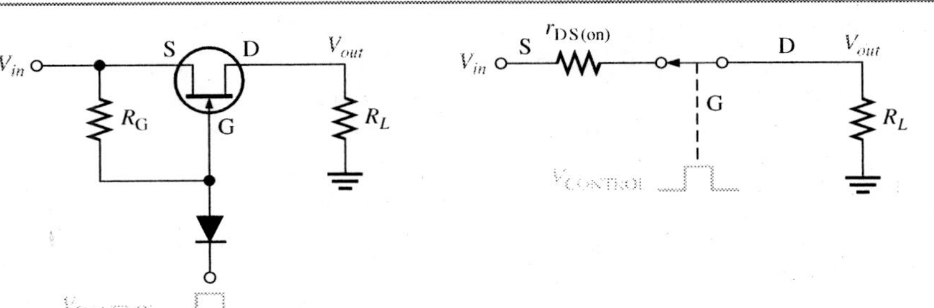

(a) A JFET analog switch.

(b) Simplified equivalent circuit for a JFET analog switch.

to connect a resistor between the source and gate terminals and add a diode in the gate circuit to prevent V_{GS} from becoming positive. To turn the JFET off, the channel must be pinched off. This is done by making the control voltage more negative than the lowest value of the signal plus the pinch-off voltage, V_P. (Recall that the magnitude of V_P is the same as $V_{GS(off)}$.)

JFETs are particularly well suited as analog switches. Because the gate draws essentially no current, they have an extremely high off resistance which provides a high degree of isolation from other signals. When the control signal is applied, the channel resistance ($r_{DS(on)}$) between the drain and source is relatively small but constant. As long as the load is much larger than $r_{DS(on)}$, the output voltage is approximately the same as the input voltage.

The on-state channel resistance is a function of the $V_{GS(off)}$ and I_{DSS} parameters, which can be found on manufacturer's specification sheets. The equation for $r_{DS(on)}$ is

$$r_{DS(on)} = -\frac{V_{GS(off)}}{2I_{DSS}} \tag{4–11}$$

EXAMPLE 4–9

Problem

Compute $r_{DS(on)}$ for a JFET with the transconductance curve shown in Figure 4–37.

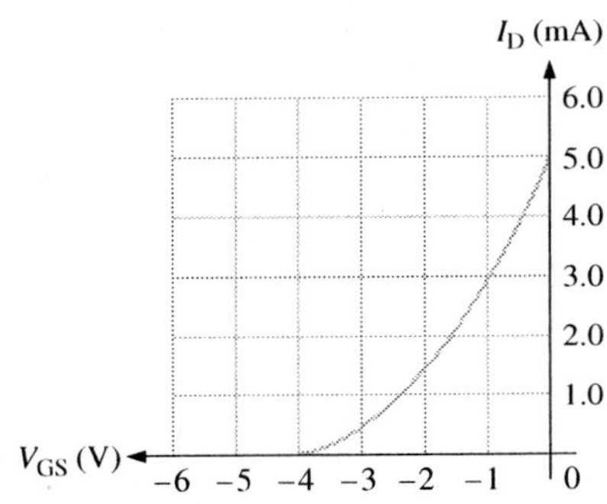

FIGURE 4–37

Solution

Reading from the graph, $V_{GS(off)} = -4$ V and $I_{DSS} = 5.0$ mA. Substituting into the equation for $r_{DS(on)}$ gives

$$r_{DS(on)} = -\frac{V_{GS(off)}}{2I_{DSS}} = -\frac{-4\text{ V}}{2(5.0\text{ mA})} = \mathbf{400\ \Omega}$$

Question

What fraction of the input voltage appears at the output if the JFET switch in this example is connected to a 10 kΩ load resistor?

JFET switches are available in IC form. They are fast and switching is accomplished from control voltages that can be obtained from standard logic families. A disadvantage of JFET switches is that they are also more prone to switching transients appearing on the signal lines when control signals change states. All in all, the advantages far outweigh the disadvantages and JFET switches are widely used in switching applications for instrumentation systems.

MOSFET Analog Switch

MOSFETs are also widely used for analog switching applications. They offer simpler circuits than JFETs and both positive and negative control. A basic MOSFET switch, using a *p*-channel E-MOSFET, is shown in Figure 4–38. A disadvantage to MOSFET switches is that the on-state resistance tends to be higher than that of JFETs. In the special case of switching a high current analog signal, power MOSFETs are available that can switch as high as 10 A with a control circuit that supplies essentially no current to the gate. There are no high current JFETs available for this application.

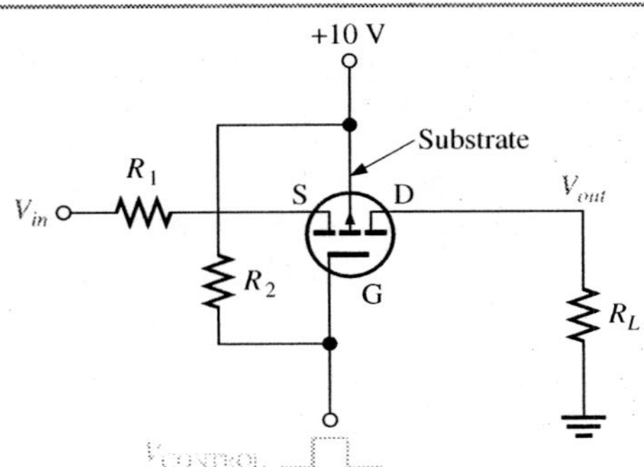

FIGURE 4–38

A *p*-channel E-MOSFET analog switch.

Solid-State Relays

A related analog switch is called a *solid-state relay (SSR).* SSRs are packaged with an optical isolator on the input circuit and power MOSFET output. In the past, mechanical relays have been necessary for applications such as low-level multiplexing of analog signals. However, because of extremely high off-state resistance, SSRs are now commonly used in these applications.

The Digital Switch

Discrete MOSFET Switches

Ideally, any digital switch is either open or closed. In practice, the operation of a FET switch can be described by a load line drawn on the transistor's characteristic curve. A circuit example, shown in Figure 4–39(a) uses a switching transistor, designed for small-power applications. The load line for the circuit is shown in Figure 4–39(b) superimposed on a typical characteristic curve. The drain resistor represents a load (such as a small light bulb) that requires approximately 200 mA of current at 12 V. The power supply is set higher than +12 V to allow for the drop across the transistor. Notice that when the transistor is on, there is a small drop (about 1 V) across it as indicated by $V_{DS(on)}$. The series resistor in the gate circuit

FIGURE 4–39

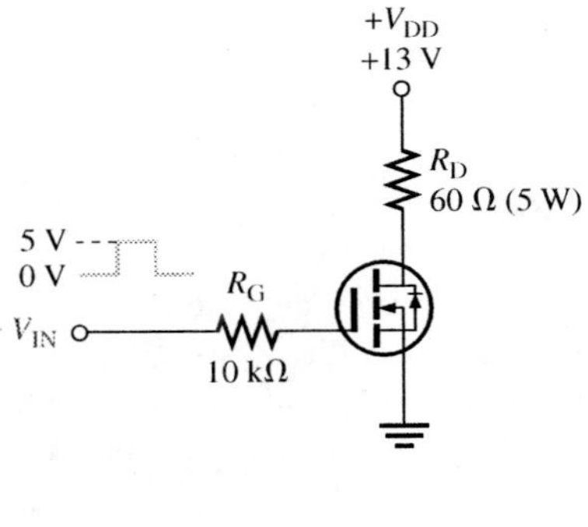

(a) E-MOSFET circuit

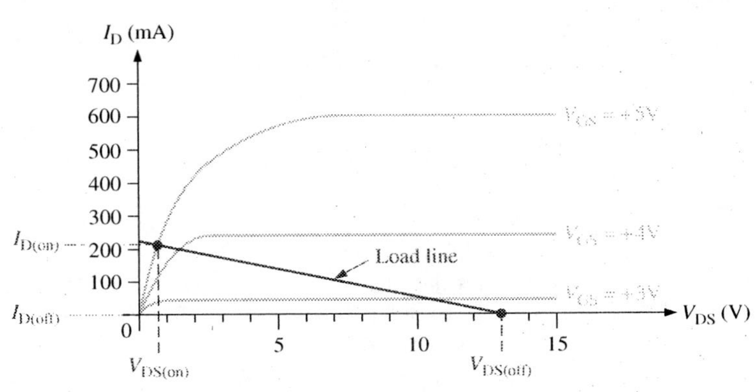

(b) Operation

is optional; it is often added by designers to protect the MOSFET. In addition, the schematic symbol shows a reverse-biased diode between the drain and source; this is because the body of the transistor is connected internally to the source, forming a diode.

As with any device, a MOSFET switch must be operated within certain limits specified by the manufacturer. In the off state, the MOSFET must be able to withstand the power supply voltage which will appear between the drain and source terminals. On a data sheet, this maximum voltage is listed as $V_{(BR)DSX}$ which is the maximum drain-source voltage with the gate shorted to the source.

MOSFET switches have several advantages over BJT switches in power applications. One advantage is that they are easier to drive because the input is a voltage (rather than a current in the case of the BJT). For small-power applications, this difference isn't too important, but when the load current exceeds a few amps, the BJT requires a hefty drive current compared to the MOSFET's simple voltage control. Another advantage is that MOSFETs in general have immunity to an effect called *thermal runaway.* Thermal runaway can happen to BJTs. As they get hot, they tend to conduct more current, which causes them to get hotter yet. With a power MOSFET, when the current reaches some level, it starts to conduct less, helping avoid the problem of thermal runaway.

IC Switching Circuits

FET switching circuits are widely used in ICs. Let's look at how MOSFETs are used in logic circuits for switching. IC logic circuits are a form of digital switch, in which the output is at one extreme of the load line. One of the most popular types of logic, called CMOS (for *C*omplementary *M*etal-*O*xide *S*emiconductors) uses both *p*-type and *n*-type E-MOSFETs. CMOS is used in computer memories and in many other IC logic circuits. The two types of transistors are typically connected in an arrangement called a *half-bridge* similar to that shown in Figure 4–40(a), a basic CMOS inverter. The function of the inverter is to cause the output to be either at a high or low voltage (called a *logic level*) which is just opposite from the input.

FIGURE 4–40 CMOS inverter circuit.

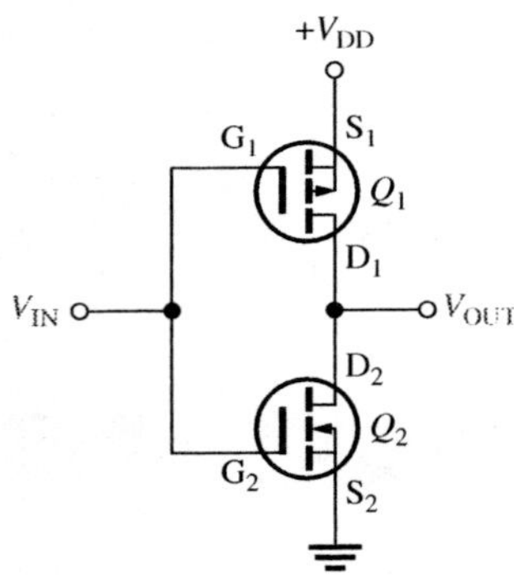

(a) Circuit

+V_{DD}

$V_{IN} = 0$ V

$V_{OUT} = V_{DD}$

(b) Switch analogy with input = 0 V

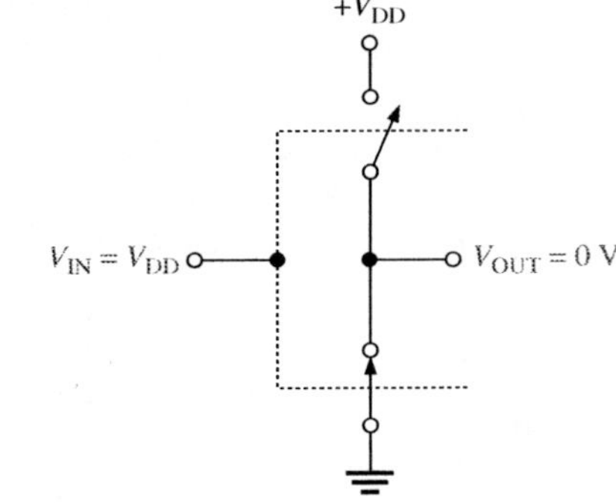

(c) Switch analogy with input = V_{DD}

Except during the very brief switching time, the two transistors do not conduct at the same time. The positive power supply, marked V_{DD}, is connected to Q_1's source (a *p*-channel device), and ground is connected to Q_2's source (an *n*-channel device). The positive supply is usually marked as a "drain" voltage, V_{DD}, (as if an *n*-type device were used). The gates of both transistors are connected together at the input, and the drains of both transistors are connected together at the output. Remember that a positive gate-to-source voltage turns on an *n*-channel E-MOSFET but turns off a *p*-channel E-MOSFET.

If the input voltage is near ground, Q_2 is off (it requires a positive gate voltage to turn on) and Q_1 is on (since the gate is negative with respect to the source). This causes the output to be nearly equal to the power-supply voltage, V_{DD}. A switch analogy of this condition is shown in Figure 4–40(b). If the input voltage is increased past the threshold voltage for Q_2, it conducts; and the increasing input voltage causes Q_1 to reduce conduction. This rapid switching action occurs at about one-half the supply voltage. At this point, the lower transistor appears as a short and the upper transistor appears as an open, as shown in the switch analogy in Figure 4–40(c). As a result, the output approaches 0 V.

Review Questions

31. What is the difference between an analog and a digital switch?
32. What are the attributes of an ideal analog switch?
33. How do you find the on-state resistance for a JFET analog switch?
34. What advantage do MOSFETs have over BJTs for digital switches?
35. How does a CMOS inverter work?

CHAPTER REVIEW

Key Terms

Common-drain (CD) A FET amplifier configuration in which the drain is the ac ground terminal.

Common-source (CS) A FET amplifier configuration in which the source is the ac ground terminal.

Constant-current region The region on the drain characteristic of a FET in which the drain current is independent of the drain-to-source voltage.

Depletion mode A class of FETs that is on with zero-gate voltage and is turned off by increasing gate voltage, which has a polarity such that it decreases channel conductivity. All JFETs and some MOSFETs are depletion-mode devices.

Drain One of the three terminals of a field-effect transistor; it is one end of the channel.

Enhancement mode A MOSFET in which the channel is formed (or enhanced) by the application of a gate voltage, thus increasing channel conductivity.

Field-effect transistor (FET) A voltage-controlled device in which the voltage at the gate terminal controls the amount of current through the device.

Gate One of the three terminals of a field-effect transistor. A voltage applied to the gate controls drain current.

Junction field-effect transistor (JFET) A type of FET that operates with a reverse-biased *pn* junction to control current in a channel. It is a depletion-mode device.

MOSFET Metal-oxide semiconductor field-effect transistor; one of two major types of FET. It uses a SiO_2 layer to insulate the gate lead from the channel. MOSFETs can be either depletion mode or enhancement mode.

Ohmic region The region on the drain characteristic of a FET with low values of V_{DS} in which the channel resistance can be changed by the gate voltage; in this region the FET can be operated as a voltage-controlled resistor.

Pinch-off voltage The value of the drain-to-source voltage of a FET at which the drain current becomes constant when the gate-to-source voltage is zero.

Source One of the three terminals of a field-effect transistor; it is one end of the channel.

Transconductance The gain of a FET; it is determined by a small change in drain current divided by a corresponding change in gate-to-source voltage. It is measured in siemens.

Important Facts

- ❑ FETs can be broadly classified into JFETs and MOSFETs. JFETs have a reverse-biased gate-source *pn* junction at the input; MOSFETs have an insulated-gate input.
- ❑ MOSFETs are classified as either depletion mode or enhancement mode. The D-MOSFET has a physical channel between the drain and the source; the E-MOSFET does not.
- ❑ All FETs are either *n*-channel or *p*-channel.
- ❑ The three terminals on a FET are the source, drain, and gate that correspond to the emitter, collector, and base of a BJT.
- ❑ JFETs have very high input resistance due to the reverse-biased gate-source *pn* junction. MOSFETs have a very high input resistance due to the insulated gate input.
- ❑ JFETs are normally *on* devices. Drain current is controlled by the amount of reverse bias on the gate-source *pn* junction.
- ❑ D-MOSFETs are normally *on* devices. Drain current is controlled by the amount of bias on the gate-source *pn* junction. A D-MOSFET can have either forward bias or reverse bias on the gate-source *pn* junction.
- ❑ E-MOSFETs are normally *off* devices. Drain current is controlled by the amount of forward bias on the gate-source *pn* junction.
- ❑ The drain characteristic curve for FETs is divided between an ohmic region and a constant-current region for normal operation.
- ❑ The transconductance curve is a plot of drain current versus gate-source voltage.
- ❑ MOSFET devices need special handling procedures to avoid destructive static electricity.
- ❑ JFETs can be biased by self-bias or a combination of self-bias and voltage-divider bias.
- ❑ A D-MOSFET can operate with a positive, negative, or zero gate-to-source voltage so it can be biased by zero bias, voltage divider with self-bias, or source bias.
- ❑ The most common method for biasing an E-MOSFET is voltage-divider bias.
- ❑ A common-source (CS) amplifier has high voltage gain and high input resistance.
- ❑ A common-drain (CD) amplifier has unity (or less) voltage gain and high input resistance.
- ❑ The voltage gain of CS and CD amplifiers can be computed by a ratio of ac resistances (including internal resistance).
- ❑ Analog switches pass or block a signal.
- ❑ Digital switches turn on or off a device.
- ❑ Digital switches are designed to operate in either saturation or cutoff.
- ❑ MOSFETs have important advantages as digital switches, particularly for high-current applications.

Formulas

Transconductance of a FET:

$$g_m = \frac{I_d}{V_{gs}} \tag{4–1}$$

Input resistance. It is the gate-source voltage divided by the gate-reverse current.

$$R_{\text{IN}} = \left|\frac{V_{\text{GS}}}{I_{\text{GSS}}}\right| \tag{4–2}$$

DC drain voltage for a FET:

$$V_{\text{D}} = V_{\text{DD}} - I_{\text{D}}R_{\text{D}} \tag{4–3}$$

DC drain-to-source voltage for a FET:

$$V_{\text{DS}} = V_{\text{DD}} - I_{\text{D}}(R_{\text{D}} + R_{\text{S}}) \tag{4–4}$$

Gate voltage with voltage-divider bias:

$$V_{\text{G}} = \left(\frac{R_2}{R_1 + R_2}\right)V_{\text{DD}} \tag{4–5}$$

Equivalent internal ac source resistance for computing voltage gain:

$$r_s' = \frac{1}{g_m} \tag{4–6}$$

Voltage gain for a CS amplifier:

$$A_v = -g_m R_d \tag{4–7}$$

Alternate voltage gain for a CS amplifier:

$$A_v = -\frac{R_d}{r_s'} \tag{4–8}$$

Voltage gain for a CD amplifier:

$$A_v = \frac{R_s}{r_s' + R_s} \tag{4–9}$$

Alternate voltage gain for a CD amplifier:

$$A_v = \frac{g_m R_s}{1 + g_m R_s} \tag{4–10}$$

Channel resistance:

$$r_{\text{DS(on)}} = -\frac{V_{\text{GS(off)}}}{2I_{\text{DSS}}} \tag{4–11}$$

Chapter Checkup

Answers are at the end of the chapter.

1. A type of transistor that is normally on when the gate-to-source voltage is zero is
   (a) JFET (b) D-MOSFET
   (c) E-MOSFET (d) answers (a) and (b)
   (e) answers (a) and (c)

2. A bias method that can be used for D-MOSFETs is
   (a) voltage divider (b) current source
   (c) self (d) all of these answers
3. In normal operation, the gate-source *pn* junction for a JFET is
   (a) reverse-biased (b) forward-biased
   (c) either (a) or (b) (d) neither (a) nor (b)
4. When the voltage between the gate and source of a JFET is zero, the drain current will be
   (a) zero (b) I_{DSS}
   (c) I_{GSS} (d) none of these answers
5. One reason an *n*-channel D-MOSFET can have zero bias is it
   (a) can operate in either depletion mode or enhancement mode
   (b) does not have an insulated gate
   (c) does not have a channel
   (d) will not have drain current when operated with zero bias
6. A feature of FETs that is superior to BJTs is their
   (a) high gain (b) low distortion
   (c) high input resistance (d) all of these answers
7. An amplifier with high voltage gain and high input resistance is a common-
   (a) drain (b) source
   (c) answers (a) and (b) (e) neither (a) nor (b)
8. An amplifier that inverts the signal between input and output is a common-
   (a) drain (b) source
   (c) answers (a) and (b) (e) neither (a) nor (b)
9. A transistor that has a closed channel unless a voltage is applied to the gate is
   (a) a JFET (b) a D-MOSFET
   (c) an E-MOSFET (d) all of these answers
   (e) none of these answers
10. The value of the drain-to-source voltage of a FET at which the drain current becomes constant when the gate-to-source voltage is zero is called the
    (a) bias voltage (b) pinch-off voltage
    (c) saturation voltage (d) cutoff voltage
11. The voltage gain of a common-drain amplifier cannot exceed
    (a) 1.0 (b) 2.0
    (c) 10 (d) 100
12. A type of electronic switching circuit which can be used to connect a given signal to the input of an analog-to-digital converter (ADC) is a(n)
    (a) analog switch (b) digital switch
    (c) logic switch (d) bipolar switch
13. Refer to Figure 4–41. The schematic symbol for a *p*-channel E-MOSFET is
    (a) a (b) b
    (c) c (d) d
    (e) e (f) f

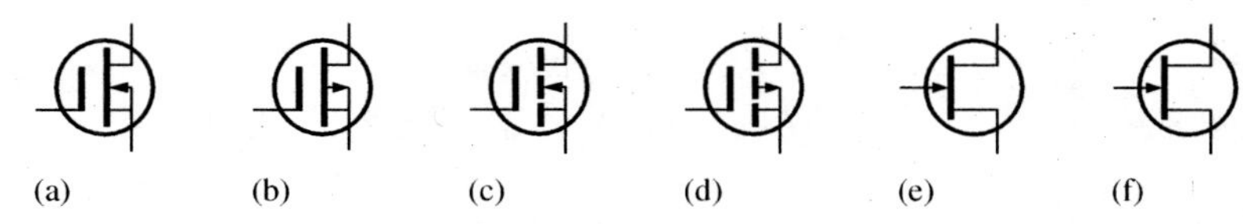

FIGURE 4–41

14. Refer to Figure 4–41. The schematic symbol for a *n*-channel D-MOSFET is
(a) a (b) b
(c) c (d) d
(e) e (f) f

15. The type of device used in a CMOS switching circuit is
(a) *n*-channel D-MOSFET
(b) *p*-channel D-MOSFET
(c) both (a) and (b)
(d) neither (a) nor (b)

16. If R_G is 1.0 MΩ instead of 10 MΩ in Figure 4–42, the gate voltage will
(a) increase (b) decrease
(c) not change

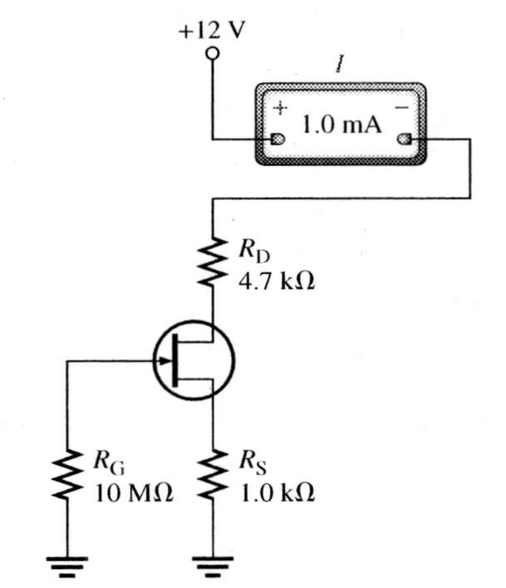

FIGURE 4–42

17. For the conditions described in Question 16, the drain current will
(a) increase (b) decrease
(c) not change

18. For the conditions described in Question 16, the input resistance will
(a) increase (b) decrease
(c) not change

Questions

Answers to odd-numbered questions are at the end of the book.

1. What is the basic difference between a JFET and a MOSFET?
2. What type of transistor conducts with a reverse-biased *pn* junction at the input?

3. What type of transistor has an insulated gate?
4. What accounts for the extremely high input resistance of JFETs and MOSFETs?
5. What internal structure are the source and drain connected to on a JFET?
6. How does a voltage applied to the gate control current in the drain circuit of a JFET?
7. What type of JFET is drawn with the arrow pointing out on the symbol?
8. What are the three regions on a JFET characteristic curve?
9. Why is a JFET considered to be a "normally-on" device?
10. Why must the gate-to-source voltage of a *p*-channel JFET always be zero or positive?
11. What is meant by I_{DSS}?
12. What is meant by I_{GSS}?
13. What is meant by $V_{GS(off)}$?
14. How does a small resistor between source and ground create the necessary bias condition for a JFET?
15. With self-bias, what is the purpose of a large gate resistor?
16. Why does dividing an average value of $|V_{GS(off)}|$ by I_{DSS} give a reasonable value for a self-bias resistor?
17. Why is it necessary to have a source resistor if voltage-divider bias is used with a JFET?
18. Why isn't it necessary to have a source resistor if voltage-divider bias is used with an E-MOSFET?
19. How does the symbol for a D-MOSFET differ from the symbol for an E-MOSFET?
20. What type of MOSFET will conduct if the gate voltage is zero?
21. What type of MOSFET can work in either depletion or enhancement mode?
22. Why is it essential that a wrist strap used for ESD protection have a high-value series resistor?
23. What parameter is obtained by dividing a change in the drain current by a change in the gate voltage for a FET?
24. How do you find the gain of a CS amplifier?
25. What is another name for a CD amplifier?
26. What is an analog switch?
27. Why is $r_{DS(on)}$ an important characteristic in a FET analog switch?
28. If a MOSFET switch is *off,* what voltage appears between the source and drain?
29. If a MOSFET switch is *on,* what voltage appears between the source and drain?
30. What are the two operating conditions for a MOSFET digital switch?

PROBLEMS

Basic Problems

Answers to odd-numbered problems are at the end of the book.

1. The V_{GS} of a *p*-channel JFET is increased from +1 V to +3 V.
   (a) Does the depletion region narrow or widen?
   (b) Does the resistance of the channel increase or decrease?
   (c) Does the transistor conduct more current or less?
2. A JFET has a specified pinch-off voltage of −5 V. When $V_{GS} = 0$, what is V_{DS} at the point where I_D becomes constant?

3. An n-channel JFET is biased such that $V_{GS} = -2$ V using self-bias. The gate resistor is connected to ground.

   (a) What is V_S?

   (b) What is the value of $V_{GS(off)}$ if V_P is specified to be 6 V?

4. A certain p-channel JFET has a $V_{GS(off)} = +6$ V. What is I_D when $V_{GS} = +8$ V?

5. The JFET in Figure 4–43 has a $V_{GS(off)} = -4$ V and an $I_{DSS} = 2.5$ mA. Assume that you increase the supply voltage, V_{DD}, beginning at 0 until the ammeter reaches a steady value. At this point,

   (a) What does the voltmeter read?

   (b) What does the ammeter read?

   (c) What is V_{DD}?

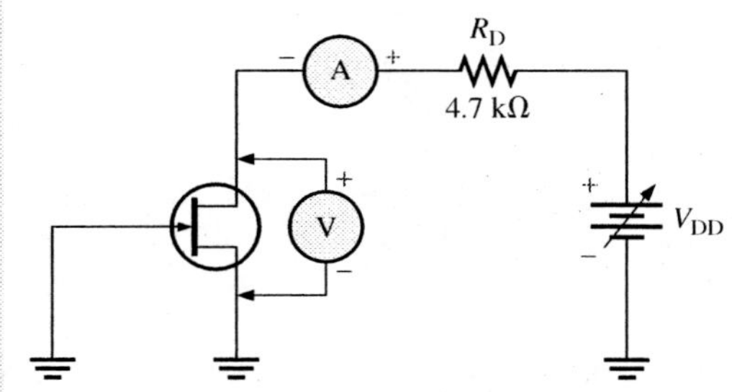

FIGURE 4–43

6. Assume a JFET has the transconductance curve shown in Figure 4–44.

   (a) What is I_{DSS}?

   (b) What is $V_{GS(off)}$?

   (c) What is the transconductance at a drain current of 2.0 mA?

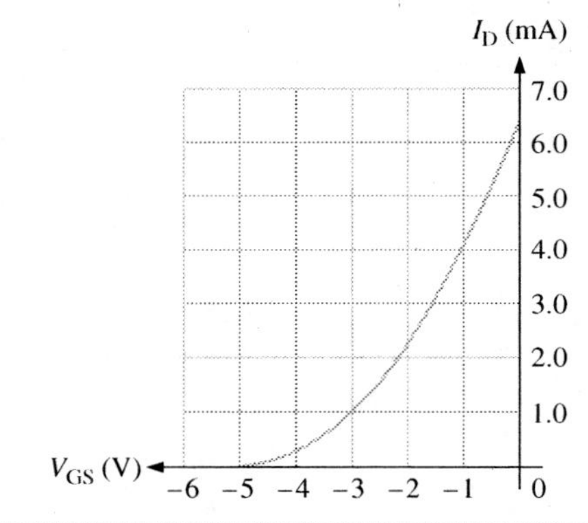

FIGURE 4–44

7. Assume the JFET with the transconductance curve shown in Figure 4–44 is connected in the circuit shown in Figure 4–45.

   (a) What is V_S?

   (b) What is I_D?

   (c) What is V_{DS}?

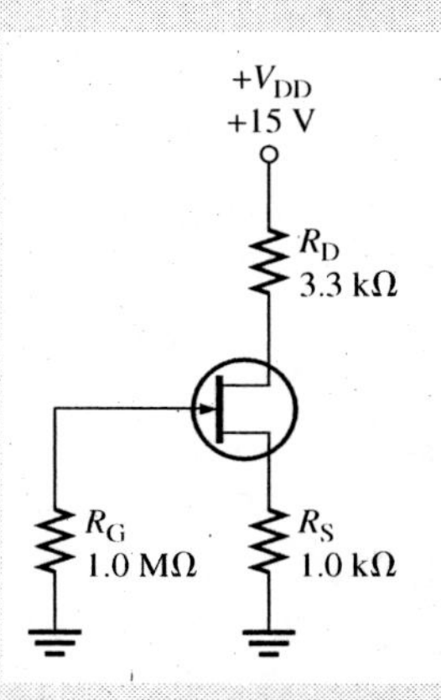

FIGURE 4–45

8. Assume the JFET in Figure 4–45 is replaced with one with a lower transconductance.

   (a) What will happen to V_{GS}?

   (b) What will happen to V_{DS}?

9. For each circuit in Figure 4–46, determine V_{DS} and V_{GS}.

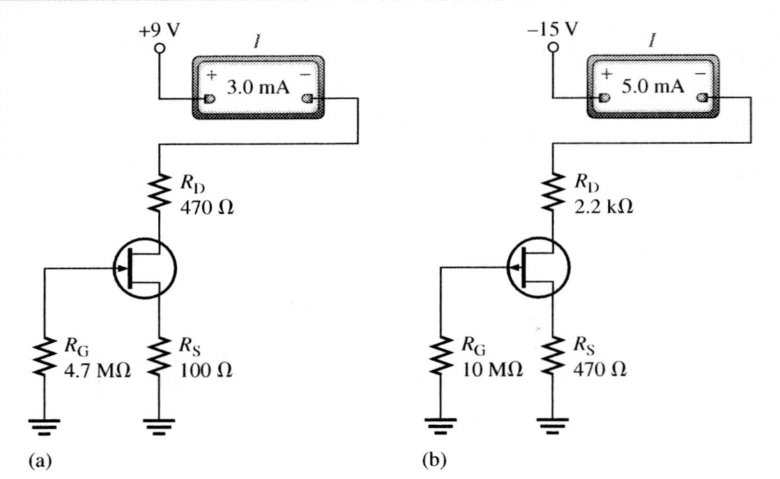

FIGURE 4–46

10. Sketch the schematic symbols for *n*-channel and *p*-channel D-MOSFETs and E-MOSFETs. Label the terminals.

11. A certain E-MOSFET has a $V_{GS(th)} = 3$ V. What is the minimum V_{GS} for the device to turn on?

12. Determine in which mode (depletion, enhancement, neither) each D-MOSFET in Figure 4–47 is biased.

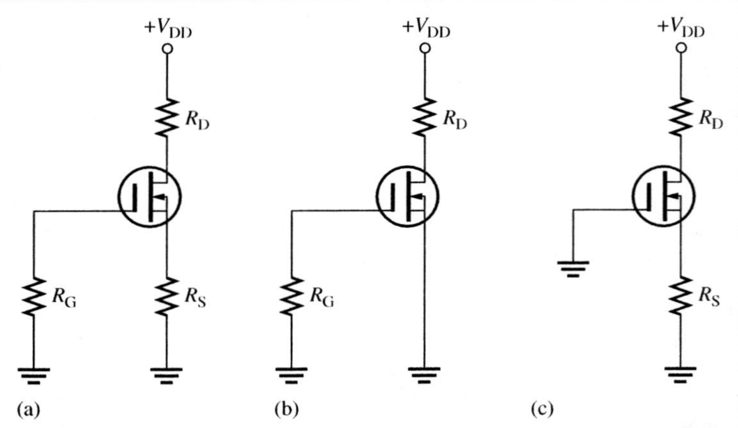

FIGURE 4–47

13. Each E-MOSFET in Figure 4–48 has a $V_{GS(th)}$ of +5 V or −5 V, depending on whether it is an *n*-channel or a *p*-channel device. Determine whether each MOSFET is on or off.

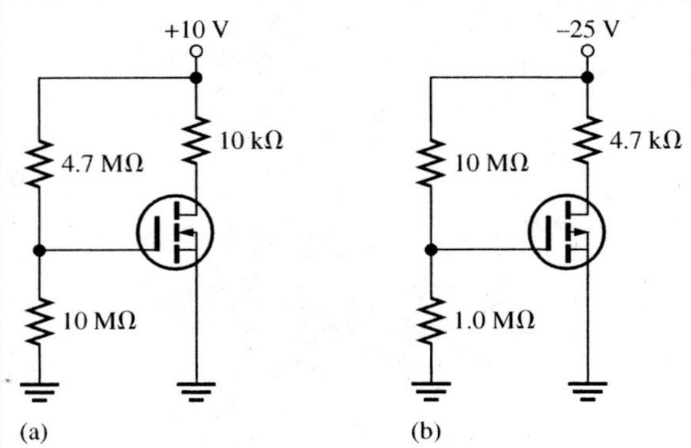

FIGURE 4–48

14. (a) Assume $V_{GS} = -2.0$ V for the circuit in Figure 4–49. Determine V_G, V_S, and V_D.
(b) If $g_m = 3000\ \mu S$, what is the voltage gain?
(c) What is V_{out}?

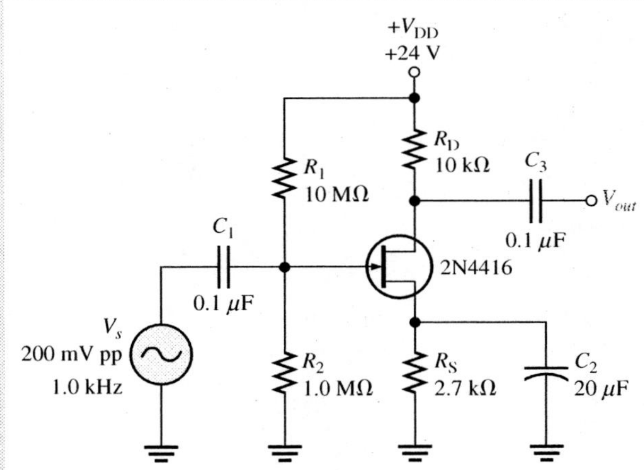

FIGURE 4–49

15. Determine the gain of the amplifier in Figure 4–49 when a 27 kΩ load is connected from the output to ground. $g_m = 3000\ \mu S$

16. For the circuit in Figure 4–49, what effect would an open R_1 have on
(a) V_G
(b) A_v
(c) I_D

17. Assume the source voltage for the D-MOSFET in Figure 4–50 is measured and found to be 1.6 V.

(a) Compute I_D and V_{DS}.

(b) If $g_m = 2000\ \mu S$, what is the voltage gain?

(c) Compute the input resistance of the amplifier.

(d) Is the D-MOSFET operating in the depletion or the enhancement mode?

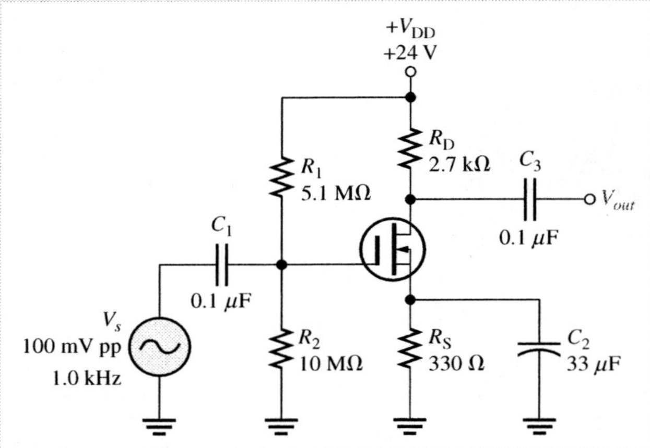

FIGURE 4–50

18. Repeat Problem 17 (a) and (b) if a 5.1 kΩ load is connected between V_{out} and ground.

Basic-Plus Problems

19. A certain JFET data sheet gives $V_{GS(off)} = -8$ V, $I_{DSS} = 10$ mA, and $I_{GSS} = 1.0$ nA.

(a) When $V_{GS} = 0$, what is I_D for values of V_{DS} above pinch-off?

(b) When $V_{GS} = -4$ V, what is R_{IN}?

20. Find I_D and V_D for the current-source biased JFET in Figure 4–51.

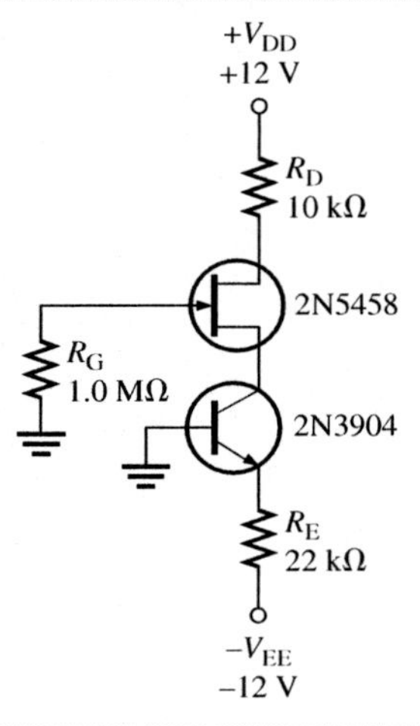

FIGURE 4–51

21. The minimum specified value of g_m for a 2N5457 is 1000 μS and the maximum specified value is 5000 μS. From these values, find the minimum and maximum gain for the CD amplifier in Figure 4–52.

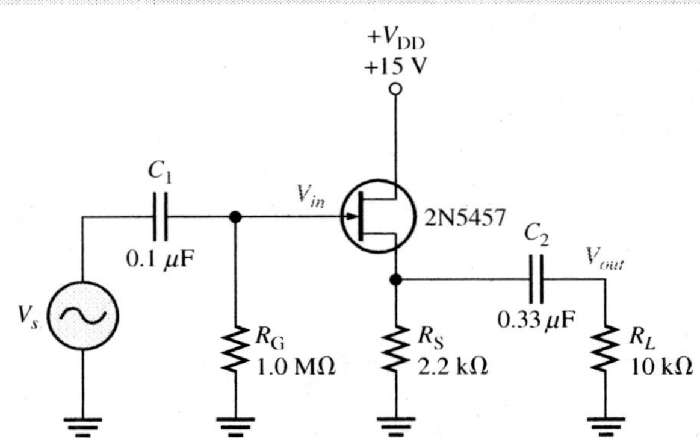

FIGURE 4–52

22. Assume the g_m of Q_1 is 1500 μS for the amplifier in Figure 4–53.
(a) Compute I_D.
(b) If $g_m = 1500\ \mu$S, what is the voltage gain?
(c) What is V_{out}?
(d) What is the purpose of C_2? What happens if it is open?

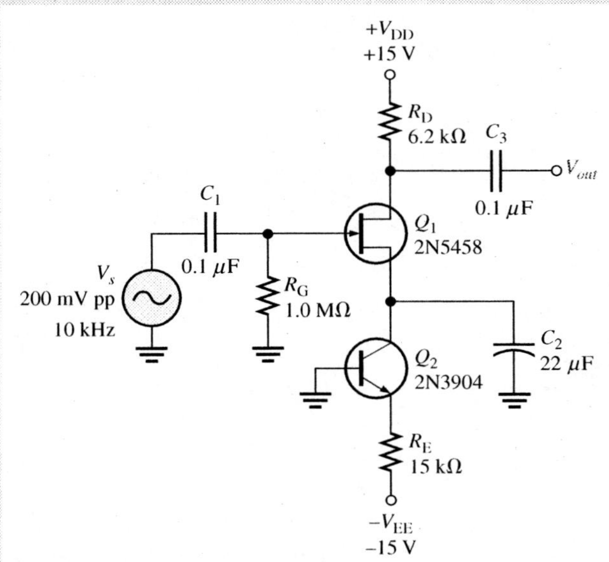

FIGURE 4–53

23. Assume the amplifier in Figure 4–53 has no output voltage. A check of the dc conditions reveals that the drain voltage is +15 V. Name at least three failures that can account for this.

24. Refer to Figure 4–53. Assume the dc voltages and the ac input voltage are correct, but V_{out} is very small. What failure can account for this?

25. Refer to Figure 4–53. What is the minimum value of I_{DSS} that Q_1 can have before the gate-source pn junction is forward-biased?

26. The 2N5555 is an n-channel JFET switching transistor. The specification sheet shows $r_{DS(on)(max)} = 150\ \Omega$ and $I_{DSS(min)} = 15$ mA. It also shows a maximum drain current of 2.0 μA when $V_{GS} = -10$ V at a temperature of 100°C. Assume it is used in the analog switch circuit of Figure 4–54 for the worst case conditions given here.

(a) What is the output voltage when $V_{GS} = 0$ V?

(b) What is the output voltage when $V_{GS} = -10$ V?

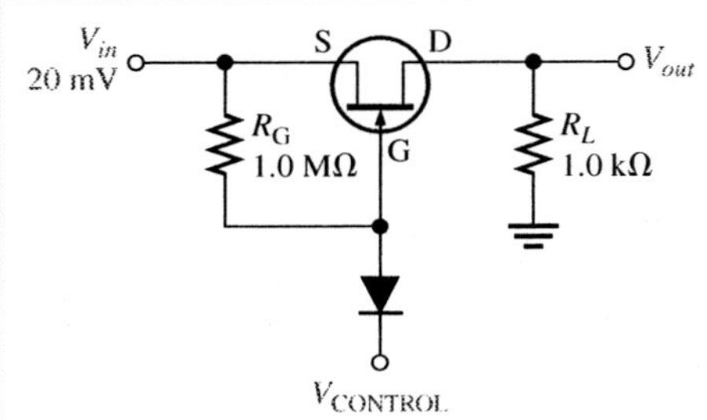

FIGURE 4–54

ANSWERS

Example Questions

4–1: I_D remains at approximately 12 mA.

4–2: ≈ 1.0 mS

4–3: $V_{DS} = 6.34$ V, $V_{GS} = -0.66$ V

4–4: $I_{D(min)} \cong 0.3$ mA; $I_{D(max)} \cong 2.3$ mA

4–5: 6.8 V

4–6: 5.3 V

4–7: A larger source resistor reduces (slightly) the transconductance. As a result, the gain will be reduced.

4–8: 10 MΩ

4–9: 0.96

Review Questions

1. Drain, source and gate
2. JFETs and MOSFETs
3. MOSFET
4. They require smaller areas than BJTs, are easy to manufacture in ICs, and produce simpler circuits.
5. BJTs are controlled by a current, FETs are controlled by a voltage. BJT circuits have higher gain but lower input resistance.
6. Transconductance curve
7. Positive
8. By the gate-to-source voltage
9. 7 V
10. Decrease

11. $V_{GS(off)}$ and I_{DSS}
12. It ties the gate to a solid 0 V.
13. The bias circuit for BJT is designed to forward-bias the base-emitter *pn* junctions; the bias circuit for a FET is designed to reverse-bias the gate-source *pn* junction.
14. -8 V
15. The source resistor in voltage-divider bias must develop sufficient voltage to overcome the positive gate voltage and make the gate-source voltage negative.
16. Depletion MOSFET and enhancement-only MOSFET. The D-MOSFET has a physical channel; the E-MOSFET does not.
17. Yes; the current is I_{DSS}.
18. No
19. Yes
20. Electrostatic discharge, which can damage sensitive devices
21. I_{DSS}
22. It is a normally off device. It must have forward bias to turn it on.
23. D-MOSFET
24. +2 V
25. A type of biasing where a transistor acts as a current-source to another transistor.
26. The gain is the transconductance (g_m) times the ac drain resistance (R_d) *or* the ratio of the ac drain resistance (R_d) to the internal ac source resistance (r'_s).
27. The gain can be calculated from the transconductance and source resistance as $A_v = g_m R_s/(1 + g_m R_s)$ or the ratio of the ac source resistance (R_s) to the sum of the ac source resistance and the internal source resistance (r'_s).
28. CD
29. It is the resistance of the voltage-divider resistors in parallel.
30. High input resistance and low noise
31. An analog switch passes or blocks an ac signal; a digital switch turns on or off a device.
32. When closed, it has no resistance to the signal; when open, it is an infinite resistance.
33. $r_{DS(on)} = -V_{GS(off)}/(2I_{DSS})$
34. They are voltage controlled and draw no drive current. They can control a large current to a device and they are immune to thermal runaway.
35. *N*-channel and *p*-channel E-MOSFETs are connected with common gates and drains, and the output is connected to the drains. The *n*-channel source is connected to ground and the *p*-channel source is connected to a positive supply voltage. When the input is greater than one-half the power supply voltage, the *n*-channel device is on, causing the output to be near ground; when the input is less than one-half the power supply voltage, the *p*-channel MOSFET is on causing the output to be near the power supply voltage.

Chapter Checkup

1. (a)	2. (d)	3. (a)	4. (b)	5. (a)
6. (c)	7. (b)	8. (b)	9. (c)	10. (b)
11. (a)	12. (a)	13. (d)	14. (a)	15. (d)
16. (c)	17. (c)	18. (b)		

MULTISTAGE, POWER, AND DIFFERENTIAL AMPLIFIERS

CHAPTER OUTLINE

Study aids for this chapter are available at

http://www.prenhall.com/SOE

INTRODUCTION

To provide useful power for a speaker or other load, most amplifiers must have several stages of voltage amplification and a power stage. (A stage is one functional part of an amplifier containing a transistor or other active device.) In this chapter, you will learn how to connect stages together using any of three methods (capacitive coupling, transformer coupling, and direct coupling). Then two types of power amplifiers are discussed—Class A and Class B. The chapter concludes with an introduction to a very useful amplifier, the differential amplifier. Differential amplifiers are widely used in instrumentation systems and are the first stage of integrated circuit (IC) operational amplifiers, introduced in Chapter 6.

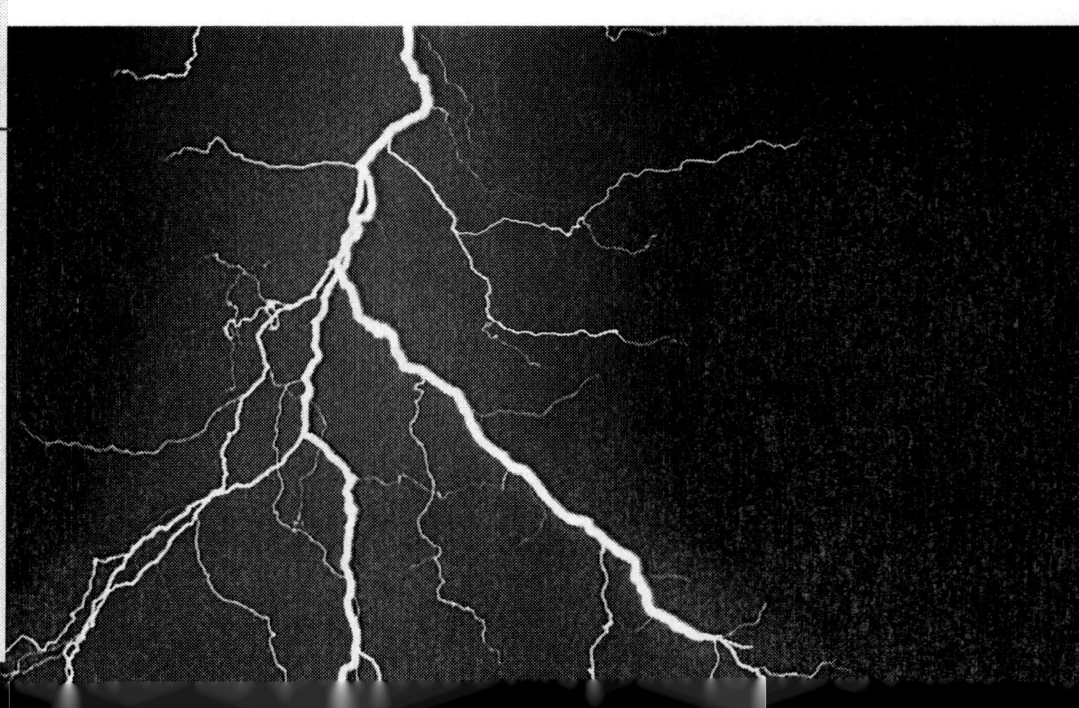

KEY TERMS

- Radio frequency
- Quality factor
- Intermediate frequency
- Mixer
- Negative feedback
- Open-loop voltage gain
- Closed-loop voltage gain
- Class A
- Efficiency
- Class B
- Push-pull
- Class AB
- Current mirror
- Differential amplifier
- Differential-mode
- Common-mode
- Common-mode rejection ratio (CMRR)

KEY OBJECTIVES

A section number is given for each objective. After completing this chapter, you should be able to

5–1 Determine the ac parameters for a capacitively coupled multistage amplifier

5–2 Describe the characteristics of transformer-coupled amplifiers, tuned amplifiers, and mixers

5–3 Determine dc and ac parameters for direct-coupled amplifiers and explain how negative feedback can be used to stabilize the bias and the gain

5–4 Compute dc and ac parameters for a class-A power amplifier and describe operation along the ac load line

5–5 Compute dc and ac parameters for a class-B and class AB power amplifier (both bipolar and FET) and describe the load lines

5–6 Explain the operation of a differential amplifier including how it can significantly reduce common-mode noise

COMPUTER SIMULATIONS DIRECTORY

The following figures have Multisim circuit files associated with them.

- Figure 5–4 Page 171
- Figure 5–15 Page 181
- Figure 5–19 Page 186
- Figure 5–27 Page 193

LABORATORY EXPERIMENTS DIRECTORY

The following exercises are for this chapter.

- **Experiment 9** A Two-Stage Feedback Amplifier
- **Experiment 10** Class B Push-Pull Amplifiers

Harold Black, an engineer for Western Electric Company, was traveling to work in 1927 on the Lackawanna ferry when he had an idea, which he sketched out on a copy of the *New York Times*. He was working on the task of improving amplifiers, and his idea was to feed some of the output of the amplifier back to the input out-of-phase to cancel some of the gain. This concept, termed *negative feedback*, turned out to be one of the most important ideas in electronics.

Negative feedback was first applied to improve long-distance telephone service but has applications in many scientific fields besides electronics. It is widely used in biomechanics, bioengineering, digital computers, and automatic controls to name a few. One basic example of negative feedback is an air conditioning system. The system measures the ambient temperature and sends back a signal when it is too warm to turn on the system. The feedback is "negative" because the action opposes the condition that triggers it.

Another scientific field that has a naturally occurring form of negative feedback is in chemistry, as illustrated by a process known as LeChatelier's principle. In certain reactions, the reaction can be changed by concentration, pressure, or temperature. LeChatelier's principle is used by chemists to predict the conditions of the system as it establishes a new equilibrium that opposes the original change—again negative feedback.

5–1 CAPACITIVELY COUPLED AMPLIFIERS

Two or more transistors can be connected together to form an amplifier called a multistage amplifier. Capacitive coupling is the most widely used method for passing the ac signal to the next stage.

In this section, you will learn to determine the ac parameters for a capacitively coupled multistage amplifier.

Amplifier Model

An **amplifier** is a device that increases the magnitude of a signal for use by a load. Although amplifiers are complicated arrangements of transistors, resistors, and other components, a simplified description is all that is necessary when the requirement is to analyze the source and load behavior. The amplifier can be thought of as the interface between the source and load, as shown in Figure 5–1. You can apply the concept of equivalent circuits, learned in dc/ac studies, to the more complicated case of an amplifier. By drawing an amplifier as an equivalent circuit, you can simplify equations related to its performance.

FIGURE 5–1

Basic amplifier model showing the equivalent input resistance and dependent output circuit.

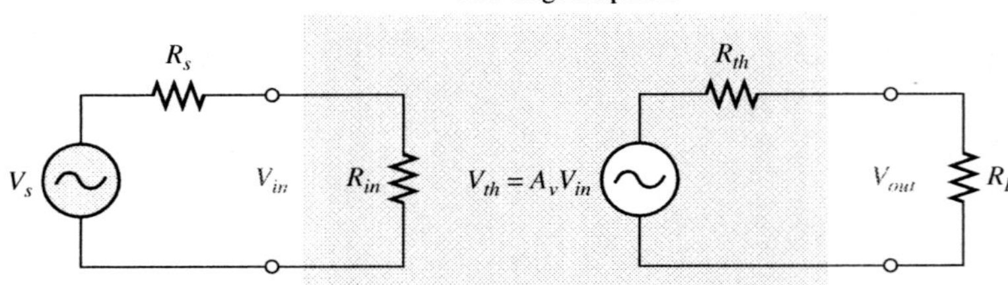

The input signal from a source is applied to the input terminals of the amplifier, and the output is taken from a second set of terminals. (Terminals are represented by open circles on a schematic.) The amplifier's input terminals present an input resistance, R_{in}, to the source. This input resistance affects the input voltage to the amplifier because it forms a voltage divider with the source resistance.

The output of the amplifier is shown as a Thevenin source in Figure 5–1. The magnitude of this source is dependent on the unloaded gain (A_v) and the input voltage; thus, the amplifier's output circuit (drawn as a Thevenin equivalent) is said to contain a *dependent source.* The value of a dependent voltage source always depends on voltage elsewhere in the circuit. The voltage values for the Thevenin case is shown in Figure 5–1.

Cascaded Stages

The Thevenin model reduces an amplifier to its "bare-bones" for analysis purposes. In addition to considering the simplified model for source and load effects, the simplified model is also useful to analyze the internal loading when two or more stages are cascaded to form a single amplifier. Each functional part that amplifies a signal is considered to be a **stage**. Consider two stages cascaded as shown in Figure 5–2. The overall gain is affected by loading effects from each of the three loops. The loops are simple series circuits, so voltages can easily be calculated with the voltage-divider rule.

FIGURE 5–2 Cascaded stages in an amplifier.

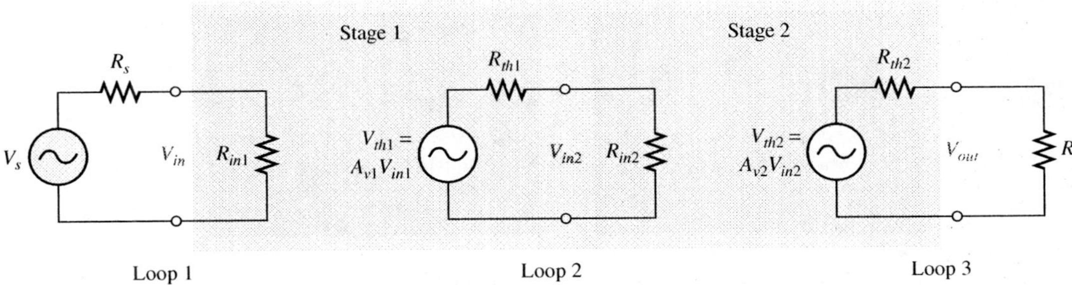

EXAMPLE 5–1

Problem

Assume a transducer with a Thevenin (unloaded) source, V_s, of 10 mV and a Thevenin source resistance, R_s, of 50 kΩ is connected to a two-stage cascaded amplifier, as shown in Figure 5–3. Compute the voltage across a 1.0 kΩ load.

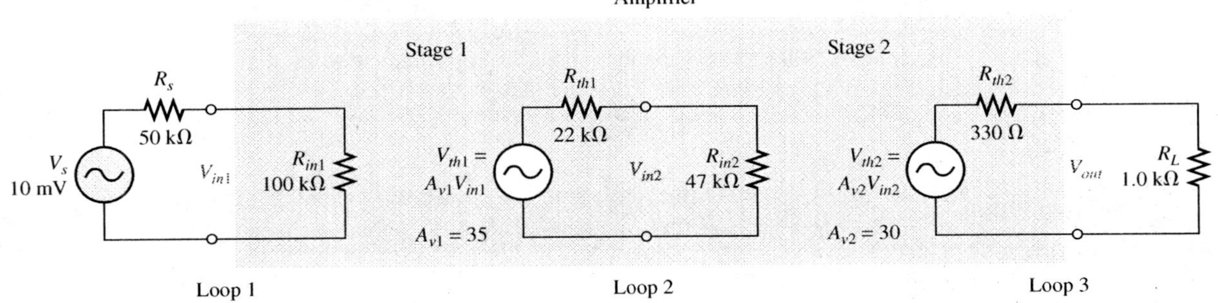

FIGURE 5–3
Two-stage cascaded amplifier.

Solution

Compute the input voltage to stage 1 from the voltage-divider rule applied to loop 1.

$$V_{in1} = V_s\left(\frac{R_{in1}}{R_{in1} + R_s}\right) = 10 \text{ mV}\left(\frac{100 \text{ k}\Omega}{100 \text{ k}\Omega + 50 \text{ k}\Omega}\right) = 6.67 \text{ mV}$$

The Thevenin voltage for stage 1 is

$$V_{th1} = A_{v1}V_{in1} = (35)(6.67 \text{ mV}) = 233 \text{ mV}$$

Compute the input voltage to stage 2 again from the voltage-divider rule, this time applied to loop 2.

$$V_{in2} = V_{th1}\left(\frac{R_{in2}}{R_{in2} + R_{th1}}\right) = 233 \text{ mV}\left(\frac{47 \text{ k}\Omega}{47 \text{ k}\Omega + 22 \text{ k}\Omega}\right) = 159 \text{ mV}$$

The Thevenin voltage for stage 2 is

$$V_{th2} = A_{v2}V_{in2} = (30)(159 \text{ mV}) = 4.77 \text{ V}$$

Apply the voltage-divider rule one more time to loop 3. The voltage across the 1.0 kΩ load is

$$V_{R_L} = V_{th2}\left(\frac{R_L}{R_L + R_{th2}}\right) = 4.77 \text{ V}\left(\frac{1.0 \text{ k}\Omega}{1.0 \text{ k}\Omega + 330 \ \Omega}\right) = \mathbf{3.59 \text{ V}}$$

Question*

Assume a transducer with a Thevenin source voltage of 5.0 mV and a source resistance of 100 kΩ is connected to the same amplifier. What is the voltage across the 1.0 kΩ load?

Transistor Amplifiers

Two or more transistors can be connected together to enhance the performance of an amplifier. Frequently, the first stage of an amplifier must have very high input resistance to avoid loading the source. In addition, the first stage needs to be designed for low noise operation because the very small signal voltage can easily be obscured by noise. Succeeding stages are designed to increase the amplitude of the signal without adding distortion.

Probably the simplest way to add gain to an amplifier is to capacitively couple two stages together as shown in Figure 5–4. In this case, both stages are identical CE amplifiers with the output of the first connected to the input of the second stage. Capacitive coupling prevents the dc bias of one stage from affecting the dc bias of another stage because capacitors block dc. Although the dc path is open, the coupling capacitor, C_3, provides almost no opposition to the ac signal and the signal passes to the next stage.

COMPUTER SIMULATION

Open the Muiltisim file F05-04DV on the website. Measure the gain for the amplifier.

* *Answers are at the end of the chapter.*

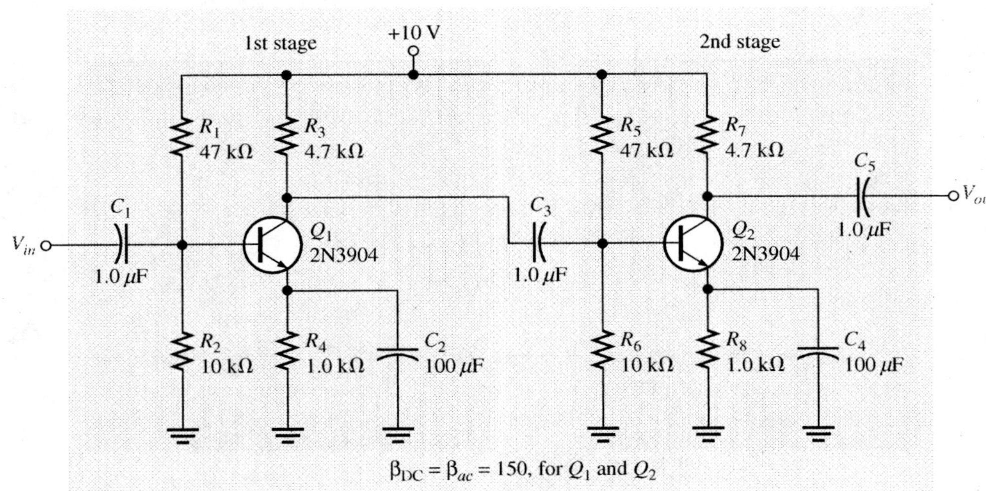

FIGURE 5–4

A two-stage CE amplifier.

The analysis of the circuit starts with the dc conditions, as explained in Section 3–2. To compute the base voltage of either stage, use the voltage-divider rule.

$$V_B \cong \left(\frac{R_2}{R_1 + R_2}\right)V_{CC} = \left(\frac{10\text{ k}\Omega}{47\text{ k}\Omega + 10\text{ k}\Omega}\right)10\text{ V} = 1.7\text{ V}$$

This estimate is slightly high because it is made for an unloaded voltage divider. After subtracting 0.7 V for the base-emitter diode, the emitter voltage is 1.0 V, which results in an emitter current of

$$I_E = \frac{V_E}{R_E} = \frac{1.0\text{ V}}{1.0\text{ k}\Omega} = 1.0\text{ mA}$$

The emitter current is also approximately equal to the collector current.

Loading Effects

Recall that amplifiers can be shown as a block diagram with essential parameters only. The ac model is simply a dependent voltage source with a series resistance (a Thevenin circuit). To compute the overall gain of the amplifier, each transistor stage in the original circuit can be modeled in a similar manner. Only three parameters need to be known: the unloaded (No-Load) voltage gain ($A_{v(NL)}$), the total input resistance, ($R_{in(tot)}$), and the output resistance (R_{out}). Notice that the unloaded output voltage is the input voltage times the unloaded gain. The two-stage CE amplifier in Figure 5–4 will be used as an example. Let's start with the model for one of the stages shown in Figure 5–5.

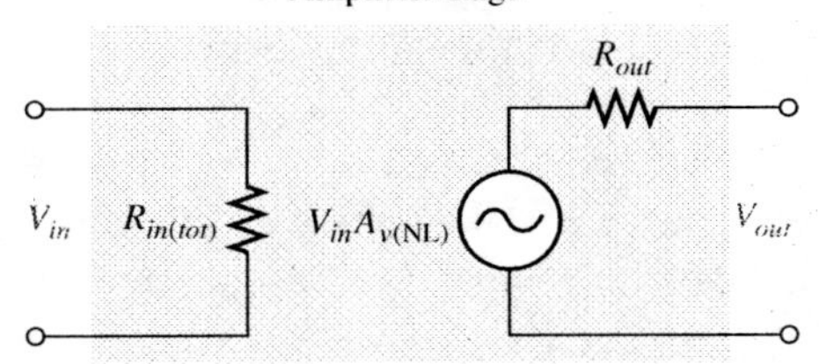

FIGURE 5–5

One-stage amplifier model.

Start by finding the unloaded gain of one stage. Because the two stages are identical, the unloaded gain is the same for both. The input resistance of the second stage acts as a load on the first stage. Thus, the loaded gain of the first stage can be found by assuming it has a load resistor equal to $R_{in(tot)}$ of stage 2. This lowers the gain of the first stage but can be considered separately from the unloaded gain calculation. An illustration of this idea should clarify how the basic amplifier model can simplify determining the overall gain.

As you know, the unloaded gain of a CE amplifier is the ratio of the ac collector resistance to the ac emitter resistance. This unloaded gain is dependent on r'_e, which in turn depends on I_E, so the calculation should be considered approximate.

Since the unloaded gain is being computed, the ac collector resistance, R_c, is the same as the actual collector resistor, R_C, which is 4.7 kΩ. The ac emitter resistance is approximately

$$r'_e \cong \frac{25\text{ mV}}{I_E} = \frac{25\text{ mV}}{1.0\text{ mA}} = 25\ \Omega$$

The unloaded gain, $A_{v(NL)}$, is approximately

$$A_{v(NL)} = -\frac{R_c}{R_e} = -\frac{R_C}{r'_e} = -\frac{4.7\text{ k}\Omega}{25\ \Omega} = -188$$

The input resistance of the CE amplifier was discussed in Section 3–4. The equation for input resistance with voltage-divider bias and no swamping resistor is

$$R_{in(tot)} = R_1 \| R_2 \| (\beta_{ac} r'_e)$$

By substitution, the input resistance of the amplifier, assuming a β_{ac} of 150, in Figure 5–4 is

$$R_{in(tot)} \cong 47\text{ k}\Omega \| 10\text{ k}\Omega \| [150(25\ \Omega)] \cong 2.58\text{ k}\Omega$$

The output resistance is the resistance looking back to the collector circuit and is simply the collector resistor.

$$R_{out} = R_C = 4.7\text{ k}\Omega$$

These values can be entered onto the model as shown in Figure 5–6.

FIGURE 5–6

Values for one stage of the amplifier in Figure 5–4.

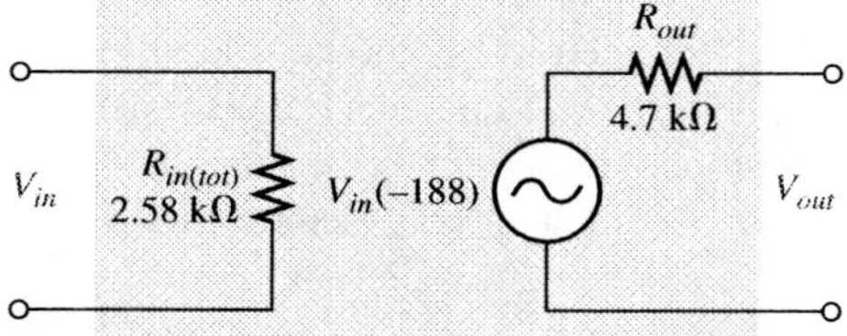

The two stages that comprise the amplifier are now connected in Figure 5–7. In this drawing, the unloaded gain for each stage is shown below the Thevenin source, and the model is used to find the overall gain. The overall gain is the product of three terms:

1. The unloaded voltage gain of the first stage
2. The gain of the voltage divider consisting of the input resistance of the second stage with the output resistance of the first stage
3. The unloaded gain of the second stage

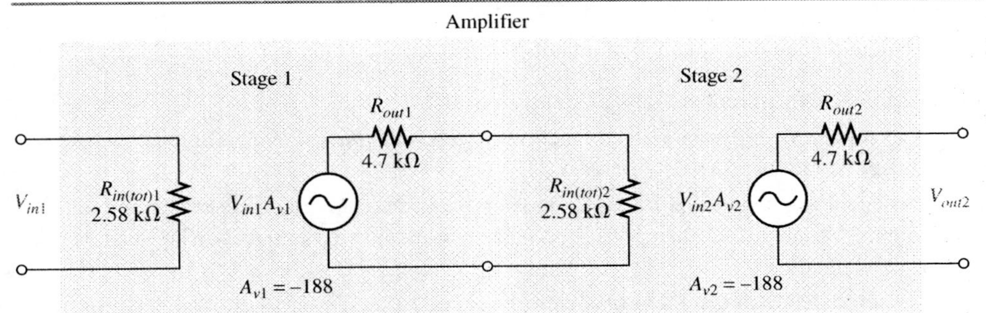

FIGURE 5–7

AC model of the complete two-stage amplifier from Figure 5–4.

If a load resistor is added to the output, it can be included as another voltage-divider term.

The unloaded gain of each stage is −188, as previously calculated. The voltage divider between the stages accounts for loading effects. It consists of $R_{in(tot)2}$ for stage 2 and R_{out1} for stage 1. The gain (attenuation) of this voltage divider is

$$A_{v(divider)} = \frac{R_{in(tot)2}}{R_{out1} + R_{in(tot)2}} = \frac{2.58\ \text{k}\Omega}{4.7\ \text{k}\Omega + 2.58\ \text{k}\Omega} = 0.35$$

The overall voltage gain is the product of the three gains.

$$A_{v(tot)} = A_{v1}A_{v(divider)}A_{v2} = (-188)(0.35)(-188) \cong 12{,}400$$

This product indicates the voltage gain is fairly large. If an input signal of 100 μV, for example, is applied to the first stage and the attenuation of the input base circuit is neglected, an output from the second stage of (100 μV)(12,400) = 1.24 V will result. Again, a factor that must be kept in mind is that this answer is approximate because the gain is very dependent on the value of r'_e and the specific transistors used. At the price of reduced gain, greater stability can be achieved by adding a swamping resistor in the emitter circuit. This will tend to make the circuit produce consistent gain that is independent of the specific transistor.

Unwanted Oscillation and Noise

Multistage amplifiers require careful design to avoid unwanted oscillations. When large signals are present in the same circuit as small signals, the large signal can have an adverse effect on the small signal due to unwanted feedback paths. This problem is compounded in high-frequency amplifiers because feedback paths tend to have lower reactance, causing more unwanted feedback. For example, protoboards have stray capacitances between rows that can lead to feedback and noise problems when constructing multistage amplifiers on them. It is usually helpful to isolate the various stages by connecting a capacitor between V_{CC} and ground at each stage; this technique is seen frequently in commercial printed circuit (PC) boards. Capacitors should be connected very close to the point that V_{CC} is applied to a stage and should have short lead lengths.

In addition to unwanted oscillations, noise voltages (unwanted electrical disturbances) can be a problem for multistage amplifiers. The ratio of signal to noise determines if the noise is sufficient to disrupt the signal. When the signal is small, a little noise voltage has a greater effect than when the signal is large. This means the first stage of an amplifier is the most important stage because of the very small signal level. FETs have the advantage for high-impedance sources; but when the source impedance is lower, (<1 MΩ), bipolar transistors can provide excellent low-noise performance.

SAFETY NOTE

If another person cannot let go of an energized conductor, switch the power off immediately. If that is not possible, use any available nonconductive material to try to separate the body from the contact point. Seek medical help right away for any electrical burns.

The following are suggestions for avoiding noise problems:

1. Keep wiring short to avoid "antennas" in circuits (particularly low-level input lines) and make signal return loops as small as possible.
2. Use capacitors between power supply and ground at each stage and make sure the power supply is properly filtered.
3. Reduce noise sources, if possible, and separate or shield the noise source and the circuit. Use shielded wiring, twisted pair, or shielded twisted pair wiring for low-level signals.
4. Ground circuits at a single point, and isolate grounds that have high currents from those with low currents by running separate ground lines back to the single point. Ground current from a high current ground can generate noise in another part of a circuit because of *IR* drops in the conductive paths.
5. Keep the bandwidth of amplifiers no larger than necessary to amplify only the desired signal, not extra noise.

Review Questions

Answers are at the end of the chapter.

1. What three parameters are needed for each stage of a multistage amplifier to determine the overall gain?
2. What is a dependent source?
3. How does a second stage affect the gain of the first stage of a two-stage amplifier?
4. Why is the first stage of a multistage amplifier the most important for reducing noise?
5. Why do multistage amplifiers require careful design to avoid noise and oscillations?

5–2 TRANSFORMER-COUPLED AMPLIFIERS

Transformers can be used to couple a signal from one stage to another. Although principally used in high-frequency designs, they are also found in some low-frequency power amplifiers. When the signal frequency is in the radio frequency (RF) range (>100 kHz), stages within an amplifier are frequently coupled with tuned transformers, which form a resonant circuit.

In this section, you will learn the characteristics of transformer-coupled amplifiers, tuned amplifiers, and mixers.

Low-Frequency Applications

Most amplifiers require that the dc signal be isolated from the ac signal. In Section 5–1 you learned how a capacitor could be used to pass the ac signal while blocking the dc signal. Transformers also block dc (because they provide no direct path) and pass ac.

Recall that *impedance* is a term used when reactance and resistance are combined and form ac opposition. With transformer coupling, it is common to refer to input and output impedance rather than resistance.

Transformers provide a useful means of matching the impedance of one part of a circuit to another. From your dc/ac studies, recall that a load on the secondary side of a transformer is changed by the transformer when looking from the primary side. A step-down transformer causes the load to look larger on the primary side as expressed by

$$R'_L = \left(\frac{N_{pri}}{N_{sec}}\right)^2 R_L \tag{5–1}$$

where R'_L is reflected resistance on the primary side, N_{pri}/N_{sec} is the ratio of primary turns to secondary turns, and R_L is the load resistance on the secondary side.

Transformers can be used at the input, the output, or between stages to couple the ac signal from one part of a circuit to another. By matching impedances in a power transformer, maximum power can be transferred to the load. Transformers can also be used for matching the impedance of a source to a line. Line-matching transformers are used primarily for low-impedance circuits (<200 Ω). For voltage amplifiers, a transformer can also step up the voltage to the next stage (but never the power).

Figure 5–8 shows examples of transformer coupling in a two-stage amplifier. Small low-frequency transformers are occasionally used in certain microphones or other transducers to couple a signal to an amplifier.

Although transformer coupling can give higher efficiency than *RC* coupling, transformer coupling is not widely applied to low-frequency designs because of two major drawbacks. First, transformers are more expensive and are much bulkier than capacitors. Second, they tend to have poorer response at high frequencies due to the reactance of the coils. For these reasons, low-frequency transformer coupling is not commonly used except in certain class A power amplifiers.

FIGURE 5–8

A basic transformer-coupled amplifier showing input, coupling, and output transformers.

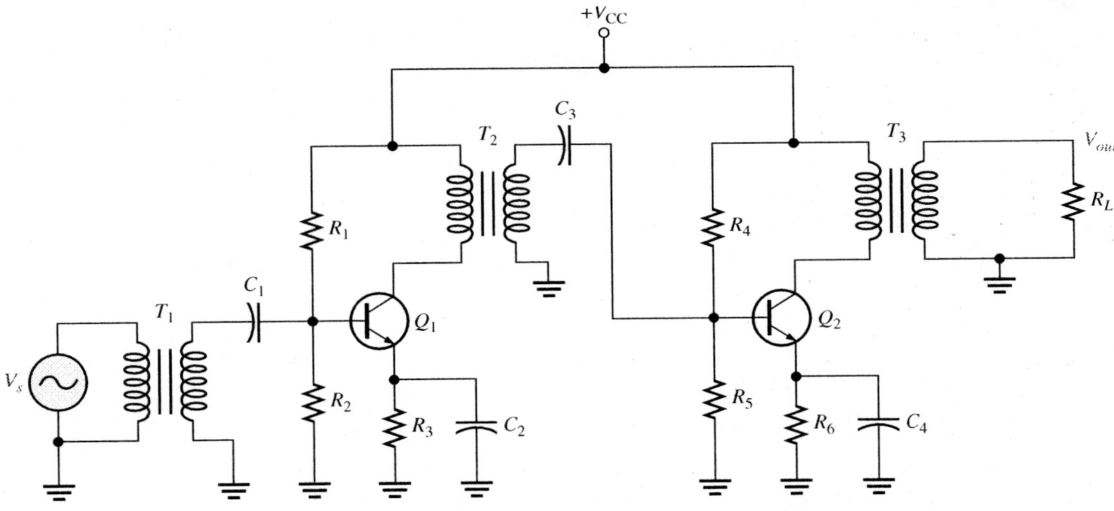

EXAMPLE 5–2

Problem

Assume the component values for the second stage of Figure 5–8 are as follows: R_4 = 5.1 kΩ, R_5 = 2.7 kΩ, $R_6 = R_E$ = 680 Ω, and R_L = 50 Ω; transformer T_3 is a 5:1 step-down transformer and V_{CC} = 12 V.

(a) Find V_{CE}.

(b) Calculate the gain of the second stage.

Solution

(a) Find the base voltage by applying the voltage-divider rule to the bias resistors.

$$V_B = \left(\frac{R_5}{R_4 + R_5}\right)V_{CC} = \left(\frac{2.7\text{ k}\Omega}{5.1\text{ k}\Omega + 2.7\text{ k}\Omega}\right)12\text{ V} = 4.2\text{ V}$$

Next, calculate the emitter voltage and current.

$$V_E = V_B - V_{BE} = 4.2\text{ V} - 0.7\text{ V} = 3.5\text{ V}$$
$$I_E = \frac{V_E}{R_E} = \frac{3.5\text{ V}}{680\ \Omega} = 5.15\text{ mA}$$

The emitter current is approximately equal to the collector current. The dc resistance of the transformer primary is small and can be ignored. With this assumption, the transformer has no effect on V_{CE}. The voltage from the collector to the emitter is the difference between V_{CC} and the drop across the emitter resistor.

$$V_{CE} \cong V_{CC} - V_E = 12\text{ V} - 3.5\text{ V} = \mathbf{8.5\ V}$$

(b) Calculate the reflected resistance of the load resistor on the primary side and the ac emitter resistance.

$$R'_L = \left(\frac{N_{pri}}{N_{sec}}\right)^2 R_L = \left(\frac{5}{1}\right)^2 50\ \Omega = 1.25\text{ k}\Omega$$
$$r'_e = \frac{25\text{ mV}}{I_E} = \frac{25\text{ mV}}{5.15\text{ mA}} = 4.85\ \Omega$$

Calculate the gain by dividing the ac resistance in the collector circuit by the ac resistance of the emitter circuit.

$$A_v = \frac{R'_L}{r'_e} = \frac{1.25\text{ k}\Omega}{4.85\ \Omega} = \mathbf{257}$$

Question

What happens to the gain if R_4 is replaced with a larger resistor?

An interesting aspect of transformer-coupled amplifiers is that the ac load line is not as steep as the dc load line. The ac saturation current is *lower* than the dc saturation current and the ac cutoff voltage is *larger* than the dc cutoff voltage (V_{CC}).

High-Frequency Applications

Frequencies greater than 100 kHz are commonly referred to as **radio frequencies (RF)**. At these frequencies, transformers are much smaller, less expensive, and offer important advantages for coupling signals over a limited bandwidth. At radio frequencies a transformer primary can be connected with a parallel capacitor to form a resonant circuit. Frequently, the secondary winding, with an appropriate capacitor across it, is also connected as a resonant circuit.

From your dc/ac studies, you may recall that a parallel resonant circuit is an *LC* combination that has an impedance maximum at the resonant frequency. This high impedance at the resonant frequency means that the gain of the amplifier can be very high at frequencies near the resonant frequency while offering little opposition to dc. This forms an efficient narrow bandwidth amplifier (typically 10 kHz) with gains as high as 1000 or so. Furthermore, the amplifier is tailored to amplify a narrow band of frequencies containing the signal of interest and not amplify other frequencies.

Tuned Amplifiers

Tuned amplifiers are different than the low-frequency amplifiers you have studied. They are designed to amplify a specific band of frequencies while eliminating any signals outside the band. They use a parallel *LC* resonant circuit for a load to provide a high impedance to the ac signal and thus produce a high gain at the resonant frequency. (At high frequencies, impedance is used instead of resistance.) The center frequency of the tuned circuit can be computed from the basic resonant frequency equation:

$$f_r = \frac{1}{2\pi\sqrt{LC}} \qquad (5\text{–}2)$$

The bandwidth of a tuned amplifier is determined by the Q (quality factor) of the resonant circuit. The **quality factor (Q)** is a dimensionless number that is the ratio of the maximum energy stored in a cycle to the energy lost in a cycle. From a practical view, the inductor almost always determines the Q; consequently Q is often expressed as the ratio of the inductive reactance, X_L, to the resistance, R. It is also the ratio of the center (resonant) frequency, f_r, to the bandwidth, BW.

$$Q = \frac{X_L}{R} = \frac{f_r}{BW} \qquad (5\text{–}3)$$

The response of a parallel resonant circuit depends on the Q of the circuit, as illustrated in Figure 5–9. The Q for an RF circuit depends on the type of inductor; it can range from 50 to 250 for ferrite-core inductors and even higher for air-core inductors.

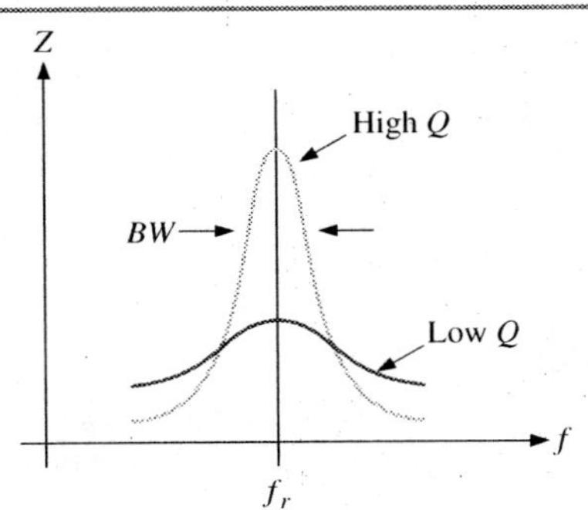

FIGURE 5–9

Impedance of a parallel resonant circuit as a function of frequency.

During signal processing, a radio frequency is usually converted to a lower frequency by mixing the RF with an oscillator. The new lower frequency that is produced is called an **intermediate frequency** or IF. Tuned transformer coupling is important in both RF and IF amplifiers.

The principal advantage to using IF is that it is a fixed frequency and requires no changes in the tuned circuit for any given RF signal (within design limits). This is accomplished by causing the oscillator to "track" the RF. Since the IF is fixed, it is easy to amplify with a fixed-resonant circuit, without need for the user to adjust any controls. This idea, first developed by Major Edwin Armstrong during World War I, is found in most communication equipment and is also used in the spectrum analyzer, an important piece of high-frequency test equipment.

Figure 5–10 shows an example of a two-stage tuned amplifier that uses resonant circuits at both the input to the first stage and the output from the second stage. Transformer coupling is used between the stages. A circuit similar to this is part of most communication equipment and consists of an RF amplifier and a mixer. The RF amplifier tunes and amplifies the high-frequency signal from a station. The **mixer** is a nonlinear circuit that combines this signal with a sine wave generated from an oscillator.

The oscillator's frequency is set to a fixed difference from the RF. When the RF and oscillator signals are mixed in a nonlinear circuit, they produce two new frequencies: the sum

FIGURE 5–10 A tuned amplifier consisting of an RF stage and a mixer.

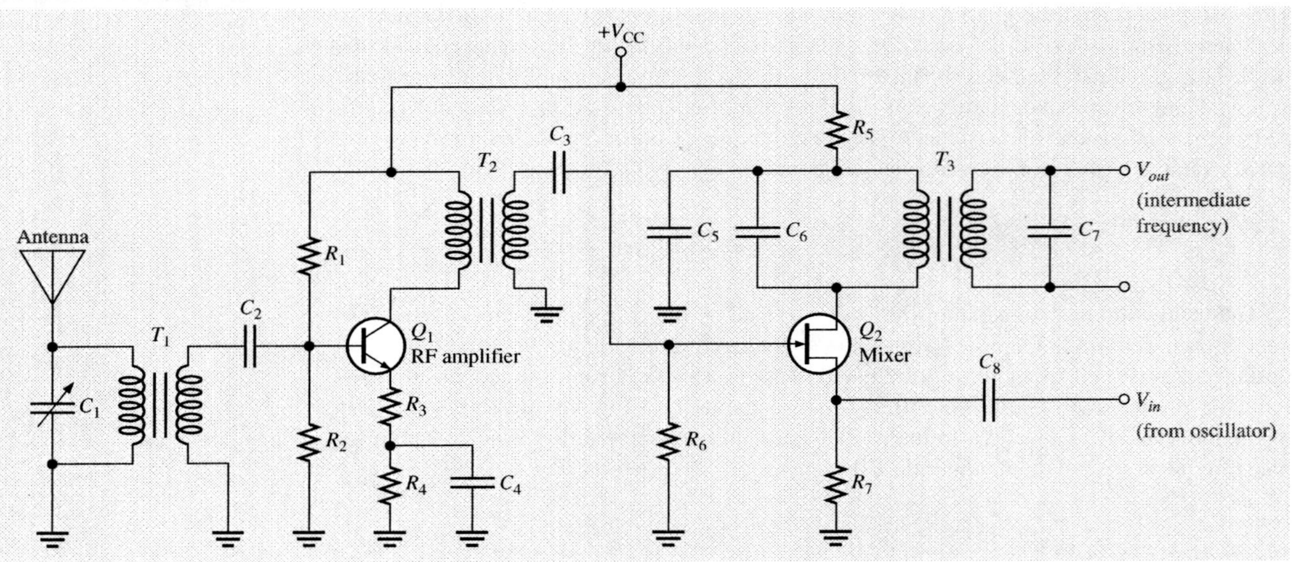

of the input signals and the difference of the input signals. The second resonant circuit is tuned to the difference frequency, while rejecting all other frequencies. This difference frequency is the IF signal that is amplified further by the IF amplifier section. The advantage of an IF section is that it is specifically designed to process a single frequency.

Let's examine the circuit in Figure 5–10 further. The first tuned circuit shown in the blue block consists of the primary of T_1 which resonates with C_1 to tune a station. Stations not at the resonant frequency are rejected by the resonant circuit. Notice that Q_1 is biased with stable voltage-divider bias. There is no collector resistor, but instead, the ac signal "sees" the primary of transformer T_2 as a load. The gain for this stage is determined by the reactance in the collector circuit divided by the ac emitter resistance consisting of R_3 and r'_e.

The RF signal is passed to the gate of Q_2 by transformer T_2 where it is combined with the signal from the oscillator in the mixer stage shown in the beige block. Note that Q_2 is a CS amplifier for the RF signal, but a CG amplifier for the oscillator signal. The resonant circuit in the output of Q_2 is tuned to the desired difference frequency. Thus, the output of Q_2 is the intermediate frequency (IF), which is sent to the next stage for further amplification. In order to generate the intermediate frequency, Q_2 must operate as a nonlinear amplifier. FETs fulfill this role nicely because they have a nonlinear characteristic curve (refer to Figure 4–6).

Notice in Figure 5–10 that a resistor, R_5, is in series with the voltage from the power supply. This resistor and C_5 form a low-pass filter called a **decoupling network** that helps isolate the circuit from other amplifiers and helps prevent unwanted oscillations. The resistor is a small value (typically 100 Ω) and the capacitor is selected to have a reactance that is <10% of this value at the operating frequency. (For example, a 100 Ω resistor can be bypassed with a capacitor that has a reactance of approximately 10 Ω.)

An IF amplifier is shown in Figure 5–11. The IF transformer is designed for the specific intermediate frequency selected. In this case, it is tuned to 455 kHz, a common IF frequency. The IF amplifier is in all respects an RF amplifier; the only difference between an IF and an RF amplifier is the function it serves in a given circuit. An IF amplifier uses a tuned input circuit and tuned output circuit to selectively amplify the intermediate frequency. The capacitor that forms the primary resonant circuit and the transformer are inside a metal enclosure that provides shielding. The exact intermediate frequency is adjusted with a tuning slug that is moved in and out of the core. Again, a decoupling network is included

FIGURE 5–11

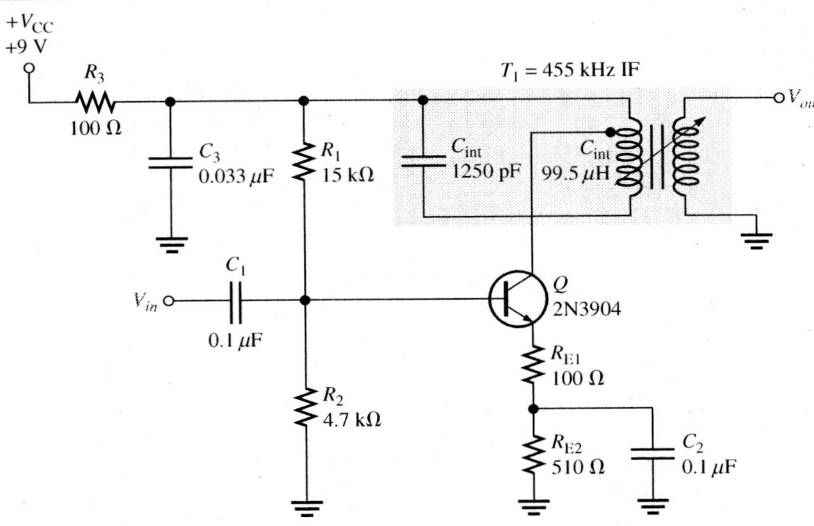

(R_3 and C_3). When tuning the IF circuit, it is important to use a high-impedance, low-capacitance test instrument to avoid changing the circuit response due to instrument loading.

Review Questions

6. What is the difference between an RF and an IF signal?
7. What is the function of a mixer?
8. What are the two signals mixed in a mixer?
9. What effect does a load resistor on the secondary of a tuned transformer have on the Q of the tuned circuit?
10. Why should a high-impedance, low-capacitance instrument be used to test an IF stage?

DIRECT-COUPLED AMPLIFIERS 5–3

Another important method for coupling signals is called direct coupling. With direct coupling, there are no coupling capacitors or transformers between stages. Depending on how the input and output signals are coupled, some amplifiers can operate with frequencies all the way down to dc.

In this section, you will learn how to determine dc and ac parameters for direct-coupled amplifiers and explain how negative feedback can be used to stabilize the bias and the gain.

Negative Feedback

Feedback is the concept of comparing the output of a system with a desired output and making a correction. In electronic systems, **negative feedback** is a correction signal that is returned to the input in a manner to cancel a fraction of the input. Figure 5–12 illustrates the concept of negative feedback. The point where the sample of the output is added (algebraically) to the input is called a summing point (in some cases this is internal to the amplifier). Although negative feedback reduces the gain, it has many positive effects, including much improved stability and reduced distortion. Negative feedback is one of the most important ideas in electronic controls and other systems (see the Science Highlight).

FIGURE 5–12

Negative feedback.

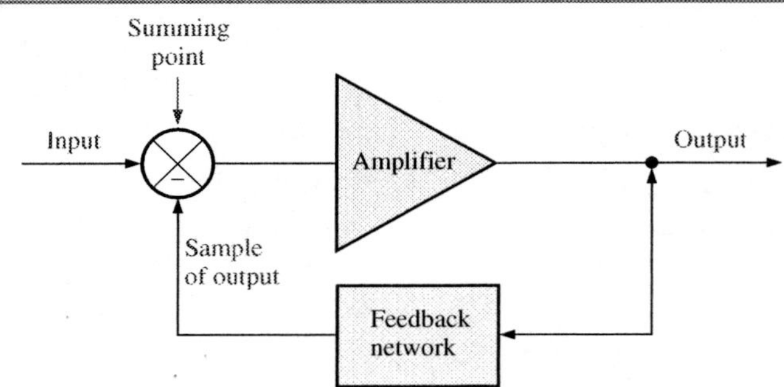

In most amplifiers some form of negative feedback is used, either for controlling gain, reducing distortion, or increasing stability. You may not have realized it, but the unbypassed emitter resistor (R_{E1}) in Figure 3–29 was a form of negative feedback. Negative feedback can be used for bias stability as well and will be applied to the direct-coupled amplifier in this section. Later, you will see many more applications for negative feedback.

Figure 5–13 shows a direct-coupled amplifier without feedback. After looking at this amplifier without feedback, you will see how feedback improves its performance. Direct coupling is from the collector of Q_1 to the base of Q_2. Since the stages are direct coupled, bias current for Q_2 is supplied by Q_1, eliminating the need for any bias resistors for Q_2 and eliminating a coupling capacitor between the stages. Although the stages are direct coupled, it is necessary in this particular amplifier to ac couple the input and output signals (through capacitors) to prevent the external signal source and the load from disturbing the dc voltages.

FIGURE 5–13

A direct-coupled amplifier without feedback.

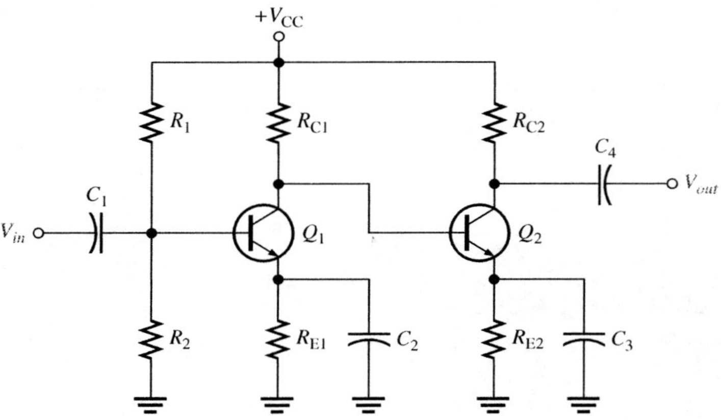

Bias for Q_2 is supplied through R_{C1}, the collector resistor for Q_1. Transistor Q_1 is relatively independent of β because it has voltage-divider bias, but Q_2 uses base bias, a method not desired in linear amplifiers because of the variation in β. In addition, thermal changes will cause the circuit to drift. Although this particular amplifier has fewer components than comparable capacitively coupled amplifiers, the drawbacks mentioned may outweigh the advantages. A relatively simple change–the addition of negative feedback–can cure the problem of β dependency and drift.

Negative Feedback for Bias Stability

The circuit in Figure 5–14 is a modification that greatly improves the bias stability of the amplifier in Figure 5–13 with an added bonus of reducing the parts count. Again, the input and output signals are capacitively coupled to avoid disturbing the bias voltages. Because there are two transistors, the feedback network, shown in red, takes advantage of the extra

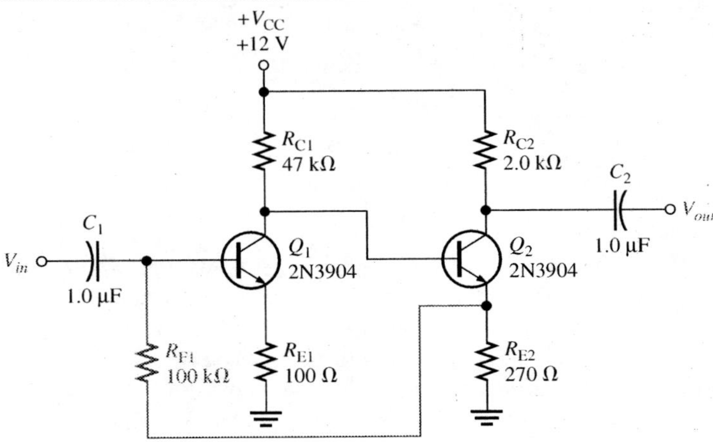

FIGURE 5–14

A direct coupled amplifier with negative feedback that produces bias stability.

gain over a single transistor and produces excellent stability for variations in β and for variations in temperature. This circuit is shown with specific values for components to give you an idea of typical values you might see.

Let's look at how the feedback works in Figure 5–14. Start with Q_2 and note that its base is forward-biased by R_{C1}, causing a collector current in Q_2 or $I_{C(Q2)}$. This current causes the emitter voltage of Q_2 to rise, turning on Q_1. As Q_1 conducts more, its collector voltage drops, reducing the bias on Q_2. This action reduces the bias on Q_2 to a stable point determined by the particular design values. $I_{C(Q1)}$ is almost completely independent of the value of β, producing a stable value of collector voltage in Q_1 and a stable base voltage in Q_2. Thus, the β dependency problem associated with base bias is no longer a factor.

Negative Feedback for Gain Stability

For the amplifier illustrated in Figure 5–14, negative feedback provided excellent bias stability that was not dependent on a particular β. It's also possible to provide excellent gain stability that is independent of β by using negative feedback with the ac signal. As you will see, negative feedback produces a self-correcting action that stabilizes the voltage gain. The modification in Figure 5–15 shows how this is achieved.

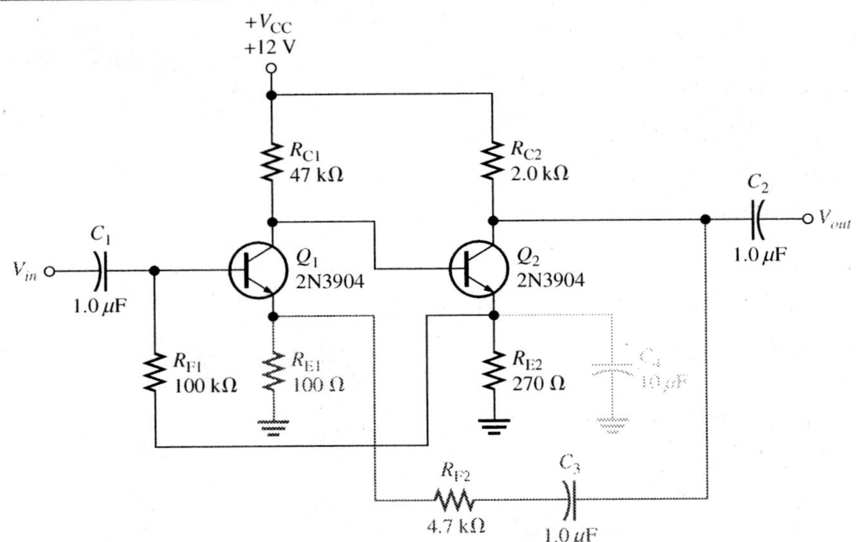

FIGURE 5–15

Modification of the circuit in Figure 5–14 to improve gain stability.

COMPUTER SIMULATION

Open the Multisim file F05-15DV on the website. Observe the input and output signals. Measure the gain.

First, a bypass capacitor, C_4, is connected in parallel with R_{E2} in order to boost the voltage gain even higher; this will produce greater gain stability when feedback is added. The gain without feedback is called **open-loop voltage gain**, which will be described further when you study operational amplifiers in Chapter 6. For the amplifier in Figure 5–15, the addition of the emitter capacitor boosts the open-loop voltage gain by a factor of approximately two.* Then a new path is added, consisting of C_3 and R_{F2}, to return a fraction of the output ac signal back to Q_1. The fraction that is returned is determined by the voltage divider consisting of R_{F2} and R_{E1}. For the amplifier in Figure 5–15, the feedback voltage, V_f, is equal to the output voltage multiplied by the feedback fraction. You will recognize the feedback voltage is derived from a simple voltage divider.

$$V_f = \left(\frac{R_{E1}}{R_{E1} + R_{F2}}\right)V_{out}$$

This feedback voltage tends to cancel the original input signal. The signal that is amplified by the open-loop voltage gain is the small *difference* in the input and negative feedback signals. As a result, the net voltage gain of the amplifier is controlled by the amount of feedback. This net gain with feedback is called the **closed-loop voltage gain**. As mentioned, the closed-loop voltage gain is determined by the amount of output signal that is returned.

An implication of a very large open-loop voltage gain is that the difference between the feedback and input signals is very small at the input to Q_1. For the amplifier in Figure 5–15, the ac signal on the base and emitter of Q_1 will have nearly the same amplitude.

Here's how the negative feedback works to achieve gain stability. Suppose the voltage gain increases due to heating (causing r'_e to be smaller). The increased open-loop gain causes the output voltage to increase and, in turn, increases the negative feedback voltage. This reduces the difference voltage at Q_1. Thus, the original change in gain is almost completely canceled by the self-correcting action of negative feedback.

Now assume a technician replaces one of the transistors with one with a lower β than in the original circuit. This causes a decrease in the open-loop gain of the amplifier. Now there will be a smaller feedback voltage that causes the difference voltage to be larger. Since there is a larger difference voltage, the original effect of a lower β has little net effect on the output voltage and, again, gain stability is achieved.

The net voltage gain of the amplifier is approximately equal to the reciprocal of the feedback fraction. For the amplifier in Figure 5–15, the net gain is

$$A_v = \left(\frac{R_{E1} + R_{F2}}{R_{E1}}\right) = \left(\frac{100\ \Omega + 4.7\ \text{k}\Omega}{100\ \Omega}\right) = 48$$

As you can see, it is easy to change the gain by simply changing the value of R_{F2}. In fact, a gain control can be easily added by using a variable resistor in place of R_{F2}. This circuit is the focus of Experiment 9 in the lab manual.

* The gain would be even higher except for the adverse loading effect on Q_1 due to lower input resistance of Q_2.

Review Questions

11. What are the main advantages of a direct-coupled amplifier?
12. How does negative feedback produce bias or gain stability?
13. Why does the addition of a bypass capacitor in the emitter circuit of a CE amplifier improve the gain stability but not the bias stability?
14. How is gain determined for the circuit in Figure 5–15?
15. For the circuit in Figure 5–15, how does R_{F2} affect the gain? Explain your answer.

CLASS A POWER AMPLIFIERS 5–4

When an amplifier is biased such that it always operates in the linear region where the output signal is an amplified replica of the input signal, it is a class A amplifier. The discussion and formulas in the previous sections apply to class A operation. Power amplifiers are those amplifiers that have the objective of delivering power to a load.

In this section, you will learn to compute dc and ac parameters for a class-A power amplifier and describe operation along the ac load line.

In a small-signal amplifier, the ac signal moves over a small percentage of the total ac load line. When the output signal is larger and approaches the limits of the ac load line, the amplifier is a **large-signal** type. Both large-signal and small-signal amplifiers are considered to be **class A** if they operate in the active region at all times. Class A power amplifiers are large-signal amplifiers with the objective of providing power (rather than voltage) to a load. As a rule of thumb, an amplifier may be considered to be a power amplifier if it is necessary to consider the problem of heat dissipation in components (> 1/4 W). One of the most useful class A power amplifiers is the common-collector amplifier, so it will be the focus in this discussion.

Heat Dissipation

Power transistors (and other power devices) must dissipate excessive internally generated heat. For bipolar power transistors, the collector terminal is the critical junction; for this reason, the transistor's case is always connected to the collector terminal. The case of all power transistors is designed to provide a large contact area between it and an external heat sink. Heat from the transistor flows through the case to the heat sink and then dissipates in the surrounding air. Heat sinks vary in size, number of fins, and type of material. Their size depends on the heat dissipation requirement and the maximum ambient temperature in which the transistor is to operate. In high-power applications, a cooling fan may be necessary.

Centered Q-Point

Recall (from Section 3–4) that the dc and ac load lines cross at the Q-point. When the Q-point is at the center of the ac load line, a maximum class A signal can be obtained. You can see this concept by examining the graph of the load line for a given amplifier in Figure 5–16(a). This graph shows the ac load line with the Q-point at its center. The collector current can vary from its Q-point value, I_{CQ}, up to its saturation value, $I_{c(sat)}$, and down to its cutoff value of zero. Likewise, the collector-to-emitter voltage can swing from its Q-point value, V_{CEQ}, up to its cutoff value, $V_{ce(cutoff)}$, and down to its saturation value of near zero. This operation is indicated in Figure 5–16(b). The peak value of the

FIGURE 5–16

Maximum class A output occurs when the Q-point is centered on the ac load line.

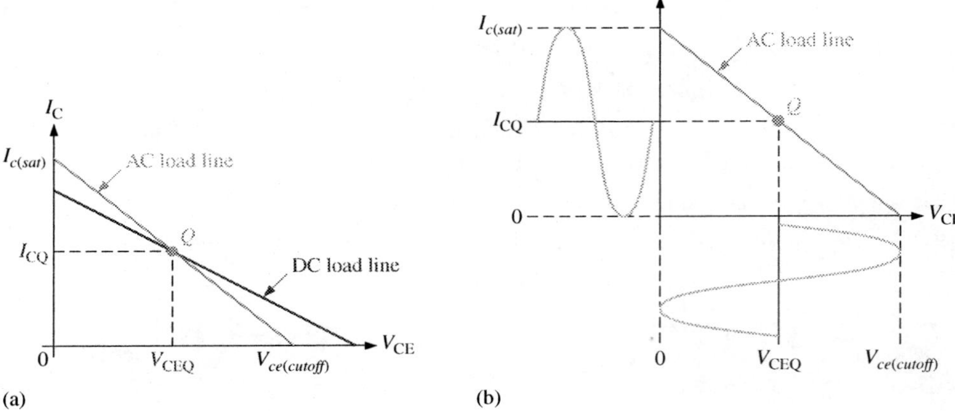

collector current equals I_{CQ}, and the peak value of the collector-to-emitter voltage equals V_{CEQ} in this case. This signal is the maximum that can be obtained from the class A amplifier. Actually, the output cannot quite reach saturation or cutoff, so the practical maximum is slightly less.

If the Q-point is not centered on the ac load line, the output signal is limited. Figure 5–17 shows a load line with the Q-point moved away from center toward cutoff. The output variation is limited by cutoff in this case. The collector current can only swing down to near zero and an equal amount above I_{CQ}. The collector-to-emitter voltage can only swing up to its cutoff value and an equal amount below V_{CEQ}. This situation is illustrated in Figure 5–17(a). If the amplifier is driven any further than this, it will "clip" at cutoff, as shown in Figure 5–17(b).

FIGURE 5–17

Q-point closer to cutoff.

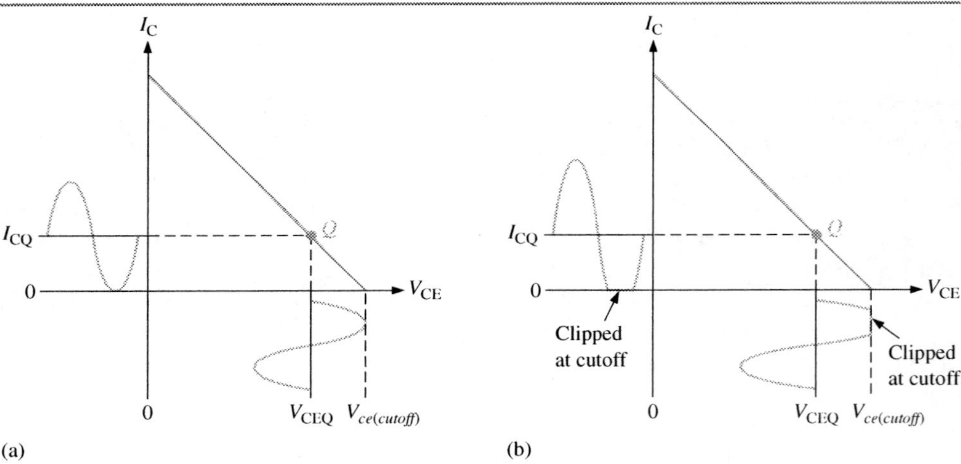

Figure 5–18 shows a load line with the Q-point moved away from center toward saturation. In this case, the output variation is limited by saturation. The collector current can only swing up to near saturation and an equal amount below I_{CQ}. The collector-to-emitter voltage can only swing down to its saturation value and an equal amount above V_{CEQ}. This situation is illustrated in Figure 5–18(a). If the amplifier is driven any further, it will "clip" at saturation, as shown in Figure 5–18(b).

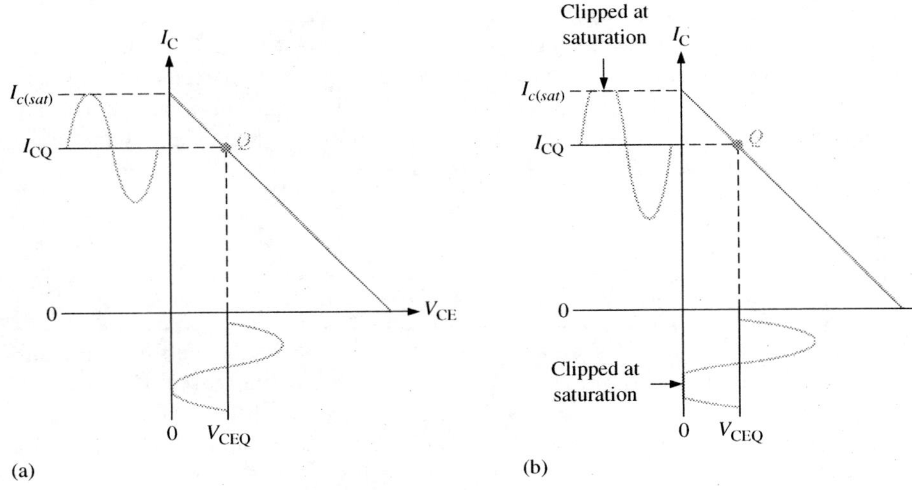

FIGURE 5–18

Q-point closer to saturation.

Power Gain

The power gain of an amplifier is the ratio of the power delivered to the load to the input power, as described in Section 3–3. Power delivered to the load is the same as output power; that is,

$$A_p = \frac{P_{out}}{P_{in}} = \frac{P_L}{P_{in}} \tag{5–4}$$

By substituting $P_L = V_L^2/R_L$ and $P_{in} = V_{in}^2/R_{in}$ into Equation 5–4, you can find power gain using resistance ratios.

$$A_p = \frac{V_L^2}{V_{in}^2}\left(\frac{R_{in}}{R_L}\right)$$

$$A_p = A_v^2\left(\frac{R_{in}}{R_L}\right) \tag{5–5}$$

Equation 5–5 says that the power gain to an amplifier is the voltage gain squared times the ratio of the input resistance to the output load resistance. It can be applied to any amplifier. For example, assume a common-collector (CC) amplifier has an input resistance of 10 kΩ and a load resistance of 100 Ω. Since a CC amplifier has a voltage gain of approximately 1, the power gain is $1^2(10\text{ k}\Omega/100\ \Omega) = 100$. For a CC amplifier, A_p is approximately equal to the ratio of the input resistance to the output load resistance. (R_{in} was discussed in Section 3–4.)

DC Quiescent Power

The power dissipation of a transistor with no signal input is the product of its Q-point current and voltage.

$$P_{DQ} = I_{CQ}V_{CEQ} \tag{5–6}$$

The only way a class A power amplifier can supply power to a load is to maintain a quiescent current that is at least as large as the peak current requirement for the load current. A signal will not increase the power dissipated by the transistor but actually causes less total power to be dissipated. The quiescent power, given in Equation 5–3, is the maximum power that a class A amplifier must handle. The transistor's power rating should normally exceed this value.

Output Power

In general, the output signal power is the product of the rms load current and the rms load voltage. The maximum unclipped ac signal occurs when the Q-point is centered on the ac load line. Assuming the Q-point is centered on the ac load line, you can find the maximum power output from the rms values of maximum current and voltage. The maximum power out from a class A amplifier is $P_{out(max)} = (0.707I_c)(0.707V_c)$.

$$P_{out(max)} = 0.5I_{CQ}V_{CEQ} \quad (5\text{–}7)$$

A Two-Stage Class A Amplifier

Figure 5–19 shows a two-stage class A amplifier. The input signal could be from a preamp or other source. The output is to a small speaker. The first stage consisting of Q_1 is a CE amplifier similar to ones you have seen; it is simply a voltage amplifier. The second stage is the power amplifier stage and is a modified CC amplifier. It is composed of two transistors, Q_2 and Q_3, that are connected in an arrangement called a **Darlington pair**. The Darlington pair is two cascaded transistors connected so that the emitter of the first drives the base of the second. The collectors are connected together, forming an equivalent high β transistor used as a CC amplifier. The voltage gain of this stage is the same as any CC amplifier, namely one; however, the power gain is considerably more. This method is a commonly used way to increase the input resistance of the amplifier and deliver a reasonable amount of power to a load.

FIGURE 5–19 A two-stage class A power amplifier.

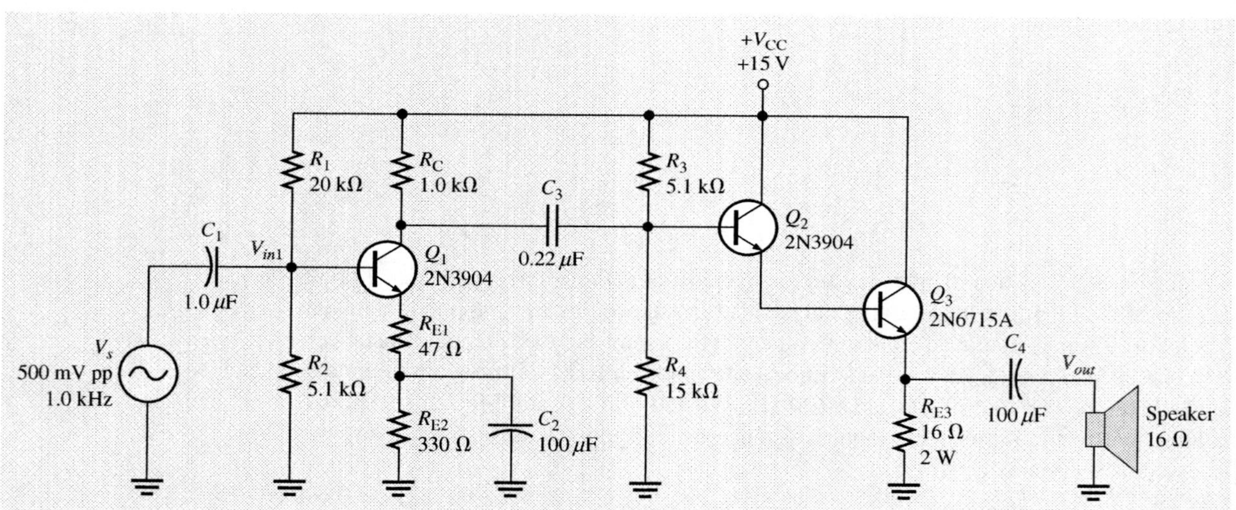

COMPUTER SIMULATION

Open the Multisim file F05-19DV on the website. Observe the signals at the input, C_3, and the output. Measure the overall gain.

The power amplifier in Figure 5–19 can be modeled as two independent amplifiers connected together as shown in Figure 5–20. The CE voltage amplifier is shown in the left box (Stage 1), and the CC power amplifier (consisting of the Darlington pair) is shown in the right box (Stage 2). The overall voltage gain for this amplifier is a modest −15, taking into account all loading effects, but the power gain is considerably more. To calculate the power gain, we can use Equation 5–5.

$$A_p = A_v^2\left(\frac{R_{in}}{R_L}\right) = (-15)^2\left(\frac{2.9\ \text{k}\Omega}{16\ \Omega}\right) = 41{,}000$$

Amplifier model. (V_{in} is shown as V_{in1} for the first stage). **FIGURE 5–20**

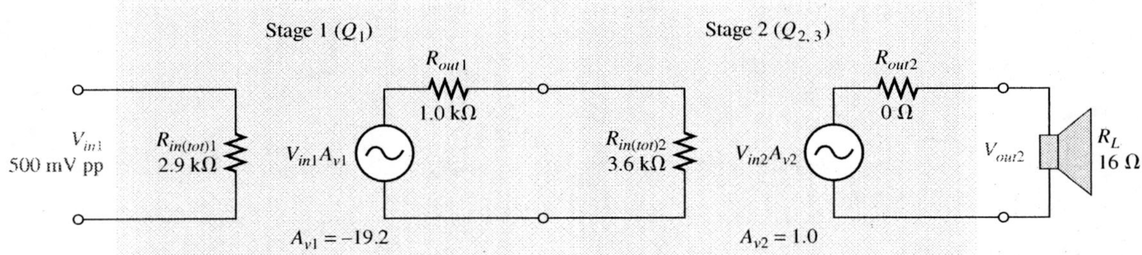

Efficiency

The **efficiency** of any amplifier is the ratio of the signal power supplied to the load to the power from the dc supply. The maximum signal power that can be obtained is given by Equation 5–7. The average power supply current, I_{CC}, is equal to I_{CQ} and the supply voltage is at least $2V_{CEQ}$. Therefore, the dc power is

$$P_{DC} = I_{CC}V_{CC} = 2I_{CQ}V_{CEQ}$$

The maximum efficiency of a capacitively coupled load is

$$eff_{max} = \frac{P_{out}}{P_{DC}} = \frac{0.5I_{CQ}V_{CEQ}}{2I_{CQ}V_{CEQ}} = 0.25$$

The maximum efficiency of a capacitively coupled class A amplifier cannot be higher than 0.25, or 25%, and, in practice, is usually considerably less (about 10%). Although the efficiency can be made higher by transformer coupling the signal to the load, there are drawbacks to transformer coupling. These drawbacks include the size and cost of transformers as well as potential distortion problems when the transformer core begins to saturate. In general, the low efficiency of class A power amplifiers limits their usefulness to small power applications that require only a few watts of load power.

EXAMPLE 5–3

Problem

Assume the dc supply for the amplifier in Figure 5–19 has a current of 0.6 A. Determine the efficiency of the power amplifier.

Solution

The efficiency is the ratio of the signal power in the load to the power supplied by the dc source. The input voltage is 500 mV peak-to-peak which is 176 mV rms. The input power is, therefore,

$$P_{in} = \frac{V_{in}^2}{R_{in}} = \frac{(176\ \text{mV})^2}{2.9\ \text{k}\Omega} = 10.7\ \mu\text{W}$$

The output power is

$$P_{out} = P_{in}A_p = (10.7\ \mu\text{W})(41{,}000) = 0.44\ \text{W}$$

The power from the power supply is

$$P_{\text{DC}} = I_{\text{CC}}V_{\text{CC}} = (0.6\ \text{A})(15\ \text{V}) = 9\ \text{W}$$

Therefore, the efficiency of the amplifier for this input is

$$eff = \frac{P_{out}}{P_{\text{DC}}} = \frac{0.44\ \text{W}}{9\ \text{W}} \cong \mathbf{0.05}$$

This represents an efficiency of 5%.

Question
What happens to the efficiency if R_{E3} were replaced with the speaker? What disadvantage does this have?

Review Questions

16. What is the purpose of a heat sink?
17. Which lead of a BJT is connected to the case?
18. What are the two types of clipping with a class A amplifier?
19. What is the maximum theoretical efficiency for a class A amplifier?
20. How can the power gain of a CC amplifier be expressed in terms of a ratio of resistances?

5–5 CLASS B POWER AMPLIFIERS

When an amplifier is biased such that it operates in the linear region for 180° of the input cycle and is in cutoff for 180°, it is a class B amplifier. The primary advantage of a class B amplifier over a class A amplifier is that the class B is much more efficient; you can get more output power for a given amount of input power. Class B amplifiers are generally configured with at least two active devices that alternately amplify the positive and negative part of the input waveform. This arrangement is called push-pull.

In this section, you will learn to compute dc and ac parameters for a class-B and class AB power amplifier (both bipolar and FET) and describe the load lines.

Class B operation refers to operation when the Q-point is located at cutoff, causing the output current to vary only during one-half of the input cycle. In a linear amplifier, two devices are required for a complete cycle; one amplifies the positive cycle and the other amplifies the negative cycle. As you will see, this arrangement has a great advantage for power amplifiers as it greatly increases the efficiency. For this reason, they are widely used as power amplifiers. We will restrict our coverage to bipolar transistors, but keep in mind that either bipolar or MOSFET transistors can be used as power amplifiers.

The Q-point Is at Cutoff

The class B amplifier is biased at cutoff so that $I_{CQ} = 0$ and $V_{CEQ} = V_{CE(cutoff)}$. Unlike the class A amplifier, there is *no dc current or power dissipated when there is no signal*. When a signal drives a class B amplifier into conduction, it then operates in its linear region. This is illustrated in Figure 5–21 with a CC amplifier (emitter-follower).

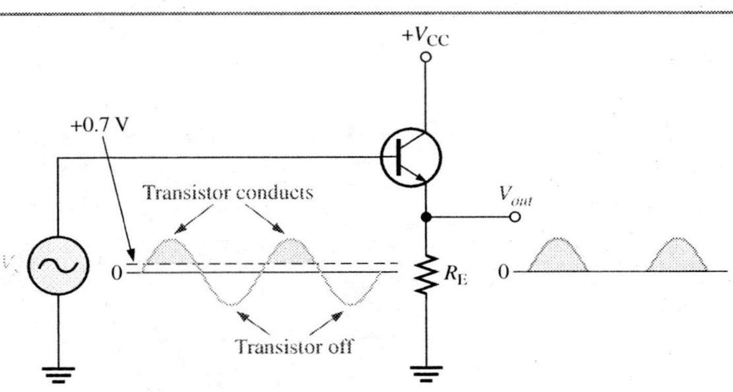

FIGURE 5–21

Common-collector class B amplifier.

Push-Pull Operation

As you can see, the circuit in Figure 5–21 only conducts for the positive half of the cycle. To amplify the entire cycle, it is necessary to add a second class B amplifier that operates on the negative half. The combination of two class B amplifiers working together is called **push-pull** operation.

The most common approach for using push-pull amplifiers together uses two **complementary symmetry transistors**; these are a matching pair of *npn/pnp* BJTs or a matching pair of *n*-channel/*p*-channel FETs. Figure 5–22 shows one of the most popular types of push-pull class B amplifiers using two emitter-followers and both positive and negative power supplies. This is a complementary amplifier because one emitter-follower uses an *npn* transistor, which conducts on the positive half of the input cycle, and the other uses a *pnp* transistor, which conducts on the negative half of the cycle. Notice that there is no dc base bias voltage

FIGURE 5–22

Class B push-pull operation.

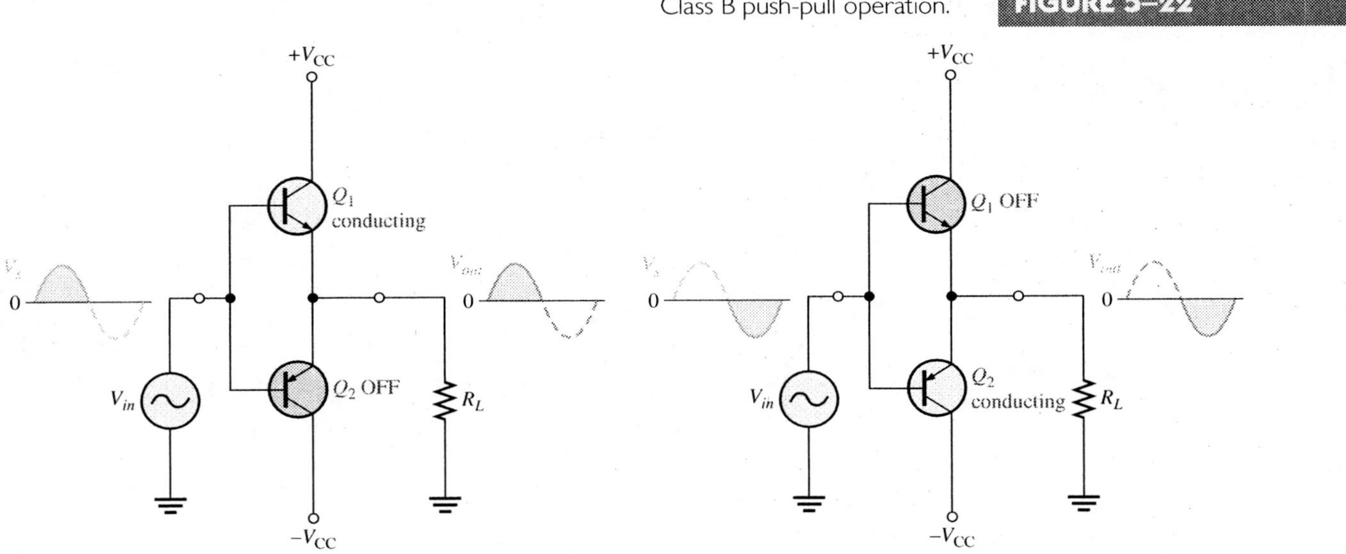

($V_B = 0$). Thus, only the signal voltage drives the transistors into conduction. Transistor Q_1 conducts during the positive half of the input cycle, and Q_2 conducts during the negative half.

Crossover Distortion

When the dc base voltage is zero, the input signal voltage must exceed V_{BE} before a transistor conducts. As a result, there is a time interval between the positive and negative alternations of the input when neither transistor is conducting, as shown in Figure 5–23. The resulting distortion in the output waveform is called **crossover distortion**.

FIGURE 5–23

Crossover distortion in a class B amplifier.

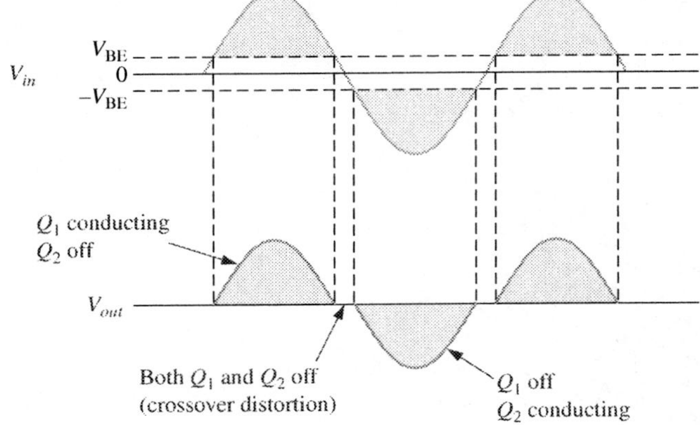

Biasing the Push-Pull Amplifier

To overcome crossover distortion, the biasing is adjusted to just overcome the V_{BE} of the transistors; this results in a modified form of operation called **class AB**. In class AB operation, the push-pull stages are biased into slight conduction, even when no input signal is present. This can be done with a voltage-divider and diode arrangement, as shown in Figure 5–24. When the diode characteristics of D_1 and D_2 are closely matched to the characteristics of the transistor base-emitter junctions, the current in the diodes and the cur-

FIGURE 5–24

Biasing the push-pull amplifier to eliminate crossover distortion.

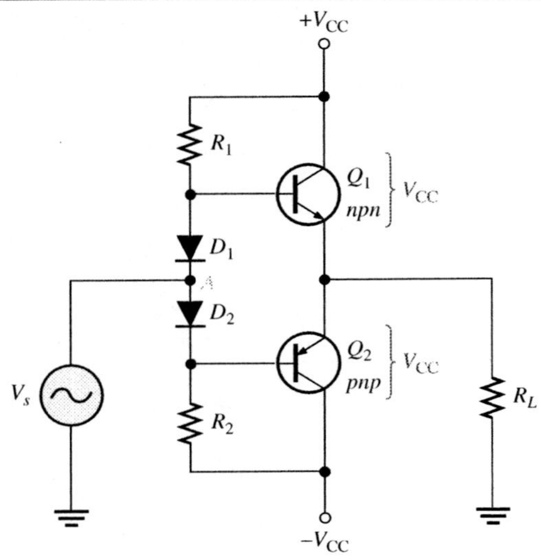

rent in the transistors is the same; this is called a **current mirror**. This current mirror produces the desired class AB operation and eliminates crossover distortion.

In the bias path, R_1 and R_2 are of equal value, as are the positive and negative supply voltages. This forces the voltage at point *A* to equal 0 V and eliminates the need for an input coupling capacitor. The dc voltage on the output is also 0 V. Assuming that both diodes and both transistors are identical, the drop across D_1 equals the V_{BE} of Q_1, and the drop across D_2 equals the V_{BE} of Q_2. Since they are matched, the diode current will be the same as I_{CQ}. The diode current and I_{CQ} can be found by applying Ohm's law to either R_1 or R_2 as follows:

$$I_{CQ} = \frac{V_{CC} - 0.7\text{ V}}{R_1}$$

This small current required of class AB operation eliminates the crossover distortion but has the potential for thermal instability if the transistor's V_{BE} drops are not matched to the diode drops or if the diodes are not in thermal equilibrium with the transistors. Heat in the power transistors decreases the base-emitter voltage and tends to increase current. If the diodes are warmed the same amount, the current is stabilized; but if the diodes are in a cooler environment, they cause I_{CQ} to increase even more. More heat is produced and thermal runaway may begin if it is not controlled. To keep this from happening, the diodes should have the same thermal environment as the transistors. In stringent cases, a small resistor in the emitter of each transistor can alleviate thermal runaway.

AC Operation

Consider the ac load line for Q_1 of the class AB amplifier in Figure 5–24. The Q-point is slightly above cutoff. (In a true class B amplifier, the Q-point is at cutoff.) The ac cutoff voltage for a two-supply operation is at V_{CC} with an I_{CQ} as given in the last equation. The ac saturation current for a two-supply operation with a push-pull amplifier is

$$I_{c(sat)} = \frac{V_{CC}}{R_L} \tag{5–8}$$

The ac and dc load lines for the *npn* transistor are as shown in Figure 5–25(a). The dc load line can be found by drawing a line that passes through V_{CEQ} and the dc saturation

FIGURE 5–25

Load analysis for a complementary symmetry push-pull amplifier. Only the load lines for the *npn* transistor are shown.

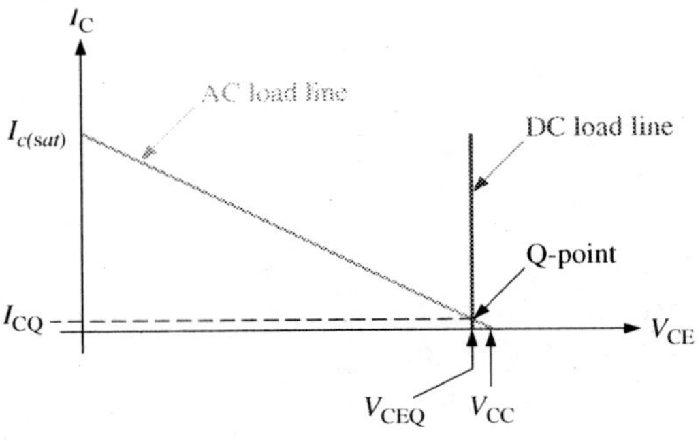

(a) DC and ac load lines for Q_1.

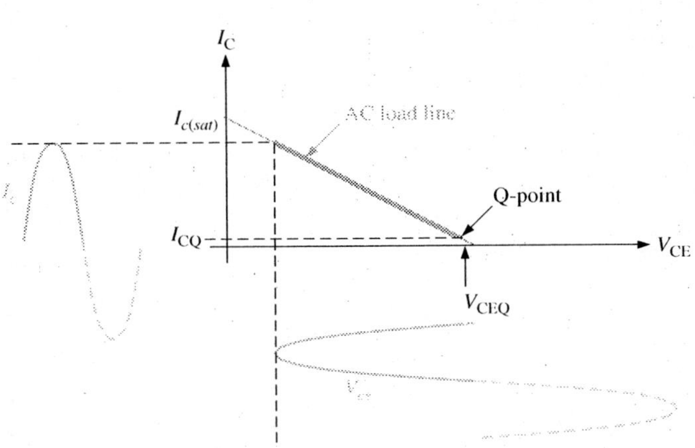

(b) AC load line operation for Q_1

current $I_{C(sat)}$. But the saturation current for dc is the current if the collector to emitter is shorted on both transistors! This assumed short across the power supplies obviously would cause maximum current from the supplies and implies the dc load line passes almost vertically through the cutoff as shown. Operation along the dc load line, such as caused by thermal runaway, could produce such a high current that the transistors are destroyed.

Figure 5–25(b) illustrates the ac load line for Q_1 of the class B amplifier. In the case illustrated, a signal is applied that swings over the region of the ac load line shown in bold. At the upper end of the ac load line, the voltage across the transistor (V_{ce}) is a minimum, and the output voltage is maximum.

Under maximum conditions, transistors Q_1 and Q_2 are alternately driven from near cutoff to near saturation. During the positive alternation of the input signal, the Q_1 emitter is driven from its Q-point value of 0 to nearly V_{CC}, producing a positive peak voltage a little less than V_{CC}. Likewise, during the negative alternation of the input signal, the Q_2 emitter is driven from its Q-point value of 0 V, to near $-V_{CC}$, producing a negative peak voltage almost equal to $-V_{CC}$. Although it is possible to operate close to the saturation current, this type of operation results in increased distortion of the signal.

The ac saturation current given in Equation 5–8 is also the peak output current. Each transistor can essentially operate over its entire load line. Recall that in class A operation, the transistor can also operate over the entire load line but with a significant difference. In class A operation, the Q-point is near the middle and there is significant current in the transistors even with no signal. In class B operation, when there is no signal, the transistors have only a very small current and therefore dissipate very little power. Thus, the efficiency of a class B amplifier can be much higher than a class A amplifier. It can be shown that the maximum theoretical efficiency of a class B amplifier is 79%.

Single-Supply Operation

Push-pull amplifiers using complementary symmetry transistors can be operated from a single voltage source as shown in Figure 5–26. The circuit operation is the same as that described previously, except the bias is set to force the output emitter voltage to be $V_{CC}/2$ instead of zero volts used with two supplies. Because the output is not biased at zero volts, capacitive coupling for the input and output is necessary to block the bias voltage from the source and the load resistor. Ideally, the output voltage can swing from zero to V_{CC}, but in practice it does not quite reach these ideal values.

FIGURE 5–26

Single ended push-pull amplifier.

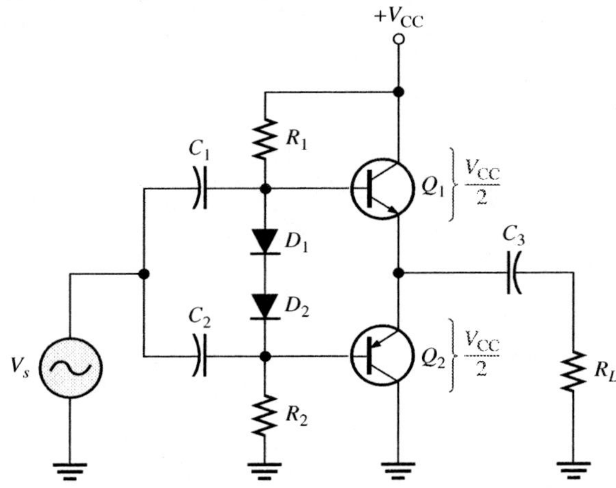

EXAMPLE 5–4

Problem

Determine the ideal maximum peak output voltage and current for the circuit shown in Figure 5–27.

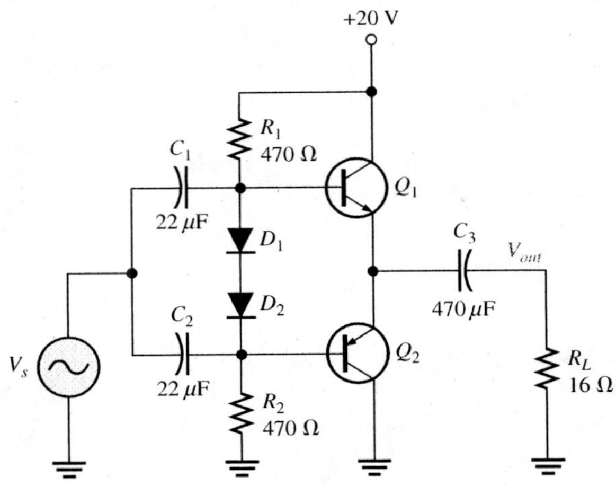

FIGURE 5–27

Solution

The ideal maximum peak output voltage is

$$V_{p(out)} \cong V_{\mathrm{CEQ}} \cong \frac{V_{\mathrm{CC}}}{2} = \frac{20\text{ V}}{2} = \mathbf{10\ V}$$

The ideal maximum peak current is

$$I_{p(out)} \cong I_{c(sat)} \cong \frac{V_{\mathrm{CEQ}}}{R_L} = \frac{10\text{ V}}{16\ \Omega} = \mathbf{0.63\ A}$$

The actual maximum values of voltage and current are slightly smaller.

Question

What is the maximum peak output voltage and current if V_{CC} is raised to +30 V?

COMPUTER SIMULATION

Open the Multisim file F05-27DV on the website. Observe the signals at the input, the base of each transistor, and at the output.

Review Questions

21. What is the difference between class B and class AB operation?
22. What is the advantage to two-supply operation with a class B complementary symmetry amplifier over single-supply operation?
23. What is crossover distortion and how is it avoided?
24. What is the maximum theoretical efficiency for a class B amplifier?
25. How is thermal runaway avoided in a circuit like that shown in Figure 5–26?

5–6 DIFFERENTIAL AMPLIFIERS

A differential amplifier is an important type of amplifier used in many applications. In particular, it is the input stage to operational amplifiers, which will be introduced in Chapter 6. The differential amplifier gets its name from its ability to amplify the difference in two input signals applied to its two inputs. Only the difference in the two signals is amplified; if identical signals are applied to the inputs, no amplification takes place.

In this section, you will learn the operation of a differential amplifier including how it can significantly reduce common-mode noise

Basic Operation

A **differential amplifier** (diff-amp) circuit and its symbol are shown in Figure 5–28. Unlike other amplifiers you have seen, there are two inputs and two outputs in a basic diff-amp. The transistors are carefully matched to have identical characteristics. Also, for the diff-amp in our example, assume the collector resistors are matched and have the same resistance. Notice that the two transistors share a common emitter resistor, R_E.

FIGURE 5–28

Basic differential amplifier.

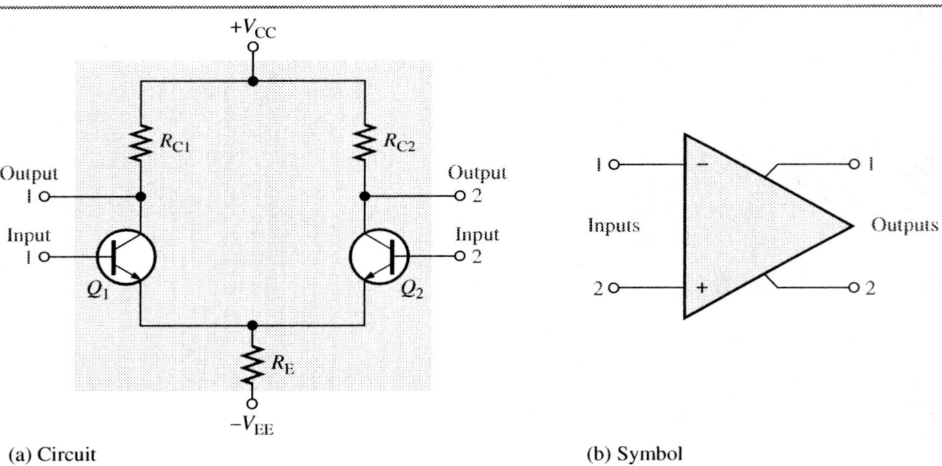

To understand the operation, assume both bases are on ground. The emitter voltage will be −0.7 V because there is one diode drop across either base-emitter junction. The emitter currents are equal ($I_{E1} = I_{E2}$) and are each one-half of the current in the common-emitter resistor. The emitter and collector currents in each transistor are approximately the same (ignoring the small base current) and one-half of the current in the emitter resistor. Because the collector currents are the same, the collector voltages are also the same.

Let's remove the base of Q_1 from ground and connect it to a small positive voltage (a few tenths of a volt) and investigate what happens. Q_1 will conduct more because of the positive voltage on its base, causing the emitter voltage to increase slightly. Although the emitter voltage is little higher, the total current in R_E is nearly the same. The emitter current is now divided so that more of it is in Q_1 and less in Q_2 (because of the slightly higher emitter voltage). As a result, the collector voltage of Q_1 will drop, and the collector voltage of Q_2 will increase. This situation is illustrated in Figure 5–29(a).

FIGURE 5–29

Basic operation of a differential amplifier showing the effect on current when a small voltage is connected to one of the bases.

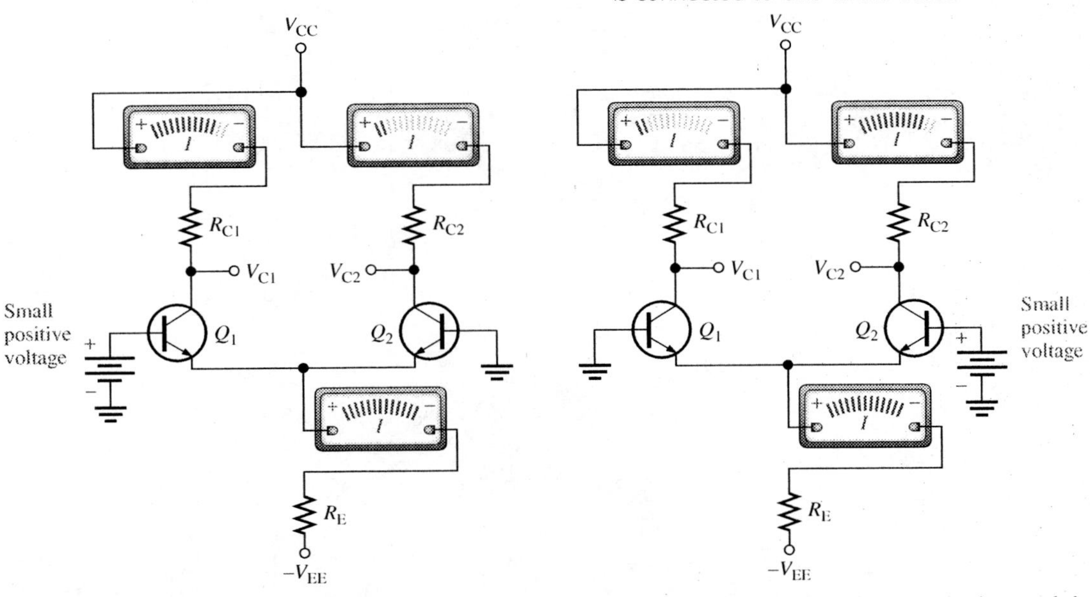

(a) A small positive voltage on Q_{B1} when Q_{B2} is grounded (b) A small positive voltage on Q_{B2} when Q_{B1} is grounded

Now assume that the base of Q_1 is placed back on ground and a small positive voltage is placed on the base of Q_2. The situation is analogous to the earlier one, but this time Q_2 will conduct more and Q_1 will conduct less. The emitter current is now divided so that more of it is in Q_2 and less in Q_1. As a result, the collector voltage of Q_1 will increase, and the collector voltage of Q_2 will decrease. This situation is illustrated in Figure 5–29(b).

Modes of Signal Operation

Single-Ended Input

In the single-ended mode, one input is grounded and the signal voltage is applied only to the other input, as shown in Figure 5–30(a). In the case where the signal voltage is applied to input 1 as in part (a), an inverted, amplified signal voltage appears at output 1 as shown. The reason that it is inverted is that Q_1 looks like a common-emitter (CE) amplifier for an input on its own base. Also, a signal voltage appears in phase at the emitter of Q_1 because it acts as a common-collector (CC) amplifier for a signal at this point in the circuit. Since the emitters of Q_1 and Q_2 are common, the emitter signal becomes the input to Q_2, which functions as a common-base (CB) amplifier, which does not invert the signal. The signal is amplified by Q_2 and appears, noninverted, at output 2. This action is illustrated in part (a).

In the case where the signal is applied to input 2 with input 1 grounded, as in Figure 5–30(b), the roles of the transistors are reversed. This time an inverted, amplified signal voltage appears at output 2 because Q_2 acts as a CE amplifier for output 2. Q_2 acts as a CC amplifier for a signal taken at its own emitter and Q_1 acts like a CB amplifier for this signal. A noninverted, amplified signal is therefore at output 1. This action is illustrated in part (b).

Differential Input

In the **differential-mode** condition, two opposite-polarity (out-of-phase) signals are applied to the inputs, as shown in Figure 5–31(a). This type of operation is also referred to as *double-ended.* As you will see, each input affects the outputs.

Figure 5–31(b) shows the output signals due to the signal on input 1 acting alone as a single-ended input. Figure 5–31(c) shows the output signals due to the signal on input 2

FIGURE 5–30

Single-ended operation of a differential amplifier.

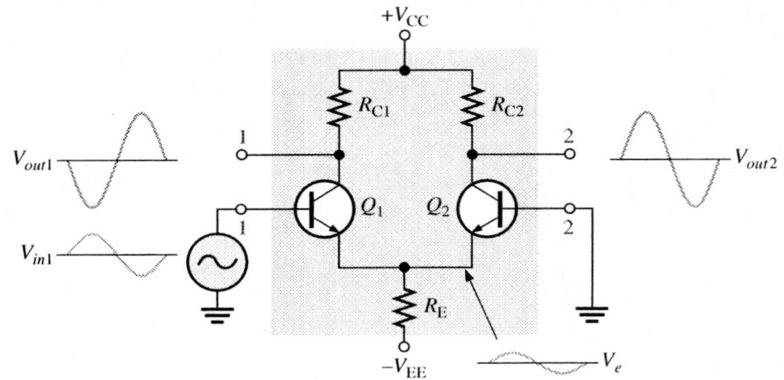

(a)

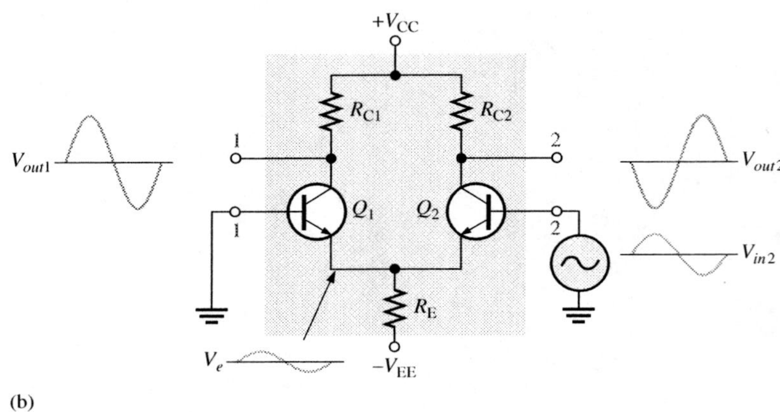

(b)

FIGURE 5–31

Differential-mode operation of a differential amplifier.

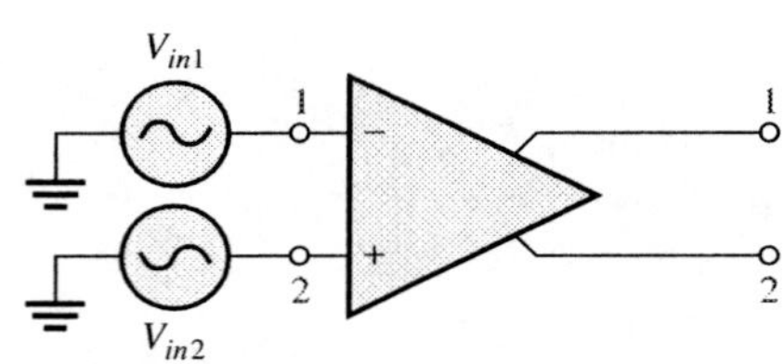

(a) Differential inputs

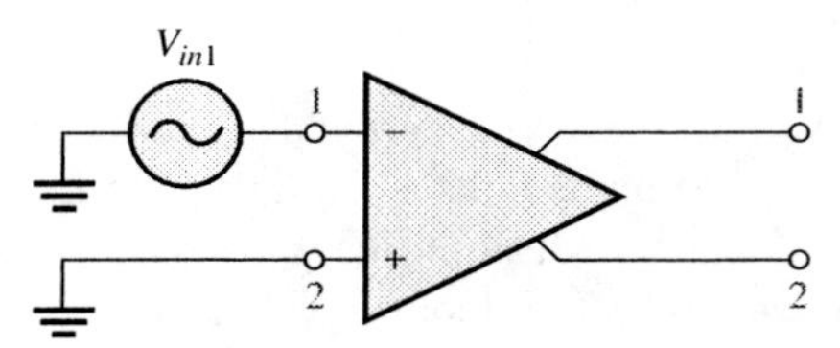

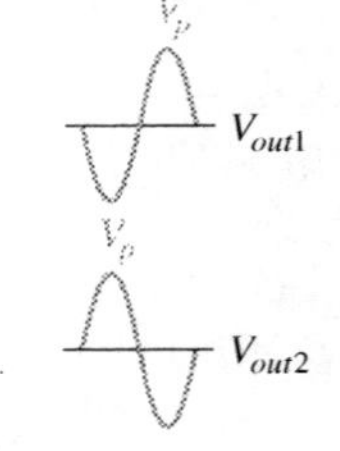

(b) Outputs due to V_{in1}

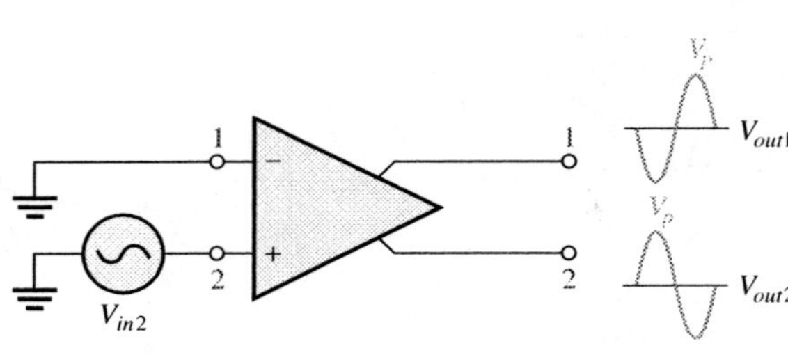

(c) Outputs due to V_{in2}

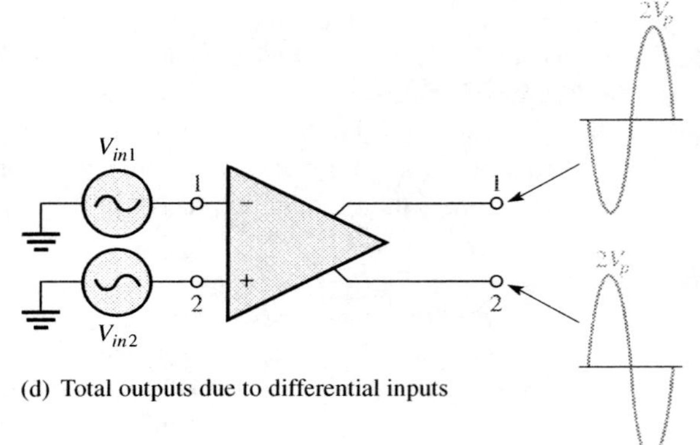

(d) Total outputs due to differential inputs

acting alone as a single-ended input. Notice in parts (b) and (c) that the signals on output 1 are of the same polarity. The same is also true for output 2. By superimposing both output 1 signals and both output 2 signals, you get the total differential operation, as shown in Figure 5–31(d).

Common-Mode Input

One of the most important aspects of the operation of a differential amplifier can be seen by considering the **common-mode** condition where two identical signals are applied to the two inputs, as shown in Figure 5–32(a). Again, by considering each input signal as acting alone, you can understand the basic operation.

Common-mode operation of a differential amplifier. **FIGURE 5–32**

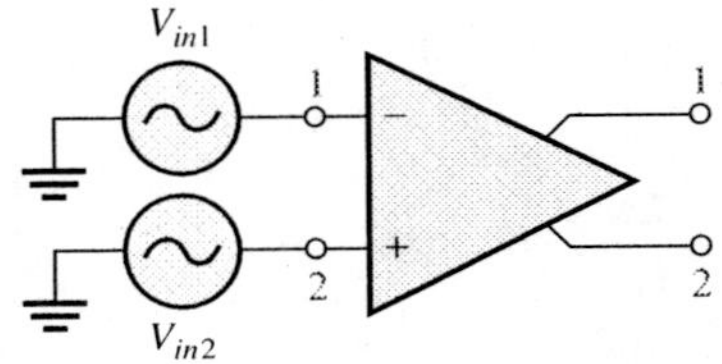

(a) Common-mode inputs

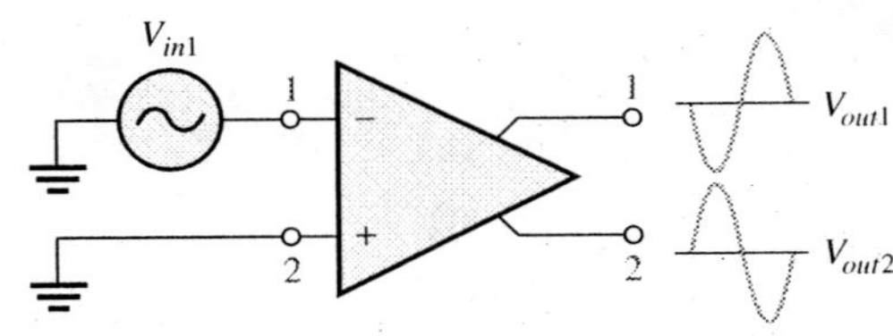

(b) Outputs due to V_{in1}

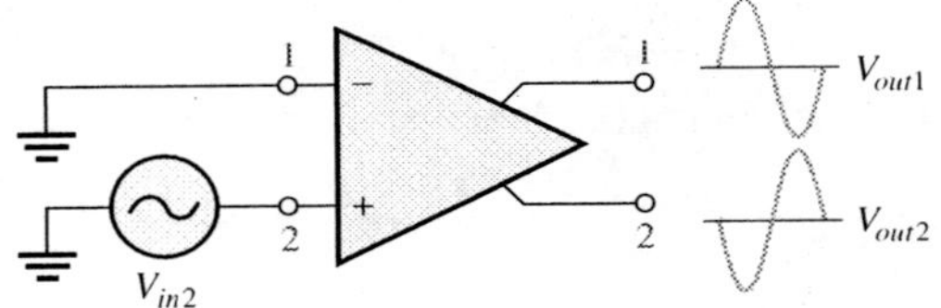

(c) Outputs due to V_{in2}

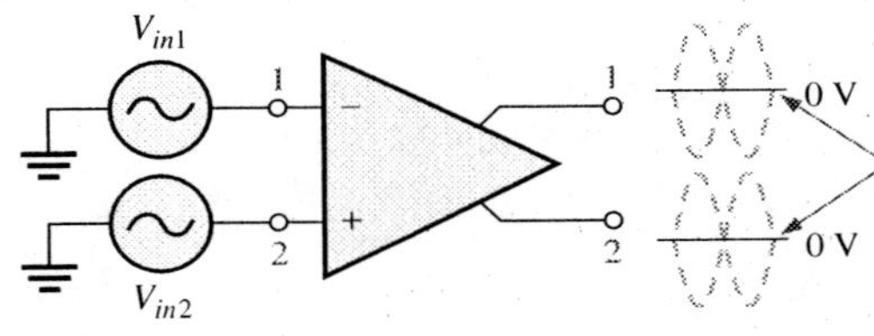

(d) Outputs cancel when common-mode signals are applied

Figure 5–32(b) shows the output signals due to the signal on only input 1, and Figure 5–32(c) shows the output signals due to the signal on only input 2. Notice that the corresponding signals on output 1 are of the opposite polarity, and so are the ones on output 2. When the input signals are applied to both inputs, the outputs are superimposed and they cancel, resulting in a zero output voltage, as shown in Figure 5–32(d).

This action is called *common-mode rejection.* Its importance lies in the situation where an unwanted signal appears commonly on both diff-amp inputs. Common-mode rejection means that this unwanted signal will not appear on the outputs to distort the desired signal. Common-mode signals (noise) generally are the result of the pick-up of radiated energy on the input lines, from adjacent lines, or the 60 Hz power line, or other sources.

Common-Mode Rejection Ratio

Wanted signals appear on only one input or with opposite polarities on both input lines. These wanted signals are amplified and appear on the outputs as previously discussed. Unwanted signals (noise) appearing with the same polarity on both input lines are essentially cancelled by the diff-amp and do not appear on the outputs. The measure of an amplifier's ability to reject common-mode signals is a parameter called the **common-mode rejection ratio (CMRR)**.

Ideally, a differential amplifier provides a very high gain for desired signals (single-ended or differential) and zero gain for common-mode signals. Practical diff-amps, however, do exhibit a very small common-mode gain (usually much less than 1), while providing a high differential voltage gain (usually several thousand). The higher the differential gain with respect to the common-mode gain, the better the performance of the diff-amp in terms of rejection of common-mode signals. This suggests that a good measure of the diff-amp's performance in rejecting unwanted common-mode signals is the ratio of the differential gain $A_{v(d)}$ to the common-mode gain, A_{cm}. This ratio is the common-mode rejection ratio, CMRR.

$$\text{CMRR} = \frac{A_{v(d)}}{A_{cm}} \tag{5–9}$$

EXAMPLE 5–5

Problem

A certain diff-amp has a differential voltage gain of 2000 and a common-mode gain of 0.2. Determine the CMRR.

Solution

$A_{v(d)} = 2000$, and $A_{cm} = 0.2$. Therefore,

$$\text{CMRR} = \frac{A_{v(d)}}{A_{cm}} = \frac{2000}{0.2} = \mathbf{10{,}000}$$

Question

Determine the CMRR for an amplifier with a differential voltage gain of 8500 and a common-mode gain of 0.25.

A CMRR of 10,000, for example, means that the desired input signal (differential) is amplified 10,000 times more than the unwanted noise (common-mode). So, as an example, if the amplitudes of the differential input signal and the common-mode noise are equal, the desired signal will appear on the output 10,000 times greater in amplitude than the noise. Thus, the noise or interference has been essentially eliminated.

Review Questions

26. What fraction of the current in the emitter resistor of a differential amplifier is in the emitter of each transistor if the inputs are both grounded?
27. If identical inputs are applied to the inputs of a differential amplifier, what happens to the output of an ideal amplifier?
28. What is the difference between a differential-mode input and a common-mode input of a diff-amp?
29. What is meant by a single-ended input signal?
30. What does the term *common-mode rejection ratio* mean?

Key Terms

Class A An amplifier that operates in the active region at all times.

Class AB An amplifier that is biased into slight conduction; the Q-point is slightly above cutoff.

Class B An amplifier that has the Q-point located at cutoff, causing the output current to vary only during one-half of the input cycle.

Closed-loop voltage gain The net voltage gain of an amplifier when negative feedback is included.

Common-mode The input condition where two identical signals are applied to the inputs of a differential amplifier.

Common-mode rejection ratio (CMRR) A measure of the ability of an amplifier to reject common-mode signals; it is the ratio of the differential gain to the common-mode gain.

Current mirror A circuit that uses matching diode junctions to form a current source. The current in a diode is reflected as a matching current in the other junction (which is typically the base-emitter junction of a transistor). Current mirrors are commonly used to bias a push-pull amplifier.

Differential amplifier An amplifier that produces an output voltage proportional to the difference of two input voltages.

Differential-mode The input condition where two opposite polarity signals are applied to the inputs of a differential amplifier.

Efficiency (power) The ratio of the signal power supplied to the load to the power from the dc supply.

Intermediate frequency A fixed frequency that is lower than the RF, produced by beating an RF signal with an oscillator frequency.

Mixer A nonlinear circuit that combines two signals and produces the sum and difference frequencies.

Negative feedback The process of returning a portion of the output back to the input in a manner to cancel a fraction of the input.

Open-loop voltage gain The voltage gain of an amplifier without external feedback.

Push-pull A type of class B amplifier with two transistors in which one transistor conducts for one half-cycle and the other conducts for the other half-cycle.

Quality factor (*Q*) A dimensionless number that is the ratio of the maximum energy stored in a cycle to the energy lost in a cycle.

Radio frequency Any frequency greater than 100 kHz.

Important Facts

- ❑ Three ways to couple amplifier stages together are capacitive coupling, transformer coupling, and direct coupling.
- ❑ Capacitive coupling and transformer coupling provide a low-impedance ac path while blocking dc. Direct coupling requires that the dc conditions from one stage are compatible with the requirements of the next stage.

- General points to alleviate noise problems in amplifiers are
  1. Keep wiring short and make signal return loops as small as possible.
  2. Use bypass capacitors between power supply and ground.
  3. Reduce noise sources and separate or shield the noise source and the circuit.
  4. Ground circuits at a single point, and isolate grounds that have high currents from those with low currents.
  5. Keep the bandwidth of amplifiers no larger than necessary.
- Tuned amplifiers use one or more resonant circuits to select a band of frequencies.
- A mixer combines a radio frequency (RF) signal with a sine wave generated from a local oscillator to produce an intermediate frequency (IF) that is amplified by an amplifier tuned to the IF.
- Negative feedback produces a self-correcting action that can produce excellent bias stability and gain stability in amplifiers.
- The voltage gain of an amplifier without feedback is called open-loop voltage gain. The voltage gain of an amplifier with negative feedback is called closed-loop voltage gain.
- A class A amplifier operates entirely in the linear region of the transistor's characteristic curves. The transistor conducts during the full 360° of the input cycle.
- The Q-point must be centered on the ac load line for maximum class A output signal swing.
- The maximum efficiency of a class A amplifier is 25%.
- A class B amplifier operates in the linear region for half of the input cycle (180°), and it is in cutoff for the other half.
- The Q-point is at cutoff for class B operation.
- Class B amplifiers are normally operated in a push-pull configuration in order to produce an output that is a replica of the input.
- The maximum efficiency of a class B amplifier is 79%.
- A class AB amplifier is biased slightly above cutoff and operates in the linear region for slightly more than 180° of the input cycle.
- Class AB eliminates the crossover distortion found in pure class B.
- A differential input voltage appears between the inverting and noninverting inputs of a differential amplifier.
- A single-ended input voltage appears between one input and ground (with the other input grounded).
- A differential output voltage appears between two output terminals of a diff-amp.
- A single-ended output voltage appears between the output and ground of a diff-amp.
- Common mode occurs when equal in-phase voltages are applied to both input terminals.

Formulas

Reflected resistance of a load resistor by a transformer:

$$R'_L = \left(\frac{N_{pri}}{N_{sec}}\right)^2 R_L \qquad \textbf{(5–1)}$$

Resonant frequency (high Q resonant circuit):

$$f_r = \frac{1}{2\pi\sqrt{LC}} \qquad \textbf{(5–2)}$$

Quality factor of a resonant circuit:

$$Q = \frac{X_L}{R} = \frac{f_r}{BW} \tag{5–3}$$

Amplifier power gain:

$$A_p = \frac{P_L}{P_{in}} \tag{5–4}$$

Alternate amplifier power gain:

$$A_p = A_v^2\left(\frac{R_{in}}{R_L}\right) \tag{5–5}$$

Power dissipation of a transistor:

$$P_{DQ} = I_{CQ}V_{CEQ} \tag{5–6}$$

Maximum power from a class A amplifier:

$$P_{out(max)} = 0.5I_{CQ}V_{CEQ} \tag{5–7}$$

AC saturation current for a two-supply operation with a push-pull amplifier:

$$I_{c(sat)} = \frac{V_{CC}}{R_L} \tag{5–8}$$

Common-mode rejection ratio (diff-amp):

$$\text{CMRR} = \frac{A_{v(d)}}{A_{cm}} \tag{5–9}$$

Chapter Checkup

Answers are at the end of the chapter:

1. If an amplifier stage with an unloaded gain of 20 is connected to an identical amplifier stage, the overall gain will be
   (a) less than 400 (b) 400
   (c) greater than 400
2. Noise can enter a circuit
   (a) by capacitive or inductive coupling
   (b) through the power supply
   (c) from within the circuit
   (d) all of these answers
3. The quality factor, Q, is a pure number that is the ratio of
   (a) X_L to X_C (b) X_L to R
   (c) R to X_C (d) none of these answers
4. In a tuned circuit, if the Q is high, the
   (a) resistance is high (b) bandwidth is small
   (c) frequency is low (d) power is high
5. Negative feedback can provide excellent
   (a) bias stability (b) gain stability
   (c) both (a) and (b) (d) neither (a) nor (b)

6. The peak current a class A amplifier can deliver to a load depends on the
   (a) maximum rating of the power supply
   (b) quiescent current
   (c) current in the bias resistors
   (d) size of the heat sink
7. An amplifier that operates in the linear region at all times is
   (a) Class A (b) Class AB
   (c) Class B (d) all of these answers
8. The efficiency of a power amplifier is the ratio of the power delivered to the load to the
   (a) input signal power
   (b) power dissipated in the last stage
   (c) power from the power supply
   (d) none of these answers
9. Crossover distortion is a problem for
   (a) Class A amplifiers (b) Class AB amplifiers
   (c) Class B amplifiers (d) all of these answers
10. A current mirror in a push-pull amplifier should give an I_{CQ} that is
   (a) equal to the current in the bias resistors and diodes
   (b) twice the current in the bias resistors and diodes
   (c) half the current in the bias resistors and diodes
   (d) zero
11. To avoid crossover distortion with an E-MOSFET push-pull amplifier, you should bias the MOSFETs with
   (a) a current mirror (b) self-bias
   (c) a voltage divider (d) a separate power supply
12. When a differential amplifier is operated single-ended,
   (a) the output is grounded
   (b) one input is grounded and a signal is applied to the other
   (c) both inputs are connected together
   (d) the output is not inverted
13. In the differential mode,
   (a) opposite polarity signals are applied to the inputs
   (b) the gain is 1
   (c) the outputs are different amplitudes
   (d) only one supply voltage is used
14. In the common mode,
   (a) both inputs are grounded
   (b) the outputs are connected together
   (c) an identical signal appears on both inputs
   (d) the output signals are in phase

15. If Q_2 in the circuit in Figure 5–33 has an open emitter, the positive side of the ac output voltage will

(a) increase (b) decrease

(c) not change

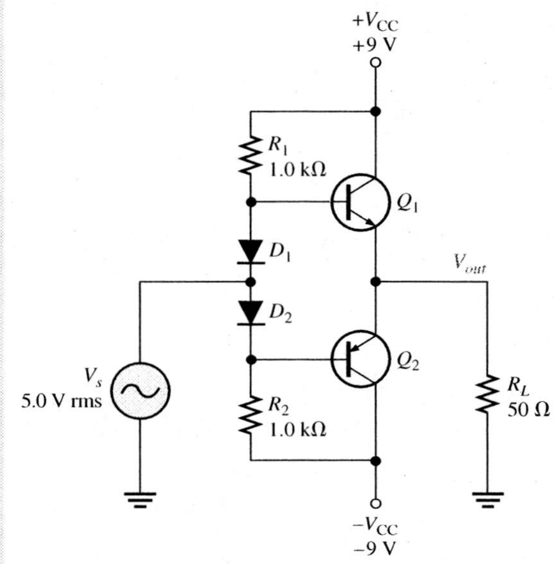

FIGURE 5–33

16. If Q_2 in the circuit in Figure 5–33 has an open emitter, the negative side of the ac output voltage will

(a) increase (b) decrease

(c) not change

17. If D_1 in the circuit in Figure 5–33, is shorted, the bias current in R_1 will

(a) increase (b) decrease

(c) not change

18. If D_1 in the circuit of Figure 5–33 is shorted, the ac output voltage will

(a) increase (b) decrease

(c) not change

Questions

Answers to odd-numbered questions are at the end of the book.

1. What is a stage of an amplifier?
2. What is the advantage of capacitively coupling two stages of an amplifier?
3. How do you find the Thevenin equivalent output voltage for a basic amplifier?
4. How does loading affect the gain of a multistage amplifier?
5. How do you find the ac resistance between the base and emitter of a transistor?
6. How does single-point grounding help avoid noise in a multistage amplifier?

7. Why is a capacitor frequently connected between V_{CC} and ground at each stage of an amplifier?
8. In what type of amplifier would you expect to see tuned transformer coupling?
9. What is the advantage of using an intermediate frequency in a radio frequency receiver?
10. What is meant by the Q of a resonant circuit?
11. What is a decoupling network?
12. What are three advantages of using negative feedback?
13. What is the difference between open-loop and closed-loop gain?
14. What type of amplifier has power gain but no voltage gain?
15. What is a Darlington pair?
16. What type of power amplifier operates best with the Q-point centered on the ac load line?
17. How does a class B amplifier differ from a class AB amplifier?
18. Why is the dc load line for a class B amplifier vertical?
19. Where does the dc load line for a class B amplifier cross the x-axis?
20. (a) What emitter voltage do you expect on a class B amplifier operated with dual supplies?
    (b) What emitter voltage do you expect on a class B amplifier operated with a single supply?
21. How does the current in the emitter resistor compare to the collector currents (in both transistors) in a differential amplifier?

PROBLEMS

Basic Problems

Answers to odd-numbered problems are at the end of the book.

1. For the two-stage amplifier modeled in Figure 5–34, determine the voltage gain.

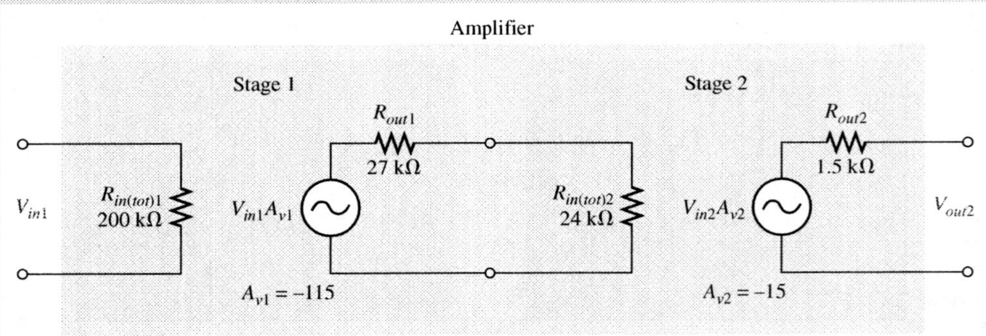

FIGURE 5–34

2. Assume a 1.0 kΩ load is connected to the amplifier modeled in Figure 5–34. What is the new gain?

3. Assume a two-stage amplifier is constructed from two identical amplifiers with the following specifications: $R_{in} = 30\ \text{k}\Omega$, $R_{out} = 2\ \text{k}\Omega$, $A_{v(\text{NL})} = 80$.

   (a) Draw the ac model of the amplifier.

   (b) What is the overall gain when the two stages are connected together?

   (c) If a 3 kΩ load resistor is connected to the amplifier, what is the overall gain?

4. Assume a source with a 600 Ω internal resistance is set to 10 mV rms, then connected to a two-stage amplifier with a 100 Ω load resistor. The following are the characteristics of each stage:

   stage 1: $R_{in} = 18\ \text{k}\Omega$, $A_{v(\text{NL})} = -40$, $R_{out} = 2.5\ \text{k}\Omega$

   stage 2: $R_{in} = 6.5\ \text{k}\Omega$, $A_{v(\text{NL})} = -30$, $R_{out} = 85\ \Omega$

   (a) Draw the equivalent circuit for the amplifier.

   (b) What is the overall gain?

   (c) What voltage is delivered to the load?

5. Assume a parallel resonant circuit is constructed from a 200 μH inductor with 9.5 Ω of resistance and a 1000 pF capacitor.

   (a) What is the resonant frequency?

   (b) What is the Q?

   (c) What is the bandwidth?

6. Assume a 10:1 step-down transformer has a load of 100 Ω connected across the secondary. What is the reflected resistance in the primary circuit?

7. The audio frequency power amplifier shown in Figure 5–35 has a 3:1 step-down transformer in the collector circuit with a 16 Ω load resistor connected to the secondary. Determine the gain of the circuit. (Since r'_e is small compared to R_{E1}, it can be ignored.)

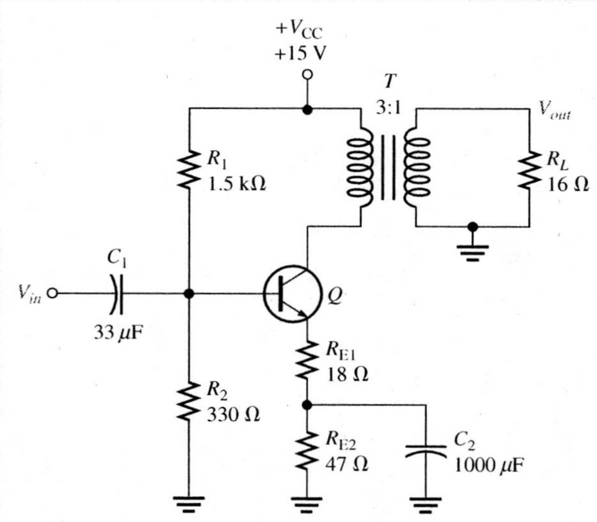

FIGURE 5–35

8. The amplifier in Figure 5–36 is a low-power audio amplifier. The transformer is a step-down impedance-matching transformer designed to give a reflected resistance in the primary of 1000 Ω when the load is 8 Ω (such as the speaker). The dc resistance of the primary winding is 66 Ω and $\beta_{ac} = 150$.
   (a) Compute V_{CE} and the I_E for the transistor.
   (b) Compute A_v, A_p, and power delivered to the load when the input is 500 mV pp.

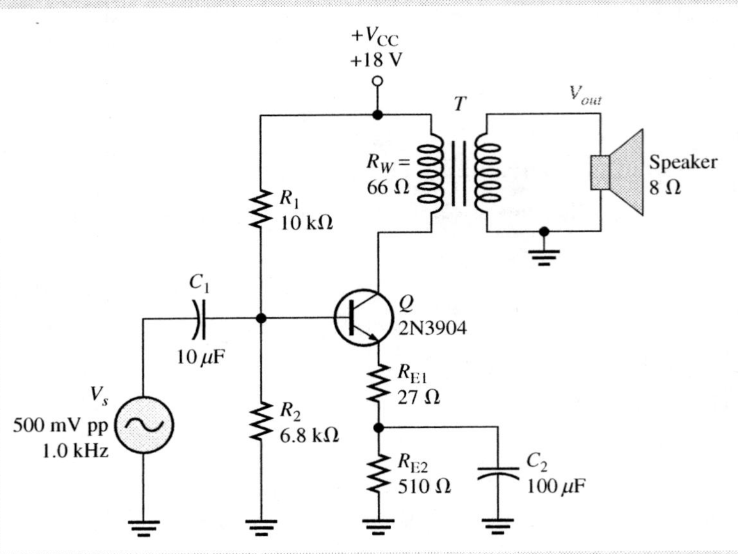

FIGURE 5–36

9. Figure 5–37 shows two dc coupled CC amplifiers (Q_2 and Q_3) with no coupling capacitors required at the input or output. Q_1 is a current source for Q_2 and produces very high input resistance for the amplifier.
   (a) Assuming the base of Q_2 is at zero volts, determine the following dc parameters: $I_{C(Q2)}$, $V_{B(Q3)}$, $I_{C(Q3)}$, $V_{E(Q3)}$.
   (b) Assuming a 5 V rms input signal, what power is delivered to the load resistor?

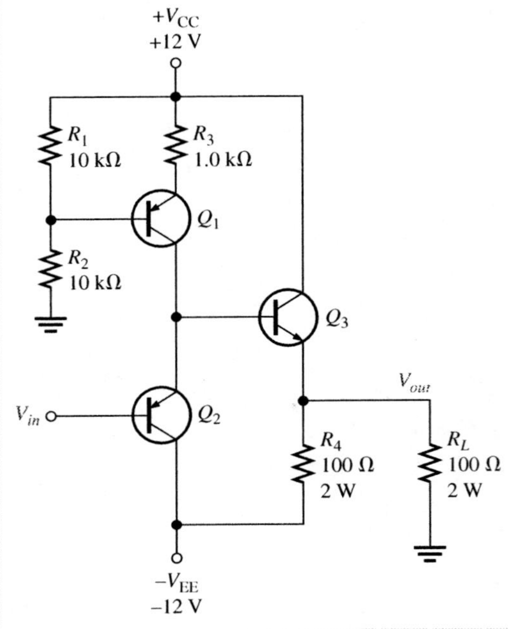

FIGURE 5–37

10. What are advantages of a dc coupled CC amplifier such as shown in Figure 5–37?

11. Assume the emitter resistor for Q_3 in Figure 5–37 is open. Will the base-emitter junction of Q_3 still be forward-biased? What happens to the collector current in Q_3?

12. For the circuit in Figure 5–15, assume a 10 kΩ resistor is substituted for R_{F2}. What effect, if any, does this change have on
   (a) the dc emitter voltage of Q_1?
   (b) the voltage gain?
   (c) the input resistance of the amplifier?

13. Figure 5–38 shows a CE power amplifier in which the collector resistor serves also as the load resistor. Assume $\beta_{DC} = \beta_{ac} = 100$.
   (a) Determine the dc Q-point (I_{CQ} and V_{CEQ}).
   (b) Determine the voltage gain and the power gain.

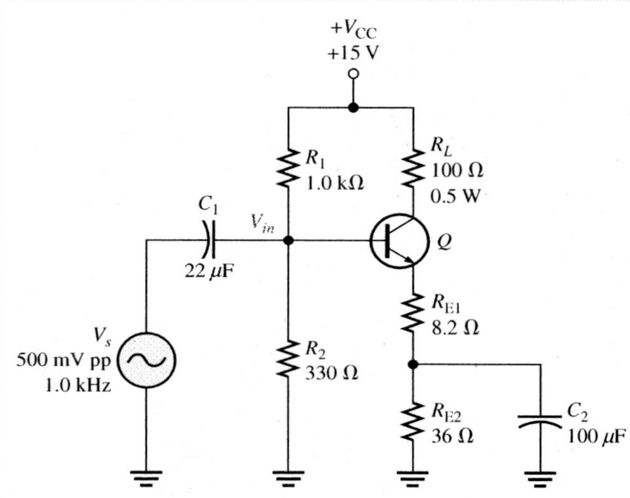

FIGURE 5–38

14. For the circuit in Figure 5–38, determine the following:
   (a) the power dissipated in the transistor with no load.
   (b) the total power from the power supply with no load.
   (c) the signal power in the load with a 500 mV pp input.

15. Refer to the circuit in Figure 5–38. What changes would be necessary to convert the circuit to a *pnp* transistor with a positive supply? What advantage would this have?

16. Assume a CC amplifier has an input resistance of 2.2 kΩ and drives an output load of 50 Ω. What is the power gain?

17. Refer to the Class AB amplifier in Figure 5–39.

(a) Determine the dc parameters $V_{B(Q1)}$, $V_{B(Q2)}$, V_E, I_{CQ}, $V_{CEQ(Q1)}$, $V_{CEQ(Q2)}$.

(b) For the 5 V rms input, determine the power delivered to the load resistor.

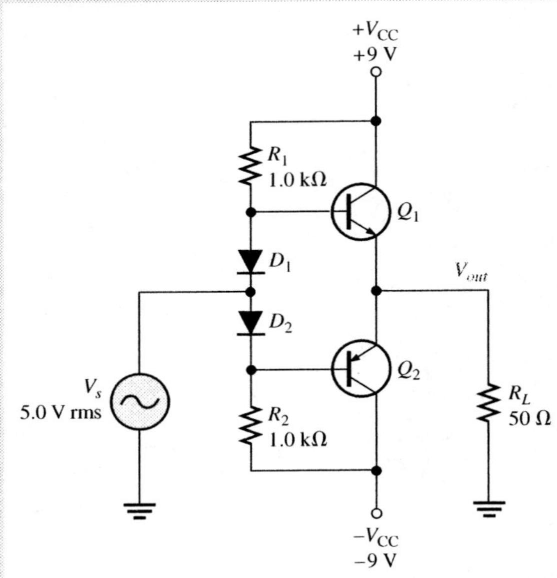

FIGURE 5–39

18. Draw the ac load line for the *npn* transistor in Figure 5–39. Label the saturation current, $I_{c(sat)}$, and show the Q-point.

19. Refer to the Class AB amplifier in Figure 5–40 operating with a single power supply.

(a) Determine the dc parameters $V_{B(Q1)}$, $V_{B(Q2)}$, V_E, I_{CQ}, $V_{CEQ(Q1)}$, $V_{CEQ(Q2)}$.

(b) Assuming the input voltage is 10 V pp, determine the power delivered to the load resistor.

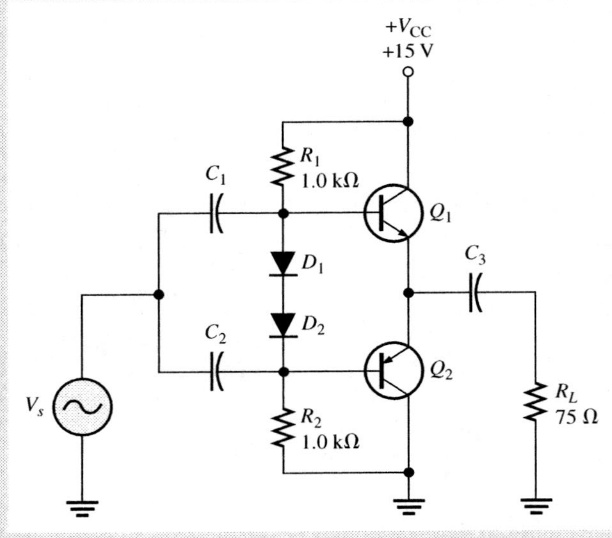

FIGURE 5–40

20. Refer to the Class AB amplifier in Figure 5–40.

(a) What is the maximum power that could be delivered to the load resistor?

(b) Assume the power supply voltage is raised to 24 V. What is the new maximum power that could be delivered to the load resistor?

21. Refer to the Class AB amplifier in Figure 5–40. What fault or faults could account for each of the following troubles?

(a) a positive half-wave output signal

(b) zero volts on both bases and the emitters

(c) no output; emitter voltage = +15 V

(d) crossover distortion observed on the output waveform

22. Identify the type of input and output configuration for each basic differential amplifier in Figure 5–41.

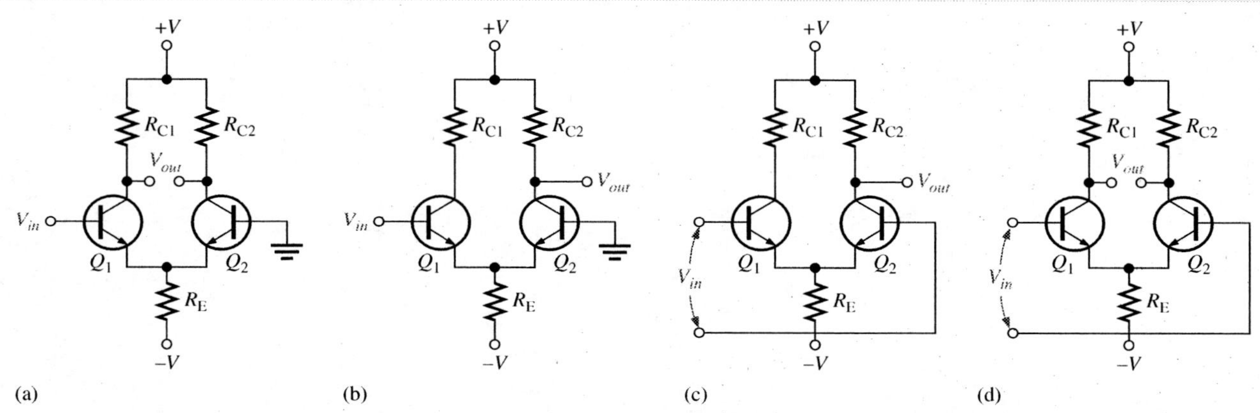

FIGURE 5–41

23. The dc base voltages in Figure 5–42 are zero. Assume the transistors are identical and have a $\beta_{DC} = 100$.

(a) What is the current in R_E?

(b) What is the current in R_{C1} and R_{C2}?

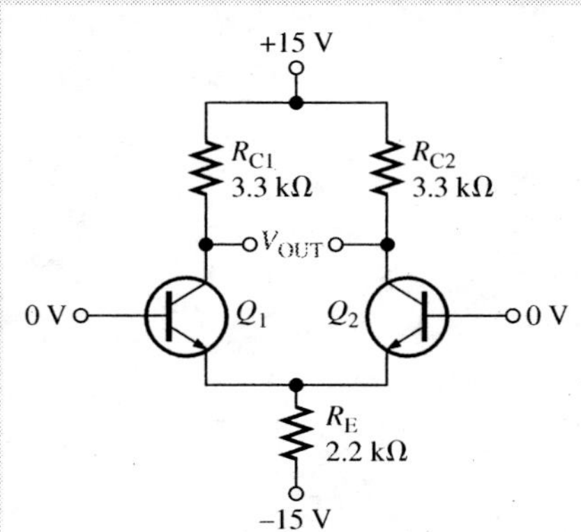

FIGURE 5–42

Basic-Plus Problems

24. For the two-stage capacitively coupled amplifier shown in Figure 5–43, compute the overall voltage gain, the input resistance, and the output resistance. Assume the g_m of the JFET is 2700 μS and β_{ac} of the BJT is 150.

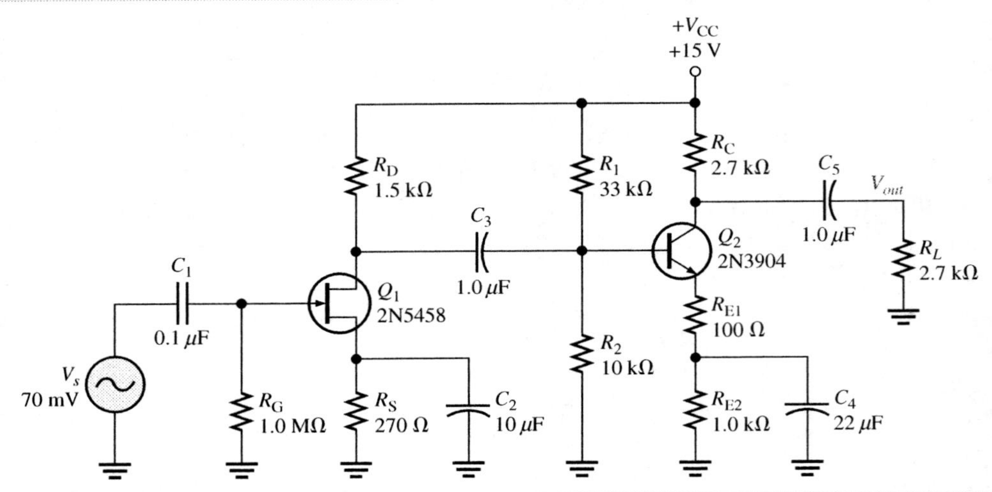

FIGURE 5–43

25. Using the results of Problem 24, draw the ac amplifier model for the two-stage amplifier in Figure 5–43. (See Figure 5–3 for an example.)

26. Assume the IF amplifier in Figure 5–44 has an IF transformer with a primary inductance of 180 μH. Internally, there is a 680 pF capacitor connected in parallel with the primary.

    (a) Find the resonant frequency.

    (b) Assume the reactance of the resonant circuit at this frequency is 40.7 kΩ. Find A_v.

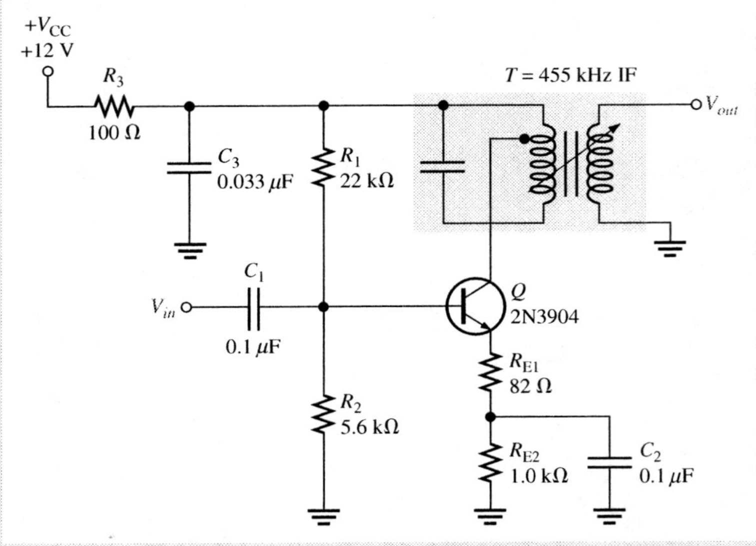

FIGURE 5–44

27. What is the purpose of R_3 and C_3 in the circuit of Figure 5–44?

28. Assume the *n*-channel E-MOSFET shown in Figure 5–45 has a threshold voltage of 2.75 V and the *p*-channel E-MOSFET has a threshold voltage of −2.75 V.

   (a) What resistance setting for R_6 will bias the output transistors to class AB operation?

   (b) Assuming the input voltage is 150 mV rms, what is the rms voltage delivered to the load?

   (c) What is the power delivered to the load with this setting?

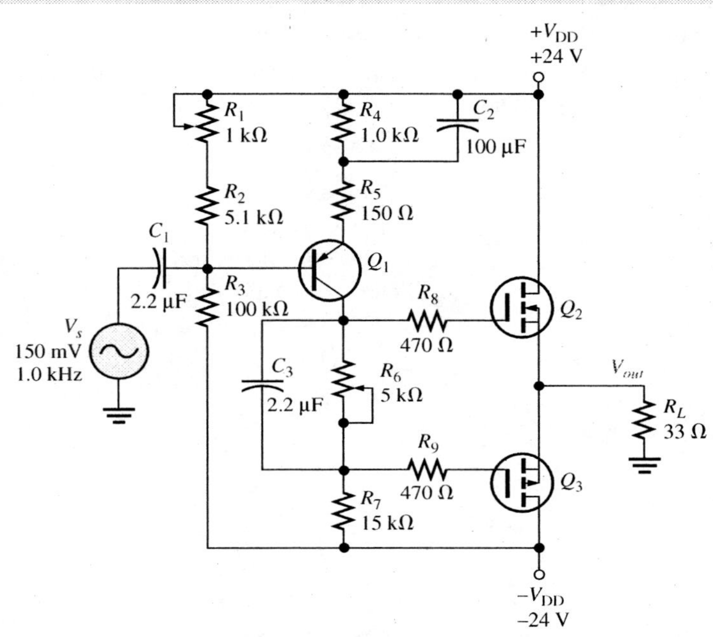

FIGURE 5–45

Example Questions

ANSWERS

5–1: 1.34 V

5–2: The gain will be lower because r'_e is larger.

5–3: The efficiency goes up because no power is wasted in R_{E3}. The disadvantage is that the speaker has a dc current (the emitter current) in the coil.

5–4: 15 V, 0.94 A

5–5: 34,000

Review Questions

1. R_{in}, R_{out}, and $A_{v(NL)}$

2. A source that has an output that depends on a voltage or current elsewhere in the circuit

3. The input resistance of the second stage forms a voltage-divider with the output resistance of the first stage. This effectively reduces the gain of the first stage by loading.

4. The signal levels are small and can easily be obscured by noise.
5. When large and small signals are present in the same amplifier, noise or signals in a later stage can affect a low-level signal adversely.
6. RF indicates *R*adio *F*requency and is any frequency useful for radio transmission. IF means *I*ntermediate *F*requency; it represents a frequency that has been shifted for processing.
7. A mixer combines two signals in a nonlinear circuit, producing the sum and difference frequencies as a result. The difference frequency becomes the IF.
8. The RF signal and a signal from the local oscillator
9. A load resistor affects the Q by reflecting a resistance into the primary side. Since Q is the ratio of X_L to R, the increase in R decreases Q.
10. Any instrument connected to the circuit can change the Q of the circuit because of resistance loading and can change the frequency because of capacitance loading.
11. Direct coupling reduces the parts count and allows for frequencies down to dc.
12. Negative feedback returns a portion of the output in a way that tends to cancel changes in the bias circuit or the gain.
13. The capacitor has no effect on the dc circuit, but it increases the open-loop gain. A higher open-loop gain means a small change in a circuit parameter will have less effect.
14. The gain is determined by the reciprocal of the feedback fraction.
15. R_{F2} along with R_{E1} form a voltage divider for ac signals that determine the amount of feedback. This in turn affects the gain.
16. To dissipate excessive heat
17. The collector
18. Cutoff and saturation clipping
19. 25%
20. It is the ratio of the input resistance to the output resistance
21. Class B is biased such that the Q-point is at cutoff. Class AB is biased such that the Q-point is slightly above cutoff.
22. The signal can be direct coupled at the input and output; parts count can be reduced.
23. Crossover distortion occurs when the input signal is less than the base-emitter drop of the push-pull amplifier. It can be avoided by biasing the class B amplifier on slightly, producing class AB operation.
24. 79%
25. Thermal runaway is generally avoided by ensuring that the diodes are in the same thermal environment as the transistors.
26. 50%
27. Nothing. There is no change.
28. A differential-mode input has opposite polarity signals on each input. A common-mode input has the same signal on each input.
29. A single-ended input is applied to only one of the inputs of a differential amplifier. The other input is at 0 V.
30. It is the ratio of the differential gain to the common-mode gain.

Chapter Checkup

1. (a)	2. (d)	3. (b)	4. (b)	5. (c)
6. (b)	7. (a)	8. (c)	9. (c)	10. (a)
11. (c)	12. (b)	13. (a)	14. (c)	15. (c)
16. (b)	17. (a)	18. (b)		

CHAPTER

OPERATIONAL AMPLIFIERS

CHAPTER OUTLINE

INTRODUCTION

Devices such as the diode and the transistor are separate devices that are individually packaged and interconnected in a circuit with other devices to form a complete, functional unit. Such devices are referred to as *discrete components*.

Now you will learn more about analog (linear) integrated circuits where many transistors, diodes, resistors, and capacitors are fabricated on a single tiny chip of semiconductive material and packaged in a single case to form a functional circuit.

The operational amplifier (op-amp), a general-purpose IC, is the most versatile and widely used of all linear integrated circuits. Although the op-amp is made up of many resistors, diodes, and transistors, it is treated as a single device. This means that you will be concerned with what the circuit does more from an external viewpoint than from an internal, component-level viewpoint.

Study aids for this chapter are available at

http://www.prenhall.com/SOE

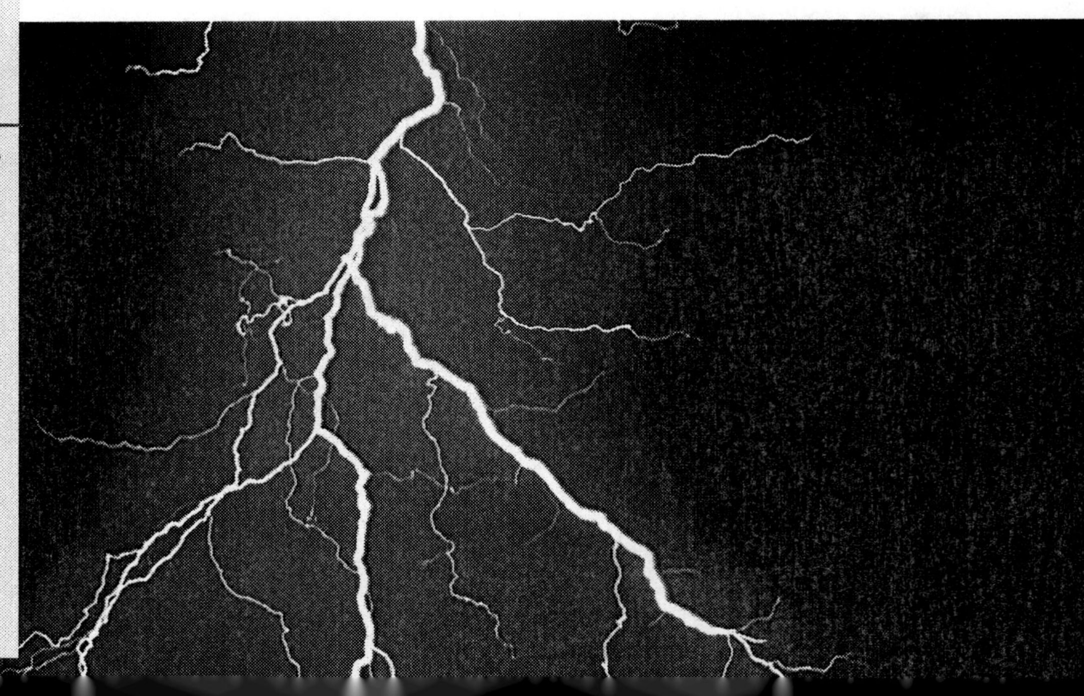

KEY OBJECTIVES

A section number is given for each objective. After completing this chapter, you should be able to

- **6–1** Describe the basic op-amp and its characteristics
- **6–2** Discuss several op-amp parameters
- **6–3** Explain negative feedback in op-amp circuits
- **6–4** Analyze three op-amp configurations and explain the closed-loop frequency response of an op-amp
- **6–5** Troubleshoot op-amp circuits

COMPUTER SIMULATIONS DIRECTORY

The following figures have Multisim circuit files associated with them.

- Figure 6–16 Page 227
- Figure 6–20 Page 229

LABORATORY EXPERIMENTS DIRECTORY

The following exercises are for this chapter.

- **Experiment 11** Op-Amp Characteristics
- **Experiment 12** Linear Op-Amp Circuits

KEY TERMS

- Operational amplifier
- Open-loop voltage gain
- Common-mode rejection ratio (CMRR)
- Slew rate
- Closed-loop voltage gain
- Noninverting amplifier
- Voltage-follower
- Inverting amplifier
- Gain-bandwidth product

ON THE JOB. . .

(Getty Images)

Social skills include your ability to properly take instruction and directions from your supervisor and to get along with the people with whom you work. It is important to foster a good working environment for all. In particular, if you are dealing with customers, you must be able to work with them in a courteous and professional manner. If a situation looks like it is getting out of hand, ask your supervisor for help. The impression that you make on customers will affect current and future business for your company.

Often, when we want to solve a problem in science, we turn to laws of physics that describe the applicable behavior. Models in science are useful for predicting the results of a new problem. A goal of any model is to describe laws of physics with simplicity, if possible. To be useful, a model must predict the new result with reasonable accuracy.

Newton attempted to describe the behavior of all moving bodies with his three laws of motion. These laws certainly met the simplicity standard, and we can use his laws to model the behavior of moving objects—from a falling object to the motion of the planets. However, it is known that in some situations, Newton's laws do not accurately depict reality. For example, when particles in a particle accelerator are traveling near the speed of light, Newton's laws break down. Despite special cases where his laws cannot be applied, they are still useful for ordinary situations that we encounter in problem solving.

An important aspect of electronics is to provide models of circuits that will allow us to predict the circuit's behavior accurately. In the study of operational amplifiers, we use ideal models. To simplify a circuit, the models leave out a lot of extraneous details. The ideal models are excellent for understanding basic behavior and in many situations enable us to predict circuit behavior quite well. If the situation requires more detail, we can use computer models that add more detail to our basic models.

6–1 INTRODUCTION TO OPERATIONAL AMPLIFIERS

Developed in the 1940s, early operational amplifiers (op-amps) were used primarily to perform mathematical operations such as addition, subtraction, integration, and differentiation, thus the term *operational*. These devices were constructed with vacuum tubes and worked with high voltages. Today, op-amps are linear integrated circuits that use relatively low supply voltages and are reliable and inexpensive.

In this section, you will learn about the basic op-amp and its characteristics.

An **operational amplifier** (op-amp) is an electronic device that amplifies the difference voltage between the two inputs. A typical op-amp is made up of three types of amplifier circuits: a *differential amplifier*, a *voltage amplifier*, and a *push-pull amplifier*, as shown in Figure 6–1. A differential amplifier is the input stage for the op-amp; it has two inputs and provides amplification of the difference voltage between the two inputs. The

FIGURE 6–1 Basic internal arrangement of an op-amp.

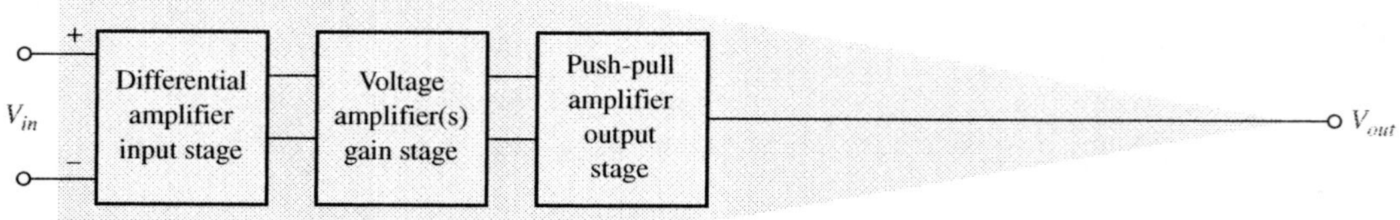

voltage amplifier is usually a class A amplifier that provides additional gain. Some op-amps may have more than one voltage amplifier stage. A push-pull class B amplifier is generally used for the output stage.

Symbol and Terminals

The standard op-amp symbol is shown in Figure 6–2(a). It has two input terminals, the inverting input (−) and the noninverting input (+), and one output terminal. The typical op-amp operates with two dc supply voltages, one positive and the other negative, as shown in Figure 6–2(b). Usually these dc voltage terminals are left off the schematic symbol for simplicity but are always understood to be there. Some typical op-amp IC packages are shown in Figure 6–2(c).

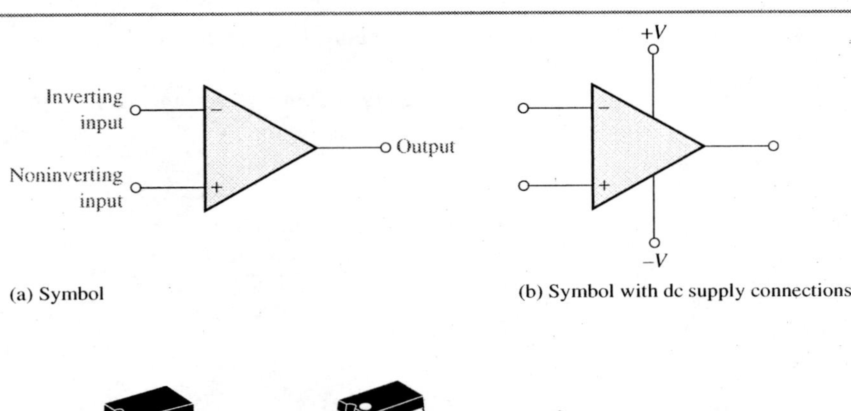

(c) Typical packages. Looking from the top, pin 1 always is to the left of the notch or dot on the DIP and SOIC packages. The dot indicates pin 1 on the plastic-leaded chip carrier (PLCC) package.

FIGURE 6–2

Op-amp symbols and packages.

The Ideal Op-Amp

The ideal op-amp has *infinite voltage gain* and an *infinite input resistance* (open), so that it does not load the driving source. Also, it has a *zero output resistance*. These characteristics are illustrated in Figure 6–3. The input voltage V_{in} appears between the two input terminals,

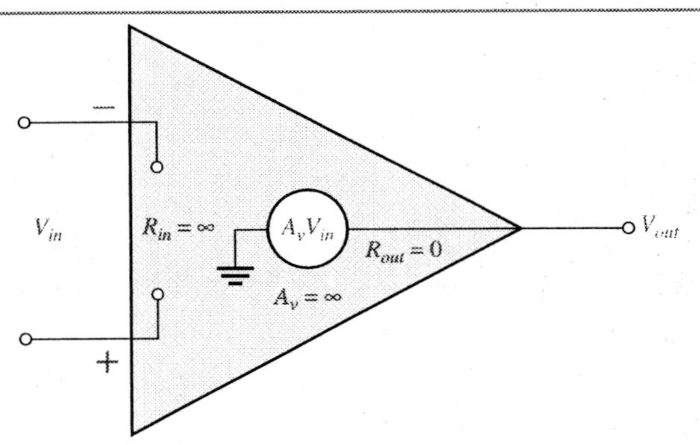

FIGURE 6–3

Ideal op-amp representation.

and the output voltage is $A_v V_{in}$, as indicated by the internal voltage source symbol. The concept of infinite input resistance is a particularly valuable analysis tool for the various op-amp configurations, which will be discussed in Section 6–4. A practical op-amp, of course, falls short of these ideal standards, but it is much easier to understand and analyze the device from an ideal point of view.

The Practical Op-Amp

Although modern integrated circuit (IC) op-amps approach parameter values that can be treated as ideal in many cases, no practical op-amp can be ideal. Any device has limitations, and the IC op-amp is no exception. Op-amps have both voltage and current limitations. Peak-to-peak output voltage, for example, is usually limited to slightly less than the difference between the two supply voltages. Output current is also limited by internal restrictions such as power dissipation and component ratings.

Characteristics of a practical op-amp are *high voltage gain, high input resistance* and *low output resistance*. Some of these are illustrated in Figure 6–4.

FIGURE 6–4

Practical op-amp representation.

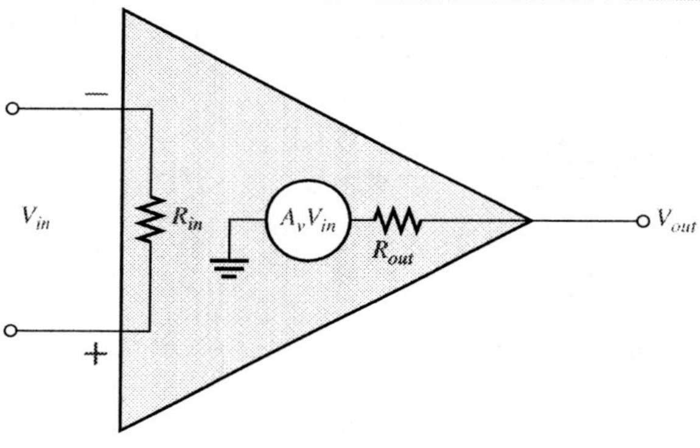

Review Questions

Answers are at the end of the chapter.

1. What are the connections to a basic op-amp?
2. What are some of the characteristics of a practical op-amp.
3. How does the voltage gain of a practical op-amp differ from an ideal op-amp?
4. What is the op-amp input indicated by +?
5. What are the three types of circuits that make up an op-amp?

OP-AMP PARAMETERS 6–2

Op-amp parameters are used to specify performance and provide for comparison of different op-amps. Open-loop voltage gain, CMRR, and slew rate are three important parameters. Other parameters include input offset voltage, input bias current, input and output resistances, and common-mode input voltage range.

In this section, you will learn about several op-amp parameters.

Input Offset Voltage

The ideal op-amp produces zero volts out for zero volts in. In a practical op-amp, however, a small dc voltage, $V_{OUT(error)}$, appears at the output when no differential input voltage is applied. Its primary cause is a slight mismatch of the transistors in the differential input stage of an op-amp, as illustrated in Figure 6–5(a).

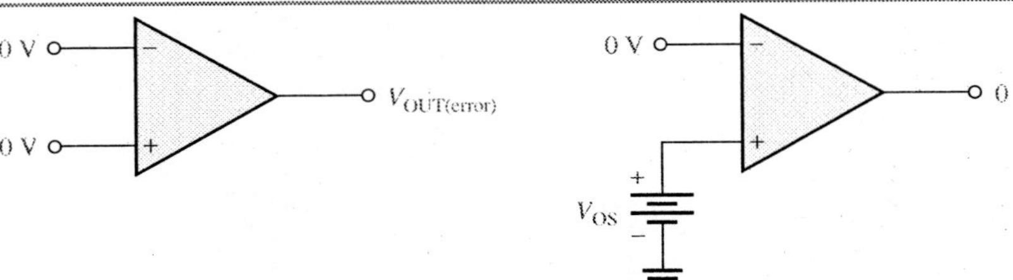

(a) A slight input mismatch (internal to the device) causes a small output error voltage with no differential input voltage.

(b) The input offset voltage is the difference in the voltage between the inputs that is necessary to eliminate the output error voltage (makes $V_{OUT} = 0$).

FIGURE 6–5

Illustration of input offset voltage, V_{OS}.

As specified on an op-amp data sheet, the **input offset voltage** (V_{OS}) is the differential dc voltage required between the inputs to force the output to zero volts. V_{OS} is demonstrated in Figure 6–5(b). Typical values of input offset voltage are in the range of 2 mV or less. In the ideal case, it is 0 V.

Input Bias Current

The input terminals of a bipolar differential amplifier are the transistor bases and, therefore, the input currents are the base currents. The **input bias current** is the dc current required by the inputs of the amplifier to properly operate the first stage. By definition, the input bias current is the *average* of both input currents. The input bias current is so small in most practical applications that it can be considered to be zero. The concept of input bias current is illustrated in Figure 6–6.

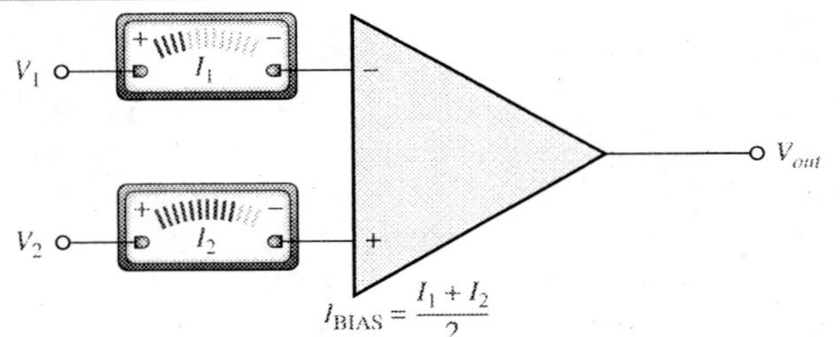

FIGURE 6–6

Input bias current is the average of the two op-amp input currents.

Input Resistance

Two basic ways of specifying the input resistance of an op-amp are the differential and the common mode. The **differential input resistance** is the total resistance between the inverting and the noninverting inputs, as illustrated in Figure 6–7(a). Differential input impedance is measured by determining the change in bias current for a given change in differential input voltage. The **common-mode input resistance** is the resistance between each input and ground and is measured by determining the change in bias current for a given change in common-mode input voltage. It is depicted in Figure 6–7(b). The input resistance is always very high.

FIGURE 6–7

Op-amp input resistance.

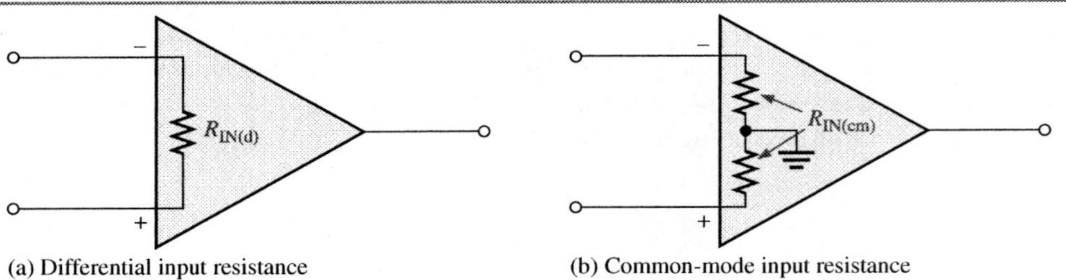

Output Resistance

Output resistance is the resistance viewed from the output terminal of the op-amp, as indicated in Figure 6–8. The output resistance is always very small.

FIGURE 6–8

Op-amp output resistance.

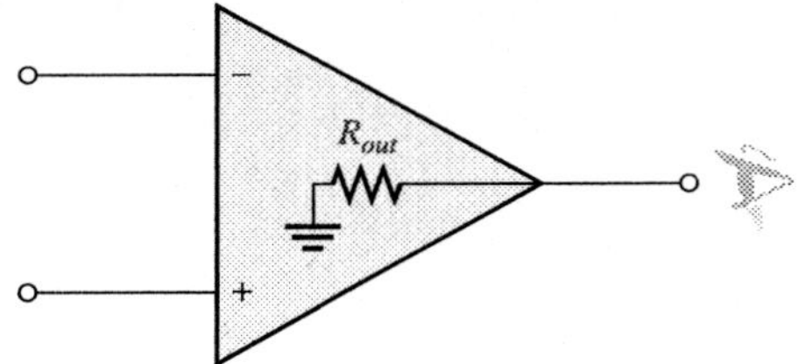

Common-Mode Input Voltage Range

All op-amps have limitations on the range of voltages over which they will operate. The **common-mode input voltage range** is the range of input voltages which, when applied to both inputs, will not cause clipping or other output distortion. Many op-amps have common-mode ranges of no more than ±10 V with dc supply voltages of ±15 V, while in others the output can go as high as the supply voltages (this is called rail-to-rail).

FIGURE 6–9

Open-loop op-amp.

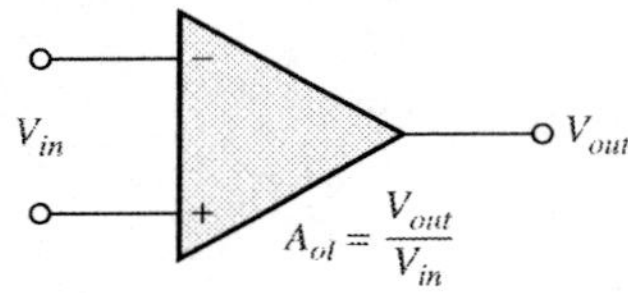

Open-Loop Voltage Gain

The **open-loop voltage gain**, A_{ol}, of an op-amp is the internal voltage gain of the device measured in the differential mode. It represents the ratio of output voltage to input voltage when there are no external components, as shown in Figure 6–9. The open-loop voltage gain is set entirely by the internal design. Open-loop voltage gain can range up to 200,000 or more and is not a well-controlled parameter. Data sheets often refer to the open-loop voltage gain as the *large-signal voltage gain.*

Common-Mode Rejection Ratio for an Op-Amp

The common-mode rejection ratio (CMRR) was discussed in conjunction with the diff-amp in Section 5–6. Similarly, for an op-amp, **CMRR** is a measure of an op-amp's ability to reject common-mode signals. An infinite value of CMRR means that the output is zero when the same signal is applied to both inputs (common-mode).

An infinite CMRR is never achieved in practice, but a good op-amp does have a very high value of CMRR. Common-mode signals are undesired interference voltages such as 60 Hz power-supply ripple and noise voltages due to pick-up of radiated energy. A high CMRR enables the op-amp to virtually eliminate these interference signals from the output.

The accepted definition of CMRR for an op-amp is the open-loop voltage gain (A_{ol}) divided by the common-mode gain. This is equivalent to the CMRR for a differential amplifier that was introduced in Chapter 5.

$$\text{CMRR} = \frac{A_{ol}}{A_{cm}} \qquad (6\text{–}1)$$

The CMRR is commonly expressed as a logarithmic ratio in decibels on op-amp data sheets. Logarithms and decibels are covered in Appendix A.

EXAMPLE 6–1

Problem

A certain op-amp has an open-loop voltage gain of 100,000 and a common-mode gain of 0.25. Determine the CMRR.

Solution

$$\text{CMRR} = \frac{A_{ol}}{A_{cm}} = \frac{100{,}000}{0.25} = \mathbf{400{,}000}$$

Question*

If a particular op-amp has an open-loop gain of 250,000 and a common-mode gain of 0.5, what is the CMRR?

Slew Rate

The maximum rate of change of the output voltage in response to a step input voltage is the **slew rate** of an op-amp. The slew rate is dependent upon the high-frequency response of the amplifier stages within the op-amp.

Slew rate is measured with an op-amp connected as shown in Figure 6–10(a). This particular op-amp connection is a unity-gain, noninverting configuration which will be discussed in Section 6–4. It gives a worst-case (slowest) slew rate. The high-frequency components of a voltage step are contained in the rising edge, and the upper critical frequency of an amplifier limits its response to a step input. The lower the upper critical frequency is, the more gradual the slope on the output for a step input.

A pulse is applied to the input as shown in Figure 6–10(b), and the ideal output voltage is measured as indicated. The width of the input pulse must be sufficient to allow the output to "slew" from its lower limit to its upper limit, as shown. As you can see, a certain time

**Answers are at the end of the chapter.*

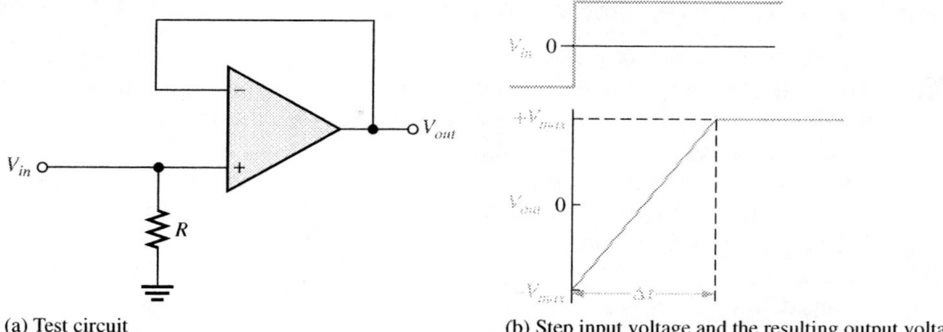

FIGURE 6–10
Slew rate measurement.

interval, Δt, is required for the output voltage to go from its lower limit $-V_{max}$ to its upper limit $+V_{max}$, once the input step is applied. The slew rate is expressed as

$$\text{Slew rate} = \frac{\Delta V_{out}}{\Delta t} \quad (6\text{–}2)$$

where ΔV_{out} is $+V_{max} - (-V_{max})$. The unit of slew rate is volts per microsecond (V/μs).

EXAMPLE 6–2

Problem

The output voltage of a certain op-amp appears as shown in Figure 6–11 in response to a step input. Determine the slew rate.

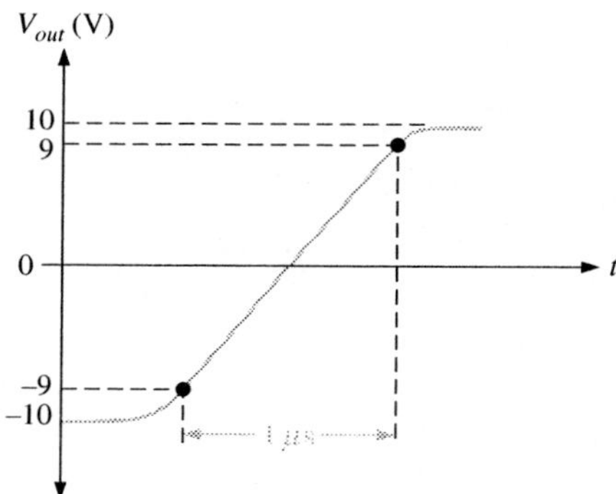

FIGURE 6–11

Solution

The output goes from the lower to the upper limit in 1 μs. Since this response is not ideal, the limits are taken at the 90% points, as indicated, So, the upper limit is +9 V and the lower limit is −9 V. The slew rate is

$$\text{Slew rate} = \frac{\Delta V}{\Delta t} = \frac{+9\text{V} - (-9\text{V})}{1\ \mu\text{s}} = \mathbf{18\ V/\mu s}$$

Question

When a pulse is applied to an op-amp, the output voltage goes from −8 V to +7 V in 0.75 μs. What is the slew rate?

Frequency Response

The internal amplifier stages that make up an op-amp have voltage gains limited by junction capacitances. Although the differential amplifiers used in op-amps are somewhat different from the basic amplifiers discussed in Chapter 5, the same principles apply. An op-amp has no internal coupling capacitors; therefore, the low-frequency response extends down to dc (0 Hz).

Comparison of Op-Amp Parameters

Table 6–1 provides a comparison of values of some of the parameters just described for several common IC op-amps. Any values not listed were not given on the manufacturer's data sheet.

TABLE 6–1

OP-AMP	Input offset voltage (mV) (max)	Input bias current (nA) (max)	Input resistance (MΩ) (min)	Open-loop gain (typ)	Slew rate (V/μs) (typ)	CMRR (dB) (min)	Comment
LM741C	6	500	0.3	200,000	0.5	70	Industry standard
LM101A	7.5	250	1.5	160,000	—	80	General-purpose
OP113E	0.075	600	—	2,400,000	1.2	100	Low noise, low drift
OP177A	0.01	1.5	26	12,000,000	0.3	130	Ultra precision
OP184E	0.065	350	—	240,000	2.4	60	Precision, rail-to-rail*
AD8009AR	5	150	—	—	5500	50	*BW* = 700 MHz, ultra fast, low distortion, current feedback
AD8041A	7	2000	0.16	56,000	160	74	*BW* = 160 MHz, rail-to-rail
AD8055A	5	1200	10	3500	1400	82	Very fast voltage feedback

* Rail-to-rail means that the output voltage can go as high as the supply voltages.

Other Features

Most available op-amps have three important features: short-circuit protection, no latch-up, and input offset nulling. Short-circuit protection keeps the circuit from being damaged if the output becomes shorted, and the no latch-up feature prevents the op-amp from hanging up in one output state (high or low voltage level) under certain input conditions. Input offset nulling is achieved by an external potentiometer that sets the output voltage at precisely zero with zero input.

Review Questions

6. What is the input offset voltage?
7. Is the input resistance of an op-amp high or low?
8. What is the open-loop voltage gain?
9. What does CMRR stand for?
10. What is the slew rate?

6–3 NEGATIVE FEEDBACK IN OP-AMPS

Negative feedback is one of the most useful concepts in electronics, particularly in op-amp applications. Recall from Chapter 5 that negative feedback is the process whereby a portion of the output voltage of an amplifier is returned to the input with a phase angle that opposes (or subtracts from) the input signal.

In this chapter, you will learn the effects of negative feedback in op-amp circuits.

Negative feedback is illustrated in Figure 6–12. The inverting input (−) effectively makes the feedback signal 180° out of phase with the input signal. The op-amp has extremely high gain and amplifies the difference in the signals applied to the inverting and noninverting inputs. A very tiny difference in these two signals is all the op-amp needs to produce the required output. *When negative feedback is present, the noninverting and inverting inputs are nearly identical.* This concept can help you figure out what signal to expect in many op-amp circuits.

FIGURE 6–12

Illustration of negative feedback.

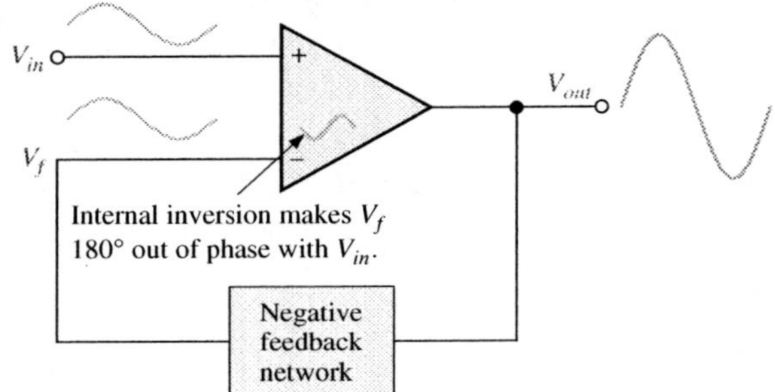

HISTORICAL NOTE

In 1921, Harold S. Black began working on the task of linearizing, stabilizing, and otherwise improving amplifiers for his employer, Western Electric Co. The problem facing Black was the requirement for an amplifier that could reduce distortion by a vast amount. In 1927, he realized that if he fed some of the output back to the input in opposite phase, he had a means of canceling the distortion. This idea, sketched on a copy of the *New York Times*, led to the development of the negative feedback amplifier.

Now let's see why the signals at the inverting and noninverting-terminals are nearly identical when negative feedback is used. Assume a 1.0 V input signal is applied to the noninverting terminal and the open-loop gain of the op-amp is 100,000. The amplifier responds to the voltage at its noninverting input terminal and moves the output toward saturation. Immediately, a fraction of this output is returned to the inverting terminal through the feedback path. But if the feedback signal ever reaches 1.0 V, there is nothing left for the op-amp to amplify! Thus, the feedback signal tries (but never quite succeeds) in matching the input signal. The gain is controlled by the amount of feedback used. When you are troubleshooting an op-amp circuit with negative feedback present, remember that the two inputs will look identical on a scope but in fact are very slightly different.

Now suppose something happens that reduces the internal gain of the op-amp. This causes the output signal to drop a small amount, returning a smaller signal to the inverting input via the feedback path. This means the difference between the signals is larger than it was. The output increases, compensating for the original drop in gain. The net change in the output is so small, it can hardly be measured. The main point is that any variation in the amplifier is immediately compensated for by the negative feedback, resulting in a very stable, predictable output.

Why Use Negative Feedback?

As you have seen, the inherent open-loop gain of a typical op-amp is very high (usually greater than 100,000). Therefore, an extremely small difference in the two input voltages drives the op-amp into its saturated output states. In fact, even the input offset voltage of the op-amp can drive it into saturation. For example, assume $V_{in} = 1$ mV and $A_{ol} = 100{,}000$. Then,

$$V_{in}A_{ol} = (1\text{ mV})(100{,}000) = 100\text{ V}$$

Since the output level of an op-amp can never reach 100 V, it is driven into saturation and the output is limited to its maximum output levels, as illustrated in Figure 6–13 for both a positive and a negative input voltage of 1 mV.

FIGURE 6–13

Without negative feedback, an extremely small difference in the two input voltages drives the op-amp to its output limits and it becomes nonlinear.

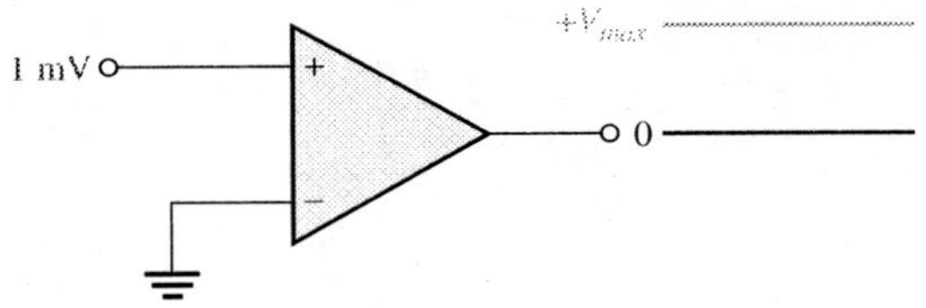

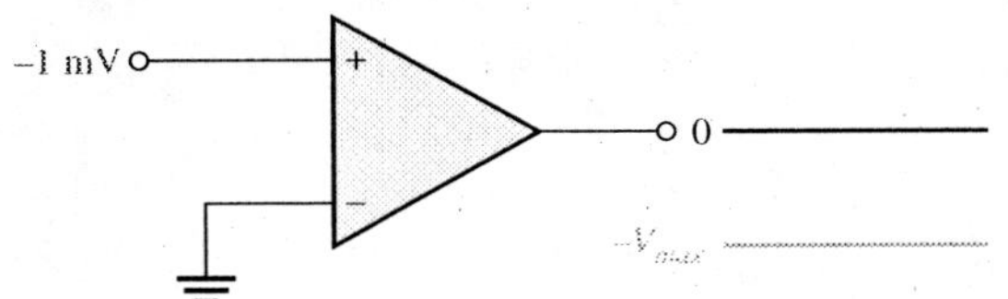

The usefulness of an op-amp operated in this manner is severely restricted and is generally limited to comparator applications (to be studied in Chapter 7). With negative feedback, the overall closed-loop voltage gain (A_{cl}) can be reduced and controlled so that the op-amp can function as a linear amplifier. In addition to providing a controlled, stable voltage gain, negative feedback also provides for control of the input and output resistances and amplifier bandwidth. Table 6–2 summarizes the general effects of negative feedback on op-amp performance.

TABLE 6–2

	Voltage gain	Input resistance	Output resistance	Bandwidth
Without negative feedback	A_{ol} is too high for linear amplifier applications	Relatively high (see Table 6–1)	Relatively low	Relatively narrow (because the gain is so high)
With negative feedback	A_{cl} is set to desired value by the feedback network	Can be increased or reduced to a desired value depending on type of circuit	Can be reduced to a desired value	Significantly wider

Review Questions

11. For negative feedback, the output is connected through a feedback network to which input?
12. What are the benefits of negative feedback in an op-amp circuit?
13. Why is it necessary to reduce the gain of an op-amp from its open-loop value?
14. Does negative feedback increase or decrease the bandwidth?
15. When troubleshooting an op-amp circuit in which negative feedback is present, what do you expect to observe on the input terminals?

6–4 OP-AMP CONFIGURATIONS

An op-amp can be connected in three basic ways using negative feedback to stabilize and reduce the gain and increase frequency response. The extremely high open-loop gain of an op-amp creates an unstable condition in which the op-amp can be driven out of its linear region or it can oscillate. In addition, the open-loop gain parameter of an op-amp can vary greatly from one device to the next.

In this section, you will learn to analyze and to explain the closed-loop frequency response of an op-amp.

Closed-Loop Voltage Gain

The **closed-loop voltage gain,** A_{cl}, is the voltage gain of an op-amp with negative feedback. The amplifier configuration consists of the op-amp and an external feedback circuit that connects the output to the inverting input. The closed-loop voltage gain is then determined by the component values in the feedback circuit and can be precisely controlled by them. We will now look at three types of closed-loop op-amp configurations.

Noninverting Amplifier

An op-amp connected in a closed-loop configuration in which the input signal is applied to the noninverting input (+) is a **noninverting amplifier**, as shown in Figure 6–14. A portion of the output is applied back to the inverting input (−) through the feedback circuit. This constitutes negative feedback.

FIGURE 6–14

Noninverting amplifier.

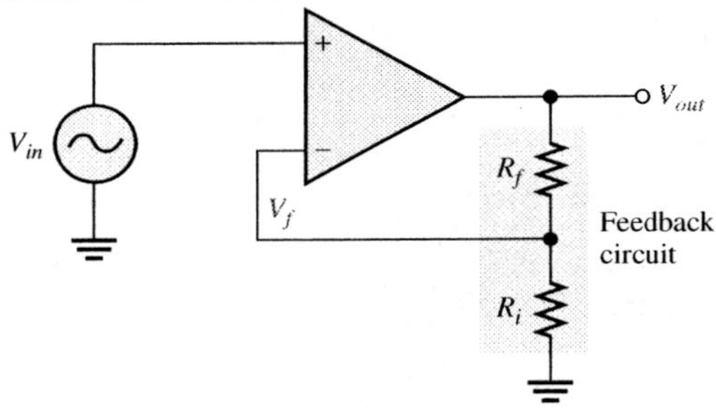

The differential voltage, V_{diff}, between the op-amp's input terminals is illustrated in Figure 6–15 and can be expressed as

$$V_{diff} = V_{in} - V_f$$

FIGURE 6–15

Differential input, $V_{in} - V_f$.

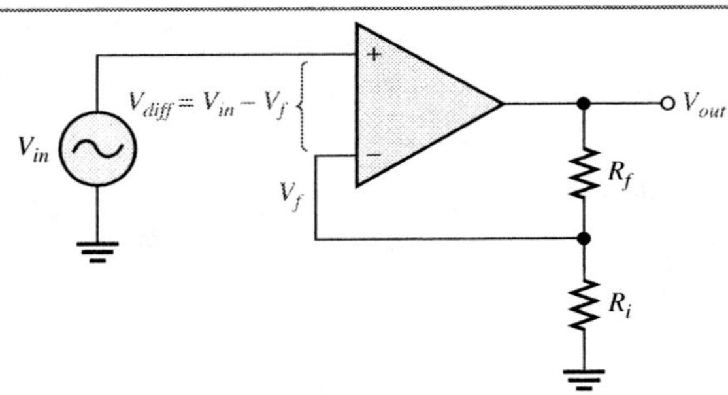

This input differential voltage is forced to be very small as a result of the negative feedback and the high open-loop gain, A_{ol}. Therefore, a close approximation is

$$V_{diff} = 0 \text{ V}$$

We can assume that $V_{in} = V_f$.

Resistors R_i and R_f form a voltage-divider network. The fraction of the output voltage, V_{out}, that is returned to the inverting input is found by applying the voltage-divider rule to the feedback circuit.

$$V_f = \left(\frac{R_i}{R_i + R_f}\right)V_{out}$$

The closed loop voltage gain is V_{out}/V_{in}. Since $V_{in} = V_f$, you can write the previous equation as

$$\frac{V_{out}}{V_{in}} = \left(\frac{R_i + R_f}{R_i}\right) = 1 + \frac{R_f}{R_i}$$

Expressing V_{out}/V_{in} as $A_{cl(\text{NI})}$,

$$A_{cl(\text{NI})} = \frac{R_f}{R_i} + 1 \qquad (6\text{–}3)$$

Equation 6–3 shows that the closed-loop voltage gain of the noninverting (NI) amplifier, $A_{cl(\text{NI})}$, is not dependent on the op-amp's open-loop gain but can be set by selecting values of R_i and R_f. This equation is based on the assumption that the open-loop gain is very high compared to the ratio of the feedback resistors, causing the input differential voltage, V_{diff}, to be zero. In nearly all practical circuits, this is an excellent assumption.

EXAMPLE 6–3

Problem

Determine the closed-loop voltage gain of the amplifier in Figure 6–16.

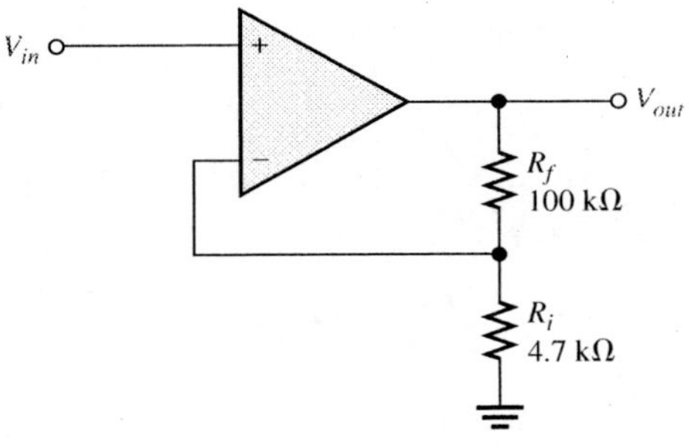

FIGURE 6–16

Solution

This is a noninverting op-amp. Therefore, the closed-loop voltage gain is

$$A_{cl(\text{NI})} = \frac{R_f}{R_i} + 1 = \frac{100 \text{ k}\Omega}{4.7 \text{ k}\Omega} + 1 = \mathbf{22.3}$$

Question

If R_f in Figure 6–16 is increased to 150 kΩ, what is the closed-loop gain?

COMPUTER SIMULATION

Open the Multisim file F06-16DV on the website. Determine the closed-loop gain.

Voltage-Follower

The **voltage-follower** is a special case of the noninverting amplifier where all of the output voltage is fed back to the inverting input by a straight connection, as shown in Figure 6–17. The straight feedback connection produces a voltage gain of approximately 1, so the closed-loop gain of the voltage-follower is

$$A_{cl(VF)} = 1 \tag{6–4}$$

FIGURE 6–17

Op-amp voltage-follower.

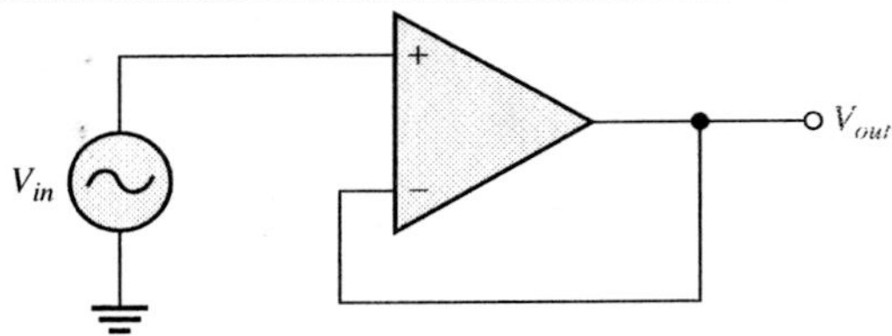

The most important features of the voltage-follower configuration are its very high input resistance and its very low output resistance. These features make it a nearly ideal buffer amplifier for interfacing high-resistance sources and low-resistance loads.

Inverting Amplifier

An op-amp connected in a closed-loop configuration in which the input signal is applied through a series resistor to the inverting input (−) is an **inverting amplifier**, as shown in Figure 6–18. The output is fed back through R_f to the inverting input. The noninverting input is grounded.

FIGURE 6–18

Inverting amplifier.

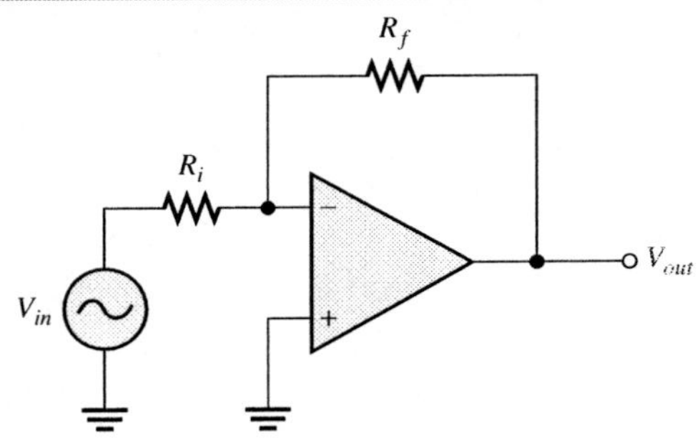

At this point, the ideal op-amp parameters mentioned earlier are useful in simplifying the analysis of this circuit. In particular, the concept of infinite input resistance is of great value. An infinite input resistance implies that there is *no* current in or out of the inverting input. If there is no current through the input resistance, then there must be *no* voltage drop between the inverting and noninverting inputs. This means that the voltage at the inverting input (−) is zero because the noninverting input (+) is grounded. This zero voltage at the inverting input terminal is referred to as *virtual ground.* This condition is illustrated in Figure 6–19(a).

FIGURE 6–19

Illustration of the virtual ground concept and closed-loop voltage gain development for the inverting amplifier.

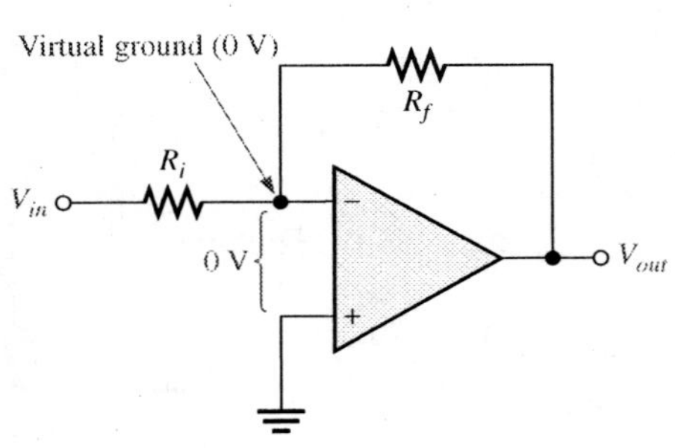

(a) Virtual ground

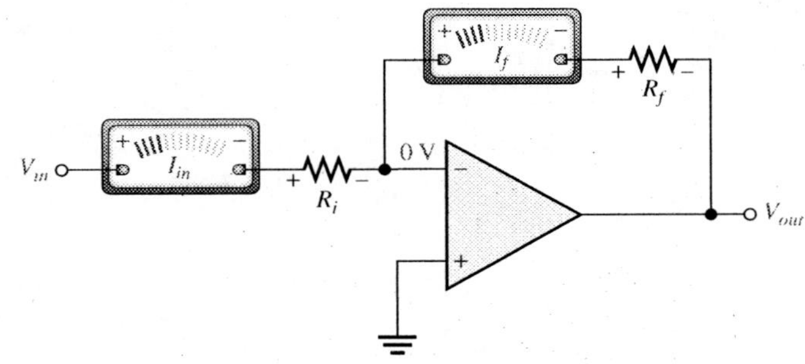

(b) $I_{in} = I_f$ and current at the inverting input (–) is 0.

Since there is no current at the inverting input, the current through R_i and the current through R_f are equal, as shown in Figure 6–19(b). The voltage across R_i equals V_{in} because of virtual ground on the other side of the resistor. Also, the voltage across R_f equals $-V_{out}$ because of virtual ground. Since $I_f = I_{in}$,

$$\frac{-V_{out}}{R_f} = \frac{V_{in}}{R_i}$$

Rearranging the terms,

$$\frac{V_{out}}{V_{in}} = -\frac{R_f}{R_i}$$

Of course, V_{out}/V_{in} is the closed-loop gain of the inverting amplifier, $A_{cl(I)}$.

$$A_{cl(I)} = -\frac{R_f}{R_i} \tag{6–5}$$

Equation 6–5 shows that the closed-loop voltage gain of the inverting amplifier, $A_{cl(I)}$, is the ratio of the feedback resistance R_f to the resistance R_i. *The closed-loop gain is independent of the op-amp's internal open-loop gain.* Thus, the negative feedback stabilizes the voltage gain. The negative sign indicates inversion.

EXAMPLE 6–4

Problem

Given the op-amp configuration in Figure 6–20, determine the value of R_f required to produce a closed-loop voltage gain of -100.

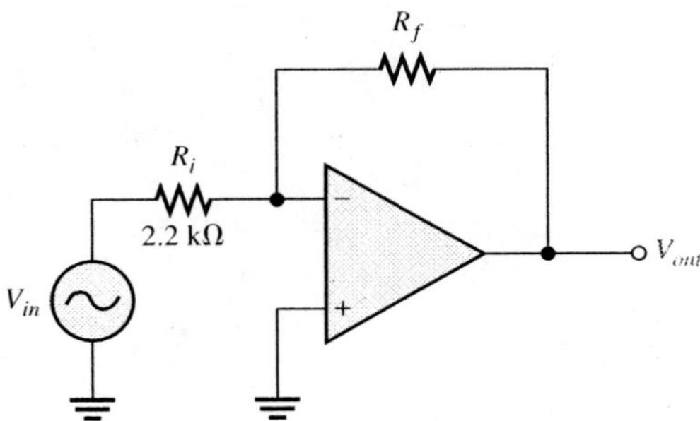

FIGURE 6–20

Solution

Knowing that $R_i = 2.2\text{ k}\Omega$ and $A_{cl(I)} = -100$, calculate R_f as follows:

$$A_{cl(I)} = -\frac{R_f}{R_i}$$

$$R_f = -A_{cl(I)}R_i = -(-100)(2.2\text{ k}\Omega) = \mathbf{220\text{ k}\Omega}$$

Question

(a) If R_i is changed to 2.7 kΩ in Figure 6–20, what value of R_f is required to produce a closed-loop gain of −25?

(b) If R_f failed open, what would you expect to see at the output?

COMPUTER SIMULATION

Open the Multisim file F06-20DV on the website. Measure the closed-loop gain.

Effect of Negative Feedback on Bandwidth

You have learned how negative feedback reduces the gain from its open-loop value. Now you will learn how it affects the amplifier's bandwidth. Figure 6–21 graphically illustrates the concept of closed-loop frequency response for an op-amp with negative feedback. When the open-loop gain of an op-amp is reduced by negative feedback, the bandwidth is increased. The closed-loop gain is independent of the open-loop gain up to the point of intersection of the two gain curves. This point of intersection is the critical frequency, $f_{c(cl)}$, for the closed-loop response, which equals the closed-loop bandwidth, BW_{cl}. Notice that beyond the closed-loop critical frequency the closed-loop gain decreases at the same rate (called the roll-off rate) as the open-loop gain.

FIGURE 6–21

Closed-loop gain compared to open-loop gain.

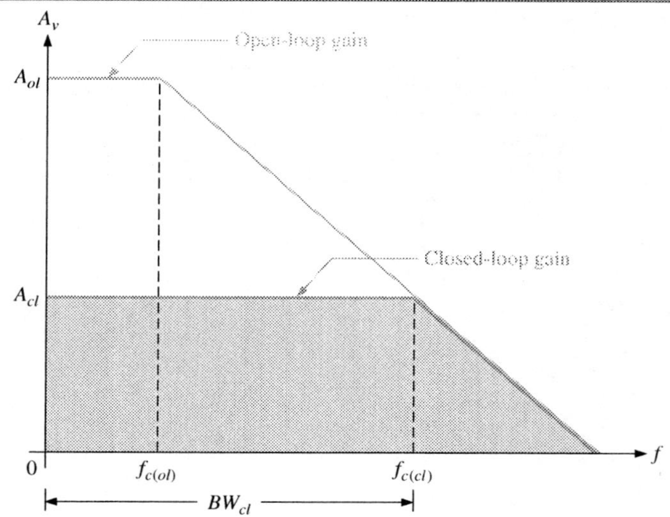

Gain-Bandwidth Product

An increase in closed-loop gain causes a decrease in the bandwidth and vice versa, such that *the product of gain and bandwidth is a constant.* This is true as long as the roll-off rate is a fixed −20 dB/decade.* If A_{cl} represents the gain of any of the noninverting closed-loop configurations and $f_{c(cl)}$ represents the closed-loop critical frequency (same as the bandwidth), then

$$A_{cl}f_{c(cl)} = A_{ol}f_{c(ol)}$$

The **gain-bandwidth product** is always equal to the frequency at which the op-amp's open-loop gain is unity (unity-gain bandwidth).

$$A_{cl}f_{c(cl)} = \text{unity-gain bandwidth} \quad (6\text{–}6)$$

EXAMPLE 6–5

Problem

Determine the bandwidth of each of the amplifiers in Figure 6–22. Both op-amps have a unity-gain bandwidth of 3 MHz.

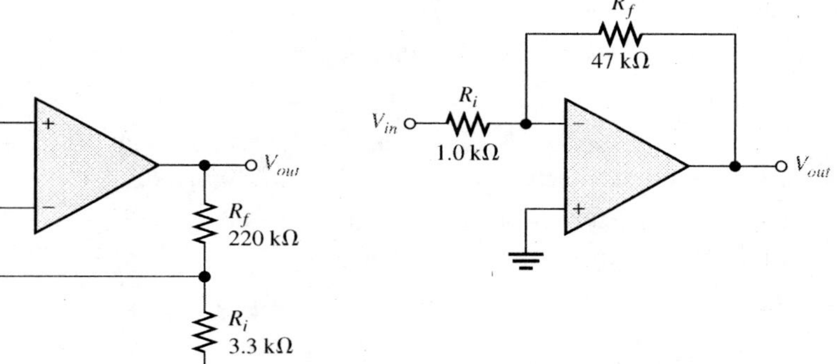

FIGURE 6–22

Solution

(a) For the noninverting amplifier in Figure 6–22(a), the closed-loop gain is

$$A_{cl(\text{NI})} = \frac{R_f}{R_i} + 1 = \frac{220\text{ k}\Omega}{3.3\text{ k}\Omega} + 1 = 67.7$$

Use Equation 6–6 and solve for $f_{c(cl)}$ (where $f_{c(cl)} = BW_{cl}$).

$$f_{c(cl)} = BW_{cl} = \frac{\text{unity-gain } BW}{A_{cl}}$$

$$BW_{cl} = \frac{3\text{ MHz}}{67.7} = \mathbf{44.3\text{ kHz}}$$

* dB stands for decibel, a logarithmic measurement. A decade is a ten times increase in frequency. See Appendix A for a discussion of logarithms and decibels.

(b) For the inverting amplifier in Figure 6–22(b), the closed-loop gain is

$$A_{cl(I)} = -\frac{R_f}{R_i} = -\frac{47\ \text{k}\Omega}{1.0\ \text{k}\Omega} = -47$$

Using the absolute value of $A_{cl(I)}$, the closed-loop bandwidth is

$$BW_{cl} = \frac{3\ \text{MHz}}{47} = \mathbf{63.8\ kHz}$$

Question
What is the bandwidth of each of the amplifiers in Figure 6–22 if both op-amps have a unity-gain bandwidth of 2 MHz?

Review Questions

16. What three op-amp configurations were discussed?
17. What is the main purpose of negative feedback?
18. Is the closed-loop voltage gain of each of the op-amp configurations discussed dependent on the internal open-loop voltage gain of the op-amp?
19. The attenuation of the negative feedback circuit of a noninverting op-amp configuration is 0.02. What is the closed-loop gain of the amplifier?
20. What is the bandwidth of a closed-loop op-amp configuration that has an $f_{c(cl)}$ of 20 kHz?

6–5 TROUBLESHOOTING

As a technician, you may encounter situations in which an op-amp or its associated circuitry has malfunctioned. The op-amp is a complex integrated circuit with many types of internal failures possible. However, since you cannot troubleshoot the op-amp internally, treat it as a single device with only a few connections to it. If it fails, replace it just as you would a resistor, capacitor, or transistor.

In this section, you will learn to troubleshoot op-amp amplifiers.

In op-amp configurations, there are only a few components that can fail. Both inverting and noninverting amplifiers have a feedback resistor, R_f, and an input resistor, R_i. Depending on the circuit, a load resistor, bypass capacitors, or a voltage compensation resistor may also be present. Any of these components can appear to be open or appear to be shorted. An open is not always due to the component itself, but may be due to a poor solder connection or a bent pin on the op-amp. Likewise, a short circuit may be due to a solder bridge.

Of course, the op-amp itself can fail. Let's examine the basic configurations, considering only the feedback and input resistor failure modes and associated symptoms.

SAFETY NOTE

Liquids and electricity don't mix! You should never place a drink or other liquid where it could be accidentally spilled onto an electrical circuit.

Faults in the Noninverting Amplifier

The first thing to do when you suspect a faulty circuit is to check for the proper power supply voltage. *The positive and negative supply voltages should be measured on the op-amp's pins with respect to a nearby circuit ground.* If either voltage is missing or incorrect, trace the power connections back toward the supply before making other checks. Check that the ground path is not open, giving a misleading power supply reading. If you have verified the supply voltages and ground path, possible faults with the basic amplifier are as follows.

Open R_f

If the feedback resistor, R_f, opens, as shown in Figure 6–23(a), the op-amp is operating with its very high open-loop gain, which causes the input signal to drive the device into nonlinear operation and results in a severely clipped output signal.

FIGURE 6–23 Faults in the noninverting amplifier.

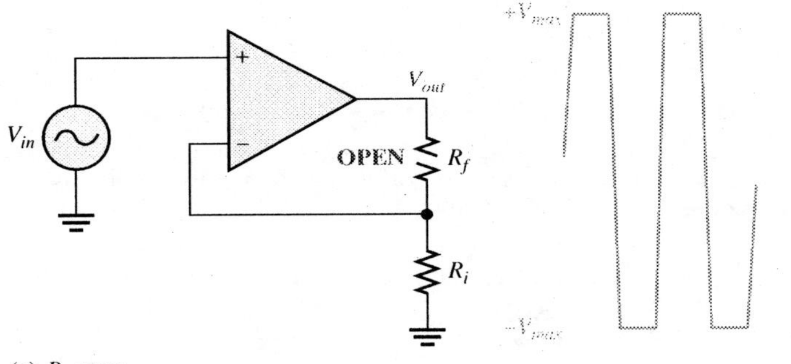

(a) R_f open

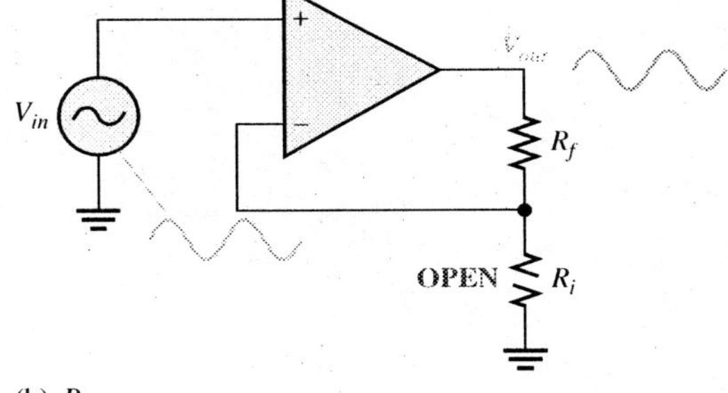

(b) R_i open

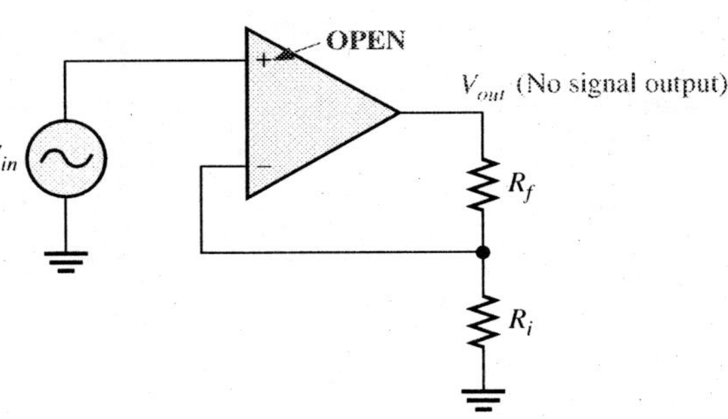

(c) Input to op-amp open internally

Open R_i

In this case, you still have a closed-loop configuration. But, since R_i is open and effectively equal to infinity, ∞, the closed-loop gain from Equation 6–3 is

$$A_{cl(NI)} = \frac{R_f}{R_i} + 1 = \frac{R_f}{\infty} + 1 = 0 + 1 = 1$$

This shows that the amplifier acts like a voltage-follower. You would observe an output signal that is the same as the input, as indicated in Figure 6–23(b).

Internally Open Noninverting Op-Amp Input

In this situation, because the input voltage is not applied to the op-amp, the output is zero. This is indicated in Figure 6–23(c).

Other Op-Amp Faults

In general, an internal failure will result in a loss or distortion of the output signal. The best approach is to first make sure that there are no external failures or faulty conditions. If everything else is good, then the op-amp must be bad.

Faults in the Voltage-Follower

The voltage-follower is a special case of the noninverting amplifier. Except for a bad power supply, a bad op-amp, or an open or short at a connection, about the only thing that can happen in a voltage-follower circuit is an open feedback loop. This would have the same effect as an open feedback resistor as previously discussed.

Faults in the Inverting Amplifier

Power Supply

As in the case of the noninverting amplifier, the power supply voltages should be checked first. Power supply voltages should be checked on the op-amp's pins with respect to a nearby ground.

Open R_f

If R_f opens as indicated in Figure 6–24(a), the input signal still feeds through the input resistor and is amplified by the high open-loop gain of the op-amp. This forces the device to be driven into nonlinear operation, and you will see an output something like that shown. This is the same result as in the noninverting configuration.

FIGURE 6–24 Faults in the inverting amplifier.

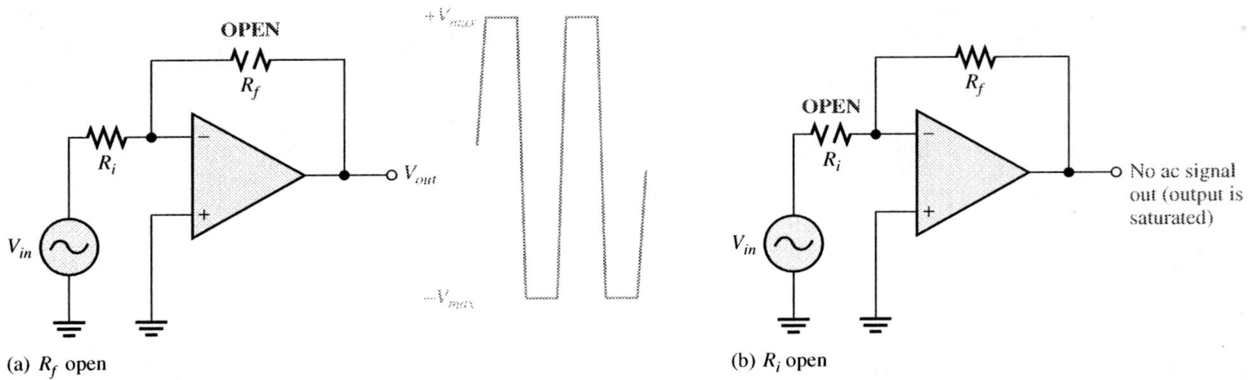

Open R_i

This prevents the input signal from getting to the op-amp input, so there will be no output signal, as indicated in Figure 6–24(b).

Failures in the op-amp itself have the same effects as previously discussed for the noninverting amplifier.

Review Questions

21. If you notice that the op-amp output is saturated, what should you check first?
22. If there is no op-amp output signal when there is a verified input signal, what should you check first?
23. What component do you suspect if the gain of a noninverting op-amp drops to 1?
24. If the feedback resistor in an op-amp circuit is the wrong value, what is the effect on the output?
25. Why should the dc supply voltages be measured directly on the op-amp pins?

CHAPTER REVIEW

Key Terms

Closed-loop voltage gain The net voltage gain of an amplifier when negative feedback is included.

Common-mode rejection ratio (CMRR) The ratio of open-loop gain to common-mode gain; a measure of an op-amp's ability to reject common-mode signals.

Gain-bandwidth product A constant which is the product of the closed-loop gain and the closed-loop critical frequency; the frequency at which the op-amp's open-loop gain is unity (1).

Inverting amplifier A closed-loop op-amp circuit in which the input signal is applied to the inverting input.

Noninverting amplifier A closed-loop op-amp circuit in which the input signal is applied to the noninverting input.

Open-loop voltage gain The internal gain of an amplifier without external feedback.

Operational amplifier An electronic device that amplifies the difference voltage between the two inputs. It has very high voltage gain, very high input resistance, very low output resistance, and good rejection of common-mode signals.

Slew rate The rate of change of the output voltage of an op-amp in response to a step input.

Voltage-follower A closed-loop, noninverting op-amp circuit with a voltage gain of one.

Important Facts

- ❑ The basic op-amp has three terminals not including power and ground: inverting input (−), noninverting input (+), and output.
- ❑ Most op-amps require both a positive and a negative dc supply voltage.
- ❑ The ideal (perfect) op-amp has infinite input resistance, zero output resistance, and infinite open-loop voltage gain.
- ❑ A practical op-amp has very high input resistance, very low output resistance, and high open-loop voltage gain.
- ❑ A differential amplifier is normally used for the input stage of an op-amp.

- Common mode occurs when equal in-phase voltages are applied to both input terminals of an op-amp.
- Input offset voltage produces an output error voltage (with no input voltage).
- Input bias current also produces an output error voltage (with no input voltage).
- Open-loop voltage gain is the gain of an op-amp with no external feedback connections.
- Closed-loop voltage gain is the gain of an op-amp with external feedback.
- The common mode rejection ratio (CMRR) is a measure of an op-amp's ability to reject common-mode inputs.
- Slew rate is the rate in volts per microsecond at which the output voltage of an op-amp can change in response to a step input.
- There are three basic op-amp negative feedback configurations: inverting, noninverting and voltage–follower.
- Negative feedback occurs when a portion of the output voltage is connected back to the inverting input such that it subtracts from the input voltage, thus reducing the voltage gain but increasing the stability and bandwidth.
- The product of gain and bandwidth is constant for most op-amps.
- The gain-bandwidth product equals the frequency at which the open-loop voltage gain is unity (1).

Formulas

Common-mode rejection ratio (op-amp):

$$\text{CMRR} = \frac{A_{ol}}{A_{cm}} \qquad \textbf{(6–1)}$$

Slew rate:

$$\text{Slew rate} = \frac{\Delta V_{out}}{\Delta t} \qquad \textbf{(6–2)}$$

Voltage gain (noninverting):

$$A_{cl(\text{NI})} = \frac{R_f}{R_i} + 1 \qquad \textbf{(6–3)}$$

Voltage gain (voltage-follower):

$$A_{cl(\text{VF})} = 1 \qquad \textbf{(6–4)}$$

Voltage gain (inverting):

$$A_{cl(\text{I})} = -\frac{R_f}{R_i} \qquad \textbf{(6–5)}$$

Unity-gain bandwidth:

$$A_{cl}f_{c(cl)} = \text{unity-gain bandwidth} \qquad \textbf{(6–6)}$$

Chapter Checkup

Answers are at the end of the chapter.

1. An integrated circuit (IC) op-amp has
   (a) two inputs and two outputs
   (b) one input and one output
   (c) two inputs and one output

2. Which of the following characteristics does not necessarily apply to an op-amp?
   (a) High gain (b) Low power
   (c) High input resistance (d) Low output resistance

3. With zero volts on both inputs, an op-amp ideally should have an output
   (a) equal to the positive supply voltage
   (b) equal to the negative supply voltage
   (c) equal to zero
   (d) equal to the CMRR

4. Of the values listed, the most realistic value for open-loop gain of an op-amp is
   (a) 1 (b) 2000
   (c) 80 dB (d) 100,000

5. The output of a particular op-amp increases 8 V in 12 μs. The slew rate is
   (a) 96 V/μs (b) 0.67 V/μs
   (c) 1.5 V/μs (d) none of these

6. For an op-amp with negative feedback, the output is
   (a) equal to the input
   (b) increased
   (c) fed back to the inverting input
   (d) fed back to the noninverting input

7. The use of negative feedback
   (a) reduces the voltage gain of an op-amp
   (b) makes the op-amp oscillate
   (c) makes linear operation possible
   (d) answers (a) and (c)

8. Negative feedback
   (a) increases the input and output impedances
   (b) increases the input impedance and the bandwidth
   (c) decreases the output impedance and the bandwidth
   (d) does not affect impedances or bandwidth

9. A certain noninverting amplifier has an R_i of 1.0 kΩ and an R_f of 100 kΩ. The closed-loop gain is
   (a) 100,000 (b) 1000
   (c) 101 (d) 100

10. If the feedback resistor in Question 9 is open, the voltage gain

(a) increases (b) decreases

(c) is not affected (d) depends on R_i

11. A certain inverting amplifier has a closed-loop gain of 25. The op-amp has an open-loop gain of 100,000. If another op-amp with an open-loop gain of 200,000 is substituted in the configuration, the closed-loop gain

(a) doubles (b) drops to 12.5

(c) remains at 25 (d) increases slightly

12. A voltage-follower

(a) has a gain of one (b) is noninverting

(c) has no feedback resistor (d) answers (a), (b), and (c)

13. When negative feedback is used, the bandwidth of an op-amp

(a) increases (b) decreases

(c) stays the same (d) fluctuates

14. If R_i is open in Figure 6–25, the closed-loop gain will

(a) increase (b) decrease

(c) not change

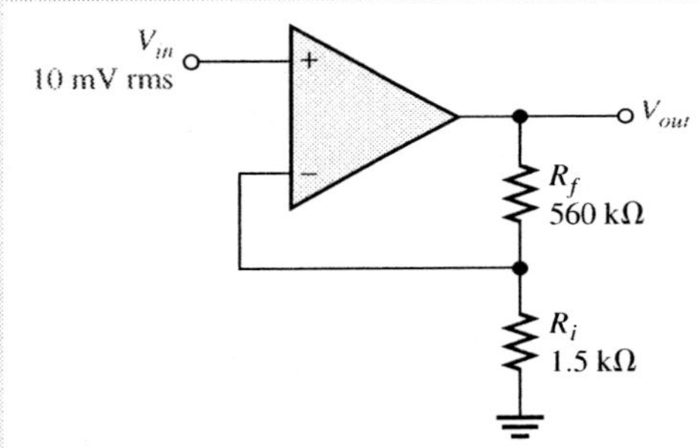

FIGURE 6–25

15. If R_i is open in Figure 6–25, for a given input signal, the output signal will

(a) increase (b) decrease

(c) not change

16. If R_f is open in Figure 6–25, the output voltage will

(a) increase (b) decrease

(c) not change

17. If R_f is open in Figure 6–25, the open-loop gain will

(a) increase (b) decrease

(c) not change

18. If R_f is open in Figure 6–25, the closed-loop gain will

(a) increase (b) decrease

(c) not change

Questions

Answers to odd-numbered questions are at the end of the book.

1. How are the inverting and noninverting inputs of an op-amp indicated?
2. What is the expression for the output voltage of an op-amp in terms of A_v and V_{in}?
3. The peak-to-peak output of an op-amp is limited to slightly less than what?
4. Ideally, when does an op-amp produce a 0 V output?
5. In practice, what is the input required to force the output to 0 V called?
6. What is the term for the average of the two input currents to an op-amp?
7. What is the difference between the differential input resistance and the common-mode input resistance?
8. What is the difference between open-loop gain and closed-loop gain?
9. Assume that two op-amps with the same differential input are compared. Op-amp B has less gain for the common-mode signals. Which op-amp has the highest CMRR?
10. Op-amp A has a slew rate of 10 V/μs and op-amp B has a slew rate of 15 V/μs. Which op-amp has the higher upper critical frequency?
11. In a noninverting op-amp configuration, if R_f is increased, what happens to the voltage gain?
12. How does the voltage-follower configuration differ from the noninverting configuration?
13. What is the voltage gain of an inverting op-amp configuration with $R_f = 100\ \text{k}\Omega$ and $R_i = 1.0\ \text{k}\Omega$?
14. In an inverting amplifier configuration, why is the inverting input considered to be 0 V?
15. If the open-loop gain of an op-amp is equal to 1 at a frequency of 15 kHz, what is the gain-bandwidth product of a noninverting configuration using the op-amp?

Basic Problems

PROBLEMS

Answers to odd-numbered problems are at the end of the book.

1. Compare a practical op-amp to the ideal.
2. Two IC op-amps are available to you. Their characteristics are listed below. Choose the one you think is more desirable and explain your choice.

   Op-amp 1: $R_{in} = 5\ \text{M}\Omega$, $R_{out} = 100\ \Omega$, $A_{ol} = 50{,}000$

   Op-amp 2: $R_{in} = 10\ \text{M}\Omega$, $R_{out} = 75\ \Omega$, $A_{ol} = 150{,}000$
3. Determine the bias current, I_{BIAS}, given that the input currents to an op-amp are 8.3 μA and 7.9 μA.
4. A certain op-amp has a CMRR of 250,000. If the common-mode gain is 0.25, what is the open-loop gain?
5. The open-loop gain of a certain op-amp is 175,000. Its common-mode gain is 0.18. Determine the CMRR.
6. An op-amp data sheet specifies a CMRR of 300,000 and an A_{ol} of 90,000. What is the common-mode gain?

7. Identify each of the op-amp configurations in Figure 6–26.

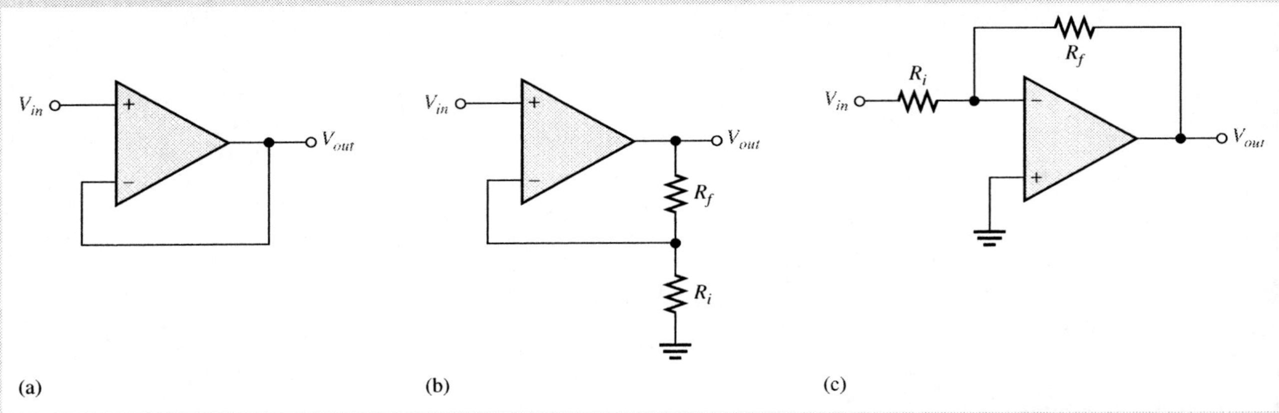

FIGURE 6–26

8. Determine the closed-loop gain of each amplifier in Figure 6–27.

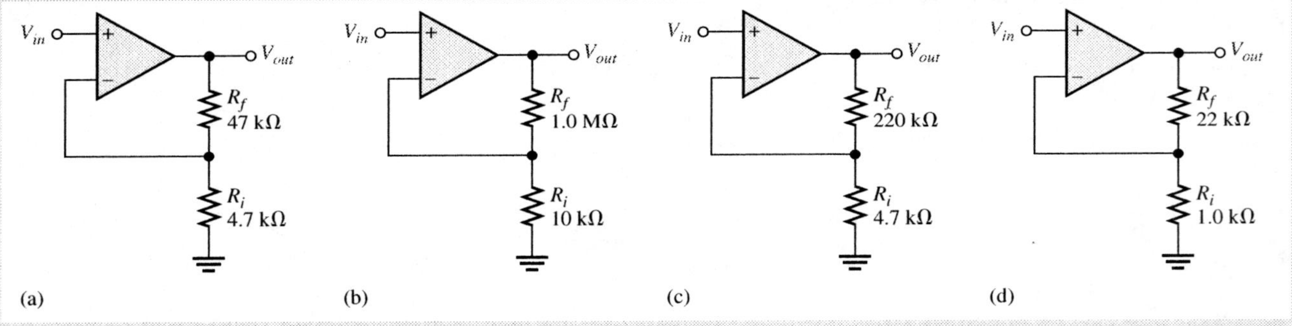

FIGURE 6–27

9. Find the gain of each amplifier in Figure 6–28.

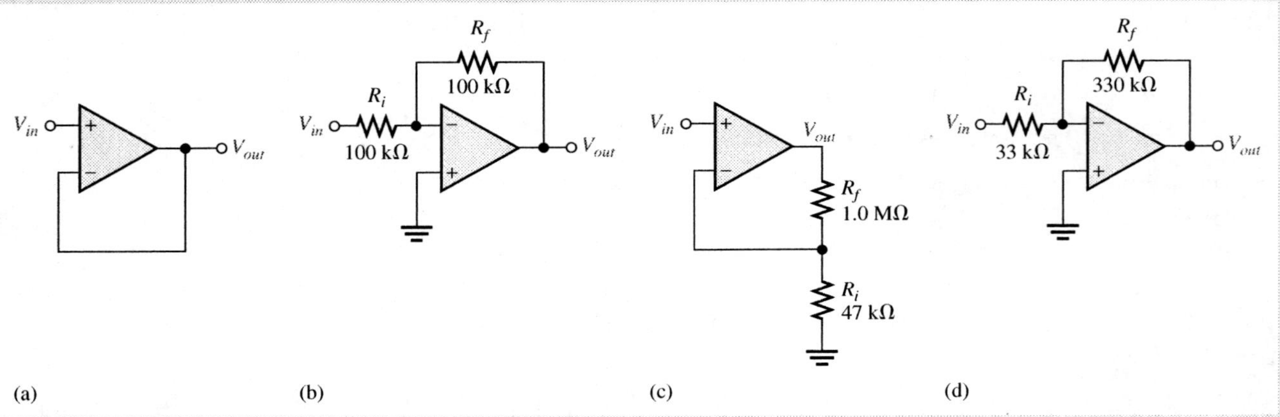

FIGURE 6–28

10. Determine the most likely fault(s) for each of the following symptoms in Figure 6–29 with a 100 mV signal applied.

(a) No output signal.

(b) Output severely clipped on both positive and negative swings.

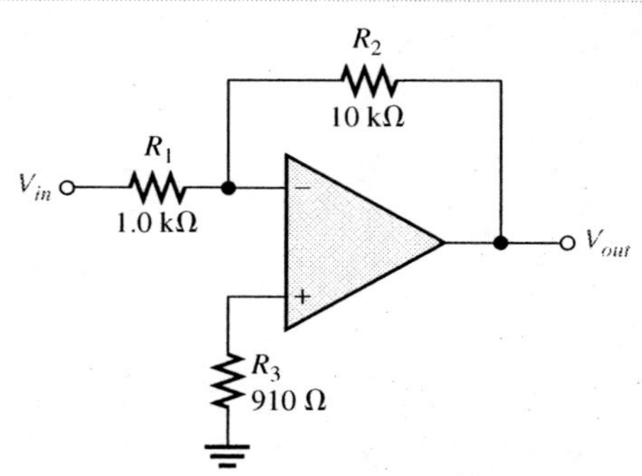

FIGURE 6–29

Basic-Plus Problems

11. Figure 6–30 shows the output voltage of an op-amp in response to a step input. What is the slew rate?

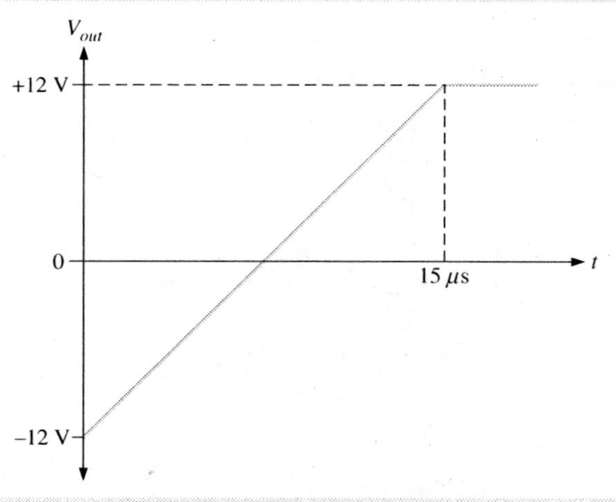

FIGURE 6–30

12. How long does it take the output voltage of an op-amp to go from −10 V to +10 V if the slew rate is 0.5 V/μs?

13. A noninverting amplifier has an R_i of 1.0 kΩ and an R_f of 100 kΩ. Determine V_f and V_{in} if $V_{out} = 5$ V.

14. For the amplifier in Figure 6–31, determine the following:

(a) $A_{cl(NI)}$ (b) V_{out}

(c) V_f

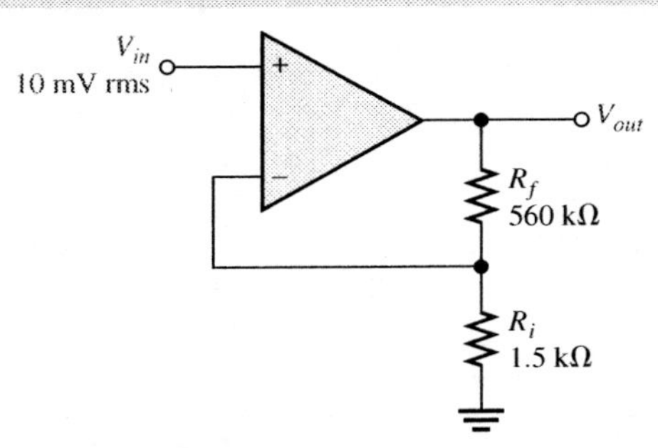

FIGURE 6–31

15. Find the value of R_f that will produce the indicated closed-loop gain in each amplifier in Figure 6–32.

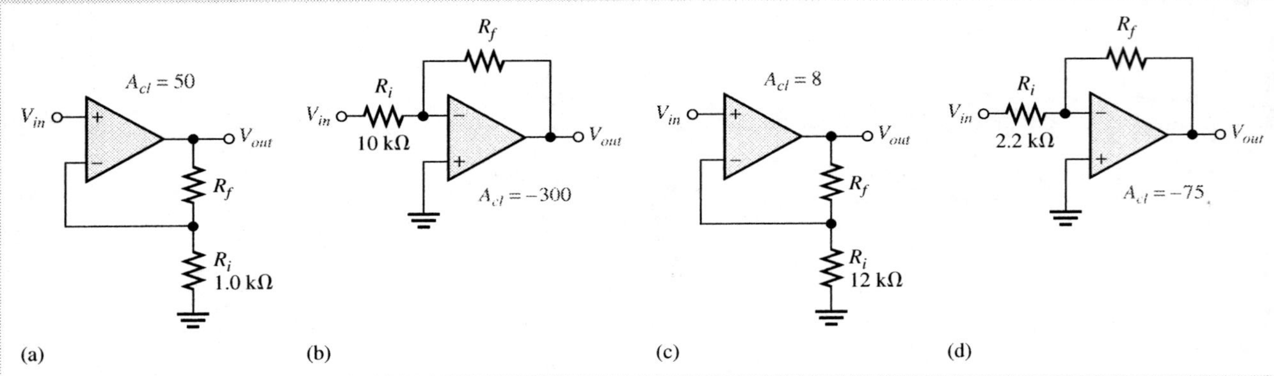

FIGURE 6–32

16. If a signal voltage of 10 mV rms is applied to each amplifier in Figure 6–28, what are the output voltages and what is their phase relationship with inputs?

17. Determine the approximate values for each of the following quantities in Figure 6–33.

(a) I_{in} (b) I_f

(c) V_{out} (d) Closed-loop gain

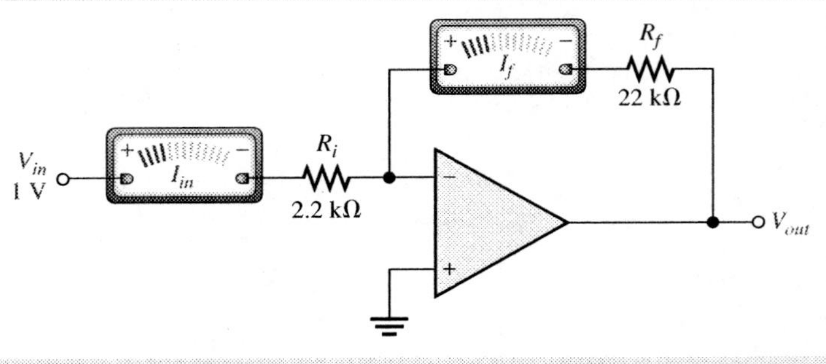

FIGURE 6–33

18. Given that $f_{c(ol)} = 75.0$ Hz, $A_{ol} = 100{,}000$, and $f_{c(cl)} = 55$ kHz, determine the approximate closed-loop gain.

19. What is the unity-gain bandwidth in Problem 18?

20. For each amplifier in Figure 6–34, determine the closed-loop gain and bandwidth. The op-amps in each circuit exhibit an open-loop gain of 150,000 and a unity-gain bandwidth of 2.8 MHz.

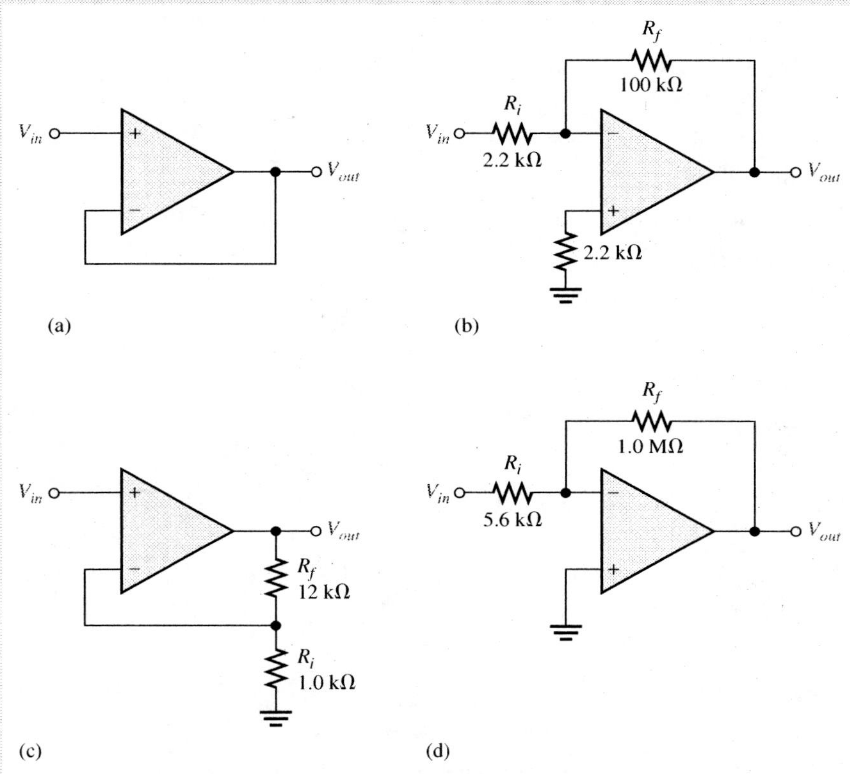

FIGURE 6–34

21. Determine the effect on the output of the circuit in Figure 6–29 for each of the following faults:

    (a) Output pin is shorted to the inverting input.

    (b) R_3 is open.

    (c) R_3 is 10 kΩ instead of 910 Ω.

    (d) R_1 and R_2 are swapped.

22. What should the output look like if a 100 mV signal is applied to the input of the circuit in Figure 6–29?

Example Questions

ANSWERS

6–1: 500,000

6–2: 20 V/μs

6–3: 32.9

6–4: (a) 67.5 kΩ

(b) The amplifier would have an open-loop gain producing a square wave.

6–5: (a) 29.5 kHz

(b) 42.6 kHz

Review Questions

1. Inverting input, noninverting input, output, positive and negative supply voltages
2. A practical op-amp has high input impedance, low output impedance and high voltage gain.
3. A practical op-amp has a very high voltage gain (not infinite).
4. Noninverting
5. Differential amplifier, voltage amplifier, and push-pull amplifier
6. The differential dc voltage required to make the output voltage zero
7. High
8. The gain without feedback
9. Common-mode rejection ratio
10. The maximum rate of change of output voltage in response to a step input
11. The inverting input
12. Negative feedback provides a stable controlled voltage gain, control of input and output impedances, and wider bandwidth.
13. The open-loop gain is so high that a very small signal on the input will drive the op-amp into saturation.
14. Increase
15. Both inputs will be the same voltage.
16. Inverting, noninverting and voltage follower
17. The main purpose of negative feedback is to stabilize the gain.
18. No
19. $A_{cl} = 1/0.02 = 50$
20. 20 kHz
21. Check power supply voltages with respect to ground. Verify ground connections. Check for an open feedback resistor.
22. Verify power supply voltages and ground leads. For inverting amplifiers, check for open R_i. For noninverting amplifiers, check that V_{in} is actually on (+) pin; if so, check (−) pin for identical signal.
23. R_i may be open.
24. The gain will be incorrect.
25. To make sure the voltage is getting to the op-amp

Chapter Checkup

1. (c)	2. (b)	3. (c)	4. (d)	5. (b)
6. (c)	7. (d)	8. (b)	9. (c)	10. (a)
11. (c)	12. (d)	13. (a)	14. (b)	15. (b)
16. (a)	17. (c)	18. (a)		

CHAPTER 7

BASIC OP-AMP CIRCUITS

CHAPTER OUTLINE

INTRODUCTION

In the last chapter, you learned about the principles, operation, and characteristics of the operational amplifier. Op-amps are used in such a wide variety of applications that it is impossible to cover all of them in one chapter, or even in one book.

Therefore, in this chapter, we will examine some of the more fundamental applications to illustrate how versatile the op-amp is and to give you a foundation in basic op-amp circuits.

Study aids for this chapter are available at

http://www.prenhall.com/SOE

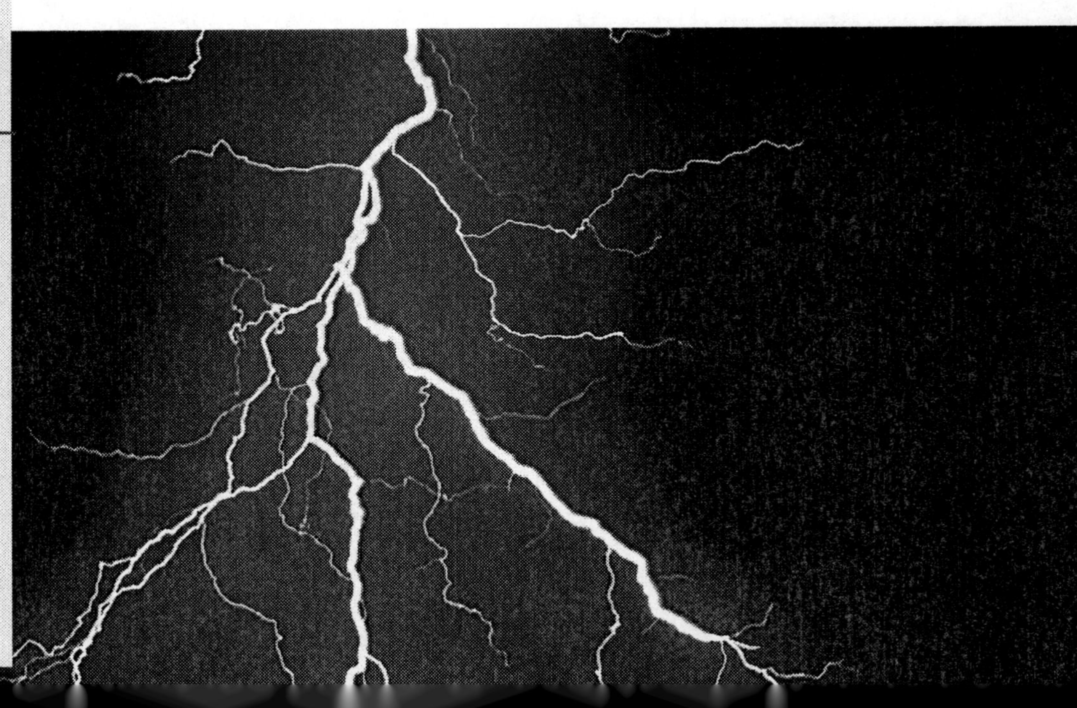

KEY TERMS

- Comparator
- Schmitt trigger
- Summing amplifier
- Integrator
- Differentiator
- Clamper
- Limiter
- Peak detector

KEY OBJECTIVES

A section number is given for each objective. After completing this chapter, you should be able to

7–1 Describe the operation of several basic comparator circuits

7–2 Describe the operation of several types of summing amplifiers

7–3 Describe the operation of integrators and differentiators

7–4 Describe the operation of clampers, limiters, and peak detectors

7–5 Troubleshoot op-amp comparators and summing amplifiers

COMPUTER SIMULATIONS DIRECTORY

The following figures have Multisim circuit files associated with them.

LABORATORY EXPERIMENTS DIRECTORY

The following exercises are for this chapter.

- ◆ **Experiment 13** Comparators and the Schmitt Trigger
- ◆ **Experiment 14** Summing Amplifiers
- ◆ **Experiment 15** The Integrator and Differentiator

The nervous system of living organisms is an interconnected structure made up of many elements, called neurons, working in connection with one another. The neuron is characterized by many inputs and one output. The output can have two states, excited or not excited. The input signals to a neuron are attenuated by the synapses, which are the junction parts of the neuron. Each individual neuron in a neural network continuously evaluates its output by looking at its inputs, calculating the weighted sum and comparing to a threshold to decide if it should fire (go to the excited state). Neural networks can "learn" by adjusting the resistance of the synapses to the incoming signals.

Artificial neural networks can be constructed using artificial neurons. A neuron can be simulated using a scaling adder followed by a comparator. The synoptic weights are determined by the values of the input resistors of the scaling adder. Using this approach a system can be implemented which learns how to recognize certain input patterns. In such a system, there is a training process where known patterns are presented and the weights are adjusted to achieve required outputs. After the system is "trained," it can recall specific patterns and produce the proper outputs.

7–1 COMPARATORS

Operational amplifiers are often used as nonlinear devices to compare the amplitude of one voltage with another. In this application, the op-amp is used in the open-loop configuration, with the input voltage on one input and a reference voltage on the other.

In this section, you will learn the operation of several basic comparator circuits.

Zero-Level Detector

A **comparator** is a circuit that compares two input voltages and produces an output in either of two states indicating the greater than or less than relationship of the inputs. One application of an op-amp used as a comparator is to determine when an input voltage exceeds a certain level. Figure 7–1(a) shows a zero-level detector. Notice that the inverting input (−) is grounded to produce a zero level and that the input signal voltage is applied to the noninverting input (+). Because of the high open-loop voltage gain, a very small difference voltage between the two inputs drives the amplifier into saturation, causing the output voltage to go to its limit.

FIGURE 7–1

The op-amp as a zero-level detector.

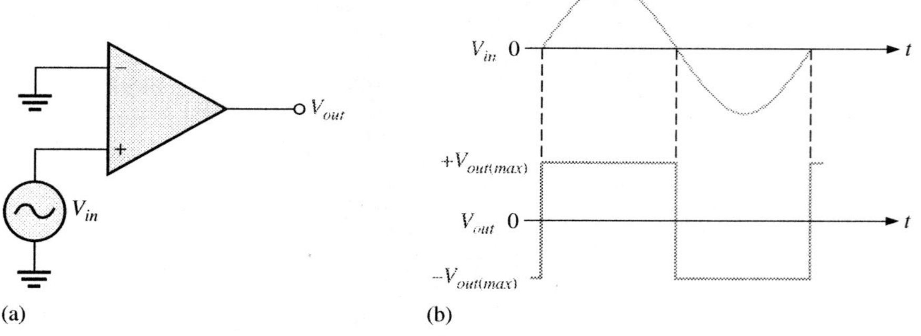

For example, consider an op-amp having A_{ol} = 100,000. A voltage difference of only 0.25 mV between the inputs could produce an output voltage of (0.25 mV)(100,000) = 25 V *if* the op-amp were capable. However, since most op-amps have output voltage limitations of ±15 V or less, the device would be driven into saturation. For many comparison applications, special op-amp comparators are used. These ICs are generally designed to maximize switching speed. In less stringent applications, a general-purpose op-amp works nicely as a comparator.

Figure 7–1(b) shows the result of a sinusoidal input voltage applied to the noninverting input of the zero-level detector. When the sine wave is negative, the output is at its maximum negative level. When the sine wave crosses 0, the amplifier is driven to its opposite state and the output goes to its maximum positive level, as shown. As you can see, the zero-level detector can be used as a squaring circuit to produce a square wave from a sine wave.

Nonzero-Level Detection

The zero-level detector in Figure 7–1 can be modified to detect positive and negative voltages by connecting a fixed reference voltage to the inverting input (−), as shown in Figure 7–2(a). A more practical arrangement is shown in Figure 7–2(b) using a voltage divider to set the reference voltage as follows:

$$V_{REF} = \frac{R_2}{R_1 + R_2}(+V) \tag{7–1}$$

where $+V$ is the positive op-amp supply voltage. The circuit in Figure 7–2(c) uses a zener diode to set the reference voltage ($V_{REF} = V_Z$). As long as the input voltage V_{in} is less than V_{REF}, the output remains at the maximum negative level. When the input voltage exceeds

Nonzero-level detectors. **FIGURE 7–2**

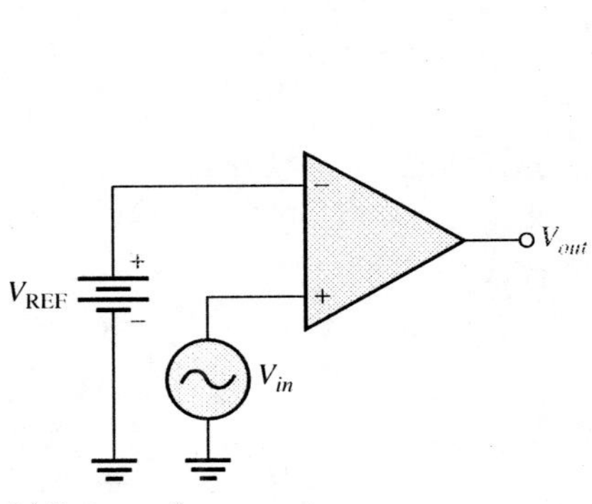

(a) Battery reference

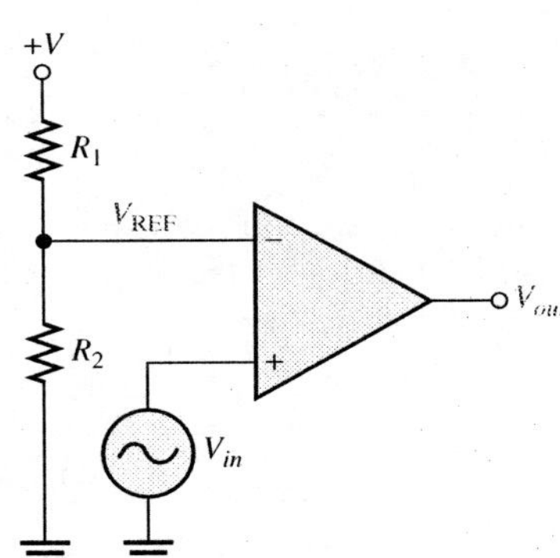

(b) Voltage-divider reference

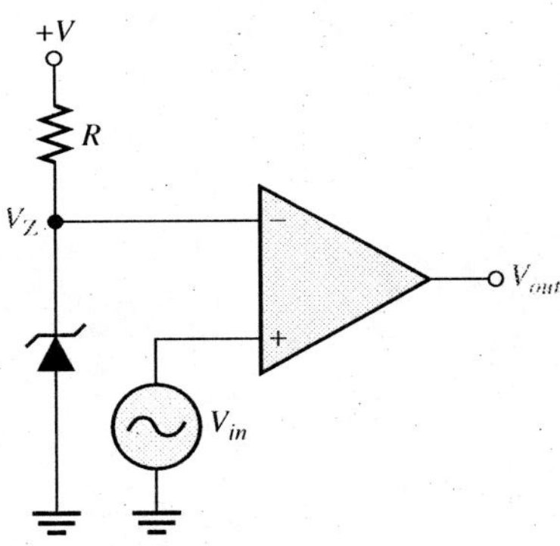

(c) Zener diode sets reference voltage

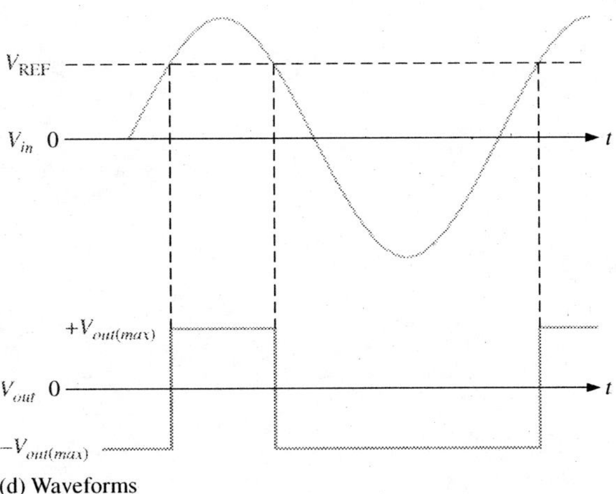

(d) Waveforms

the reference voltage, the output goes to its maximum positive state, as shown in Figure 7–2(d) with a sinusoidal input voltage.

EXAMPLE 7–1

Problem

The input signal in Figure 7–3(a) is applied to the comparator circuit in Figure 7–3(b). Make a sketch of the output showing its proper relationship to the input signal. Assume the maximum output levels of the op-amp are ±12 V.

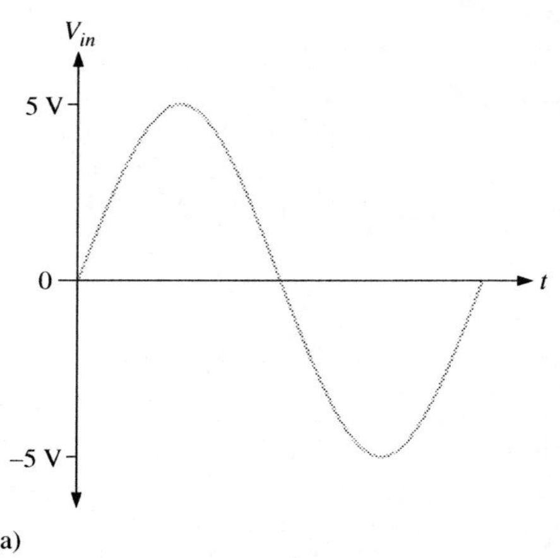

FIGURE 7–3 (a) (b)

Solution

The reference voltage is set by R_1 and R_2 as follows:

$$V_{REF} = \frac{R_2}{R_1 + R_2}(+V) = \frac{1.0\text{ k}\Omega}{8.2\text{ k}\Omega + 1.0\text{ k}\Omega}(+15\text{ V}) = 1.63\text{ V}$$

As shown in Figure 7–4, each time the input exceeds +1.63 V, the output voltage switches to its +12 V level, and each time the input goes below +1.63 V, the output switches back to its −12 V level.

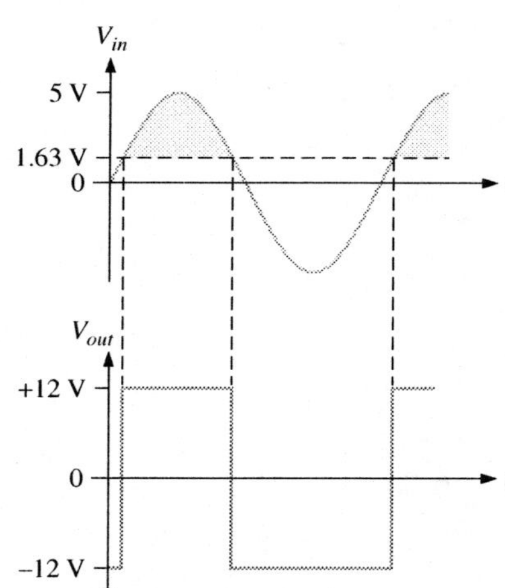

FIGURE 7–4

Question*

What is the reference voltage in Figure 7–3 if R_1 = 22 kΩ and R_2 = 3.3 kΩ?

* *Answers are at the end of the chapter.*

COMPUTER SIMULATION

Open the Multisim file F07-03DV on the website. Apply the input shown, and observe the output waveform.

Effects of Input Noise on Comparator Operation

In many practical situations, **noise** (unwanted voltage or current fluctuations) may appear on the input line. This noise voltage becomes superimposed on the input voltage, as shown in Figure 7–5, and can cause a comparator to erratically switch output states.

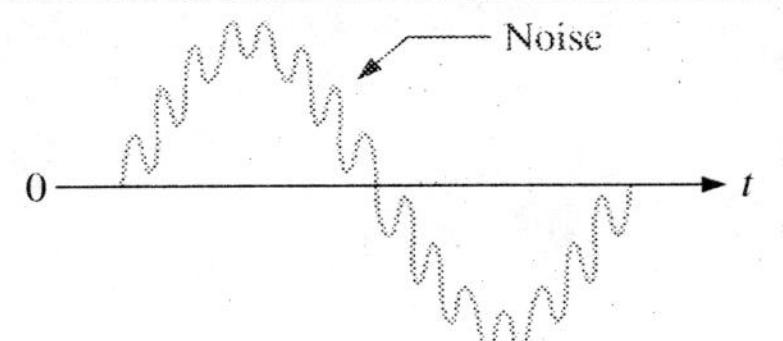

FIGURE 7–5

Sine wave with superimposed noise.

In order to understand the potential effects of noise voltage, consider a low-frequency sinusoidal voltage applied to the noninverting input (+) of an op-amp comparator used as a zero-level detector, as shown in Figure 7–6(a). Part (b) of the figure shows the input sine wave plus noise and the resulting output. As you can see, when the sine wave approaches 0, the fluctuations due to noise cause the total input to vary above and below 0 several times, thus producing an erratic output voltage.

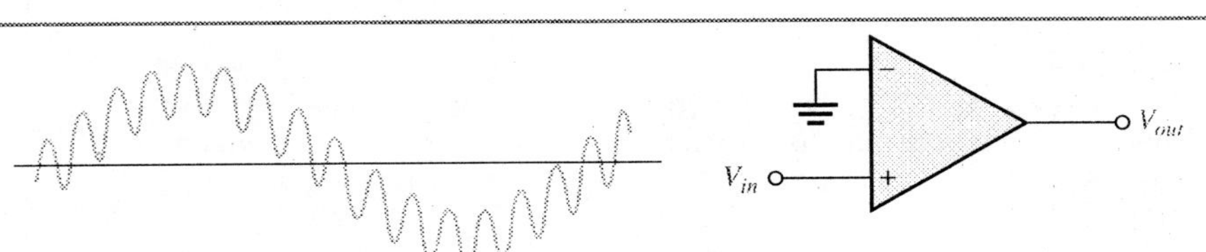

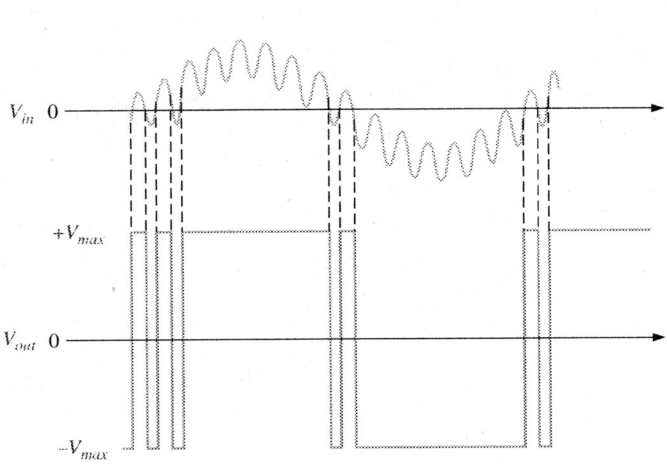

FIGURE 7–6

Effects of noise on comparator circuit.

Reducing Noise Effects with Hysteresis

An erratic output voltage caused by noise on the input occurs because the op-amp comparator switches from its negative output state to its positive output state at the same input voltage level that causes it to switch in the opposite direction, from positive to negative. This unstable condition occurs when the input voltage hovers around the reference voltage, and any small noise fluctuations cause the comparator to switch first one way and then the other.

In order to make the comparator less sensitive to noise, a technique incorporating positive feedback, called *hysteresis*, can be used, as described in Section 3–5 for transistors. Recall that hysteresis uses two reference levels. A good example of hysteresis is a common household thermostat that turns the furnace on at one temperature and off at another.

The two reference levels are referred to as the upper trigger point (UTP) and the lower trigger point (LTP). This two-level hysteresis is established with a positive feedback arrangement, as shown in Figure 7–7. Notice that the noninverting input (+) is connected to a resistive voltage divider such that a portion of the output voltage is fed back to the input. The input signal is applied to the inverting input (−) in this case.

FIGURE 7–7

Comparator with positive feedback for hysteresis.

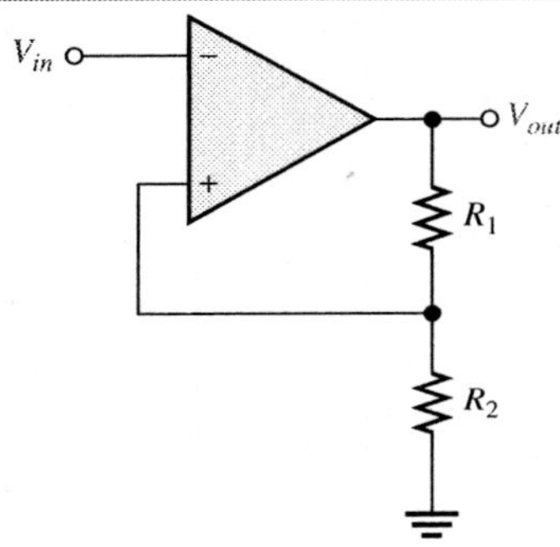

The basic operation of the comparator with hysteresis is as follows with reference to Figure 7–7. Assume that the output voltage is at its positive maximum, $+V_{out(max)}$. The voltage fed back to the noninverting input is V_{UTP} and is expressed as

$$V_{\text{UTP}} = \frac{R_2}{R_1 + R_2}(+V_{out(max)})$$

When the input voltage V_{in} exceeds V_{UTP}, the output voltage drops to its negative maximum, $-V_{out(max)}$. Now the voltage fed back to the noninverting input is V_{LTP} and is expressed as

$$V_{\text{LTP}} = \frac{R_2}{R_1 + R_2}(-V_{out(max)})$$

The input voltage must now fall below V_{LTP} before the device will switch back to its other voltage level. This means that a small amount of noise voltage has no effect on the output, as illustrated by Figure 7–8.

A comparator with hysteresis is sometimes known as a **Schmitt trigger**. The amount of hysteresis is defined by the difference of the two trigger levels.

$$V_{\text{HYS}} = V_{\text{UTP}} - V_{\text{LTP}} \quad (7\text{–}2)$$

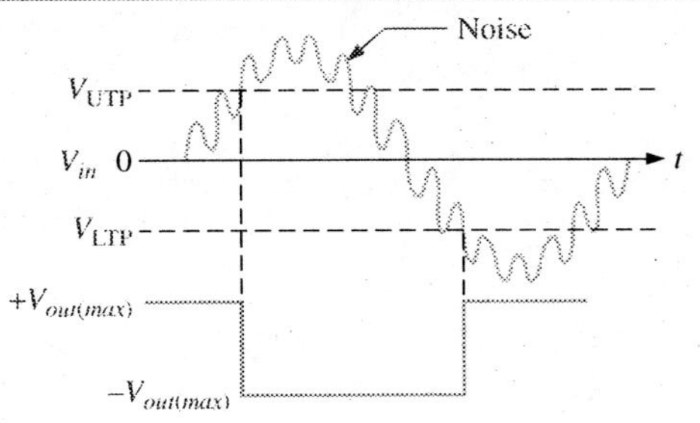

FIGURE 7–8

Operation of a comparator with hysteresis. The device triggers only once when the UTP or the LTP is reached; thus, there is immunity to noise that is riding on the input signal.

EXAMPLE 7–2

Problem

Determine the upper and lower trigger points and the hysteresis for the comparator circuit in Figure 7–9. Assume that $+V_{out(max)} = +5$ V and $-V_{out(max)} = -5$ V.

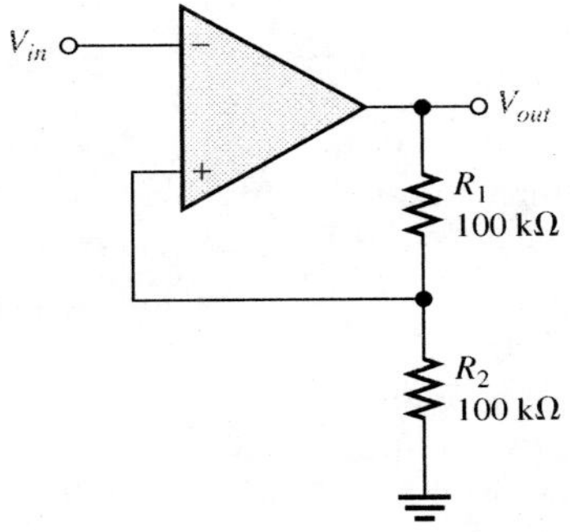

FIGURE 7–9

Solution

$$V_{UTP} = \frac{R_2}{R_1 + R_2}(+V_{out(max)}) = 0.5(5\text{ V}) = \mathbf{+2.5\text{ V}}$$

$$V_{LTP} = \frac{R_2}{R_1 + R_2}(-V_{out(max)}) = 0.5(-5\text{ V}) = \mathbf{-2.5\text{ V}}$$

$$V_{HYS} = V_{UTP} - V_{LTP} = 2.5\text{ V} - (-2.5\text{ V}) = \mathbf{5\text{ V}}$$

Question

What are the upper and lower trigger points and the hysteresis in Figure 7–9 for $R_1 = 68$ kΩ and $R_2 = 82$ kΩ? The maximum output voltage levels are ±7 V.

COMPUTER SIMULATION

Open the Multisim file F07-09DV on the website. Observe the output voltage.

A Comparator Application: Over-Temperature Sensing Circuit

Figure 7–10 shows an op-amp comparator used in a precision over-temperature sensing circuit to determine when the temperature reaches a certain critical value. The circuit consists of a Wheatstone bridge with the op-amp used to detect when the bridge is balanced. One leg of the bridge contains a thermistor (R_1), which is a temperature-sensing resistor with a

FIGURE 7–10

An over-temperature sensing circuit.

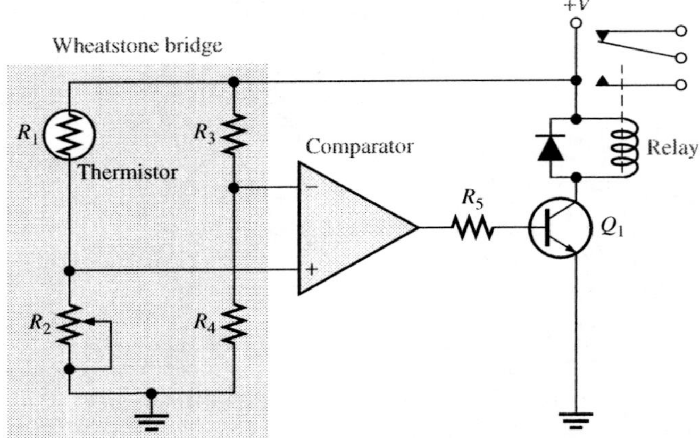

negative temperature coefficient (its resistance decreases as temperature increases and vice versa). The potentiometer (R_2) is set at a value equal to the resistance of the thermistor at the critical temperature. At normal temperatures (below critical), R_1 is greater than R_2, thus creating an unbalanced condition that drives the op-amp to its low saturated output level and keeps transistor Q_1 off.

As the temperature increases, the resistance of the thermistor decreases. When the temperature reaches the critical value, R_1 becomes equal to R_2, and the bridge becomes balanced (since $R_3 = R_4$). At this point the op-amp switches to its high saturated output level, turning Q_1 on. This energizes the relay, which can be used to activate an alarm or initiate an appropriate response to the over-temperature condition.

Review Questions

Answers are at the end of the chapter.

1. What is the reference voltage for each comparator in Figure 7–11?
2. What is noise in a circuit?
3. What is the purpose of hysteresis in a comparator?
4. What is a comparator with hysteresis commonly called?
5. Hysteresis is what type of feedback?

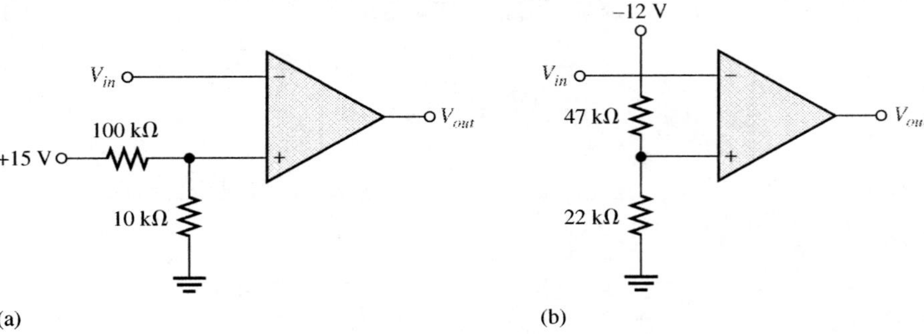

FIGURE 7–11

SUMMING AMPLIFIERS 7–2

The summing amplifier is a variation of the inverting op-amp configuration covered in Chapter 6. The **summing amplifier** has two or more inputs, and its output voltage is proportional to the negative of the algebraic sum of its input voltages.

In this section, you will learn the operation of several types of summing amplifiers.

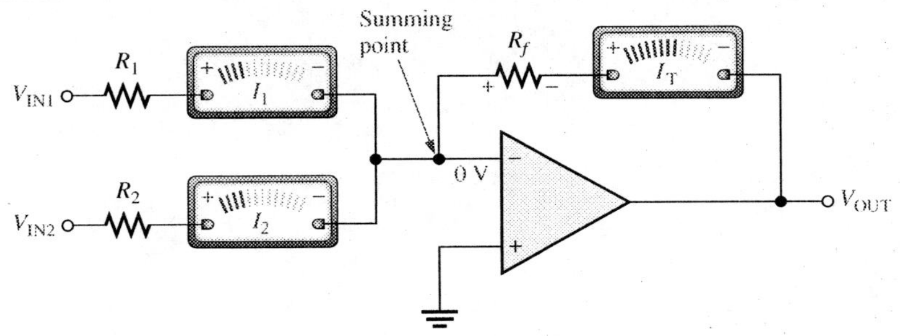

FIGURE 7–12 Two-input inverting summing amplifier.

A two-input summing amplifier is shown in Figure 7–12, but any number of inputs can be used. The operation of the circuit and derivation of the output expression are as follows. Two voltages, V_{IN1} and V_{IN2}, are applied to the inputs and produce currents I_1 and I_2, as shown. From the concepts of infinite input resistance and virtual ground, the voltage at the inverting input (−) of the op-amp is approximately 0 V, and therefore there is no current at the inverting input. This means that both input currents I_1 and I_2 combine at this summing point and form the total current, which is through R_f, as indicated ($I_T = I_1 + I_2$). Since $V_{OUT} = -I_T R_f$, the following steps apply.

$$V_{OUT} = -(I_1 + I_2)R_f = -\left(\frac{V_{IN1}}{R_1} + \frac{V_{IN2}}{R_2}\right)R_f$$

If all three of the resistors are equal in value ($R_1 = R_2 = R_f = R$), then

$$V_{OUT} = -\left(\frac{V_{IN1}}{R} + \frac{V_{IN2}}{R}\right)R = -(V_{IN1} + V_{IN2})$$

The previous equation shows that the output voltage has the same magnitude as the sum of the two input voltages but with a negative sign. A general expression is given in Equation 7–3 for a summing amplifier with n inputs, as shown in Figure 7–13 where all resistors are equal in value.

$$V_{OUT} = -(V_{IN1} + V_{IN2} + \cdots + V_{INn}) \quad (7\text{–}3)$$

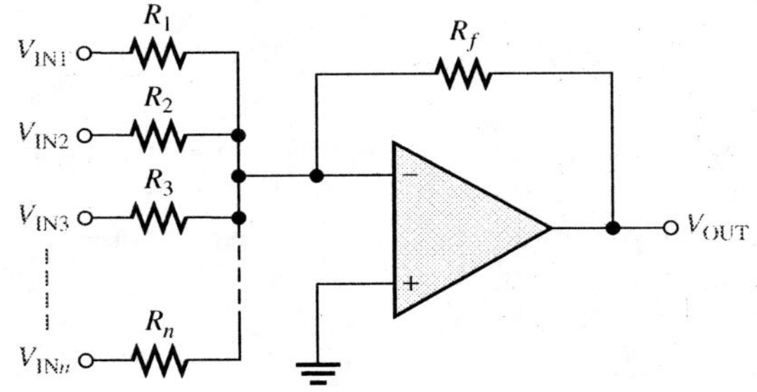

FIGURE 7–13 Summing amplifier with n inputs.

EXAMPLE 7–3

Problem

Determine the output voltage in Figure 7–14.

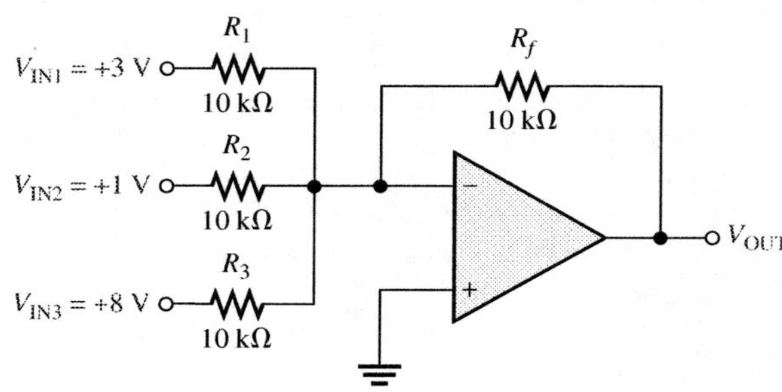

FIGURE 7–14

Solution

$$V_{OUT} = -(V_{IN1} + V_{IN2} + V_{IN3}) = -(3\text{ V} + 1\text{ V} + 8\text{ V}) = \mathbf{-12\text{ V}}$$

Question

If a fourth input of +0.5 V is added to Figure 7–14 with a 10 kΩ resistor, what is the output voltage?

COMPUTER SIMULATION

Open the Multisim file F07-14DV on the website. Measure the output voltage.

Averaging Amplifier

A summing amplifier can be made to produce the mathematical average of the input voltages. The amplifier has a gain of R_f/R, where R is the value of each input resistor. The general expression for the output of an averaging amplifier is

$$V_{OUT} = -\frac{R_f}{R}(V_{IN1} + V_{IN2} + \cdots + V_{INn}) \qquad (7\text{–}4)$$

Averaging is done by setting the ratio R_f/R equal to the reciprocal of the number of inputs (n); that is, $R_f/R = 1/n$.

You obtain the average of several numbers by first adding the numbers and then dividing by the quantity of numbers you have. Examination of Equation 7–4 and a little thought will convince you that a summing amplifier will do this. The next example illustrates this idea.

EXAMPLE 7–4

Problem

Show that the amplifier in Figure 7–15 produces an output whose magnitude is the mathematical average of the input voltages.

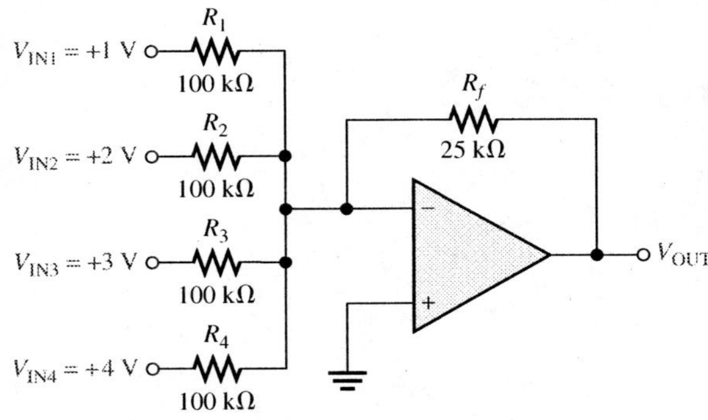

FIGURE 7–15

Solution

Since the input resistors are equal, $R = 100\text{ k}\Omega$. The output voltage is

$$V_{OUT} = -\frac{R_f}{R}(V_{IN1} + V_{IN2} + V_{IN3} + V_{IN4})$$

$$= -\frac{25\text{ k}\Omega}{100\text{ k}\Omega}(1\text{ V} + 2\text{ V} + 3\text{ V} + 4\text{ V}) = -\frac{1}{4}(10\text{ V}) = \mathbf{-2.5\text{ V}}$$

A simple calculation shows that the average of the input values is the same magnitude as V_{OUT} but of opposite sign.

$$V_{IN(avg)} = \frac{1\text{ V} + 2\text{ V} + 3\text{ V} + 4\text{ V}}{4} = \frac{10\text{ V}}{4} = 2.5\text{ V}$$

Question

What are the changes required in the averaging amplifier in Figure 7–15 in order to handle five inputs?

COMPUTER SIMULATION

Open the Multisim file F07-15DV on the website. Measure the output voltage.

Scaling Adder

A different weight can be assigned to each input of a summing amplifier by simply adjusting the values of the input resistors. The output voltage can be expressed as

$$V_{OUT} = -\left(\frac{R_f}{R_1}V_{IN1} + \frac{R_f}{R_2}V_{IN2} + \cdots + \frac{R_f}{R_n}V_{INn}\right) \tag{7–5}$$

The weight of a particular input is set by the ratio of R_f to the resistance for that input. For example, if an input voltage is to have a weight of 1, then $R = R_f$. Or, if a weight of 0.5 is required, $R = 2R_f$. The smaller the value of R, the greater the weight, and vice versa.

EXAMPLE 7–5

Problem

Determine the weight of each input voltage for the scaling adder in Figure 7–16 and find the output voltage.

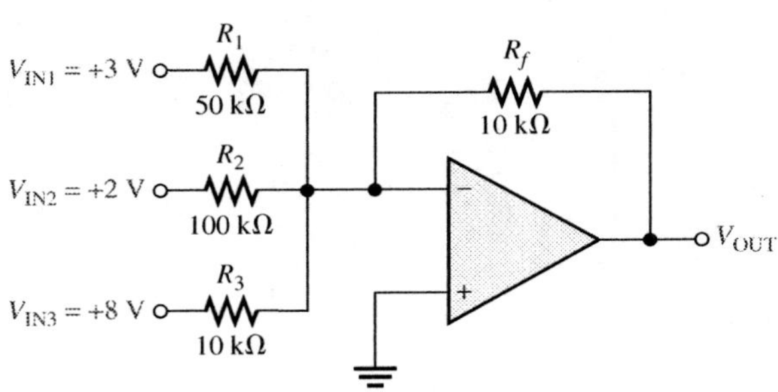

FIGURE 7–16

Solution

Weight of input 1: $\dfrac{R_f}{R_1} = \dfrac{10\text{ k}\Omega}{50\text{ k}\Omega} = \mathbf{0.2}$

Weight of input 2: $\dfrac{R_f}{R_2} = \dfrac{10\text{ k}\Omega}{100\text{ k}\Omega} = \mathbf{0.1}$

Weight of input 3: $\dfrac{R_f}{R_3} = \dfrac{10\text{ k}\Omega}{10\text{ k}\Omega} = \mathbf{1}$

The output voltage is

$$V_{OUT} = -\left(\frac{R_f}{R_1}V_{IN1} + \frac{R_f}{R_2}V_{IN2} + \frac{R_f}{R_3}V_{IN3}\right)$$
$$= -[0.2(3\text{ V}) + 0.1(2\text{ V}) + 1(8\text{ V})] = -(0.6\text{ V} + 0.2\text{ V} + 8\text{ V}) = \mathbf{-8.8\text{ V}}$$

Question

What is the weight of each input voltage in Figure 7–16 if $R_1 = 22$ kΩ, $R_2 = 82$ kΩ, $R_3 = 56$ kΩ, and $R_f = 10$ kΩ? Also find V_{OUT}.

COMPUTER SIMULATION

Open the Multisim file F07-16DV on the website. Measure the output voltage.

A Scaling Adder Application: Digital-to-Analog (D/A) Conversion

D/A conversion is an important interface process for converting digital signals to analog (linear) signals. An example is a voice signal that is digitized for storage, processing, or transmission and must be changed back into an approximation of the original audio signal in order to drive a speaker.

One method of D/A conversion uses a scaling adder with input resistor values that represent the binary weights of the digital input code. Figure 7–17 shows a four-digit digital-to-analog converter (DAC) of this type (called a *binary-weighted resistor DAC*). The switch symbols represent transistor switches for applying each of the four binary digits to the inputs.

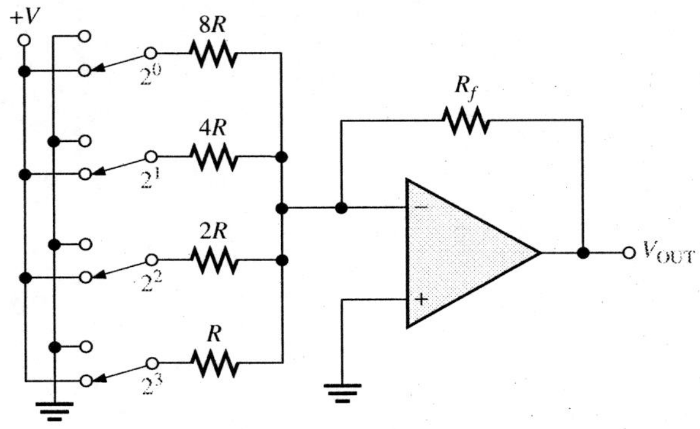

FIGURE 7–17

A scaling adder as a four-digit digital-to-analog converter (DAC).

The inverting input (−) is at virtual ground, so the output voltage is proportional to the current through the feedback resistor R_f (sum of input currents). The lowest-value resistor R corresponds to the highest weighted binary input (2^3). All of the other resistors are multiples of R and correspond to the binary weights 2^2, 2^1, and 2^0.

SAFETY NOTE

Always be careful when you work with electricity. Electric shock can cause muscle spasms, weakness, shallow breathing, rapid pulse, severe burns, unconsciousness, or death. In a shock incident, the path that electric current takes through the body gets very hot. Burns occur all along that path, including the places on the skin where the current enters and leaves the body. It's not only power lines that are dangerous if you contact them. You can also be injured by a shock from an appliance, instrument, or power cord in your home or in the lab.

Review Questions

6. What is a summing point?
7. What is the output voltage of a unity gain ($A_{cl} = 1$) summing amplifier with 1 V, 2 V, and 4 V on its inputs?
8. What is the value of R_f/R for a five-input averaging amplifier?
9. A certain scaling adder has two inputs, one having twice the weight of the other. If the resistor value for the lower weighted input is 10 kΩ, what is the value of the other input resistor?
10. What does DAC stand for?

INTEGRATORS AND DIFFERENTIATORS 7–3

An op-amp integrator simulates mathematical integration, which is basically a summing process that determines the total area under the curve of a function. An op-amp differentiator simulates mathematical differentiation, which is a process of determining the instantaneous rate of change of a function. The integrators and differentiators shown in this section are idealized to show basic principles. Practical integrators often have an additional resistor or other circuitry in parallel with the feedback capacitor to prevent saturation.

In this section, you will learn the operation of integrators and differentiators.

The Op-Amp Integrator

An **integrator** is a circuit that produces an inverted output that approximates the area under the curve of the input function. An ideal integrator is shown in Figure 7–18. Notice that the feedback element is a capacitor that forms an RC circuit with the input resistor.

FIGURE 7–18

An ideal op-amp integrator.

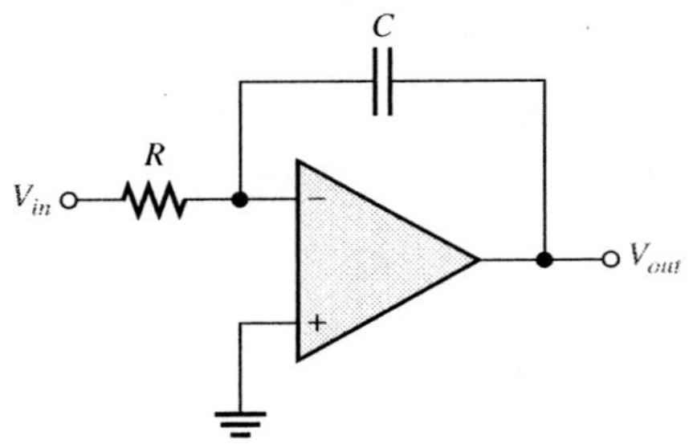

How a Capacitor Charges

To understand how the integrator works, it is important to review how a capacitor charges. Recall that the charge Q on a capacitor is proportional to the charging current (I_C) and the time (t).

$$Q = I_C t$$

Also, in terms of the voltage, the charge on a capacitor is

$$Q = CV_C$$

From these two relationships, the capacitor voltage can be expressed as

$$V_C = \left(\frac{I_C}{C}\right)t$$

This expression is an equation for a straight line which begins at zero with a constant slope of I_C/C. (Remember that the general formula for a straight line is $y = mx + b$. In this case, $y = V_C$, $m = I_C/C$, $x = t$, and $b = 0$.)

The capacitor voltage in a simple RC circuit is not linear but is exponential. This is because the charging current continuously decreases as the capacitor charges and causes the rate of change of the voltage to continuously decrease. The advantage of using an op-amp with an RC circuit to form an integrator is that the capacitor's charging current is made constant, thus producing a straight-line (linear) voltage rather than an exponential voltage. Now let's see why this is true.

In Figure 7–19 the inverting input of the op-amp is at virtual ground (0 V), so the voltage across R_i equals V_{in}. Therefore, the input current is

$$I_{in} = \frac{V_{in}}{R_i}$$

FIGURE 7–19

Currents in an integrator.

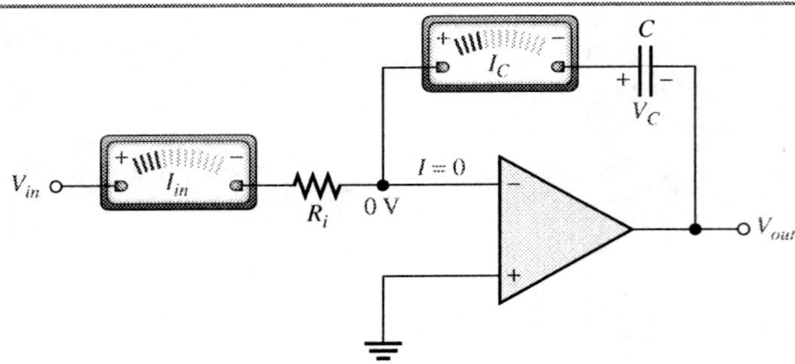

If V_{in} is a constant voltage, then I_{in} is also a constant because the inverting input always remains at 0 V, keeping a constant voltage across R_i. Because of the very high input impedance of the op-amp, there is negligible current at the inverting input. This makes all of the input current charge the capacitor, so

$$I_C = I_{in}$$

The Capacitor Voltage

Since I_{in} is constant, so is I_C. The constant I_C charges the capacitor linearly and produces a linear voltage across C. The positive side of the capacitor is held at 0 V by the virtual ground of the op-amp. The voltage on the negative side of the capacitor decreases linearly from zero as the capacitor charges, as shown in Figure 7–20. This voltage is called a *negative ramp* and is the consequence of a constant positive input.

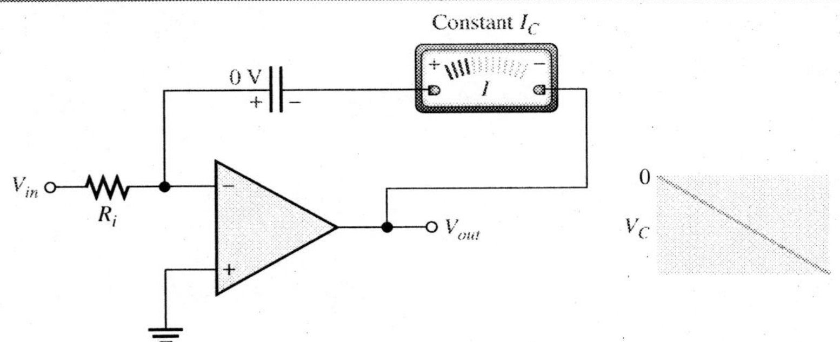

FIGURE 7–20

A linear ramp voltage is produced across C by the constant charging current.

The Output Voltage

V_{out} is the same as the voltage on the negative side of the capacitor. When a constant positive input voltage in the form of a step or pulse (a pulse has a constant amplitude when high) is applied, the output ramp decreases negatively until the op-amp saturates at its maximum negative level. This is indicated in Figure 7–21.

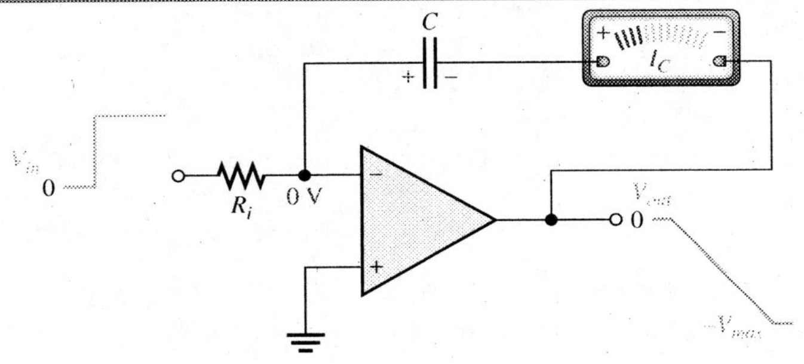

FIGURE 7–21

A constant input voltage produces a ramp on the output of the integrator.

Rate of Change of the Output

The rate at which the capacitor charges, and therefore the slope of the output ramp, is set by the ratio I_C/C, as you have seen. Since $I_C = V_{in}/R_i$, the rate of change or slope of the integrator's output voltage is

$$\text{Output rate of change} = -\frac{V_{in}}{R_iC} \qquad (7\text{–}6)$$

Integrators are especially useful in triangular-wave generators.

EXAMPLE 7–6

Problem

(a) Determine the rate of change of the output voltage in response to the input square wave, as shown for the ideal integrator in Figure 7–22(a). The output voltage is initially zero. The pulse width is 100 μs.

(b) Describe the output and draw the waveform.

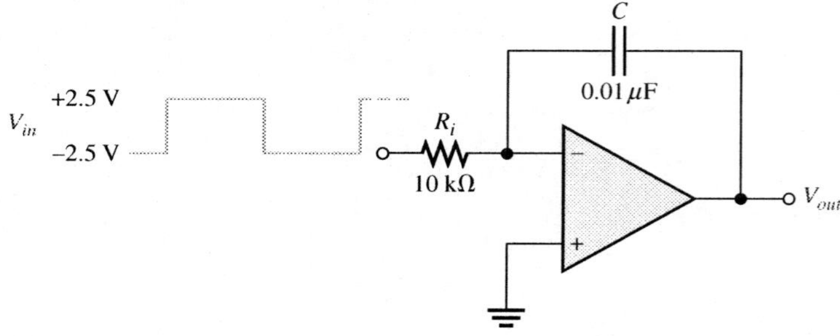

(a)

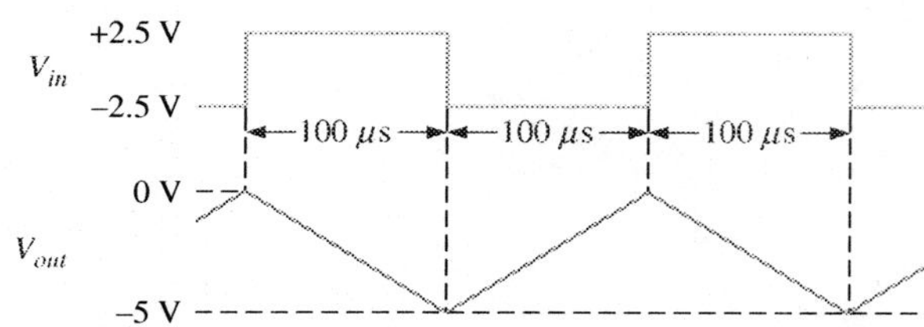

FIGURE 7–22 (b)

Solution

(a) During the time that the input is positive, the capacitor is charging such that the rate of change of the output voltage is negative.

$$\text{Output rate of change} = -\frac{V_{in}}{R_iC} = -\frac{5\text{ V}}{(10\text{ k}\Omega)(0.01\ \mu\text{F})} = -50\text{ kV/s}$$
$$= \mathbf{-50\ mV/\mu s}$$

During the time that the input is negative, the capacitor is discharging; so the rate of change of the output voltage is the same as during charging but it is positive.

$$\text{Output rate of change} = \mathbf{+50\ mV/\mu s}$$

(b) When the input is at +2.5 V, the output is a negative-going ramp. When the input is at −2.5 V, the output is a positive-going ramp.

$$\text{Change in output voltage} = (50\text{ mV}/\mu\text{s})(100\ \mu\text{s}) = 5\text{ V}$$

During the time the input is at +2.5 V, the output will go from 0 V to − 5 V. During the time the input is at −2.5 V, the output will go from −5 V to 0 V. Therefore, the output is a triangular wave with peaks at 0 V and −5 V, as shown in Figure 7–22(b).

Question

How would you change the integrator in Figure 7–22 to make the output change from 0 V to −10 V with the same input?

COMPUTER SIMULATION

Open the Multisim file F07-22DV on the website. Notice that the output waveform slowly drifts in this ideal circuit because of the high gain at dc. To cure this, place a 1 MΩ resistor in parallel with the capacitor. Although an initial drift will be observed, it will stabilize because the dc gain has been reduced.

The Op-Amp Differentiator

A **differentiator** is a circuit that produces an inverted output that approximates the rate of change of the input function. An ideal differentiator is shown in Figure 7–23. Notice how the placement of the capacitor and resistor differ from that in the integrator. The capacitor is now the input element. A differentiator produces an output that is proportional to the rate of change of the input voltage. Although a small-value resistor is normally used in series with the capacitor to limit the gain, it does not affect the basic operation and is not shown for purposes of this analysis.

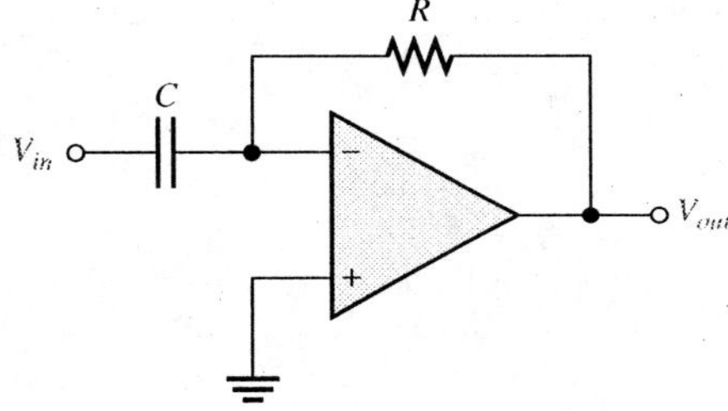

FIGURE 7–23

An ideal op-amp differentiator.

To see how the differentiator works, let's apply a positive-going ramp voltage to the input as indicated in Figure 7–24. In this case, $I_C = I_{in}$ and the voltage across the capacitor is equal to V_{in} at all times ($V_C = V_{in}$) because of virtual ground on the inverting input.

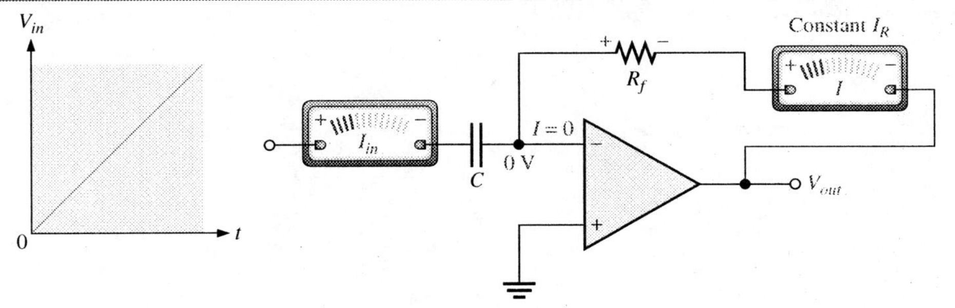

FIGURE 7–24

A differentiator with a ramp input.

From the basic formula, which is $V_C = (I_C/C)t$,

$$I_C = \left(\frac{V_C}{t}\right)C$$

Since the current at the inverting input is negligible, $I_R = I_C$. Both currents are constant because the slope of the capacitor voltage (V_C/t) is constant. The output voltage is also constant and equal to the voltage across R_f because one side of the feedback resistor is always 0 V (virtual ground).

$$V_{out} = I_R R_f = I_C R_f$$

Substituting $(V_C/t)C$ for I_C,

$$V_{out} = -\left(\frac{V_C}{t}\right)R_fC \quad (7\text{–}7)$$

The output is negative when the input is a positive-going ramp and positive when the input is a negative-going ramp, as illustrated in Figure 7–25. During the positive slope of the input, the capacitor is charging from the input source with constant current through the feedback resistor. During the negative slope of the input, the constant current is in the opposite direction because the capacitor is discharging.

FIGURE 7–25

Output of a differentiator with a series of positive and negative ramps (triangle wave) on the input.

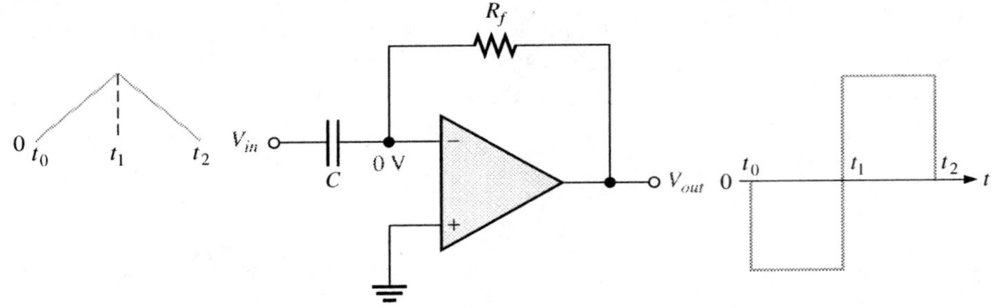

Notice in Equation 7–7 that the term V_C/t is the slope of the input. If the slope increases, V_{out} becomes more negative. If the slope decreases, V_{out} becomes more positive. So, the output voltage is proportional to the negative slope (rate of change) of the input. The constant of proportionality is the time constant, R_fC.

EXAMPLE 7–7

Problem

Determine the output voltage of the ideal op-amp differentiator in Figure 7–26 for the triangular-wave input shown.

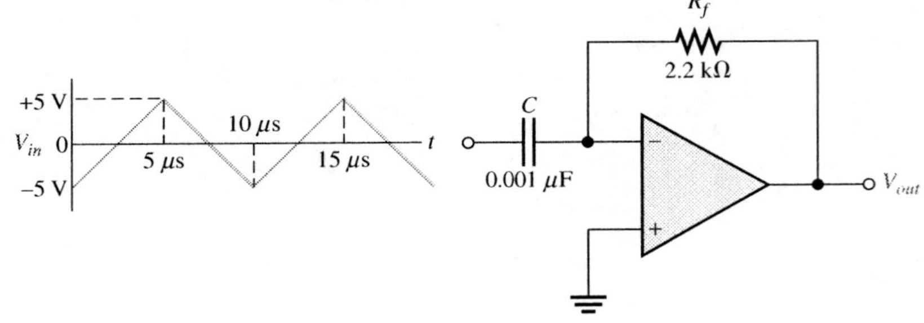

FIGURE 7–26

Solution

Starting at $t = 0$, the input voltage is a positive-going ramp ranging from −5 V to +5 V (a + 10 V change) in 5 μs. Then it changes to a negative-going ramp ranging from +5 V to −5 V (a −10 V change) in 5 μs.

Substituting into Equation 7–7, the output voltage for the positive-going ramp is

$$V_{out} = -\left(\frac{V_C}{t}\right)R_fC = -\left(\frac{10\text{ V}}{5\ \mu\text{s}}\right)(2.2\text{ k}\Omega)(0.001\ \mu\text{F}) = \mathbf{-4.4\text{ V}}$$

The output voltage for the negative-going ramp is calculated the same way.

$$V_{out} = -\left(\frac{V_C}{t}\right)R_f C = -\left(\frac{-10\ \text{V}}{5\ \mu\text{s}}\right)(2.2\ \text{k}\Omega)(0.001\ \mu\text{F}) = \mathbf{+4.4\ V}$$

Finally, the output voltage waveform is graphed relative to the input as shown in Figure 7–27.

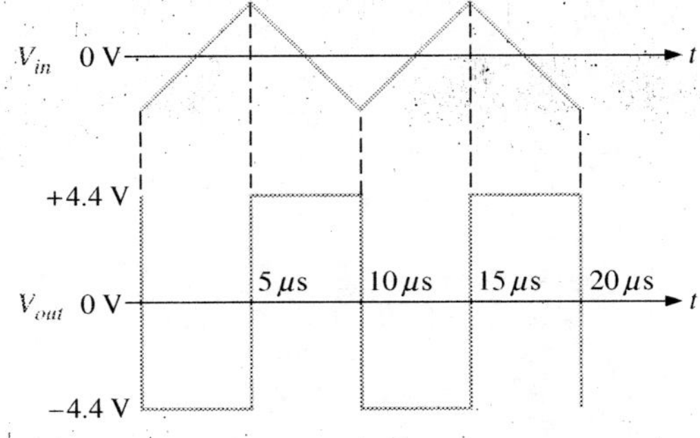

FIGURE 7–27

Question

What would the output voltage be if the feedback resistor in Figure 7–26 is changed to 3.3 kΩ?

COMPUTER SIMULATION

Open the Multisim file F07-26DV on the website. Notice that the output waveform initially has a small instability. A tiny amount of noise appears on the output waveform due to the very high gain at high frequencies. Connect a 100 Ω resistor in series with the input to eliminate the noise. The output amplitude will be slightly lower.

Review Questions

11. What is the feedback element in an op-amp integrator?
12. For a constant input voltage to an integrator, why is the voltage across the capacitor linear?
13. What is the feedback element in an op-amp differentiator?
14. How is the output of a differentiator related to the input?
15. What type of waveform does a differentiator produce when the input is a triangular waveform?

ACTIVE DIODE CIRCUITS 7–4

The term *active* in relation to a circuit indicates that a gain element is used, in this case, an op-amp. Some of the circuits introduced in this section use both op-amps and diodes to provide certain functions. Circuits that shift the dc level of a signal are called clampers, and circuits that limit the amplitude of a signal are called limiters. Another type of active diode circuit that is introduced is the peak detector.

In this section you will learn the operation of clampers, limiters, and peak detectors.

Clamping Circuits

The Basic Diode Clamper

A **clamper** is a circuit used to add a dc level to a signal voltage. Clampers are often referred to as *dc restorers* because they are used to restore a dc level to a signal that has been processed through capacitively coupled amplifiers. To illustrate the basic principle of clamper operation, Figure 7–28 shows a simple passive diode clamping circuit that adds a positive dc level to the input signal. To understand the operation of this circuit, start with the first negative half-cycle of the input voltage. When the input initially goes negative, the diode is forward-biased, allowing the capacitor to charge to *near* the peak of the input, as shown in Figure 7–28(a). Just past the negative peak, the diode becomes reverse-biased because the cathode is held to $V_{p(in)} - 0.7$ V by the charge on the capacitor.

FIGURE 7–28

Positive clamping operation with a passive clamper.

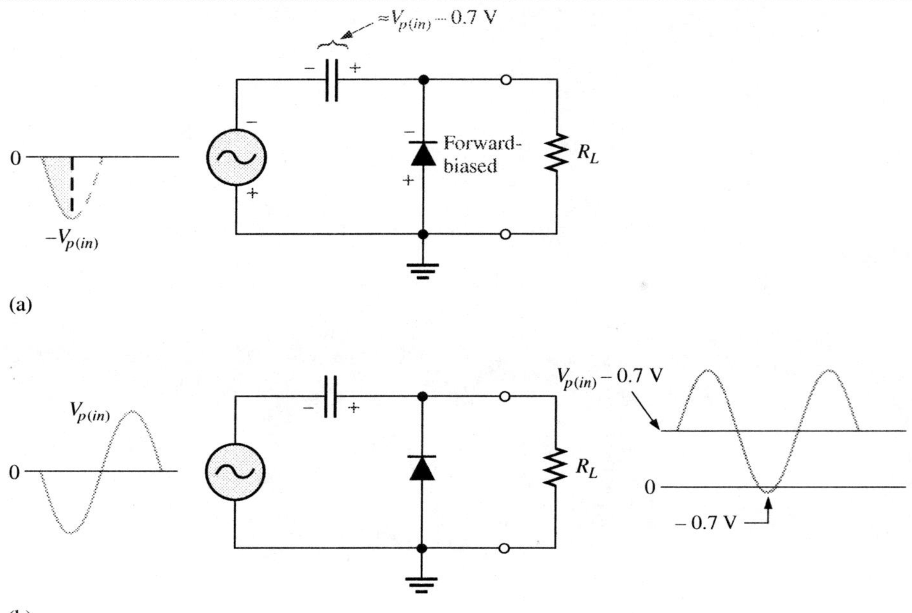

The capacitor can discharge only through R_L. Thus, from the peak of one negative half-cycle to the next, the capacitor discharges very little. The amount that is discharged depends on the value of R_L and the period of the input signal. For good clamping action, the *RC* time constant should be at least ten times the period of the input. The net effect of the clamping action is that the capacitor retains a charge approximately equal to the peak value of the input less the diode drop. The dc voltage of the capacitor adds to the input voltage by superposition, as shown in Figure 7–28(b). If the diode is turned around, a negative dc voltage is added to the input signal, as shown in Figure 7–29.

FIGURE 7–29

Negative clamping.

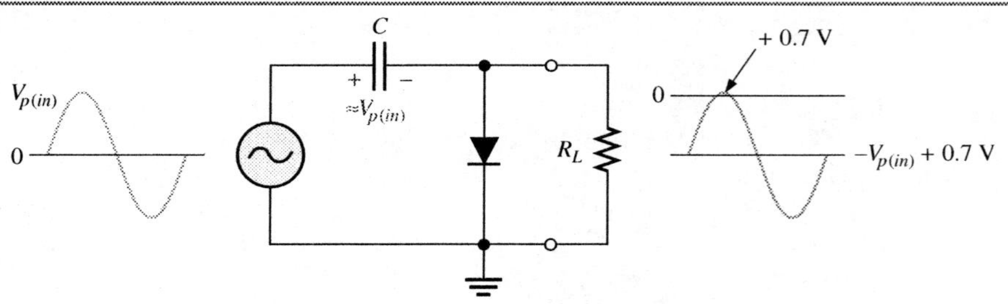

An Active Clamping Circuit

A positive clamper with an op-amp and a diode is shown in Figure 7–30. This circuit overcomes a couple of disadvantages of the passive clamper previously discussed. The use of the op-amp eliminates the −0.7 V peak found in the positive passive clamper output, and it prevents loading the input source when the diode is forward-biased.

FIGURE 7–30 An active clamping circuit and its operation.

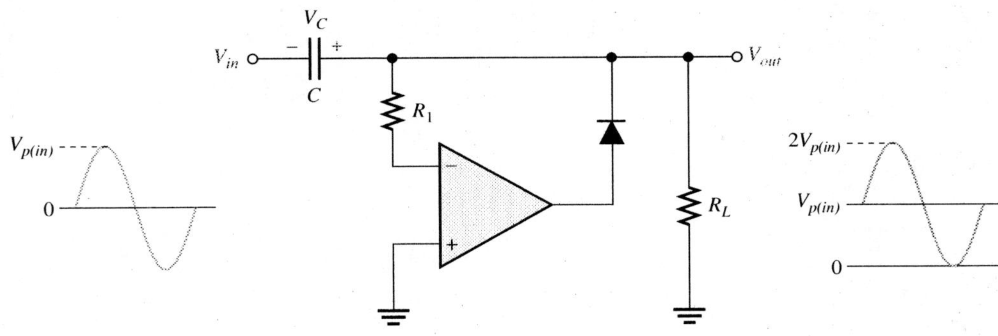

The operation is as follows. On the first negative half-cycle of the input voltage, V_{in}, the differential input is positive, which produces a positive output voltage. Because of the feedback loop, the positive op-amp output voltage forward-biases the diode, allowing the capacitor to quickly charge. The maximum voltage across the capacitor occurs at the negative peak of the input with the polarity shown in Figure 7–30. This capacitor voltage adds to the input voltage so that the minimum peak of the output voltage, V_{out}, is at 0 V as indicated.

During the time between the minimum output peaks of V_{out} and after the capacitor is charged, the differential input voltage to the op-amp becomes negative (inverting input more positive than noninverting input). As a result, the output of the op-amp becomes negative and reverse-biases the diode, thus breaking the feedback path. The only change in the capacitor voltage during this time is due to a very small discharge through R_L. At each minimum peak of the signal, the diode is forward-biased for a very short time to replenish the voltage across the capacitor.

The positive clamper can be converted to a negative clamper by reversing the diode. In this case, the output waveform would occur below 0 V with its maximum peaks at zero, as illustrated in Figure 7–31(a). Also, the clamping level can be changed to a value other than 0 V by connecting a reference voltage source at the + input of the op-amp, as shown in Figure 7–31(b) and illustrated in Example 7–8.

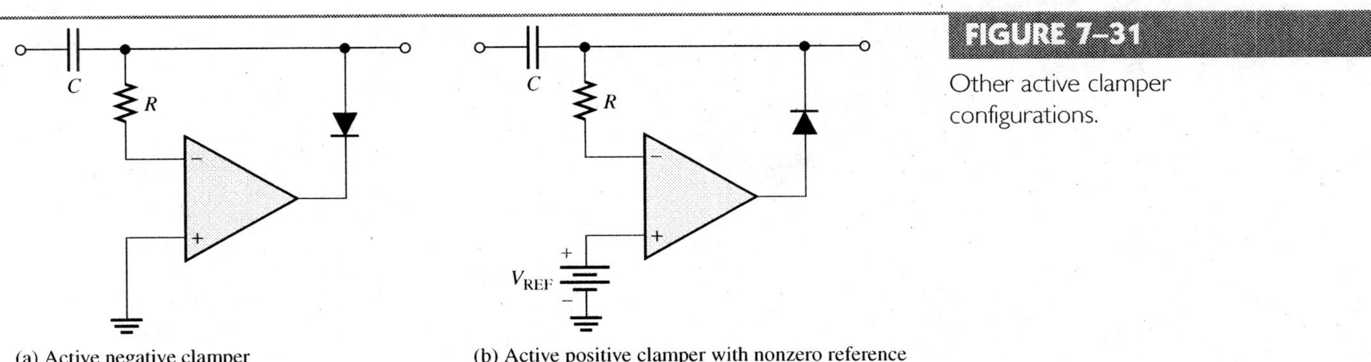

(a) Active negative clamper

(b) Active positive clamper with nonzero reference

FIGURE 7–31 Other active clamper configurations.

EXAMPLE 7–8

Problem

Determine the output voltage for the clamping circuit in Figure 7–32 for the input voltage shown.

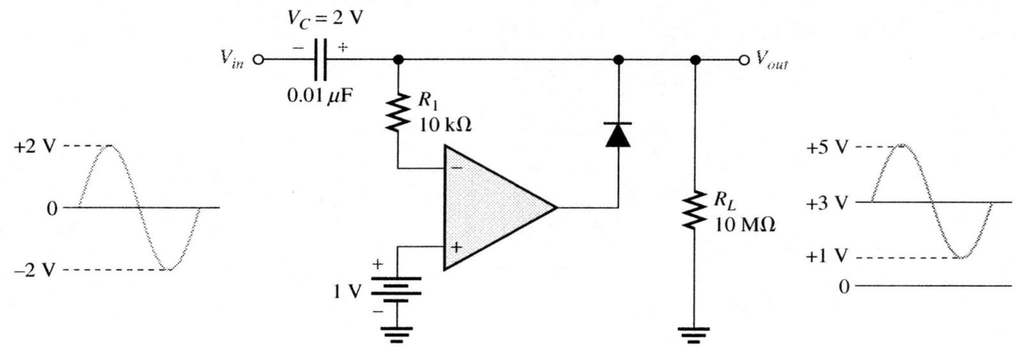

FIGURE 7–32

Solution

This is a positive clamping circuit and the reference voltage is +1 V, so the minimum peak value of the output voltage is also +1 V. The voltage is effectively shifted by 3 V, as indicated in the figure.

Question

What is the peak value of the output in Figure 7–32 if the reference voltage is +2.5 V?

COMPUTER SIMULATION

Open the Multisim file F07-32DV on the website. Observe the output waveform and dc level.

A Clamper Application

A clamping circuit is often used in television receivers as a dc restorer. The incoming composite video signal is normally processed through capacitively coupled amplifiers that eliminate the dc component in the signal. This results in loss of the black and white reference levels and the blanking level. Before being applied to the picture tube, these reference levels must be restored. Figure 7–33 illustrates this process in a general way.

FIGURE 7–33

Clamping application (dc restorer) in a TV receiver.

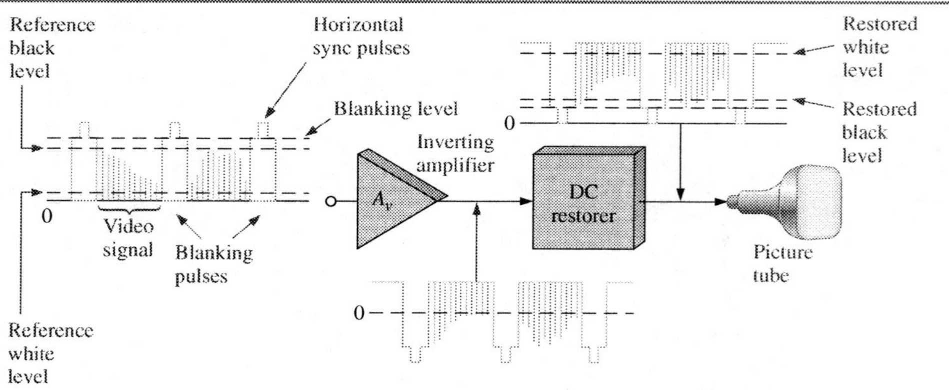

Limiting Circuits

The Basic Diode Limiter

Diode **limiters** (also called clippers) cut off or limit voltage above or below specified voltage levels. To understand how a limiter works, let's first look at a simple passive positive limiter such as the one shown in Figure 7–34(a). When the input signal is positive, the diode is reverse-biased and the output voltage looks like the input voltage. When the input signal is negative, the diode is forward-biased and the output is limited to −0.7 V, which is the diode drop. If you turn the diode around, you get a negative limiter, as shown in Figure 7–34(b). To change the limiting level, a reference voltage source can be used in series with the diode or a zener diode can be used in place of the rectifier diode.

FIGURE 7–34

Basic diode limiters. For good limiting action $R_L \gg R_s$.

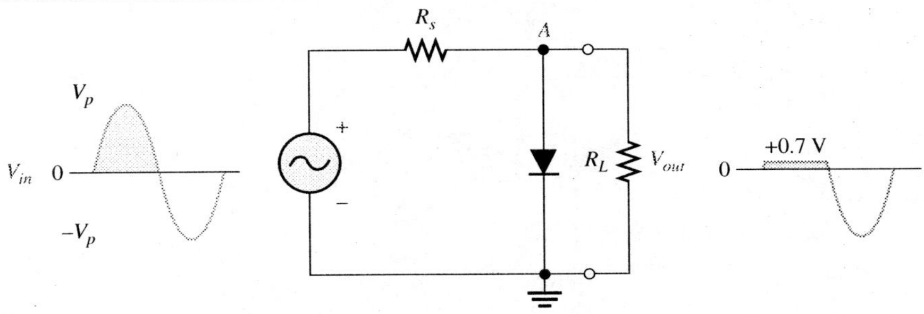

(a) Limiting of the positive alternation

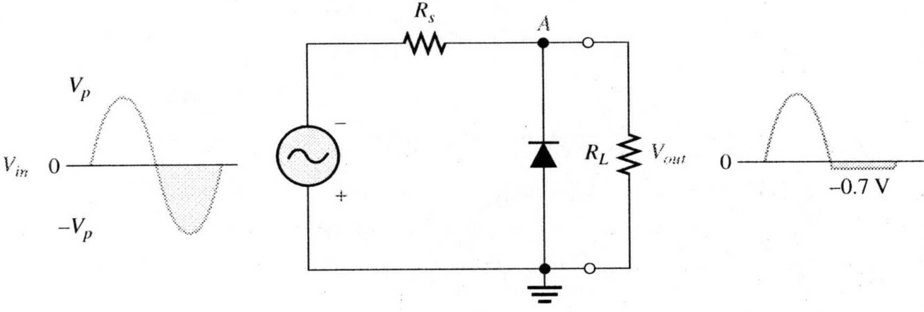

(b) Limiting of the negative alternation

Active Limiting Circuits

One type of op-amp limiting circuit that uses an op-amp and a diode is shown in Figure 7–35. The operation is as follows, assuming negligible output loading. When the input voltage, V_{in}, is less than the reference voltage, V_{REF}, the op-amp differential input voltage is positive. This produces a positive voltage at the op-amp output that forward-biases the diode. When the diode is forward-biased, the op-amp operates as a voltage-follower and the output voltage, V_{out}, is limited to V_{REF} so that $V_{out} = V_{REF}$. When the input voltage, V_{in}, is greater than V_{REF}, the op-amp differential voltage is negative. This produces a negative voltage at the op-amp output that reverse-biases the diode. With the diode effectively open, the input voltage is coupled directly to the output through R so that $V_{out} = V_{in}$.

Another example of an active limiter uses two zener diodes to limit the output voltage both positively and negatively, as shown in Figure 7–36. The limiting voltage is set by the zener diodes connected in the feedback loop of an inverting amplifier to a value of $\pm(V_Z + 0.7\text{ V})$. Of course, when the input voltage is so small that the limiting voltage is not reached, one of the zener diodes is reverse-biased and acts as an open, so the output of the op-amp is linear and equal to $V_{out} = (R_f/R_i)V_{in}$. When the output reaches $\pm(V_Z + 0.7\text{ V})$, one of the zeners goes into reverse breakdown and the other is forward-biased.

FIGURE 7–35 An active negative limiting circuit.

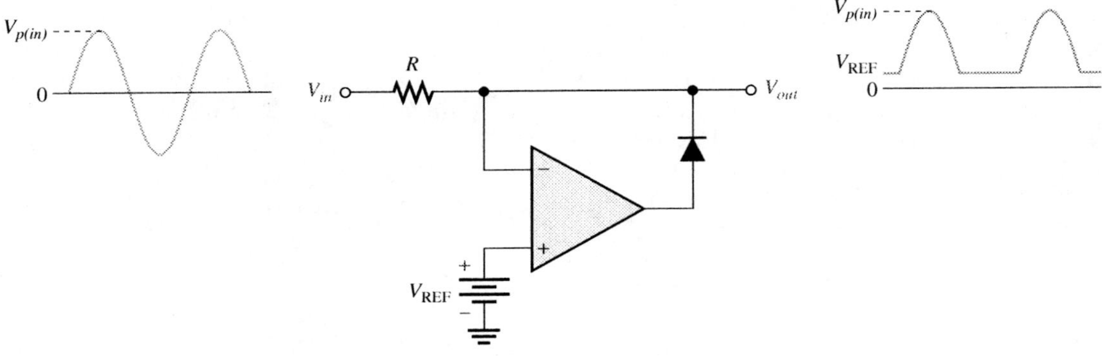

FIGURE 7–36 Limiting circuit using zener diodes in an inverting op-amp configuration.

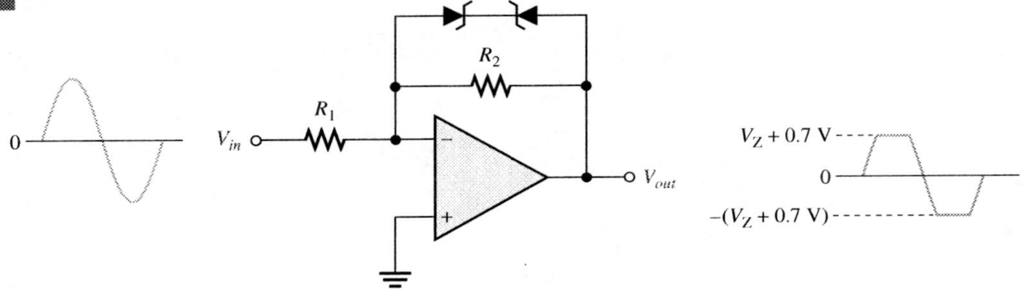

EXAMPLE 7–9

Problem

(a) Describe the output waveform for the limiting circuit in Figure 7–37 for a 1 kHz sine wave with a peak value of 100 mV.

(b) Repeat part (a) for a 1 kHz sine wave with a peak value of 1 V.

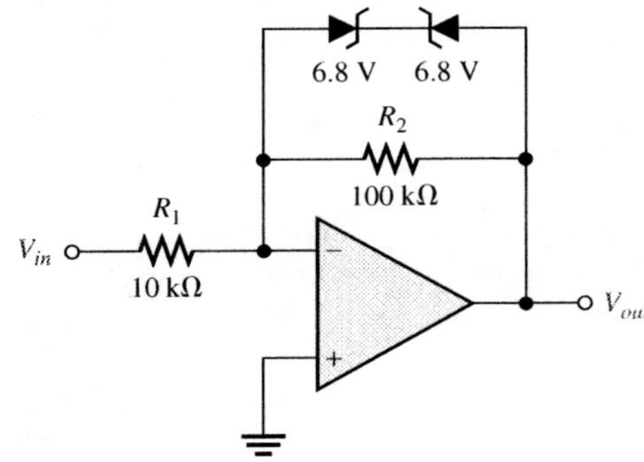

FIGURE 7–37

Solution

(a) The peak value of the output is

$$V_{p(out)} = (R_2/R_1)V_{p(in)} = (10)100 \text{ mV} = 1 \text{ V}$$

This value is less than the limiting level as set by the 6.8 V zener diodes, so the output is a sine wave that is not limited on its peaks.

(b) Since the peak output value exceeds the limiting level, the output is limited at $\pm(6.8\text{ V} + 0.7\text{ V})$, as shown in Figure 7–38.

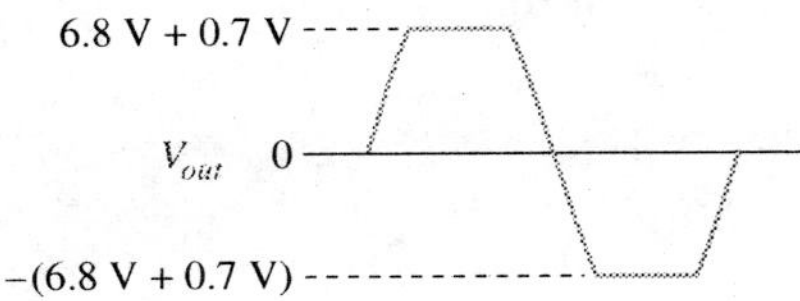

FIGURE 7–38

Question

What happens to each of the outputs in parts (a) and (b) if R_2 in Figure 7–37 is decreased to 68 kΩ?

COMPUTER SIMULATION

Open the Multisim File F07-37DV on the website. Observe the output waveform.

Peak Detector

Another application of the op-amp is in a peak detector circuit such as the one shown in Figure 7–39. In this case the op-amp is used as a comparator. A **peak detector** is a circuit used to detect the peak of the input voltage and store that peak voltage on a capacitor. For example, this circuit can be used to detect and store the maximum value of a voltage surge; this value can then be measured at the output with a voltmeter or recording device. The basic operation is as follows. When a positive voltage is applied to the noninverting input of the op-amp through R_i, the high-level output voltage of the op-amp forward-biases the diode and charges the capacitor. The capacitor continues to charge until its voltage reaches a value equal to the input voltage and thus both op-amp inputs are at the same voltage. At this point, the op-amp comparator switches and its output goes to the low level. The diode is now reverse-biased, and the capacitor stops charging. It has reached a voltage equal to the peak of V_{in} and will hold this voltage until the charge eventually leaks off. If a greater input peak occurs, the capacitor charges to the new peak.

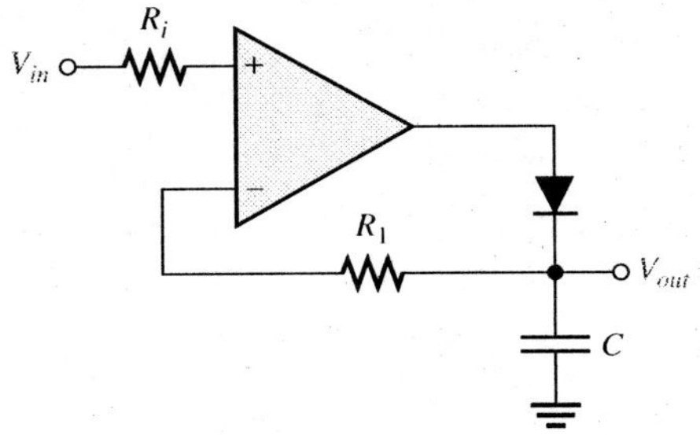

FIGURE 7–39

A basic peak detector.

Review Questions

16. What does a clamping circuit do?
17. What does a limiting circuit do?

18. What does a peak detector do?
19. By what other name is a clamping circuit sometimes known?
20. By what other name is a limiting circuit sometimes known?

7–5 TROUBLESHOOTING

Although integrated circuit op-amps are extremely reliable and trouble-free, failures do occur from time to time. One type of internal failure mode is a condition where the op-amp output is "stuck" in a saturated state resulting in a constant high or constant low level, regardless of the input. Also, external component failures will produce various types of failure modes in op-amp circuits.

In this section, you will learn to troubleshoot op-amp comparators and summing amplifiers.

Failures in Comparators

Figure 7–40 illustrates an internal failure of a comparator circuit that results in a "stuck" output.

FIGURE 7–40 Internal comparator failures typically result in the output being "stuck" in the HIGH or LOW state.

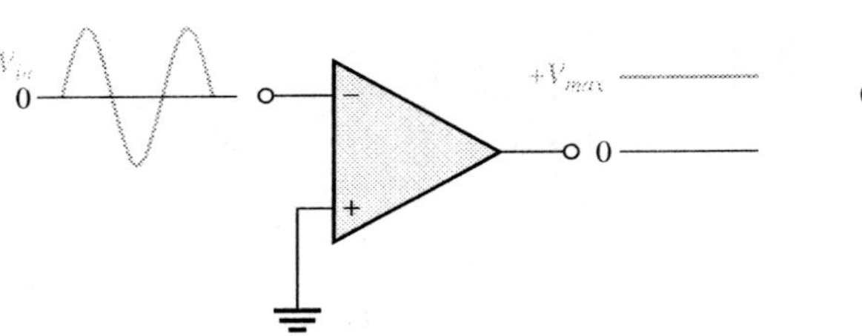

(a) Output failed in the HIGH state

(b) Output failed in the LOW state

A comparator with hysteresis is shown in Figure 7–41. In addition to a failure of the op-amp itself, one of the resistors could be faulty. Recall that R_1 and R_2 set the UTP and LTP

FIGURE 7–41 Examples of comparator circuit failures and their effects.

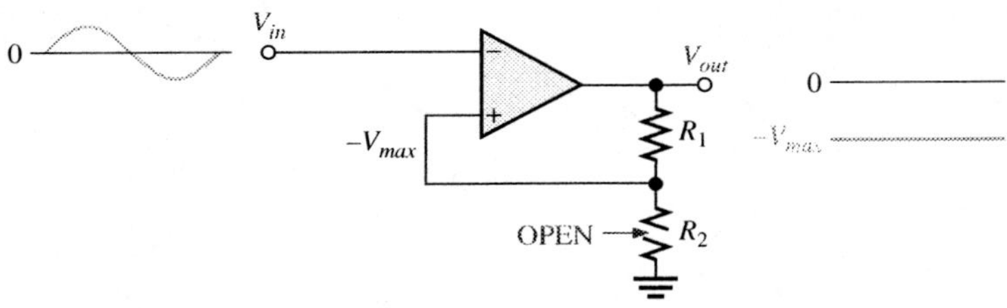

(a) Open R_2 causes output to "stick" in one state (either positive or negative)

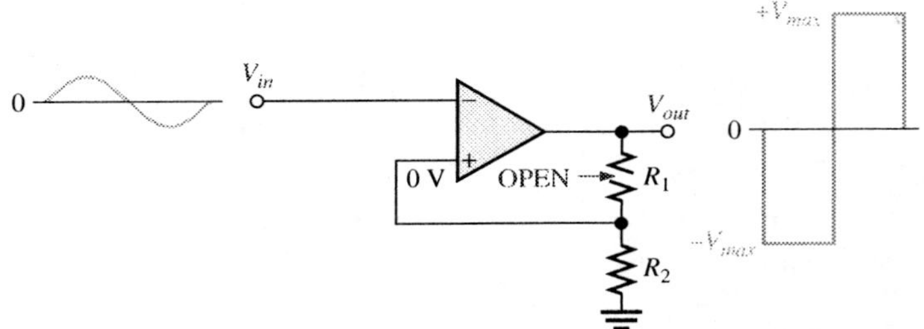

(b) Open R_1 forces the circuit to operate as a zero-level detector for the hysteresis comparator. Now, suppose that R_2 opens. Essentially all of the output voltage is fed back to the noninverting input, and, since the input voltage will never exceed the output, the device will remain in one of its saturated states. This symptom can also indicate a faulty op-amp, as mentioned before. Now, assume that R_1 opens. This leaves the noninverting input near ground potential and causes the circuit to operate as a zero-level detector. These conditions are shown in parts (a) and (b) of Figure 7–41.

Symptoms of Component Failures in Summing Amplifiers

If one of the input resistors in a unity-gain summing amplifier opens, the output will be less than the normal value by the amount of the voltage applied to the open input. Stated another way, the output will be the sum of the remaining input voltages.

If the summing amplifier has a nonunity gain, an open input resistor causes the output to be less than normal by an amount equal to the gain times the voltage at the open input.

EXAMPLE 7–10

Problem

(a) What is the normal output voltage in Figure 7–42?

(b) What is the output voltage if R_2 opens?

(c) What happens if R_5 opens?

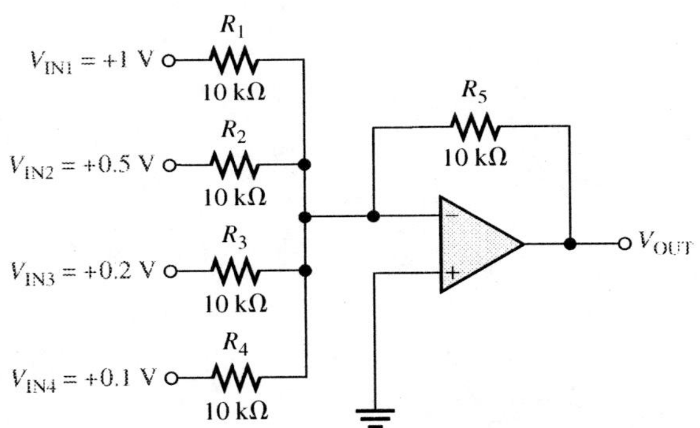

FIGURE 7–42

Solution

(a) $V_{OUT} = -(V_{IN1} + V_{IN2} + \cdots + V_{INn}) = -(1\text{ V} + 0.5\text{ V} + 0.2\text{ V} + 0.1\text{ V}) = \mathbf{-1.8\ V}$

(b) $V_{OUT} = -(1\text{ V} + 0.2\text{ V} + 0.1\text{ V}) = \mathbf{-1.3\ V}$

(c) If R_5 opens, the circuit becomes a comparator and the output goes to $-V_{max}$.

Question

In Figure 7–42, assume $R_5 = 47\text{ k}\Omega$. What is the output voltage if R_1 opens?

COMPUTER SIMULATION

Open the Multisim file 07-42DV on the website. Check the output voltage for each condition stated in the example.

As another example, let's look at an averaging amplifier. An open input resistor will result in an output voltage that is the average of all the inputs with the open input averaged in as a zero.

EXAMPLE 7–11

Problem

(a) What is the normal output voltage for the averaging amplifier in Figure 7–43?

(b) If R_4 opens, what is the output voltage? What does the output voltage represent?

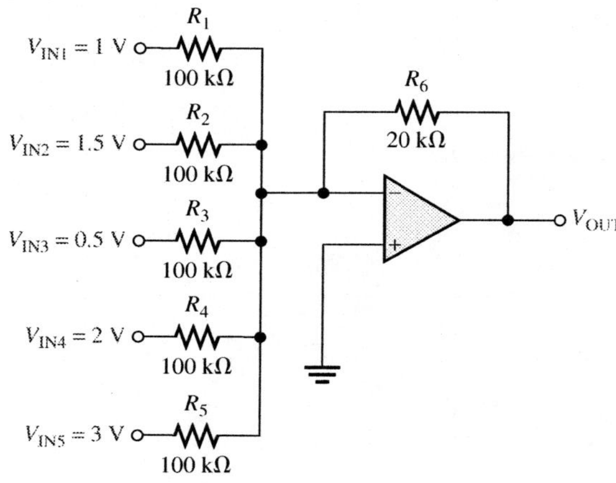

FIGURE 7–43

Solution

Since the input resistors are equal, $R = 100\ \text{k}\Omega$. $R_f = R_6$.

$$\text{(a)}\quad V_{OUT} = -\frac{R_f}{R}(V_{IN1} + V_{IN2} + \cdots + V_{INn})$$

$$= -\frac{20\ \text{k}\Omega}{100\ \text{k}\Omega}(1\ \text{V} + 1.5\ \text{V} + 0.5\ \text{V} + 2\ \text{V} + 3\ \text{V}) = -0.2(8\ \text{V}) = \mathbf{-1.6\ V}$$

$$\text{(b)}\quad V_{OUT} = -\frac{20\ \text{k}\Omega}{100\ \text{k}\Omega}(1\ \text{V} + 1.5\ \text{V} + 0.5\ \text{V} + 3\ \text{V}) = -0.2(6\ \text{V}) = \mathbf{-1.2\ V}$$

The 1.2 V result is the average of five voltages with the 2 V input replaced by 0 V. Notice that the output is not the average of the four remaining input voltages.

Question

If R_4 is open, as was the case in this example, what would you have to do to make the output equal to the average of the remaining four input voltages?

COMPUTER SIMULATION

Open the Multisim file 07-43DV on the website. Check the output for the conditions stated in the example.

Review Questions

21. What is one type of internal op-amp failure?
22. If a certain malfunction is attributable to more than one possible component that could fail open, what would you do to isolate the problem?
23. What is a "stuck" output?
24. If one of the input resistors in a summing amplifier fails open, what is the output?
25. If the feedback resistor in a summing amplifier opens, what is the output?

CHAPTER REVIEW

Key Terms

Clamper A circuit used to add a dc level to a signal voltage.

Comparator A circuit that compares two input voltages and produces an output in either of two states indicating the greater than or less than relationship of the inputs.

Differentiator A circuit that produces an inverted output that approximates the rate of change of the input function.

Integrator A circuit that produces an inverted output which approximates the area under the curve of the input function.

Limiter A circuit that cuts off or limits voltage above and/or below a specified level.

Peak detector A circuit used to detect the peak of the input voltage and store that peak value on a capacitor.

Schmitt trigger A comparator with hysteresis.

Summing amplifier A variation of a basic comparator circuit that is characterized by two or more inputs and an output voltage that is proportional to the magnitude of the algebraic sum of the input voltages.

Important Facts

- ❑ In an op-amp comparator, when the input voltage exceeds a specified reference voltage, the output changes state.
- ❑ Hysteresis gives an op-amp noise immunity.
- ❑ A comparator switches to one state when the input reaches the upper trigger point (UTP) and back to the other state when the input drops below the lower trigger point (LTP).
- ❑ The difference between the UTP and the LTP is the hysteresis voltage.
- ❑ The output voltage of a summing amplifier is proportional to the sum of the input voltages.
- ❑ An averaging amplifier is a summing amplifier with a closed-loop gain equal to the reciprocal of the number of inputs.

- In a scaling adder, a different weight can be assigned to each input, thus making the input contribute more or contribute less to the output.
- Integration is a mathematical process for determining the area under a curve.
- Integration of a step produces a ramp with a slope proportional to the amplitude.
- Differentiation is a mathematical process for determining the rate of change of a function.
- Differentiation of a ramp produces a step with an amplitude proportional to the slope.
- A clamper adds a dc level to an ac signal.
- A limiter cuts off (clips) voltage above and/or below a specified level.

Formulas

Comparator reference:

$$V_{REF} = \frac{R_2}{R_1 + R_2}(+V) \quad \text{(7–1)}$$

Hysteresis voltage:

$$V_{HYS} = V_{UTP} - V_{LTP} \quad \text{(7–2)}$$

n-input adder:

$$V_{OUT} = -(V_{IN1} + V_{IN2} + \cdots + V_{INn}) \quad \text{(7–3)}$$

Averaging amplifier:

$$V_{OUT} = -\frac{R_f}{R}(V_{IN1} + V_{IN2} + \cdots + V_{INn}) \quad \text{(7–4)}$$

Scaling adder:

$$V_{OUT} = -\left(\frac{R_f}{R_1}V_{IN1} + \frac{R_f}{R_2}V_{IN2} + \cdots + \frac{R_f}{R_n}V_{INn}\right) \quad \text{(7–5)}$$

Integrator output rate of change:

$$\text{Output rate of change} = -\frac{V_{in}}{R_iC} \quad \text{(7–6)}$$

Differentiator output voltage with ramp input:

$$V_{out} = -\left(\frac{V_C}{t}\right)R_fC \quad \text{(7–7)}$$

Chapter CheckUp

Answers are at the end of the chapter.

1. In a zero-level detector, the output changes state when the input
   (a) is positive (b) is negative
   (c) crosses zero (d) has a zero rate of change

2. The zero-level detector is one application of a

(a) comparator (b) differentiator

(c) summing amplifier (d) diode

3. Noise on the input of a comparator can cause the output to

(a) hang up in one state

(b) go to zero

(c) change back and forth erratically between two states

(d) produce the amplified noise signal

4. The effects of noise can be reduced by

(a) lowering the supply voltage

(b) using positive feedback

(c) using negative feedback

(d) using hysteresis

(e) answers (b) and (d)

5. A comparator with hysteresis

(a) has one trigger level

(b) has two trigger levels

(c) has a variable trigger level

(d) is like a magnetic circuit

6. In a comparator with hysteresis,

(a) a bias voltage is applied between the two inputs

(b) only one supply voltage is used

(c) a portion of the output is fed back to the inverting input

(d) a portion of the output is fed back to the noninverting input

7. A summing amplifier can have

(a) only one input (b) only two inputs

(c) any number of inputs

8. If the voltage gain for each input of a summing amplifier with a 4.7 kΩ feedback resistor is unity, the input resistors must have a value of

(a) 4.7 kΩ

(b) 4.7 kΩ divided by the number of inputs

(c) 4.7 kΩ times the number of inputs

9. An averaging amplifier has five inputs. The ratio R_f/R_{in} must be

(a) 5 (b) 0.2

(c) 1

10. In a scaling adder, the input resistors are

(a) all the same value

(b) equal to R_f divided by the number of inputs

(c) each inversely proportional to the weight of its inputs

(d) related by a factor of two

11. In an integrator, the feedback element is a
    (a) resistor (b) capacitor
    (c) zener diode (d) voltage divider
12. For a step input, the output of an integrator is a
    (a) pulse (b) triangular waveform
    (c) spike (d) ramp
13. The rate of change of an integrator's output voltage in response to a step input is set by
    (a) the RC time constant
    (b) the amplitude of the step input
    (c) the current through the capacitor
    (d) all of these
14. In a differentiator, the feedback element is a
    (a) resistor (b) capacitor
    (c) zener diode (d) voltage divider
15. The output of a differentiator is proportional to
    (a) the RC time constant
    (b) the rate at which the input is changing
    (c) the amplitude of the input
    (d) answers (a) and (b)
16. When you apply a triangular waveform to the input of a differentiator, the output is
    (a) a dc level
    (b) an inverted triangular waveform
    (c) a square waveform
    (d) the first harmonic of the triangular waveform
17. For a given input signal, the minimum output signal voltage of an active positive clamper is
    (a) equal to the dc value of the input
    (b) equal to the peak value of the input
    (c) equal to the voltage connected to the + input of the op-amp
    (d) equal to the voltage connected to the − input of the op-amp
18. When the + input of the clamper op-amp is connected to ground, the dc value of the output voltage is
    (a) zero
    (b) equal to the dc value of the input
    (c) equal to the average value of the input
    (d) equal to the peak value of the input
19. In an active positive limiter (similar to the one in Figure 7–35 except for the direction of the diode) with a sine wave input voltage, the output signal is
    (a) limited to values above V_{REF}
    (b) limited to values below V_{REF}
    (c) always 0.7 V
    (d) like the input signal

20. An active limiter with zener diodes across the feedback path limits the
(a) positive peaks only
(b) positive and negative peaks
(c) negative peaks only
(d) positive peak to the zener voltage and the negative peak to -0.7 V

21. If the value of R_1 in Figure 7–44 is less than specified, the output voltage will
(a) increase (b) decrease
(c) not change

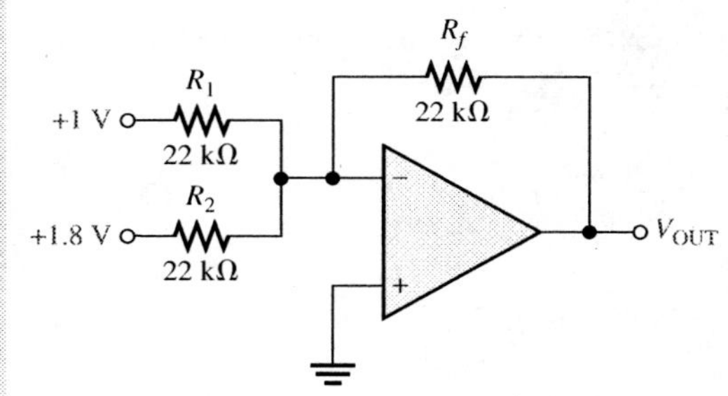

FIGURE 7–44

22. If the value of R_1 in Figure 7–44 is less than specified, the voltage at the inverting input will
(a) increase (b) decrease
(c) not change

Questions

Answers to odd-numbered questions are at the end of the book.

1. What is the reference voltage of a nonzero level detector with a voltage-divider reference if $R_1 = 10\text{ k}\Omega$, $R_2 = 10\text{ k}\Omega$, and $+V = 9$ V?
2. What is the upper trigger point (UTP) of a comparator with hysteresis if $R_1 = 47\text{ k}\Omega$, $R_2 = 68\text{ k}\Omega$, and the maximum positive output voltage is 10 V?
3. In the over-temperature sensing circuit in Figure 7–10, what happens to the thermistor resistance when the temperature increases?
4. If the two input voltages to a 2-input summing amplifier are $+2$ V and $+5$ V, what is the output voltage?
5. A certain averaging amplifier has four inputs. If the gain (R_f/R) is 0.25, and the input voltages are 5 V, 1.5 V, 2 V, and 3 V, what is the output voltage?
6. How does a scaling adder differ from a summing amplifier?
7. The rate of change of the output voltage of a certain integrator is $-10\text{ mV}/\mu\text{s}$. How many volts will the output change in 50 μs?
8. How does a differentiator differ from an integrator?
9. If a differentiator has a triangular input at a frequency of 10 kHz, what is the frequency of the output and what type of waveform is it?

10. If R_3 opens in Figure 7–42, what is the output voltage?
11. What are three components in an active clamper?
12. What types of diode can be used in an active limiter?

PROBLEMS

Basic Problems

Answers to odd-numbered problems are at the end of the book.

1. A certain op-amp has an open-loop gain of 80,000. The maximum saturated output levels of this particular device are ±12 V when the dc supply voltages are ±15 V. If a differential sine wave of 0.15 mV rms is applied between the inputs, what is the peak-to-peak value of the output?
2. Determine the output level (maximum positive or maximum negative) for each comparator in Figure 7–45.

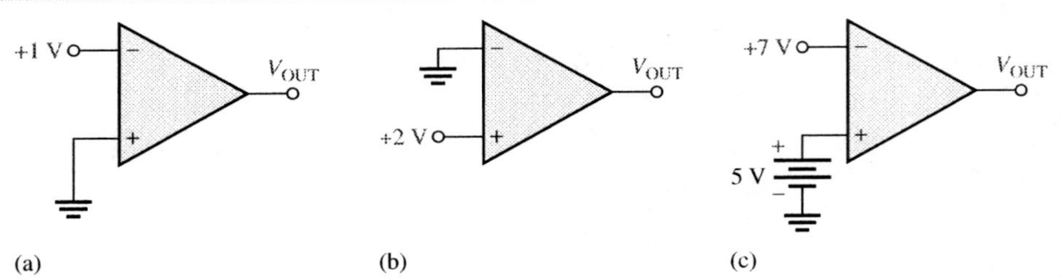

FIGURE 7–45

3. Calculate the V_{UTP} and V_{LTP} in Figure 7–46. $V_{out(max)} = -10$ V.

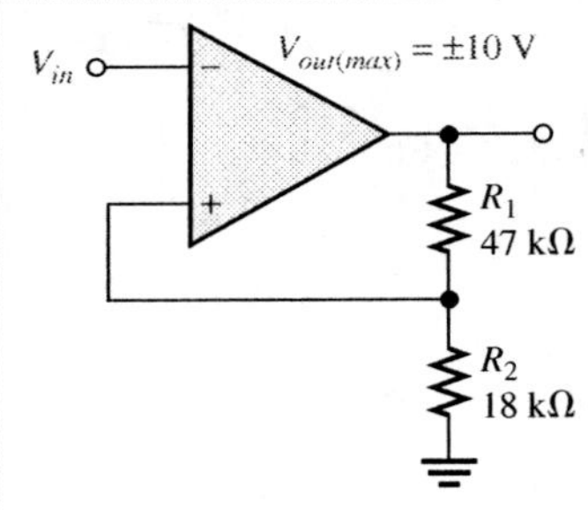

FIGURE 7–46

4. What is the hysteresis voltage in Figure 7–46?

5. Sketch the output voltage waveform for each circuit in Figure 7–47 with respect to the input. Show voltage levels.

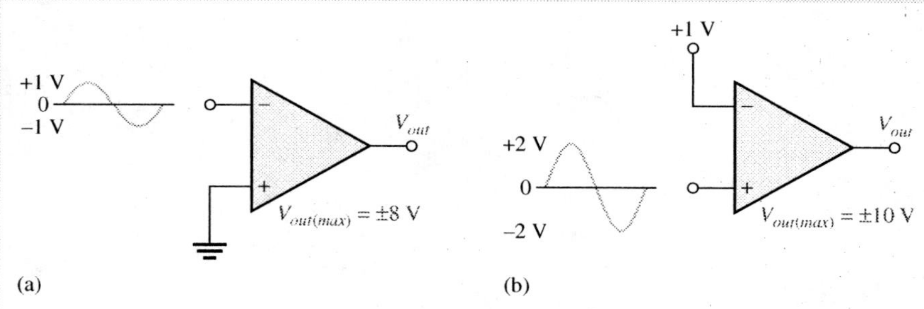

FIGURE 7–47

6. Determine the output voltage for each circuit in Figure 7–48.

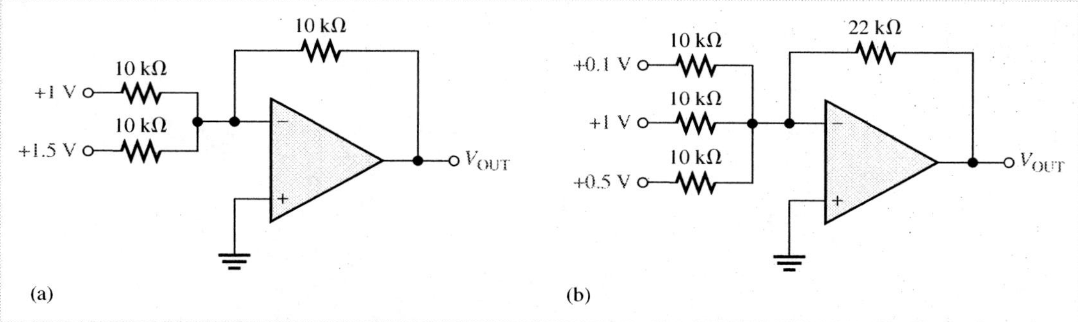

FIGURE 7–48

7. Find the output voltage when the input voltages shown in Figure 7–49 are applied to the scaling adder. What is the current through R_f?

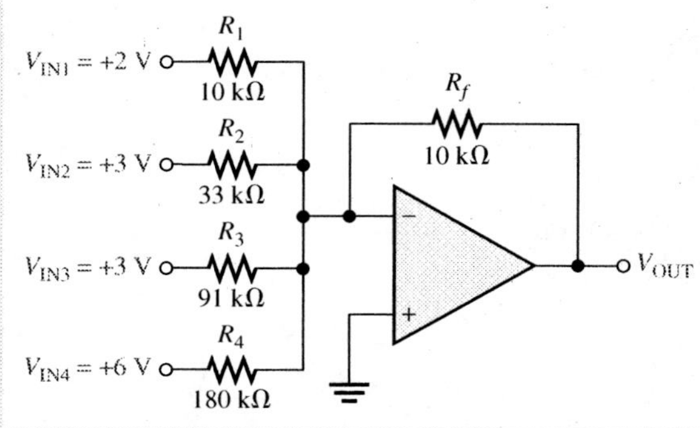

FIGURE 7–49

8. Determine the rate of change of the output voltage in response to the step input to the ideal integrator in Figure 7–50.

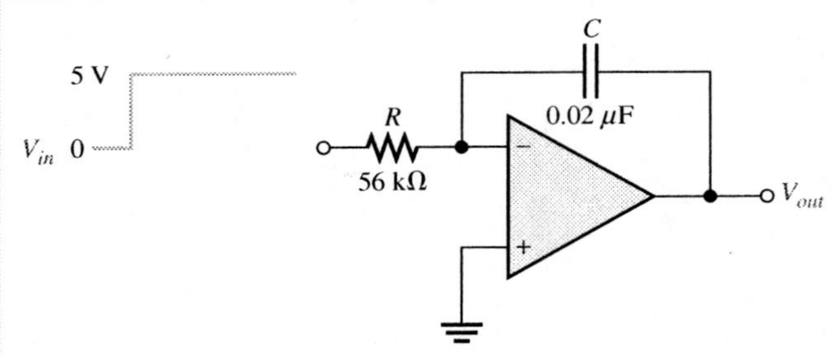

FIGURE 7–50

9. Describe the output waveform of each circuit in Figure 7–51. Assume that the R_LC time constant is much greater than the period of the input signal.

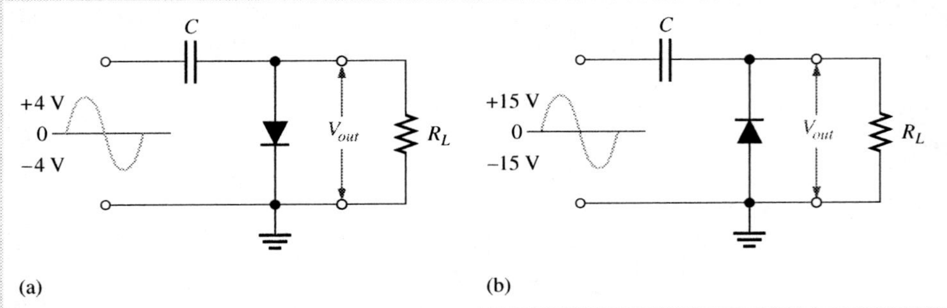

FIGURE 7–51

10. Describe the output waveform for the circuit in Figure 7–52.

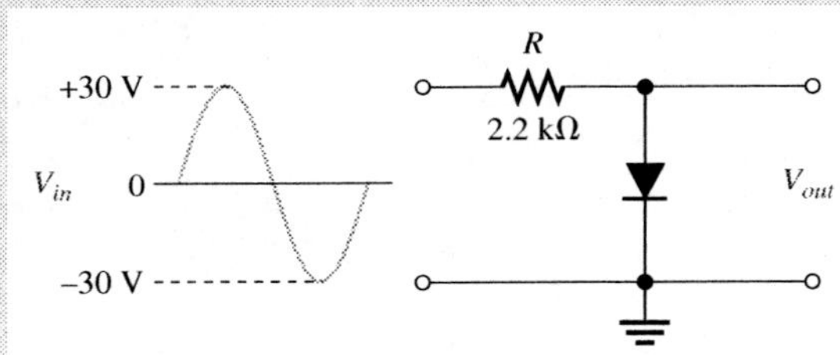

FIGURE 7–52

11. Determine the output voltage for the clamping circuit in Figure 7–53 for the input shown.

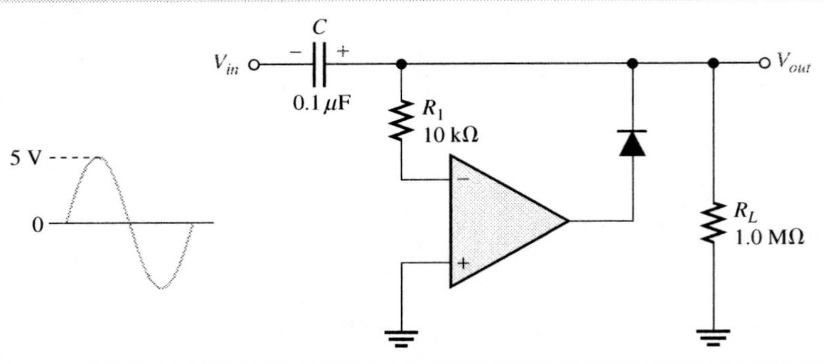

FIGURE 7–53

12. Determine the output waveform for the active limiter in Figure 7–54.

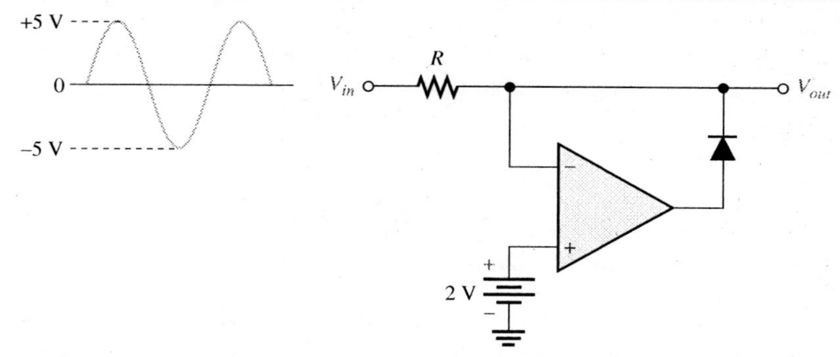

FIGURE 7–54

Basic-Plus Problems

13. Determine the hysteresis voltage for each comparator in Figure 7–55. The maximum output levels are ±11 V.

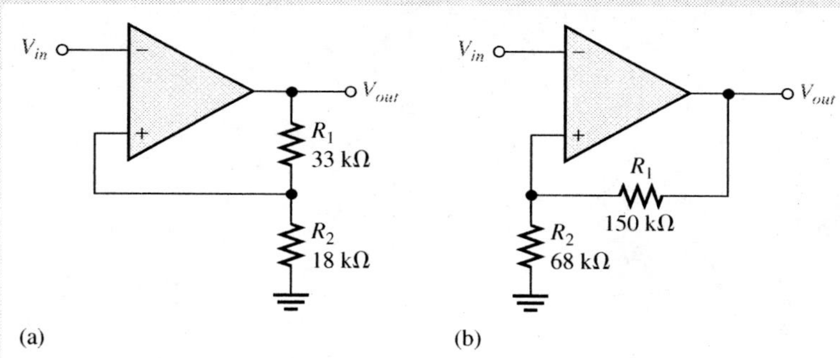

FIGURE 7–55

14. Refer to Figure 7–56. Determine the following:
    (a) I_{R1} and I_{R2}
    (b) Current through R_f
    (c) V_{OUT}

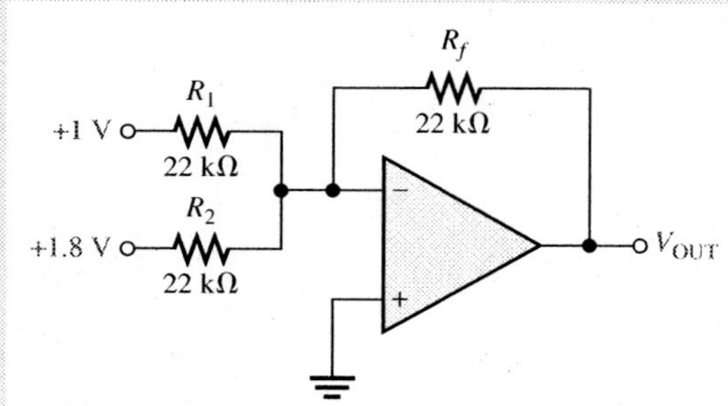

FIGURE 7–56

15. Find the value of R_f necessary to produce an output that is five times the sum of the inputs in Figure 7–56.

16. Design a summing amplifier that will average eight input voltages. Use input resistances of 10 kΩ each.

17. Determine the values of the input resistors required in a six-input scaling adder so that the lowest weighted input is 1 and each successive input has a weight twice the previous one. Use $R_f = 100\ \text{k}\Omega$.

18. A triangular waveform is applied to the input of the circuit in Figure 7–57 as shown. Determine what the output should be and sketch its waveform in relation to the input.

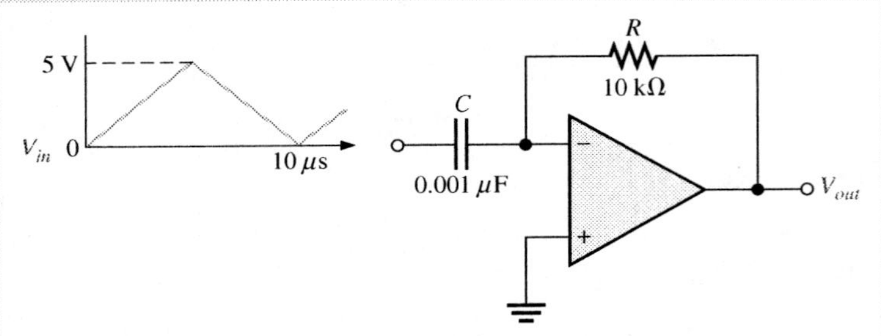

FIGURE 7–57

19. What is the magnitude of the capacitor current in Problem 18?

20. A triangular waveform with a peak-to-peak voltage of 2 V and a period of 1 ms is applied to the differentiator in Figure 7–58(a). What is the output voltage?

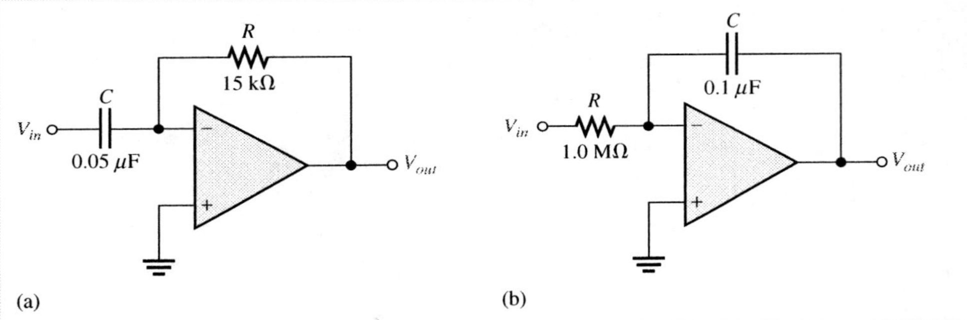

FIGURE 7–58

21. In Figure 7–58(b), a square wave with a period of 20 ms and an amplitude of ±5 V is applied to the input. Sketch the resulting output waveform. The saturated output levels of the op-amp are ±12 V.

22. Determine the load current in each circuit of Figure 7–59. (Hint: In part (b), thevenize the circuit to the left of R_i.)

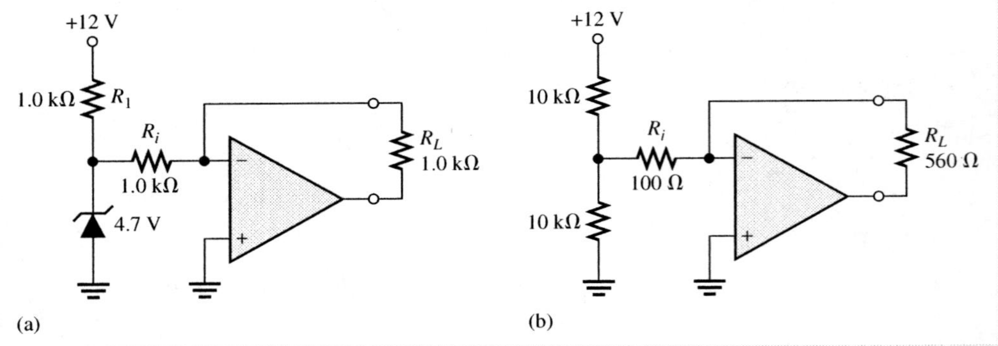

FIGURE 7–59

23. Devise a circuit for remotely sensing temperature and producing a proportional voltage that can then be converted to digital form for display. A thermistor can be used as the temperature-sensing element.

24. The sequences of voltage levels shown in Figure 7–60 are applied to the summing amplifier and the indicated output is observed. First, determine if this output is correct. If it is not correct, determine the fault. Each voltage value is to the left of the associated level.

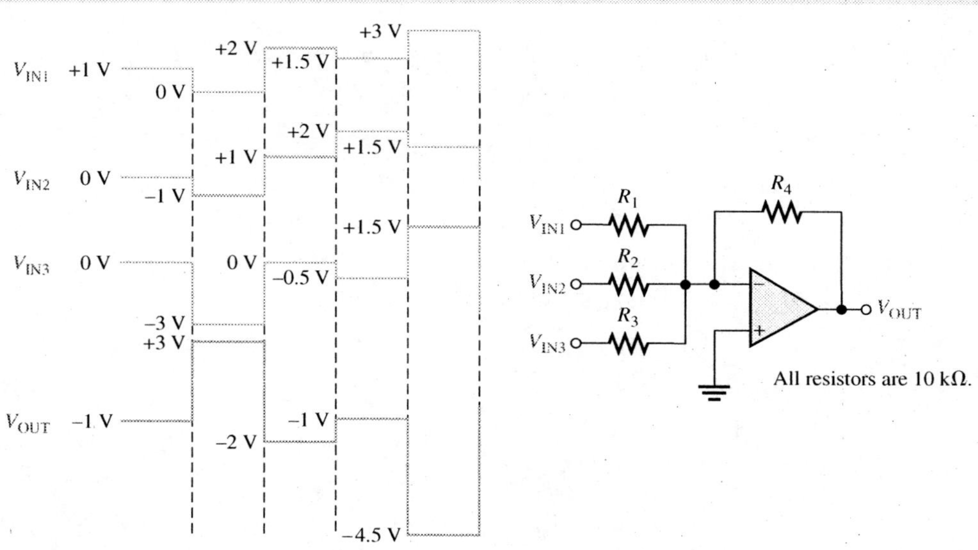

FIGURE 7–60

25. The given ramp voltages are applied to the op-amp circuit in Figure 7–61. Is the given output correct? If it isn't, what is the problem?

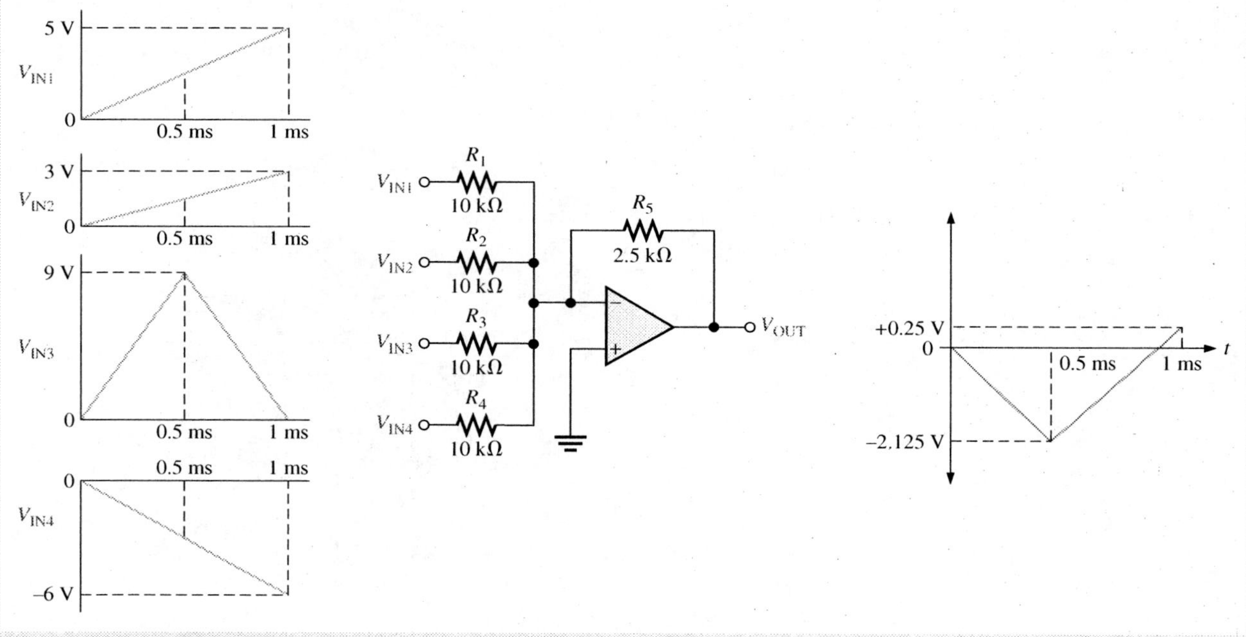

FIGURE 7–61

26. Assume you are given the responsibility of constructing 1000 circuits like the one in Figure 7–49. Consider the problems you will face in carrying out this job. Write

a summary discussing the resources you would need (people, parts, materials, equipment, space, and time).

ANSWERS

Example Questions

7–1: 1.96 V

7–2: +3.83 V, −3.83 V, V_{HYS} = 7.65 V

7–3: −12.5 V

7–4: Changes require an additional 100 kΩ input resistor and a change of R_f to 20 kΩ.

7–5: 0.45, 0.12, 0.18; V_{OUT} = −3.03 V

7–6: Change C to 5000 pF or change R to 5.0 kΩ.

7–7: Same waveform with peak voltages of ±6.6 V

7–8: 4.5 V

7–9: (a) $V_{p(out)}$ decreases to 0.68 V with no limiting.

(b) $V_{p(out)}$ is 6.8 V with no limiting.

7–10: −3.76 V

7–11: Change R_6 to 25 kΩ.

Review Questions

1. (a) $V = (10\text{ k}\Omega/110\text{ k}\Omega)15\text{ V} = 1.36\text{ V}$

   (b) $V = (22\text{ k}\Omega/69\text{ k}\Omega)(-12\text{ V}) = -3.83\text{ V}$
2. Unwanted voltage or current fluctuations
3. Hysteresis makes the output noise-free.
4. Schmitt trigger
5. Positive feedback
6. The summing point is the point where the input resistors are commonly connected.
7. −7 V
8. $R_f/R = 1/5 = 0.2$
9. 5 kΩ
10. Digital-to-analog converter
11. The feedback element in an integrator is a capacitor.
12. The capacitor voltage is linear because the capacitor current is constant.
13. The feedback element in a differentiator is a resistor.
14. The output of a differentiator is proportional to the rate of change of the input.
15. Square wave
16. A clamping circuit effectively restores the dc of an input signal.
17. A limiting circuit clips the input signal at a specified level.
18. A peak detector "stores" the peak value of an input signal.
19. A clamping circuit is called a dc restorer.
20. A limiting circuit is called a clipper.
21. An op-amp can fail with a shorted output.
22. Connect an identical component across the component that is suspected to be open.

23. The output will not change from either a HIGH or LOW state.
24. The output is the sum of the remaining inputs.
25. The output is at one of its maximum (saturated) levels.

Chapter Checkup

1. (c)	2. (a)	3. (c)	4. (e)	5. (b)
6. (d)	7. (c)	8. (a)	9. (b)	10. (c)
11. (b)	12. (d)	13. (d)	14. (a)	15. (d)
16. (c)	17. (c)	18. (d)	19. (b)	20. (b)
21. (a)	22. (c)			

ACTIVE FILTERS

CHAPTER OUTLINE

Study aids for this chapter are available at

http://www.prenhall.com/SOE

INTRODUCTION

Power supply filters were introduced in Chapter 2. In this chapter, active filters used for signal processing are introduced. Filters are circuits that are capable of passing input signals with certain selected frequencies through to the output while rejecting signals with other frequencies. This property is called *selectivity*.

Active filters use devices such as transistors or op-amps combined with passive *RC*, *RL*, or *RLC* circuits. The active devices provide voltage gain and the passive circuits provide frequency selectivity. In terms of general response, there are four basic categories of active filters: low-pass, high-pass, band-pass, and band-stop. In this chapter, you will study active filters using op-amps and *RC* circuits.

Logarithms and decibels are used in this chapter. If you are not familiar with these quantities, you should refer to Appendix A prior to or during the study of this chapter.

KEY TERMS

- Filter
- Critical frequency
- Low-pass filter
- Bandwidth
- Pole
- Roll-off
- High-pass filter
- Band-pass filter
- Band-stop filter
- Damping factor
- Order

KEY OBJECTIVES

A section number is given for each objective. After completing this chapter, you should be able to

8–1 Describe the gain-versus-frequency responses of the basic filters

8–2 Describe the three basic filter response characteristics and other filter parameters

8–3 Explain how active low-pass filters work

8–4 Explain how active high-pass filters work

8–5 Explain how active band-pass and band-stop filters work

8–6 Discuss two methods for measuring frequency response

COMPUTER SIMULATIONS DIRECTORY

The following figure has a Multisim circuit file associated with it.

- Figure 8–18
  Page 306

LABORATORY EXPERIMENTS DIRECTORY

The following exercises are for this chapter.

- **Experiment 16**
  Low-Pass and High-Pass Active Filters
- **Experiment 17**
  Multiple-Feedback Band-Pass Filters

The first music synthesizers developed in the 1960s, used many of the circuits developed for early analog computers. These synthesizers consisted of a large number of separate modules, each with a particular audio or control function. The modules performed functions such as envelope and waveform generators, oscillators, mixers, multiple filters, and amplifiers along with methods to control the sequence of sounds. All of these functions can be done with modern operational amplifiers (op-amps).

As digital techniques improved, the computing power and versatility of digital computers offered spectacular improvements in signal processing capability. Synthesizers soon incorporated digital techniques in their design. Today, many of the analog methods have been replaced by digital methods, but it is worth noting that analog techniques have also improved dramatically along with their digital counterparts.

One analog method of particular interest is the filter, used to select certain frequencies from others. Digital filters require a lot of memory to implement and require a lot of "number crunching." Analog filters, on the other hand, are simpler circuits. The Operational Transconductance Amplifier (OTA) is particularly useful in implementing filters. A combination of digital and analog techniques may ultimately produce a faster and cheaper solution to filters used in electronic music and other applications.

8–1 BASIC FILTER RESPONSES

Filters are usually categorized by the manner in which the output voltage varies with the frequency of the input voltage. The categories of active filters are low-pass, high-pass, band-pass, and band-stop.

In this section, you will learn the gain-versus-frequency responses of the basic filters.

Frequency Response of Low-Pass Filters

A **filter** is a circuit that passes certain frequencies and attenuates or rejects all other frequencies. The **passband** of a filter is the region of frequencies that are allowed to pass through the filter with minimum attenuation, usually defined as less than -3 decibels (dB)* of attenuation. The **critical frequency**, f_c, (also called the *cutoff frequency*) defines the end of the passband and is normally specified at the point where the response drops -3 dB (70.7%) from the passband response. Following the passband of a low-pass filter is a region called the *transition region* or *skirt* that leads into a region called the *stopband*. There is no precise point between the transition region and the stopband.

A **low-pass filter** is one that passes frequencies from dc (0 Hz) to f_c and significantly attenuates all other frequencies. The passband of the ideal low-pass filter is shown in the blue-shaded area of Figure 8–1(a); the response drops to zero at frequencies beyond the passband. This ideal response is sometimes referred to as a "brick-wall" because nothing gets through beyond the wall.

In general, the **bandwidth** of a filter is a measure of its passband and is defined as the difference between the upper and lower cutoff (critical) frequencies of the passband. In the case of a low-pass filter, the lower critical frequency is 0 Hz, so the bandwidth is equal to f_c.

$$BW = f_c \tag{8–1}$$

* The decibel (dB) is a logarithmic measurement. See Appendix A.

Low-pass filter responses. **FIGURE 8–1**

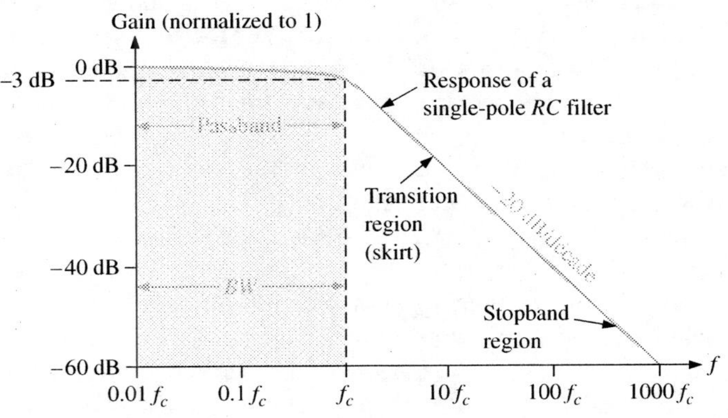

(a) Comparison of an ideal low-pass filter response with actual response

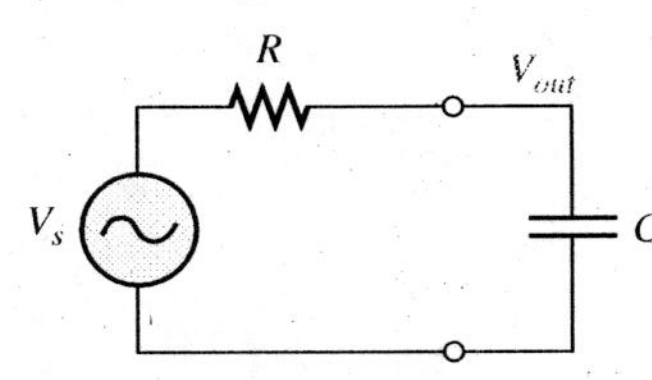

(b) Basic low-pass circuit

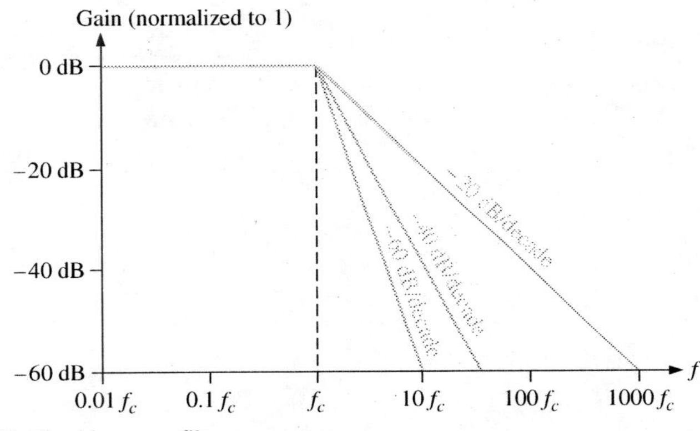

(c) Idealized low-pass filter responses

The ideal response shown in Figure 8–1(a) is not attainable by any practical filter. Actual filter responses depend on the number of **poles**, a term used with filters to describe the number of bypass circuits contained in the filter. The most basic low-pass filter is a simple *RC* circuit consisting of just one resistor and one capacitor; the output is taken across the capacitor as shown in Figure 8–1(b). This basic *RC* filter has a single pole and it decreases at −20 dB/decade beyond the critical frequency. (A decade is a ten times change in frequency.) The actual response is indicated by the blue line in Figure 8–1(a). The response is plotted on a standard log plot that is used for filters to show details of the curve as the gain drops. Notice that the gain is almost constant until the frequency is at the critical frequency; after this, the gain drops rapidly at a fixed rate called the **roll-off** rate.

The −20 dB/decade roll-off rate for the gain of a basic *RC* filter means that at a frequency of $10f_c$, the output will be −20 dB (10%) of the input. This rather gentle roll-off is not a particularly good filter characteristic because too much of the unwanted frequencies (beyond the passband) are allowed through the filter.

The critical frequency of the simple low-pass *RC* filter occurs when $X_C = R$, where

$$f_c = \frac{1}{2\pi RC}$$

The output at the critical frequency is 70.7% of the input. This response is equivalent to an attenuation of −3 dB.

Figure 8–1(c) illustrates several idealized low-pass response curves including the basic one pole response (−20 dB/decade). The approximations show a *flat* response to the cutoff frequency and a roll-off at a constant rate after the cutoff frequency. Actual filters do not have a perfectly flat response to the cutoff frequency but have dropped to −3 dB at this point as described previously.

In order to produce a filter that has a steeper transition region, (and hence form a more effective filter), it is necessary to add additional circuitry to the basic filter. Responses that are steeper than −20 dB/decade in the transition region cannot be obtained by simply cascading identical *RC* stages (due to loading effects). However, by combining an op-amp with frequency-selective feedback circuits, filters can be designed with roll-off rates of −40, −60, or more dB/decade. Filters that include one or more op-amps in the design are called **active filters**. These filters can optimize the roll-off rate or other attribute (such as phase response) with a particular filter design. In general, the more poles the filter uses, the steeper its transition region will be. The exact response depends on the type of filter and the number of poles.

Frequency Response of High-Pass Filters

A **high-pass filter** is one that significantly attenuates or rejects all frequencies below f_c and passes all frequencies above f_c. The critical frequency is, again, the frequency at which the output is 70.7% of the input (or −3 dB) as shown in Figure 8–2(a). The ideal response, in-

FIGURE 8–2 High-pass filter responses.

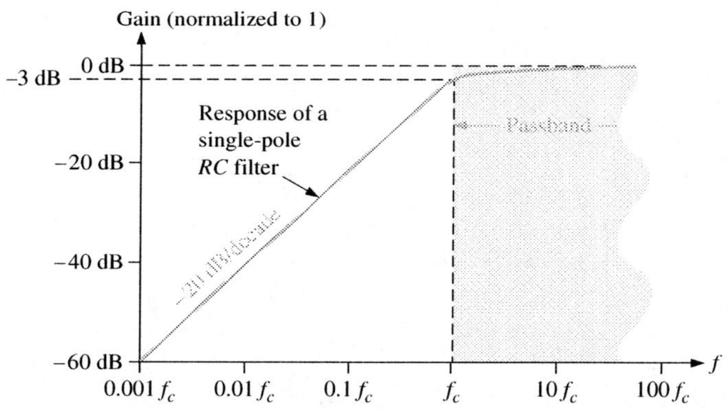

(a) Comparison of an ideal high-pass filter response with actual response

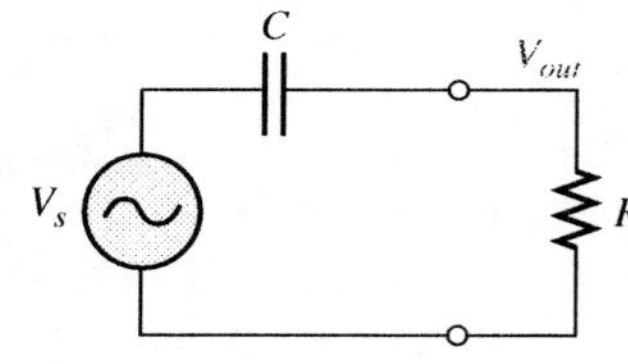

(b) Basic high-pass circuit

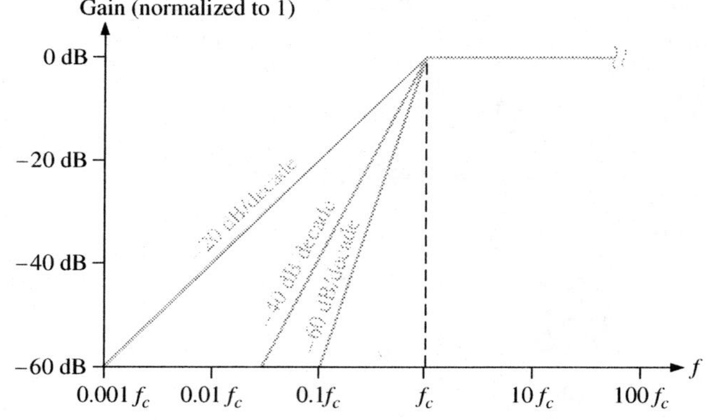

(c) Idealized high-pass filter responses

dicated by the blue-shaded area, has an instantaneous drop at f_c, which, of course, is not achievable. Ideally, the passband of a high-pass filter is all frequencies above the critical frequency. The high-frequency response of practical circuits is limited by the op-amp or other components that make up the filter.

A simple *RC* circuit consisting of a single resistor and capacitor can be configured as a high-pass filter by taking the output across the resistor as shown in Figure 8–2(b). As in the case of the low-pass filter, the basic *RC* circuit has a roll-off rate of −20 dB/decade as indicated by the blue line in Figure 8–2(a). Also, the critical frequency for the basic high-pass filter occurs when $X_C = R$, where

$$f_c = \frac{1}{2\pi RC}$$

Figure 8–2(c) illustrates several idealized high-pass response curves including the basic one pole response (−20 dB/decade) for a basic *RC* circuit. As in the case of the low-pass filter, the approximations show a *flat* response to the cutoff frequency and a roll-off at a constant rate after the cutoff frequency. Actual high-pass filters do not have the perfectly flat response indicated or the precise roll-off rate shown. Responses that are steeper than −20 dB/decade in the transition region are also possible with active high-pass filters; the particular response depends on the type of filter and the number of poles.

Frequency Response of Band-Pass Filters

A **band-pass filter** passes all signals lying within a band between a lower-frequency limit and an upper-frequency limit and essentially rejects all other frequencies that are outside this specific band. A generalized band-pass response curve is shown in Figure 8–3. The bandwidth (*BW*) of a band-pass filter is the difference between the upper critical frequency (f_{c2}) and the lower critical frequency (f_{c1}).

$$BW = f_{c2} - f_{c1} \tag{8–2}$$

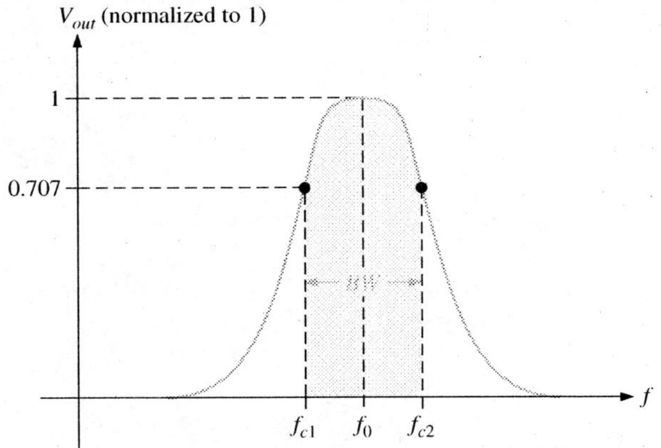

FIGURE 8–3

General band-pass response curve.

As you know, the critical frequencies are the points at which the response curve is 70.7% of its maximum. These critical frequencies are also called *3 dB frequencies*. The frequency about which the passband is centered is called the *center frequency*, f_0, defined as the geometric mean of the critical frequencies.

$$f_0 = \sqrt{f_{c1}f_{c2}} \tag{8–3}$$

Quality Factor

The quality factor (Q) of a band-pass filter is the ratio of the center frequency to the bandwidth.

$$Q = \frac{f_0}{BW} \tag{8–4}$$

The value of Q is an indication of the selectivity of a band-pass filter. The higher the value of Q, the narrower the bandwidth and the better the selectivity for a given value of f_0. Band-pass filters are sometimes classified as narrow-band ($Q > 10$) or wide-band ($Q < 10$).

EXAMPLE 8–1

Problem

A certain band-pass filter has a center frequency of 15 kHz and a bandwidth of 1 kHz. Determine the Q and classify the filter as narrow-band or wide-band.

Solution

$$Q = \frac{f_0}{BW} = \frac{15\text{ kHz}}{1\text{ kHz}} = \mathbf{15}$$

Because $Q > 10$, this is a **narrow-band** filter.

Question*

If the Q of the filter is doubled, what will the bandwidth be?

Frequency Response of Band-Stop Filters

Another category of active filter is the **band-stop filter**, also known as the *notch*, *band-reject*, or *band-elimination filter*. A general response curve for a band-stop filter is shown in Figure 8–4. Notice that the bandwidth is the band of frequencies between the 3 dB points, just as in the case of the band-pass filter response. You can think of the operation as opposite to that of the band-pass filter because frequencies within a certain bandwidth are rejected, and frequencies outside the bandwidth are passed.

FIGURE 8–4

General band-stop filter response.

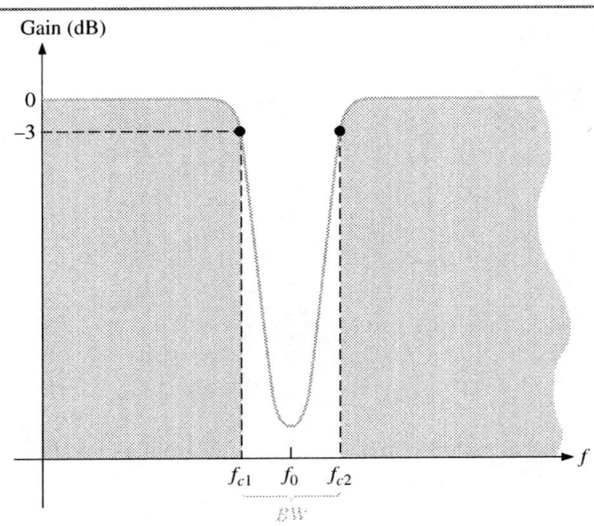

* *Answers are at the end of the chapter.*

Review Questions

Answers are at the end of the chapter.

1. What determines the bandwidth of a low-pass filter?
2. What limits the bandwidth of an active high-pass filter?
3. How are the Q and the bandwidth of a band-pass filter related?
4. How is the selectivity affected by the Q of a filter?
5. Is the passband of a low-pass filter above or below its critical frequency?

FILTER RESPONSE CHARACTERISTICS 8–2

Each type of filter (low-pass, high-pass, band-pass, or band-stop) can be tailored by circuit component values to have either a Butterworth, Chebyshev, or Bessel characteristic. Each of these characteristics is identified by the shape of the response curve, and each has an advantage in certain applications.

In this section, you will learn the three basic filter response characteristics and other filter parameters.

Butterworth, Chebyshev, or Bessel response characteristics can be realized with most active filter circuit configurations by proper selection of certain component values. A general comparison of the three response characteristics for a low-pass filter response curve is shown in Figure 8–5. High-pass, band-pass, and band-stop filters can also be designed to have any one of the characteristics.

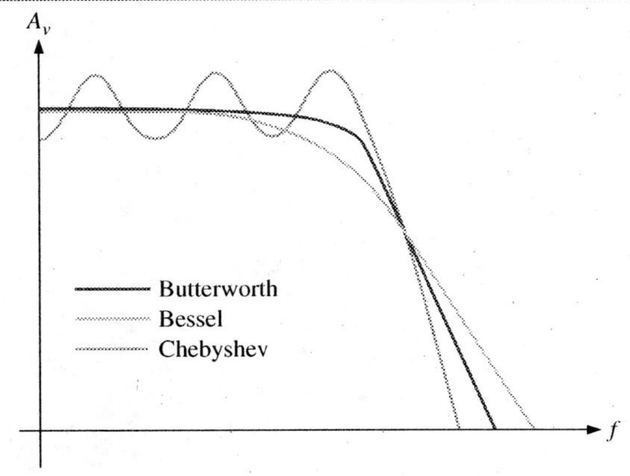

FIGURE 8–5

Comparative plots of three types of filter response characteristics.

The Butterworth Characteristic

The **Butterworth** characteristic provides a very flat amplitude response in the passband and a roll-off rate of −20 dB/decade/pole. The phase response is not linear, however, and the phase shift (thus, time delay) of signals passing through the filter varies nonlinearly with frequency. Therefore, a pulse applied to a filter with a Butterworth response will cause overshoots on the output because each frequency component of the pulse's rising and falling edges experiences a different time delay. Filters with the Butterworth response are normally used when all frequencies in the passband must have the same gain. The Butterworth response is often referred to as a *maximally flat response*.

The Chebyshev Characteristic

Filters with the **Chebyshev** response characteristic are useful when a rapid roll-off is required because it provides a roll-off rate greater than −20 dB/decade/pole. This is a greater rate than that of the Butterworth, so filters can be implemented with the Chebyshev response with fewer poles and less complex circuitry for a given roll-off rate. This type of filter response is characterized by overshoot or ripples in the passband (depending on the number of poles) and an even less linear phase response than the Butterworth.

The Bessel Characteristic

The **Bessel** response exhibits a linear phase characteristic, meaning that the phase shift increases linearly with frequency. The result is almost no overshoot on the output with a pulse input. For this reason, filters with the Bessel response are used for filtering pulse waveforms without distorting the shape of the waveform.

The Damping Factor

As mentioned, an active filter can be designed to have either a Butterworth, Chebyshev, or Bessel response characteristic regardless of whether it is a low-pass, high-pass, band-pass, or band-stop type. The **damping factor (DF)** of an active filter circuit determines which response characteristic the filter exhibits. To explain the basic concept, a generalized active filter is shown in Figure 8–6. It includes an amplifier, a negative feedback circuit, and a filter section. The amplifier and feedback circuit are connected in a noninverting configuration. The damping factor is determined by the negative feedback circuit and is defined by the following equation:

$$DF = 2 - \frac{R_1}{R_2} \qquad (8\text{–}5)$$

FIGURE 8–6

General diagram of an active filter. Note that R_1 corresponds to R_f and R_2 corresponds to R_i as defined in Section 6–4.

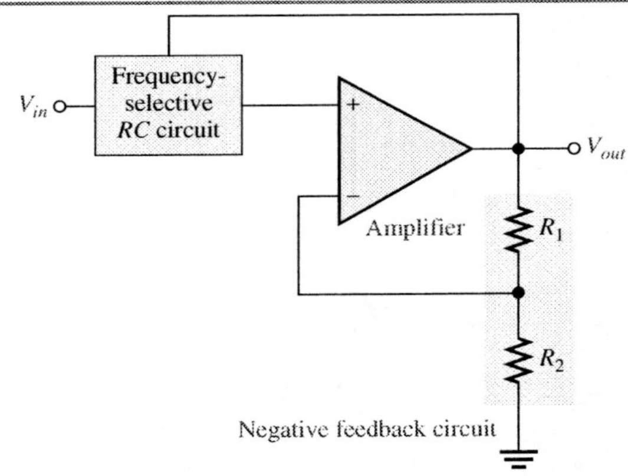

Basically, the damping factor affects the filter response by negative feedback action. Any attempted increase or decrease in the output voltage is offset by the opposing effect of the negative feedback. This tends to make the response curve flat in the passband of the filter if the value for the damping factor is precisely set. By advanced mathematics, which we will not cover, values for the damping factor have been derived for various orders of filters to achieve the maximally flat response of the Butterworth characteristic.

The value of the damping factor required to produce a desired response characteristic depends on the **order** (number of poles) of the filter. Recall that the more poles a filter has, the faster its roll-off rate is. To achieve a second-order Butterworth response, for example, the damping factor must be 1.414. To implement this damping factor, the feedback resistor ratio must be

$$\frac{R_1}{R_2} = 2 - DF = 2 - 1.414 = 0.586$$

This ratio gives the closed-loop gain of the noninverting filter amplifier, $A_{cl(NI)}$, a value of 1.586, derived as follows:

$$A_{cl\,(NI)} = \frac{R_1}{R_2} + 1 = 0.586 + 1 = 1.586$$

EXAMPLE 8–2

Problem

If resistor R_2 in the feedback circuit of an active filter of the type in Figure 8–6 is 10 kΩ, what value must R_1 be to obtain a maximally flat Butterworth response?

Solution

$$\frac{R_1}{R_2} = 0.586$$

$$R_1 = 0.586R_2 = 0.586(10\text{ k}\Omega) = \mathbf{5.86\text{ k}\Omega}$$

Using the nearest standard 5 percent value of 5600 Ω will get very close to the ideal Butterworth response.

Question

What is the damping factor for $R_2 = 10$ kΩ and $R_1 = 5.6$ kΩ?

Critical Frequency and Roll-Off Rate

The critical frequency is determined by the values of the resistor and capacitors in the *RC* circuit, as shown in Figure 8–6. For a single-pole (first-order) filter, as shown in Figure 8–7, the critical frequency is

$$f_c = \frac{1}{2\pi RC}$$

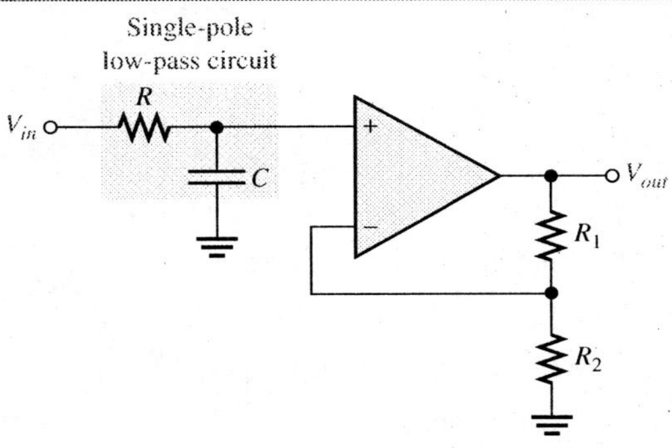

FIGURE 8–7

First-order (one-pole) low-pass filter.

Although we show a low-pass configuration, the same formula is used for the f_c of a single-pole high-pass filter. The number of poles determines the roll-off rate of the filter. A Butterworth response produces −20 dB/decade/pole. So, a first-order (one-pole) filter has a roll-off of −20 dB/decade; a second-order (two-pole) filter has a roll-off rate of −40 dB/decade; a third-order (three-pole) filter has a roll-off rate of −60 dB/decade; and so on.

Generally, to obtain a filter with three poles or more, one-pole or two-pole filters are cascaded, as shown in Figure 8–8. To obtain a third-order filter, for example, cascade a second-order and a first-order filter; to obtain a fourth-order filter, cascade two second-order filters; and so on. Each filter in a cascaded arrangement is called a *stage* or *section*.

FIGURE 8–8 The number of filter poles can be increased by cascading filters.

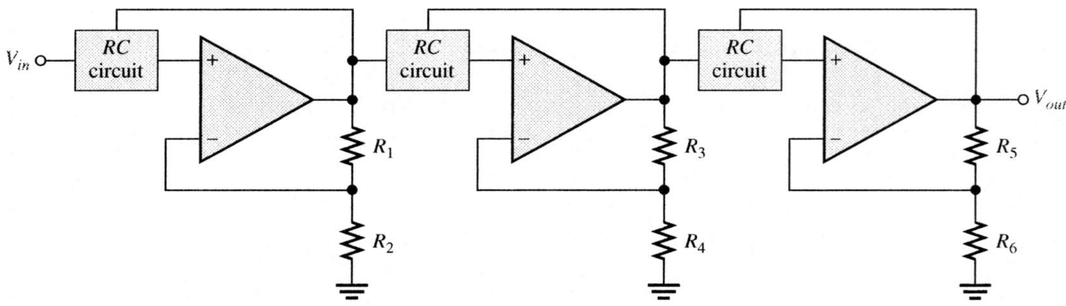

Because of its maximally flat response, the Butterworth characteristic is the most widely used. Therefore, we will limit our coverage to the Butterworth response to illustrate basic filter concepts. Table 8–1 lists the roll-off rates, damping factors, and feedback resistor ratios for up to sixth-order Butterworth filters. Resistor designations correspond to the gain-setting resistors in Figure 8–8 and may be different in other circuit diagrams.

TABLE 8–1 Values for the Butterworth response.

		1st stage			2nd stage			3rd stage		
Order	**Roll-off dB/decade**	**Poles**	**DF**	**R_1/R_2**	**Poles**	**DF**	**R_3/R_4**	**Poles**	**DF**	**R_5/R_6**
1	−20	1	Optional							
2	−40	2	1.414	0.586						
3	−60	2	1.00	1	1	1.00	1			
4	−80	2	1.848	0.152	2	0.765	1.235			
5	−100	2	1.00	1	2	1.618	0.382	1	0.618	1.382
6	−120	2	1.932	0.068	2	1.414	0.586	2	0.518	1.482

Review Questions

6. How do Butterworth, Chebyshev, and Bessel responses differ?
7. What determines the response characteristic of a filter?
8. What are the basic parts of an active filter?
9. What determines the critical frequency of a filter?
10. What determines the roll-off rate of a filter?

ACTIVE LOW-PASS FILTERS 8–3

Filters that use op-amps as the active element provide several advantages over passive filters (R, L, and C elements only). The op-amp provides gain, so that the signal is not attenuated as it passes through the filter. The high input resistance of the op-amp prevents excessive loading of the driving source, and the low output resistance of the op-amp prevents the filter from being affected by the load that it is driving. Active filters are also easy to adjust over a wide frequency range without altering the desired response.

In this section, you will learn how active low-pass filters work.

A Single-Pole Filter

Figure 8–9(a) shows an active filter with a single low-pass *RC* circuit that provides a roll-off of −20 dB/decade above the critical frequency, as indicated by the response curve in Figure 8–9(b). The critical frequency of the single-pole filter is $f_c = 1/(2\pi RC)$. The op-amp in this filter is connected as a noninverting amplifier with the closed-loop voltage gain in the passband set by the values of R_1 and R_2.

$$A_{cl(NI)} = \frac{R_1}{R_2} + 1$$

FIGURE 8–9 Single-pole active low-pass filter and response curve.

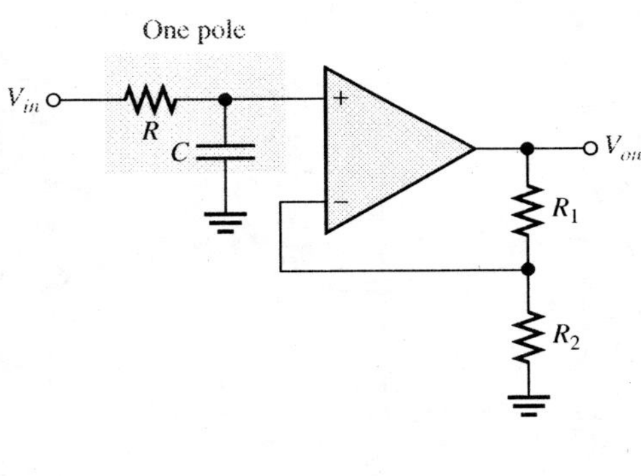

(a)

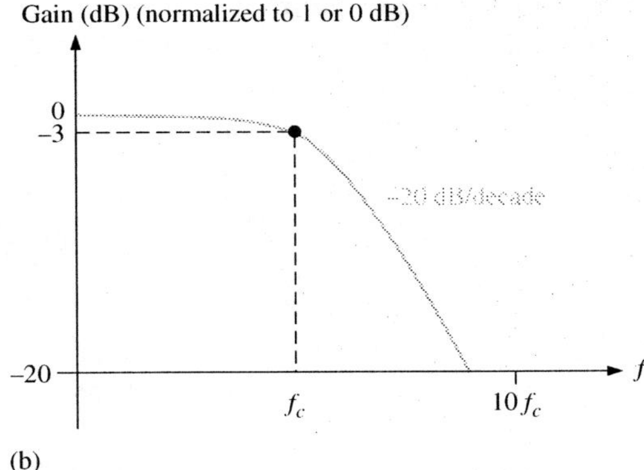

(b)

A Two-Pole Filter

The Sallen-Key is one of the most common configurations for a second-order (two-pole) filter. It is also known as a VCVS (voltage-controlled voltage source) filter. A low-pass version of the Sallen-Key filter is shown in Figure 8–10. Notice that there are two low-pass *RC* circuits that provide a roll-off of −40 dB/decade above the critical frequency (assuming a Butterworth characteristic). One *RC* circuit consists of R_A and C_A, and the second *RC* circuit consists of R_B and C_B. A unique feature is the capacitor C_A that provides feedback for shaping the response near the edge of the passband. The critical frequency for the second-order Sallen-Key filter is

$$f_c = \frac{1}{2\pi\sqrt{R_A R_B C_A C_B}} \quad (8\text{–}6)$$

FIGURE 8–10

Basic Sallen-Key second-order low-pass filter.

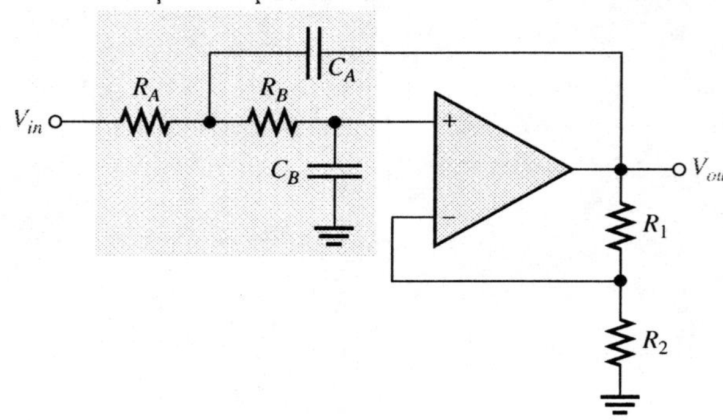

For simplicity, the component values can be made equal so that $R_A = R_B = R$ and $C_A = C_B = C$. In this case, the expression for the critical frequency simplifies to

$$f_c = \frac{1}{2\pi RC}$$

As in the single-pole filter, the op-amp in the second-order Sallen-Key filter acts as a noninverting amplifier with the negative feedback provided by the R_1/R_2 circuit. As you have learned, the damping factor is set by the values of R_1 and R_2, thus making the filter response either Butterworth, Chebyshev, or Bessel. For example, from Table 8–1, the R_1/R_2 ratio must be 0.586 to produce the damping factor of 1.414 required for a second-order Butterworth response.

EXAMPLE 8–3

Problem

Determine the critical frequency of the low-pass filter in Figure 8–11, and set the value of R_1 for an approximate Butterworth response.

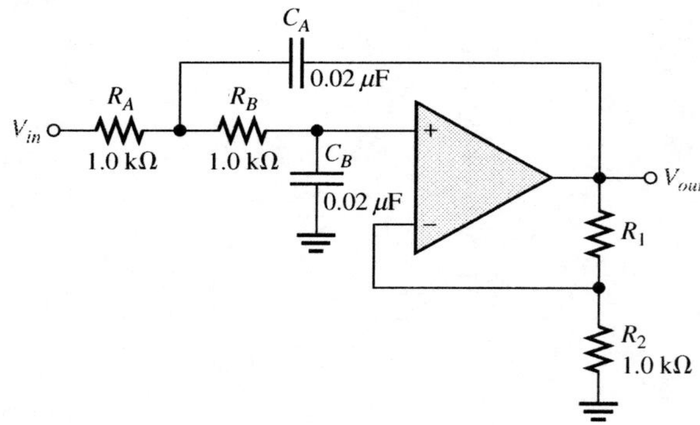

FIGURE 8–11

Solution

Since $R_A = R_B = 1.0\text{ k}\Omega$ and $C_A = C_B = 0.02\ \mu\text{F}$,

$$f_c = \frac{1}{2\pi RC} = \frac{1}{2\pi(1.0\text{ k}\Omega)(0.02\ \mu\text{F})} = \mathbf{7.96\text{ kHz}}$$

For a Butterworth response, $R_1/R_2 = 0.586$ from Table 8–1.

$$R_1 = 0.586R_2 = 0.586(1.0\text{ k}\Omega) = \mathbf{586\ \Omega}$$

Select a standard value as near as possible to this calculated value.

Question

What is the value of f_c for Figure 8–11 if $R_A = R_B = R_2 = 2.2\text{ k}\Omega$ and $C_A = C_B = 0.01\ \mu\text{F}$?

Cascaded Low-Pass Filters Achieve a Higher Roll-Off Rate

As an example, a four-pole filter is required to get a fourth-order low-pass response (−80 dB/decade). This is done by cascading two two-pole low-pass filters, as shown in Figure 8–12. The two stages must have different gains, as illustrated in Example 8–4.

FIGURE 8–12 A fourth-order filter using cascaded low-pass filters.

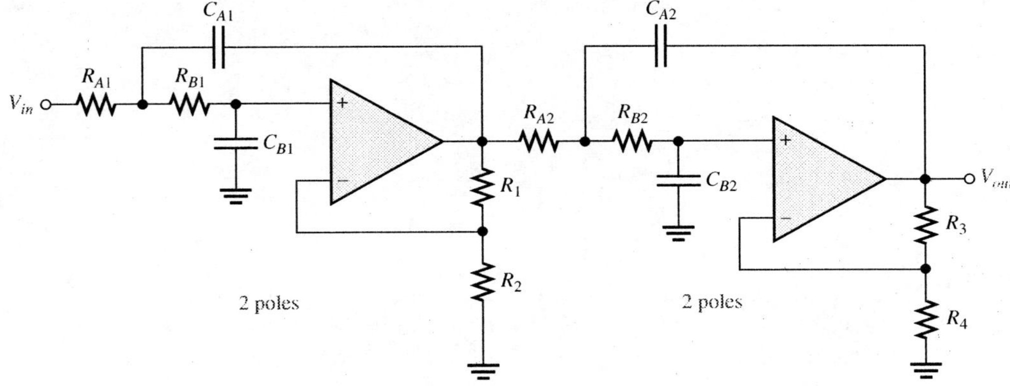

EXAMPLE 8–4

Problem

For the four-pole filter in Figure 8–12, determine the capacitance values required to produce a critical frequency of 2680 Hz if all the resistors in the *RC* low-pass circuits are 1.8 kΩ. Also select values for the feedback resistors to get a Butterworth response.

Solution

Both stages must have the same f_c. Assuming equal-value capacitors,

$$f_c = \frac{1}{2\pi RC}$$

$$C = \frac{1}{2\pi R f_c} = \frac{1}{2\pi(1.8\text{ k}\Omega)(2680\text{ Hz})} = 0.033\ \mu\text{F}$$

$$C_{A1} = C_{B1} = C_{A2} = C_{B2} = \mathbf{0.033\ \mu F}$$

Also select $R_2 = R_4 = 1.8$ kΩ for simplicity. Refer to Table 8–1. For a Butterworth response in the first stage, $DF = 1.848$ and $R_1/R_2 = 0.152$. Therefore,

$$R_1 = 0.152R_2 = 0.152(1800\ \Omega) = \mathbf{274\ \Omega}$$

Choose $R_1 = 270\ \Omega$.

In the second stage, $DF = 0.765$ and $R_3/R_4 = 1.235$. Therefore,

$$R_3 = 1.235R_4 = 1.235(1800\ \Omega) = \mathbf{2.22\ k\Omega}$$

Choose $R_3 = 2.2$ kΩ.

Question

For the filter in Figure 8–12, what are the capacitance values for $f_c = 1$ kHz if all the filter resistors are 680 Ω?

Review Questions

11. How many poles does a second-order low-pass filter have?
12. Why is the damping factor of a filter important?
13. What is the primary purpose of cascading low-pass filters?
14. How many capacitors does a two-pole filter have?
15. What roll-off rate is produced by cascading two single-pole filters?

8–4 ACTIVE HIGH-PASS FILTERS

In high-pass filters, the roles of the capacitor and resistor are reversed in the *RC* circuits. Otherwise, the basic parameters are the same as for the low-pass filters.

In this section, you will learn how active high-pass filters work.

A Single-Pole Filter

A high-pass active filter with a −20 dB/decade roll-off is shown in Figure 8–13(a). Notice that the input circuit is a single high-pass *RC* circuit. The negative feedback circuit is the same as for the low-pass filters previously discussed. The high-pass response curve is shown in Figure 8–13(b).

FIGURE 8–13 Single-pole active high-pass filter and response curve.

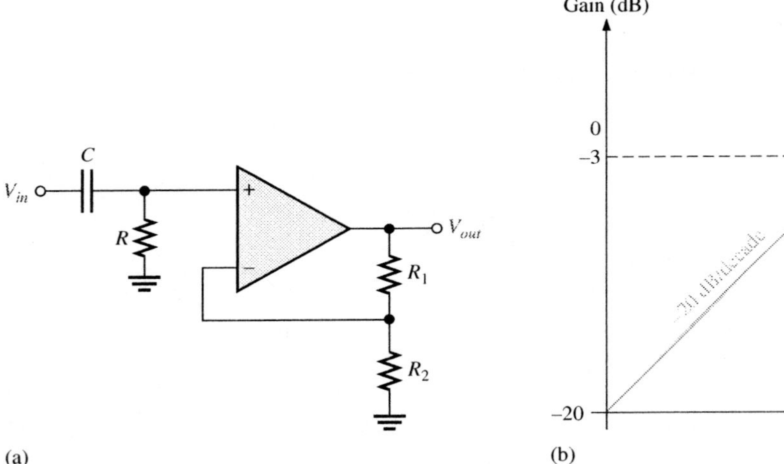

Ideally, a high-pass filter passes all frequencies above f_c without limit, as indicated in Figure 8–14(a), although in practice, this is not the case. As you have learned, all op-amps inherently have internal *RC* circuits that limit the amplifier's response at very high frequencies. Therefore, there is an upper-frequency limit on the high-pass filter's response which, in effect, makes it a band-pass filter with a very wide bandwidth. In the majority of applications, the internal high-frequency limitation is greater than the filter's f_c and the limitation can be neglected. In some applications, special current-feedback op-amps or discrete transistors are used for the gain element to increase the high-frequency limitation beyond that realizable with standard op-amps.

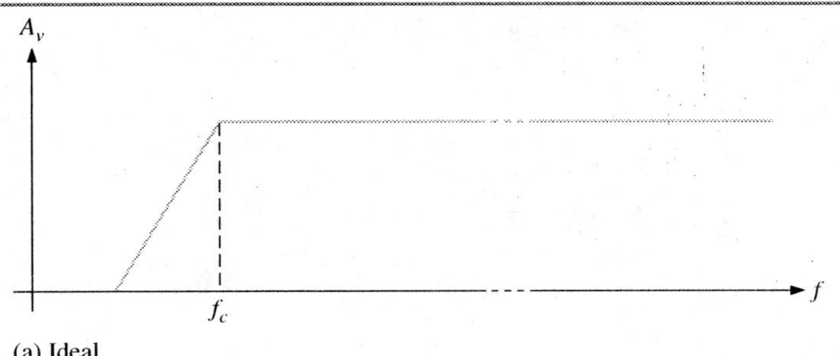

(a) Ideal

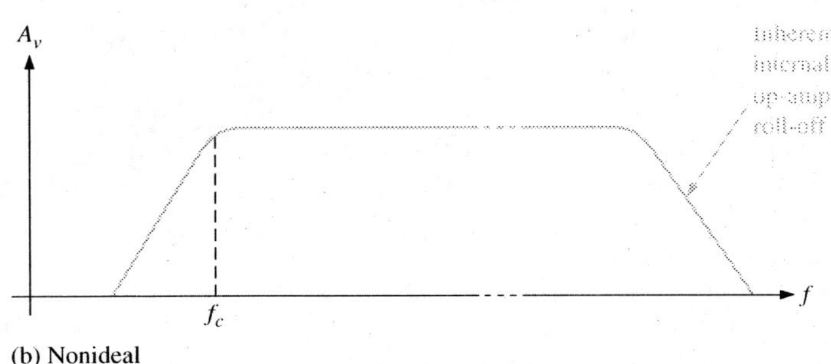

(b) Nonideal

FIGURE 8–14

High-pass filter response.

A Two-Pole Filter

A high-pass second-order Sallen-Key configuration is shown in Figure 8–15. The components R_A, C_A, R_B, and C_B form the two-pole frequency-selective circuit. Notice that the positions of the resistors and capacitors in the frequency-selective circuit are opposite to those in the low-pass configuration. As with the other filters, the circuit response characteristic can be optimized by proper selection of the feedback resistors, R_1 and R_2.

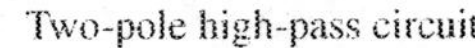

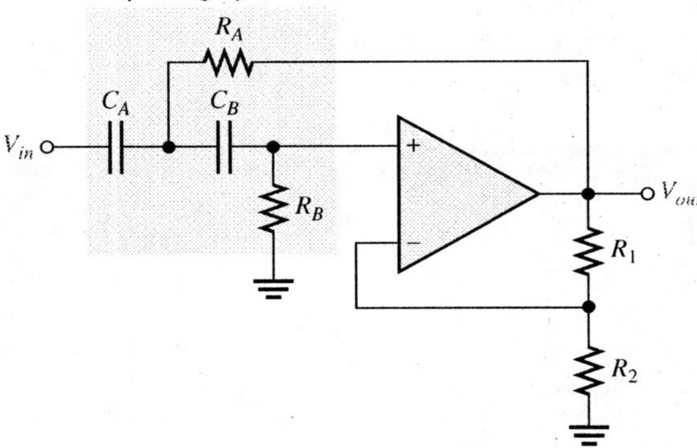

FIGURE 8–15

Basic Sallen-Key second-order high-pass filter.

EXAMPLE 8–5

Problem

Choose values for the Sallen-Key high-pass filter in Figure 8–15 to implement an equal-value second-order Butterworth response with a critical frequency of approximately 10 kHz.

Solution

Start by selecting a value for R_A and R_B (R_1 or R_2 can also be the same value as R_A and R_B for simplicity).

$$R = R_A = R_B = R_2 = \mathbf{3.3\ k\Omega} \quad \text{(an arbitrary selection)}$$

Next, calculate the capacitance value from $f_c = 1/(2\pi RC)$.

$$C = C_A = C_B = \frac{1}{2\pi Rf_c} = \frac{1}{2\pi(3.3\text{ k}\Omega)(10\text{ kHz})} = \mathbf{0.0048\ \mu F}$$

For a Butterworth response, the damping factor must be 1.414 and $R_1/R_2 = 0.586$.

$$R_1 = 0.586R_2 = 0.586(3.3\text{ k}\Omega) = \mathbf{1.93\ k\Omega}$$

If you had chosen $R_1 = 3.3\text{ k}\Omega$, then

$$R_2 = \frac{R_1}{0.586} = \frac{3.3\text{ k}\Omega}{0.586} = 5.63\text{ k}\Omega$$

Either way, an approximate Butterworth response is realized by choosing the nearest standard values.

Question
What are the values for all the components in the high-pass filter of Figure 8–15 to obtain an $f_c = 300$ Hz? Use equal-value components and optimize for a Butterworth response.

Review Questions

16. How does a high-pass Sallen-Key filter differ from the low-pass configuration?
17. To increase the critical frequency of a high-pass filter, would you increase or decrease the resistor values?
18. If three two-pole high-pass filters and one single-pole high-pass filter are cascaded, what is the resulting roll-off?
19. What limits the frequency response of a high-pass filter at very high frequencies?
20. What is a stage in a cascaded filter?

8–5 ACTIVE BAND-PASS AND BAND-STOP FILTERS

As mentioned, band-pass filters pass all frequencies bounded by a lower-frequency limit and an upper-frequency limit and reject all others lying outside this specified band. A band-pass response can be thought of as the overlapping of a low-frequency response curve and a high-frequency response curve. Band-stop filters reject a specified band of frequencies and pass all others. The response is opposite that of a band-pass filter.

In this section, you will learn how active band-pass and band-stop filters work.

Cascaded Low-Pass and High-Pass Filters Achieve a Band-Pass Response

One way to implement a band-pass filter is a cascaded arrangement of a high-pass filter and a low-pass filter, as shown in Figure 8–16(a), as long as the critical frequencies are sufficiently separated. Each of the filters shown is a two-pole Sallen-Key Butterworth configuration so that the roll-off rates are −40 dB/decade, indicated-in-the composite response curve of Figure 8–16(b). The critical frequency of each filter is chosen so that the response curves overlap sufficiently, as indicated. The critical frequency of the high-pass filter must be sufficiently lower than that of the low-pass stage.

The lower frequency, f_{c1}, of the passband is the critical frequency of the high-pass filter. The upper frequency, f_{c2}, is the critical frequency of the low-pass filter. This filter is generally limited to wide bandwidth applications.

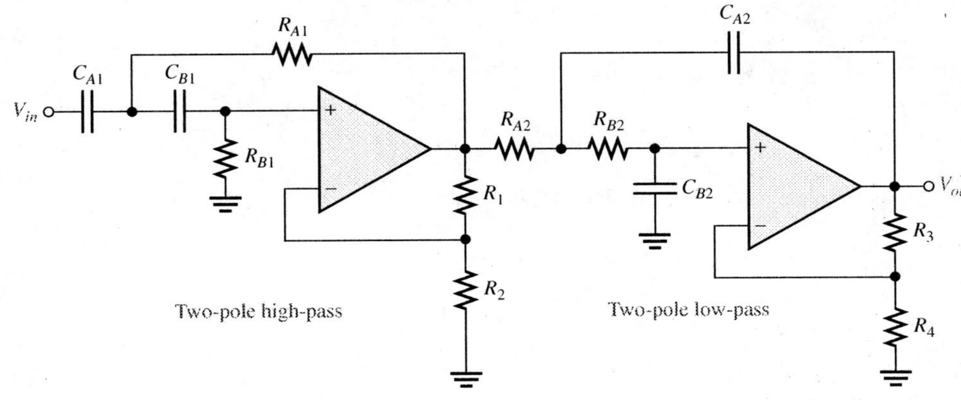

(a)

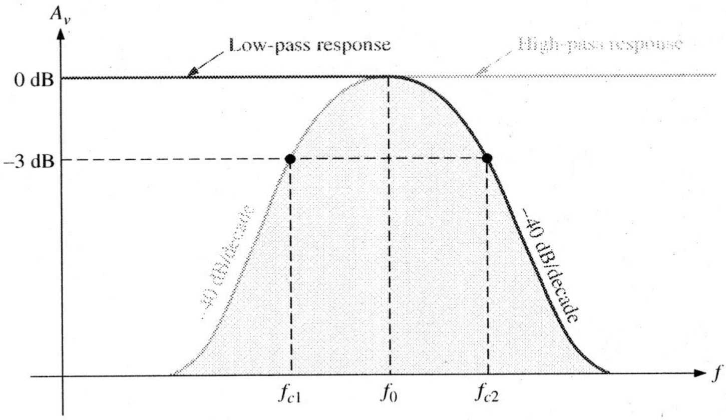

(b)

FIGURE 8–16

Band-pass filter formed by cascading a two-pole high-pass and a two-pole low-pass filter (it does not matter in which order the filters are cascaded).

Multiple-Feedback Band-Pass Filter

Another type of filter configuration, shown in Figure 8–17, is a multiple-feedback band-pass filter. This filter has an advantage over the cascaded band-pass filter just discussed because

FIGURE 8–17

Multiple-feedback band-pass filter.

it is simpler and can achieve narrower bandwidths. Notice that there are two feedback paths, one through C_2 and the other through R_3. The formula for the center frequency is

$$f_0 = \frac{1}{2\pi C}\sqrt{\frac{R_1 + R_2}{R_1 R_2 R_3}} \tag{8–7}$$

where $C = C_1 = C_2$. The gain at the center frequency is determined by R_1 and R_3 as follows.

$$A_0 = \frac{R_3}{2R_1}$$

The bandwidth is determined by the center frequency, f_0, and the Q. First find Q using the formula

$$Q = \pi f_0 C R_3$$

The bandwidth can then be found with the formula

$$BW = \frac{f_0}{Q}$$

EXAMPLE 8–6

Problem

Find the center frequency, the gain at the center frequency, and the bandwidth for the filter in Figure 8–18.

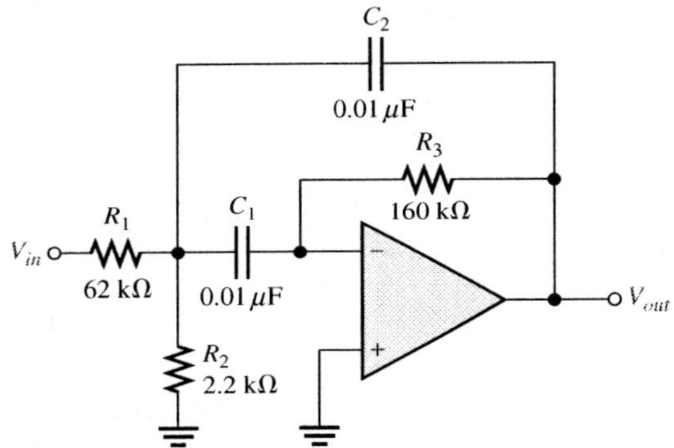

FIGURE 8–18

Solution

The center frequency is

$$f_0 = \frac{1}{2\pi C}\sqrt{\frac{R_1 + R_2}{R_1 R_2 R_3}} = \frac{1}{2\pi(0.01\ \mu\text{F})}\sqrt{\frac{62\ \text{k}\Omega + 2.2\ \text{k}\Omega}{(62\ \text{k}\Omega)(2.2\ \text{k}\Omega)(160\ \text{k}\Omega)}} = \mathbf{863\ Hz}$$

The gain is

$$A_0 = \frac{R_3}{2R_1} = \frac{160\ \text{k}\Omega}{2(62\ \text{k}\Omega)} = \mathbf{1.29}$$

To get the bandwidth, first calculate the Q.

$$Q = \pi f_0 C R_3 = \pi(863\ \text{Hz})(0.01\ \mu\text{F})(160\ \text{k}\Omega) = 4.34$$

Then the bandwidth is

$$BW = \frac{f_0}{Q} = \frac{863\ \text{Hz}}{4.34} = \mathbf{199\ Hz}$$

Question

If the value of the capacitors in the filter are increased, what happens to the center frequency?

COMPUTER SIMULATION

Open the Multisim file F08-18DV on the website. Measure the bandwidth of the cascaded band-pass filter.

Multiple-Feedback Band-Stop Filter

Figure 8–19 shows a multiple-feedback band-stop filter. Notice that this configuration is similar to the band-pass version.

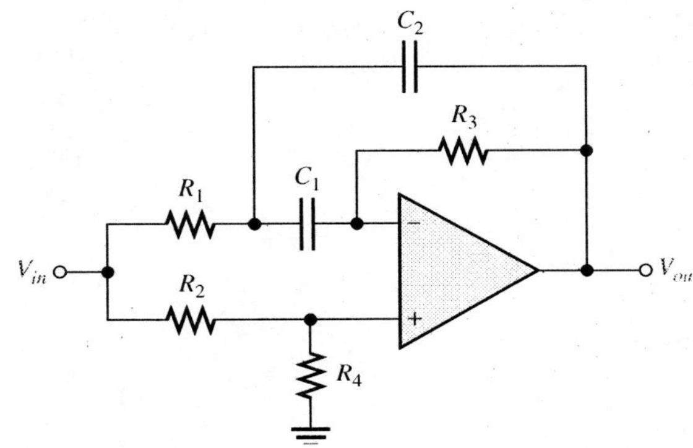

FIGURE 8–19

Multiple-feedback band-stop filter.

State-Variable Filter

The state-variable or universal active filter is widely used for band-pass applications. As shown in Figure 8–20, it consists of a summing amplifier and two op-amp integrators (which act as single-pole low-pass filters) that are combined in a cascaded arrangement to form a second-order filter. Although used primarily as a band-pass (BP) filter, the state-variable configuration also provides low-pass (LP) and high-pass (HP) outputs. The center frequency is set by the *RC* circuits in both integrators. When used as a band-pass filter, the critical frequencies of the integrators are usually made equal, thus setting the center frequency of the passband.

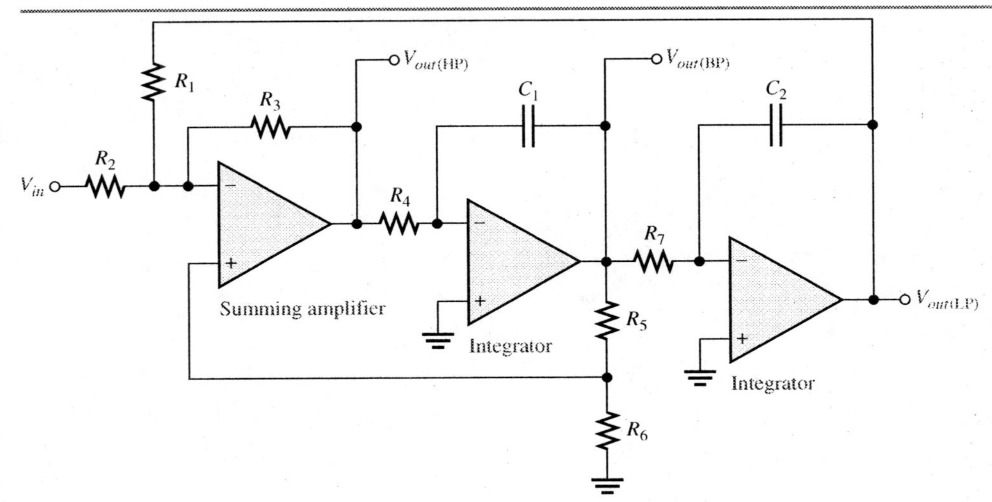

FIGURE 8–20

State-variable filter.

Basic Operation

At input frequencies below f_c, the input signal passes through the summing amplifier and integrators and is fed back 180° out-of-phase. Thus, the feedback signal and input signal cancel for all frequencies below approximately f_c. As the low-pass response of the integrators rolls off, the feedback signal diminishes, thus allowing the input to pass through to the band-pass output. Above f_c, the low-pass response disappears, thus preventing the input signal from passing through the integrators. As a result, the band-pass output peaks sharply at f_c, as indicated in Figure 8–21. Stable Qs up to 100 can be obtained with this type of filter.

FIGURE 8–21

General state-variable response curves.

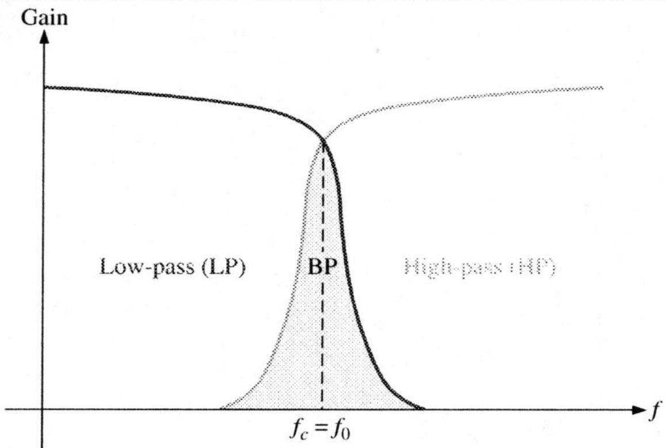

State-Variable Band-Stop Filter

Summing the low-pass (LP) and the high-pass (HP) responses of the state-variable filter shown in Figure 8–20 creates a band-stop filter as shown in Figure 8–22.

FIGURE 8–22

State-variable band-stop filter.

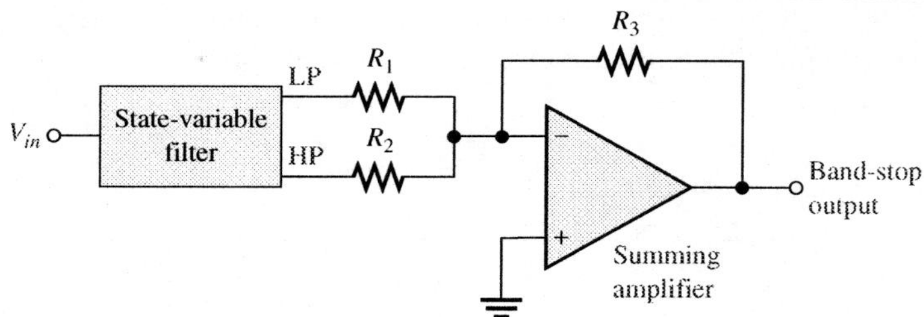

Review Questions

21. What determines selectivity in a band-pass filter?
22. One filter has a $Q = 5$ and another has a $Q = 25$. Which has the narrower bandwidth?
23. What are the active elements that make up a state-variable filter?
24. What are the two feedback components in a multiple-feedback band-stop filter?
25. In a cascaded band-pass filter, which filter stage has the highest critical frequency?

FILTER RESPONSE MEASUREMENTS 8–6

Two methods of determining a filter's response by measurement are discrete point measurement and swept frequency measurement.

In this section, you will learn about two methods for measuring frequency response.

Discrete Point Measurement

Figure 8–23 shows an arrangement for taking filter output voltage measurements at discrete values of input frequency using common laboratory instruments. The general procedure is as follows:

1. Set the amplitude of the sine wave generator to a desired voltage level.
2. Set the frequency of the sine wave generator to a value well below the expected critical frequency of the filter under test. For a low-pass filter, set the frequency as near as possible to 0 Hz. For a band-pass filter, set the frequency well below the expected lower critical frequency.
3. Increase the frequency in predetermined steps sufficient to allow enough data points for an accurate response curve.
4. Maintain a constant input voltage amplitude while varying the frequency.
5. Record the output voltage at each value of frequency.
6. After recording a sufficient number of points, plot a graph of output voltage versus frequency.

If the frequencies to be measured exceed the frequency response of the DMM, an oscilloscope may have to be used instead.

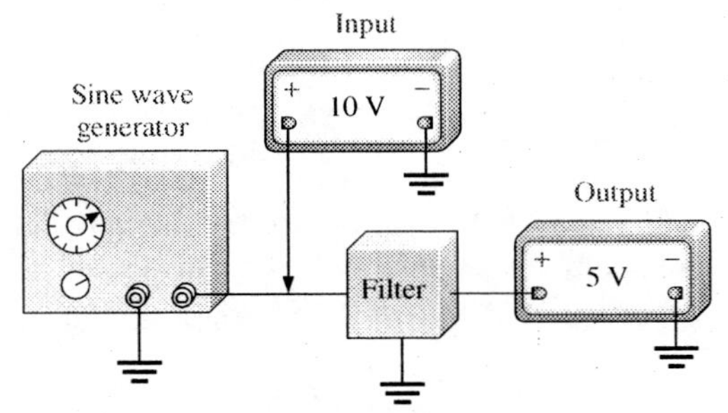

FIGURE 8–23

Test setup for discrete point measurement of the filter response. (Readings are arbitrary and for display only.)

Swept Frequency Measurement

The swept frequency method requires more elaborate test equipment than does the discrete point method, but it is much more efficient and can result in a more accurate response curve. A general test setup is shown in Figure 8–24(a) using a swept frequency generator and a spectrum analyzer. Figure 8–24(b) shows how the test can be made with an oscilloscope instead of a spectrum analyzer.

The swept frequency generator produces a constant amplitude output signal whose frequency increases linearly between two preset limits, as indicated in Figure 8–24. In part (a), the spectrum analyzer is an instrument that can be calibrated for a desired *frequency*

SAFETY NOTE

When working with electrical circuits, always wear shoes and keep them dry. Do not stand on metal or wet floors. Never handle instruments when your hands are wet.

FIGURE 8–24

Test setup for swept frequency measurement of the filter response. For clarity, only the envelope of the response is shown.

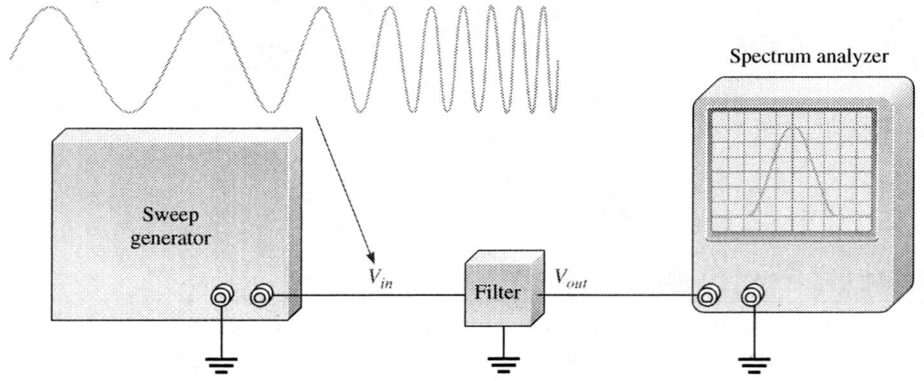

(a) Test setup for a filter response using a spectrum analyzer

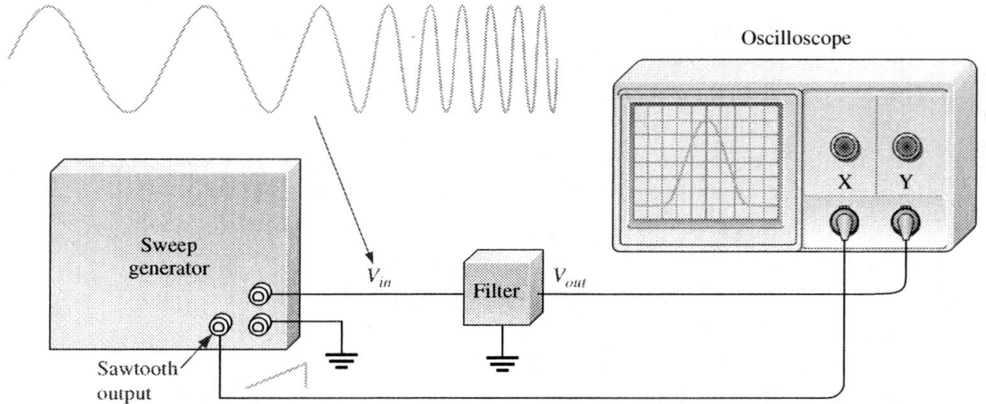

(b) Test setup for a filter response using an oscilloscope. The scope is put in X-Y mode. The sawtooth waveform from the sweep generator drives the X-channel of the oscilloscope.

span/division rather than for the usual *time/division* setting. Therefore, as the input frequency to the filter sweeps through a preselected range, the response is traced out on the screen of the spectrum analyzer. The test setup for using an oscilloscope to display the response is shown in part (b).

Review Questions

26. What are the two test methods discussed in this section?
27. What is the purpose of the two tests?
28. What are one disadvantage and one advantage of each test method?
29. What instruments are used in the swept frequency measurement?
30. In the discrete point measurement, what could prohibit the use of a DMM?

Key Terms

Band-pass filter A type of filter that passes a range of frequencies lying between a certain lower frequency and a certain higher frequency.

Band-stop filter A type of filter that blocks or rejects a range of frequencies lying between a certain lower frequency and a certain higher frequency.

Bandwidth A measure of a filter's passband; the difference between the upper and lower cutoff (critical) frequencies of the passband.

Critical frequency (f_c) The frequency that defines the end of the passband of a filter; also called *cutoff frequency*.

Damping factor (*DF*) A filter characteristic that determines the type of response.

Filter A circuit that passes certain frequencies and attenuates or rejects all other frequencies.

High-pass filter A type of filter that passes frequencies above a certain frequency while rejecting lower frequencies.

Low-pass filter A type of filter that passes frequencies below a certain frequency while rejecting higher frequencies.

Order The number of poles in a filter.

Pole A circuit containing one resistor and one capacitor that contributes −20 dB/decade to a filter's roll-off rate.

Roll-off The rate of decrease in gain below or above the critical frequencies of a filter.

Important Facts

- The bandwidth in a low-pass filter equals the critical frequency because the response extends to 0 Hz.
- The bandwidth in a high-pass filter extends above the critical frequency and is limited only by the inherent frequency limitation of the active circuit.
- A band-pass filter passes all frequencies within a band between a lower and an upper critical frequency and rejects all others outside this band.
- The bandwidth of a band-pass filter is the difference between the upper critical frequency and the lower critical frequency.
- A band-stop filter rejects all frequencies within a specified band and passes all those outside this band.
- Filters with the Butterworth response characteristic have a very flat response in the passband, exhibit a roll-off of −20 dB/decade/pole, and are used when all the frequencies in the passband must have the same gain.
- Filters with the Chebyshev characteristic have ripples or overshoot in the passband and exhibit a faster roll-off per pole than filters with the Butterworth characteristic.
- Filters with the Bessel characteristic are used for filtering pulse waveforms. Their linear phase characteristic results in minimal waveshape distortion. The roll-off rate per pole is slower than for the Butterworth.
- In filter terminology, a single *RC* circuit is called a *pole*.
- Each pole in a Butterworth filter causes the output to roll off at a rate of −20 dB/decade.

- The quality factor Q of a band-pass filter determines the filter's selectivity. The higher the Q, the narrower the bandwidth and the better the selectivity.
- The damping factor determines the filter response characteristic (Butterworth, Chebyshev, or Bessel).

Formulas

Low-pass bandwidth:

$$BW = f_c \tag{8–1}$$

Filter bandwidth of a band-pass filter:

$$BW = f_{c2} - f_{c1} \tag{8–2}$$

Center frequency of a band-pass filter:

$$f_0 = \sqrt{f_{c1} f_{c2}} \tag{8–3}$$

Quality factor of a band-pass filter:

$$Q = \frac{f_0}{BW} \tag{8–4}$$

Damping factor:

$$DF = 2 - \frac{R_1}{R_2} \tag{8–5}$$

Critical frequency for a second-order Sallen-Key filter:

$$f_c = \frac{1}{2\pi\sqrt{R_A R_B C_A C_B}} \tag{8–6}$$

Center frequency for a multiple-feedback band-pass filter:

$$f_0 = \frac{1}{2\pi C}\sqrt{\frac{R_1 + R_2}{R_1 R_2 R_3}} \tag{8–7}$$

Chapter Checkup

Answers are at the end of the chapter.

1. The term *pole* in filter terminology refers to
   (a) a high-gain op-amp (b) one complete active filter
   (c) a single *RC* circuit (d) the feedback circuit
2. A single resistor and a single capacitor can be connected to form a filter with a roll-off rate of
   (a) −20 dB/decade (b) −40 dB/decade
   (c) −6 dB/octave (d) answers (a) and (c)
3. A band-pass response has
   (a) two critical frequencies
   (b) one critical frequency
   (c) a flat curve in the passband
   (d) a wide bandwidth

4. The lowest frequency passed by a low-pass filter is
   (a) 1 Hz   (b) 0 Hz
   (c) 10 Hz   (d) dependent on the critical frequency
5. The Q of a band-pass filter depends on
   (a) the critical frequencies
   (b) only the bandwidth
   (c) the center frequency and the bandwidth
   (d) only the center frequency
6. The damping factor of an active filter determines
   (a) the voltage gain
   (b) the critical frequency
   (c) the response characteristic
   (d) the number of poles
7. A maximally flat frequency response is known as
   (a) Chebyshev   (b) Butterworth
   (c) Bessel   (d) Colpitts
8. The damping factor of a filter is set by
   (a) the negative feedback circuit
   (b) the positive feedback circuit
   (c) the frequency-selective circuit
   (d) the bandwidth of the op-amp
9. The number of poles in a filter affect the
   (a) voltage gain   (b) bandwidth
   (c) center frequency   (d) roll-off rate
10. Sallen-Key filters are always
   (a) single-pole filters   (b) VCVS filters
   (c) Butterworth filters   (d) band-pass filters
11. When filters are cascaded, the roll-off rate
   (a) increases   (b) decreases
   (c) does not change
12. When a low-pass and a high-pass filter are cascaded to get a band-pass filter, the critical frequency of the low-pass filter must be
   (a) equal to the critical frequency of the high-pass filter
   (b) less than the critical frequency of the high-pass filter
   (c) greater than the critical frequency of the high-pass filter
13. A state-variable filter consists of
   (a) one op-amp with multiple-feedback paths
   (b) a summing amplifier and two integrators
   (c) a summing amplifier and two differentiators
   (d) three Butterworth stages
14. When the gain of a filter is minimum at its center frequency, it is a
   (a) band-pass filter   (b) a band-stop filter
   (c) a notch filter   (d) answers (b) and (c)

15. If C_1 and C_2 in Figure 8–25(a) are replaced with 0.15 μF capacitors, the bandwidth will
(a) increase (b) decrease
(c) not change

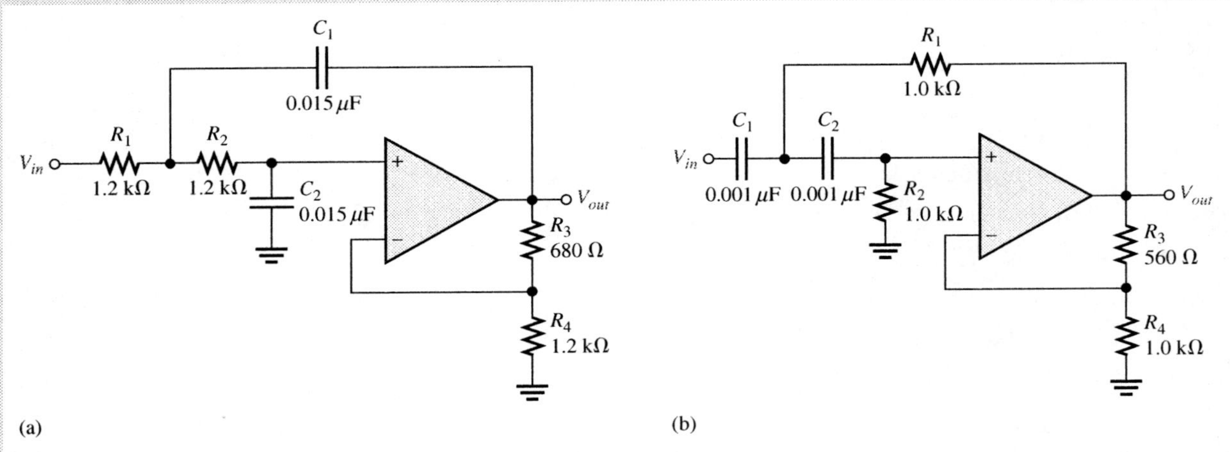

FIGURE 8–25

16. If C_1 and C_2 Figure 8–25(a) are replaced with 0.15 μF capacitors, the number of poles will
(a) increase (b) decrease
(c) not change
17. If C_1 and C_2 Figure 8–25(a) are replaced with 0.15 μF capacitors, the roll-off rate will
(a) increase (b) decrease
(c) not change
18. If C_1 is open in Figure 8–25(b), the ac output for a given ac input will
(a) increase (b) decrease
(c) not change
19. If R_4 is 10 kΩ in Figure 8–25(b), the damping factor will
(a) increase (b) decrease
(c) not change
20. If R_4 is increased to 1.2 kΩ in Figure 8–25(b), the critical frequency will
(a) increase (b) decrease
(c) not change

Questions

Answers to odd-numbered questions are at the end of the book.

1. What are the three regions of a low-pass filter response curve?
2. If the critical frequency of a low-pass filter is 5 kHz, what is the filter bandwidth?
3. If a filter produces an output voltage of 1 V at its critical frequency, what is its maximum output voltage?
4. If a band-pass filter has critical frequencies of 10 kHz and 12 kHz, what is the center frequency?
5. What is the bandwidth of the filter described in Question 4?

6. Which filter has better selectivity, one with a Q of 20 or one with a Q of 15?
7. How does a band-stop filter differ from a band-pass filter?
8. What is the damping factor for a second-order Butterworth response?
9. What determines the damping factor?
10. To achieve better selectivity, would you use two cascaded single-pole filters or two cascaded 2-pole filters?

Basic Problems

PROBLEMS

Answers to odd-numbered problems are at the end of the book.

1. Identify each type of filter response (low-pass, high-pass, band-pass, or band-stop) in Figure 8–26.

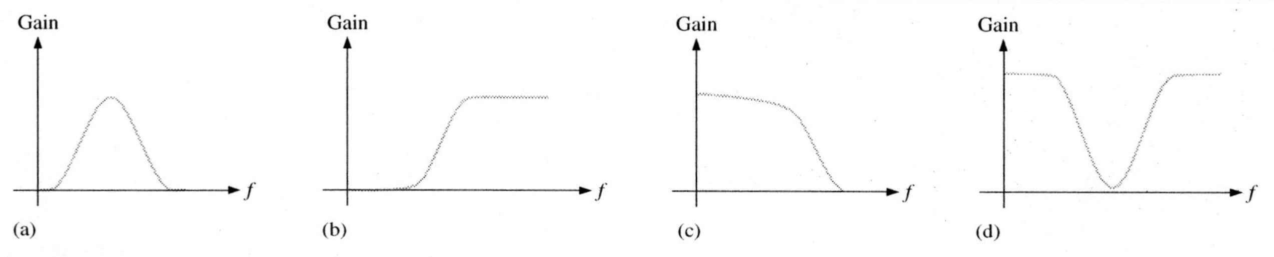

FIGURE 8–26

2. A certain low-pass filter has a critical frequency of 800 Hz. Find the bandwidth.
3. A single-pole high-pass filter has a frequency-selective network with $R = 2.2\ \text{k}\Omega$ and $C = 0.0015\ \mu\text{F}$. Determine the critical frequency.
4. What is the roll-off rate of the filter described in Problem 3?
5. What is the bandwidth of a band-pass filter whose critical frequencies are 3.2 kHz and 3.9 kHz? What is the Q of this filter?
6. Determine the center frequency of a filter with a Q of 15 and a bandwidth of 1.0 kHz.
7. What is the damping factor in each active filter shown in Figure 8–27?

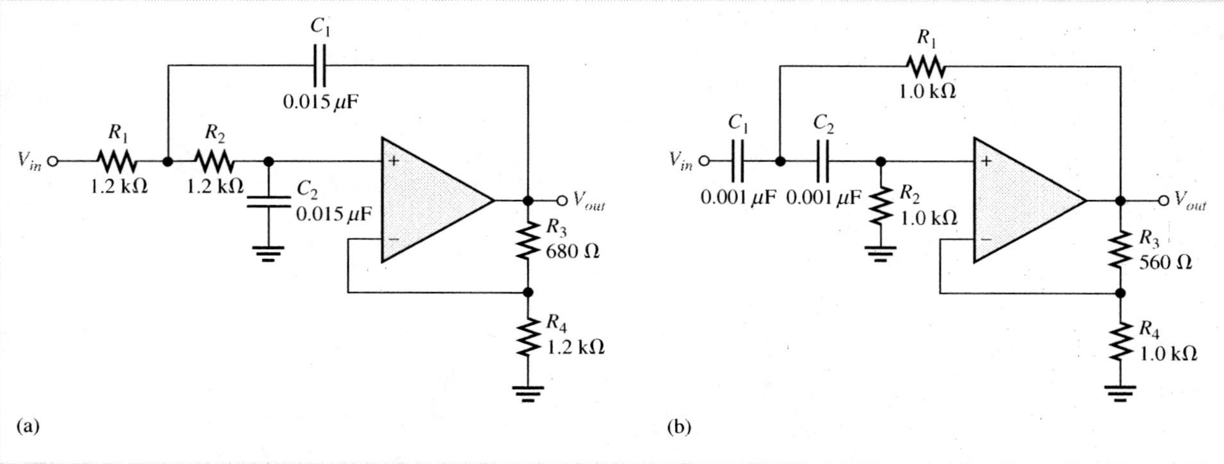

FIGURE 8–27

8. Response curves for second-order filters are shown in Figure 8–28. Identify each as Butterworth or Chebyshev.

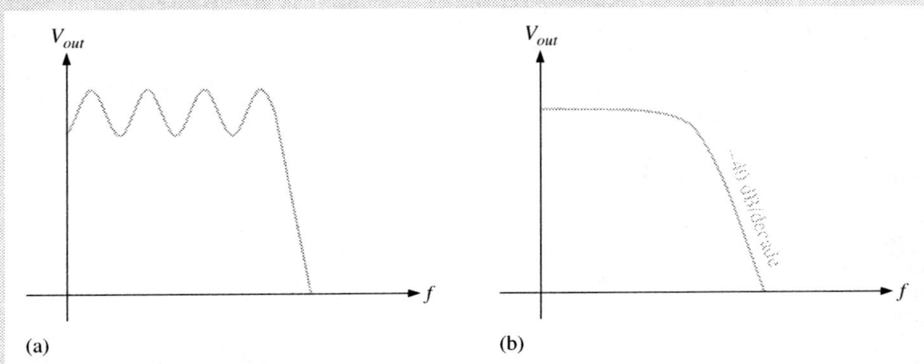

FIGURE 8–28

9. Is the four-pole filter in Figure 8–29 approximately optimized for a Butterworth response? What is the roll-off rate?

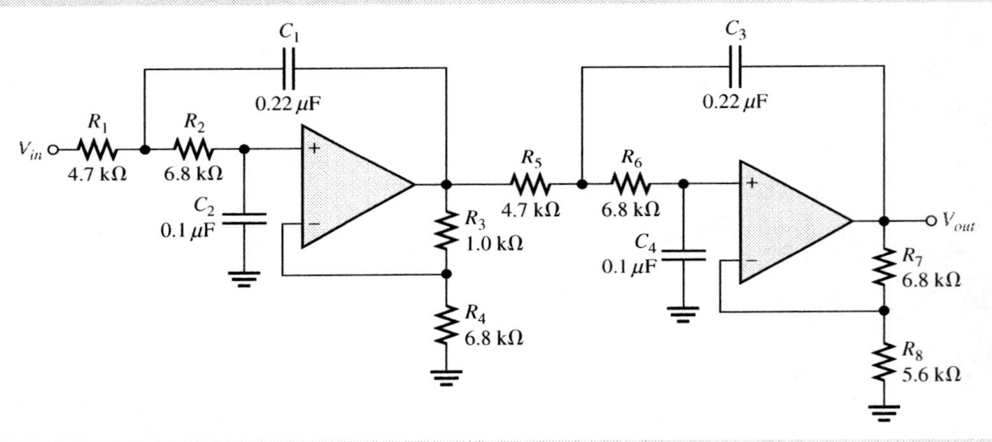

FIGURE 8–29

10. Determine the critical frequency in Figure 8–29.
11. Using a block diagram format, show how to implement the following roll-off rates using single-pole and two-pole low-pass filters with Butterworth responses.
    (a) −40 dB/decade
    (b) −20 dB/decade
    (c) −60 dB/decade
    (d) −100 dB/decade
    (e) −120 dB/decade
12. For the filter in Figure 8–30.
    (a) How would you increase the critical frequency?
    (b) How would you change the filter to a low-pass?

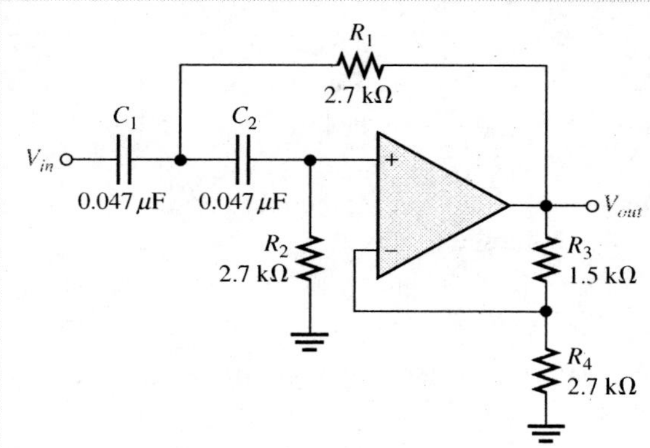

FIGURE 8–30

13. Identify each band-pass filter configuration in Figure 8–31.

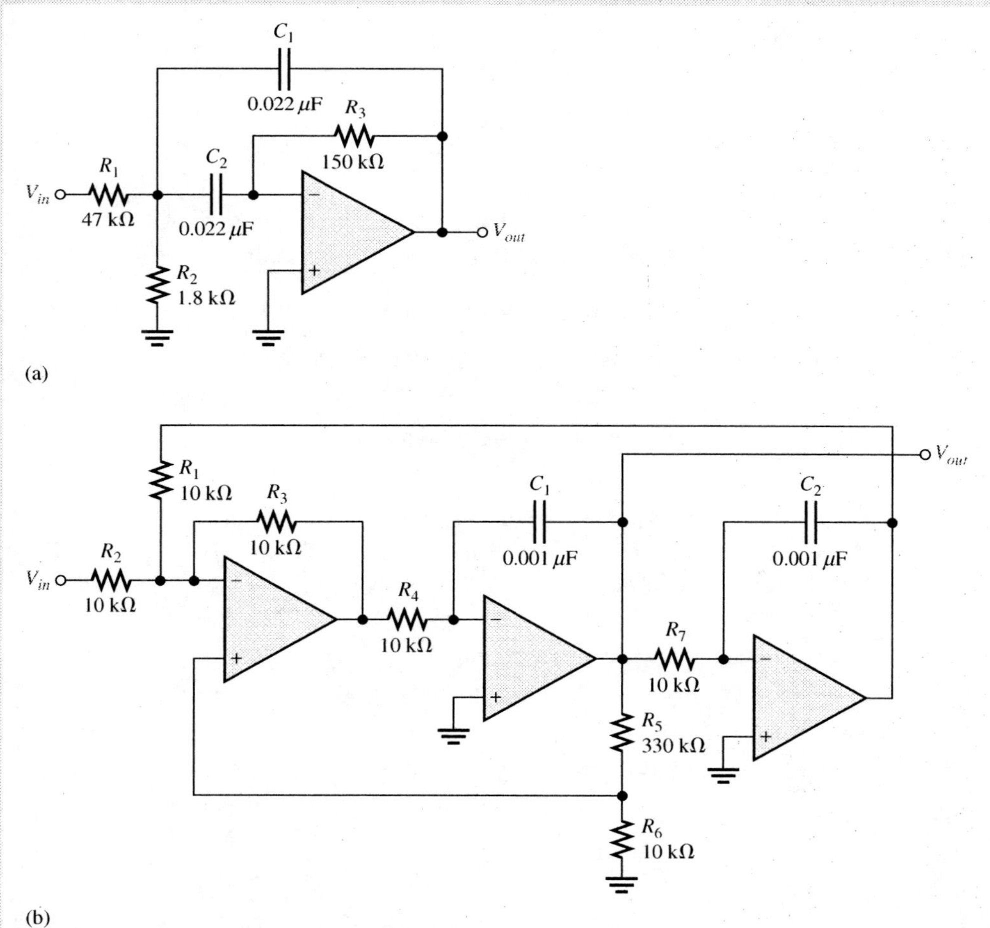

FIGURE 8–31

14. Determine the center frequency and bandwidth for the filter in Figure 8–31(a).

Basic Plus Problems

15. For the filter in Figure 8–27 that does not have a Butterworth response, specify the changes necessary to convert it to a Butterworth response. (Use nearest standard values.)
16. Without changing the response curve, adjust the component values in the filter of Figure 8–29 to make it an equal-value filter.
17. Modify the filter in Figure 8–29 to increase the roll-off rate to −120 dB/decade while maintaining an approximate Butterworth response.
18. Convert the equal-value filter from Problem 16 to a high-pass with the same critical frequency and response characteristic.
19. Make the necessary circuit modification to reduce by half the critical frequency in Problem 18.

ANSWERS

Example Questions

8–1: 500 Hz

8–2: 1.44

8–3: 7.23 kHz

8–4: $C_{A1} = C_{A2} = C_{B1} = C_{B2} = 0.234\ \mu F$

8–5: $R_A = R_B = R_2 = 10\ k\Omega$, $C_A = C_B = 0.053\ \mu F$, $R_1 = 5.86\ k\Omega$

8–6: Decreases

Review Questions

1. The critical frequency determines the bandwidth.
2. The inherent frequency limitation of the op-amp limits the bandwidth.
3. Q and BW are inversely related.
4. The higher the Q, the better the selectivity, and vice versa.
5. Below
6. Butterworth is very flat in the passband and has a −20 dB/decade/pole roll-off. Chebyshev has ripples in the passband and has greater than −20 dB/decade/pole roll-off. Bessel has a linear phase characteristic and less than −20 dB/decade/pole roll-off.
7. The damping factor determines the response characteristic.
8. Frequency-selective gain element (amplifier), and negative feedback circuit are the parts of an active filter.
9. The values of R and C determine f_c.
10. The number of poles determines the roll-off.
11. A second-order filter has two poles.
12. The damping factor sets the response characteristic.
13. Cascading increases the roll-off rate.
14. Two capacitors
15. −40 dB/decade

16. The positions of the *R*s and *C*s in the frequency-selection circuit are opposite for low-pass and high-pass configurations.
17. Decrease the *R* values to increase f_c.
18. −140 dB/decade
19. The op-amp's internal *RC* circuits
20. A stage is one filter in a cascaded arrangement.
21. *Q* determines selectivity.
22. $Q = 25$. Higher *Q* gives narrower *BW*.
23. A summing amplifier and two integrators make up a state-variable filter.
24. A resistor and a capacitor
25. The low-pass filter
26. Discrete point measurement and swept frequency measurement
27. To check the frequency response of a filter
28. Discrete point measurement—tedious and less complete; simpler equipment. Swept frequency measurement—uses more expensive equipment; more efficient, can be more accurate and complete.
29. Sweep generator and oscilloscope
30. The DMM's frequency response

Chapter Checkup

1. (c)	2. (d)	3. (a)	4. (b)	5. (c)
6. (c)	7. (b)	8. (a)	9. (d)	10. (b)
11. (a)	12. (c)	13. (b)	14. (d)	15. (b)
16. (c)	17. (c)	18. (b)	19. (d)	20. (c)

CHAPTER

SPECIAL-PURPOSE AMPLIFIERS

CHAPTER OUTLINE

Study aids for this chapter are available at

http://www.prenhall.com/SOE

INTRODUCTION

A general-purpose op-amp, such as the 741, is an extremely versatile and widely used device. However, many specialized IC amplifiers have been designed with certain types of applications in mind or with certain special features or characteristics. Most of these devices are actually derived from the basic op-amp. These special amplifiers include the instrumentation amplifier (IA) that is used in high-noise environments, the isolation amplifier that is used in high-voltage and medical applications, and the operational transconductance amplifier (OTA) that is used as a voltage-to-current amplifier. In this chapter, you will learn about each of these devices and some of their basic applications.

KEY TERMS

- Instrumentation amplifier
- Isolation amplifier
- Operational transconductance amplifier
- Transconductance

KEY OBJECTIVES

A section number is given for each objective. After completing this chapter, you should be able to

9–1 Explain how instrumentation amplifiers work and how they are applied

9–2 Explain how one type of isolation amplifier works and how it is applied

9–3 Explain how operational transconductance amplifiers work and how they are applied

LABORATORY EXPERIMENTS DIRECTORY

The following exercise is for this chapter.

- **Experiment 18**
  The Instrumentation Amplifier

ON THE JOB. . .

(Getty Images)

All valuable employees are reliable. It is important to arrive at your workplace on time and be ready to work. If there is a valid reason you can't be at work or will be late, call your supervisor and explain the situation. Most supervisors will appreciate knowing about a problem before you are absent or late rather than afterward. Honesty is a key factor in establishing a positive relationship between you and your employer. If an employer finds that you are honest, he/she will be more willing to entrust you with more responsibilities. Loyalty is also vital because it lets the employer know that you can be relied upon.

The goal of research in the field of molecular electronics is to create electronic devices from single molecules joined together. Recent developments include an organic molecule transistor and a technique for accurately measuring the flow of electrons through a single molecule.

To build the transistor, a team at Bell Labs allowed thousands of organic molecules to assemble themselves onto a gold film similar to bristles on a brush. By adding another layer of gold on top and applying an electric field with a silicon electrode, a transistor with a channel just one molecule wide was created.

The challenges ahead are to determine which shape of molecule makes the best transistors and to find ways to scale the devices down. Also, researchers are trying to find which molecules make the best wires because transistors must be interconnected with wires to form circuits. One method is to attach a gold "tip" to both ends of a small carbon chain.

9–1 INSTRUMENTATION AMPLIFIERS

Instrumentation amplifiers (IAs) are commonly used in environments with high common-mode noise such as in data acquisition systems where remote sensing of input variables is required.

In this section, you will learn how instrumentation amplifiers work and how they are applied.

One of the most common problems in measuring systems is the contamination of the signal from a transducer with unwanted noise (such as 60 Hz power line interference). The transducer signal is typically a small differential signal carrying the desired information. Noise that is added to both signal conductors in the same amount is called a common-mode noise. Ideally, the differential signal should be amplified and the common-mode noise should be rejected.

A second problem for measuring systems is that many transducers have a high output resistance and can easily be loaded down when connected to an amplifier. An amplifier for small transducer signals needs to have a very high input resistance to avoid this loading effect.

The solution to these measurement problems is the instrumentation amplifier (IA), a specially designed differential amplifier with ultra-high input resistance and extremely good common-mode rejection as well as being able to achieve high, stable gains. Instrumentation amplifiers can faithfully amplify low-level signals in the presence of high common-mode noise. They are used in a variety of signal-processing applications where accuracy is important and where low drift, low bias currents, precise gain, and very high CMRR are required. CMRR is often specified in decibels (dB), which is designated as CMRR′. IAs are available with CMRR′s of up to 130 dB.

An **instrumentation amplifier** is a differential voltage-gain device that amplifies the difference between the voltages existing at its two input terminals. The main purpose of an instrumentation amplifier is to amplify small signals that are riding on large common-mode voltages. The key characteristics are high input resistance, high common-mode rejection, low output offset, and low output resistance.

Figure 9–1 shows one way to implement a basic instrumentation amplifier (IA). Op-amps A1 and A2 are voltage-followers. The voltage-followers provide high input resistance with a gain of 1. Op-amp A3 is a differential amplifier that amplifies the difference between V_{out1} and V_{out2}. Although this circuit has the advantage of high input resistance, it requires extremely high precision matching of the gain resistors to achieve a high CMRR (R_1 must

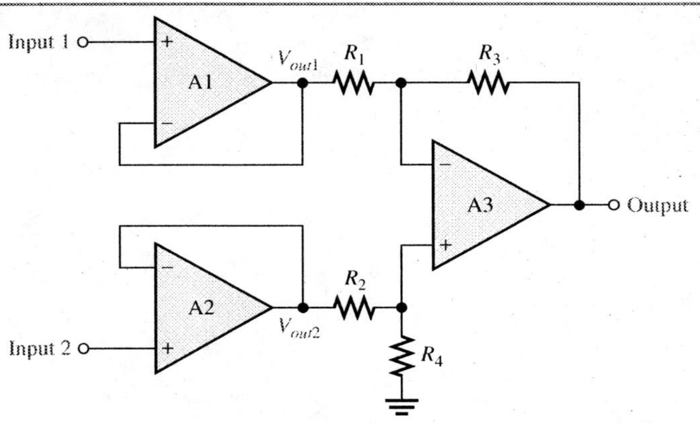

FIGURE 9–1

The basic instrumentation amplifier using three op-amps.

match R_2 and R_3 must match R_4). Further, two resistors must be changed if variable gain is desired (typically R_1 and R_2), and they must track each other with high precision over the operating temperature range.

A simple change that solves the difficulties of the circuit in Figure 9–1 and provides high gain is the IA shown in Figure 9–2. The inputs are buffered by op-amp A1 and op-amp A2, providing a very high input resistance. Resistors R_5 and R_6 are added to make op-amps A1 and A2 noninverting amplifiers with gain rather than voltage-followers. The gain is adjustable using an external resistor R_G. The entire assembly (except for R_G) is contained in a single IC, as indicated in the gray block in Figure 9–2. In this design, the common-mode gain still depends on very precisely matched resistors. However, these resistors can be critically matched during manufacture (by laser trimming) within the IC. Resistors R_1, R_2, R_3, and R_4 are generally set by the manufacturer for a gain of 1.0 for the differential amplifier. Resistors R_5 and R_6 are precision-matched resistors set equal to each other. The overall differential gain can be controlled by the value of R_G, connected as shown. If R_G is not connected, A1 and A2 become voltage-followers and the overall gain will be 1. In this event, R_5 and R_6 have no effect because there is no current in the feedback path. The equation for determining the output voltage is

$$V_{out} = \left(1 + \frac{2R}{R_G}\right)(V_{in2} - V_{in1}) \tag{9–1}$$

where the closed-loop gain is

$$A_{cl} = 1 + \frac{2R}{R_G}$$

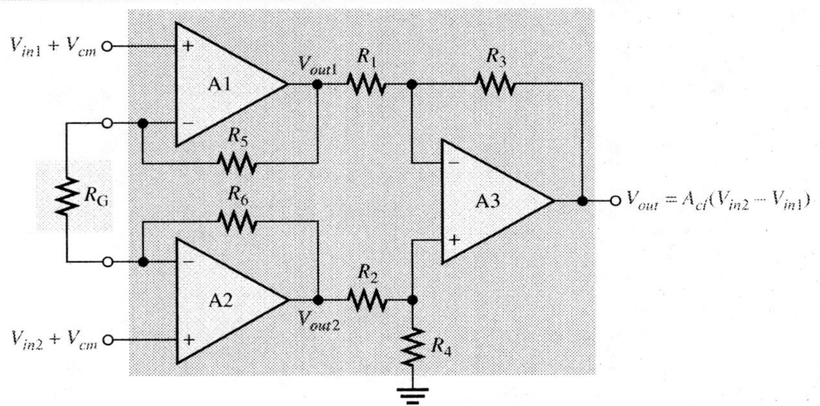

FIGURE 9–2

The instrumentation amplifier with the external gain-setting resistor R_G. Differential and common-mode signals are indicated.

and where $R_5 = R_6 = R$. The last equation shows that the gain of the instrumentation amplifier can be set by the value of the external resistor R_G when R_5 and R_6 have known fixed values.

The external gain-setting resistor R_G can be calculated for a desired voltage gain by using the following formula:

$$R_G = \frac{2R}{A_{cl} - 1} \quad (9\text{–}2)$$

Instrumentation amplifiers in which the gain is set to specific values using a binary input instead of a resistor are also available.

EXAMPLE 9–1

Problem

Determine the value of the external gain-setting resistor R_G for the IA in Figure 9–2 with $R_5 = R_6 = 25\text{ k}\Omega$. The voltage gain is to be 500.

Solution

$$R_G = \frac{2R}{A_{cl} - 1} = \frac{50\text{ k}\Omega}{500 - 1} \cong \mathbf{100\ \Omega}$$

Question*

What value of external gain-setting resistor is required for an instrumentation amplifier with $R_5 = R_6 = 39\text{ k}\Omega$ to produce a gain of 325?

Applications

The instrumentation amplifier is normally used to measure small differential signal voltages that are superimposed on a common-mode noise voltage often much larger than the signal voltage. Applications include situations where a quantity is sensed by a remote device, such as a temperature- or pressure-sensitive transducer, and the resulting small electrical signal is sent over a long line subject to electrical noise that produces common-mode voltages in the line. The instrumentation amplifier at the end of the line must amplify the small signal from the remote sensor and reject the large common-mode voltage. Figure 9–3 illustrates this.

FIGURE 9–3

Illustration of the rejection of large common-mode voltages and the amplification of smaller signal voltages by an instrumentation amplifier.

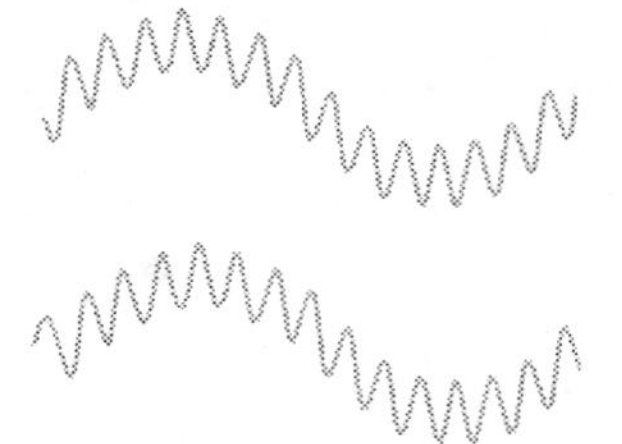

Small differential high-frequency signal riding on a larger low-frequency common-mode signal

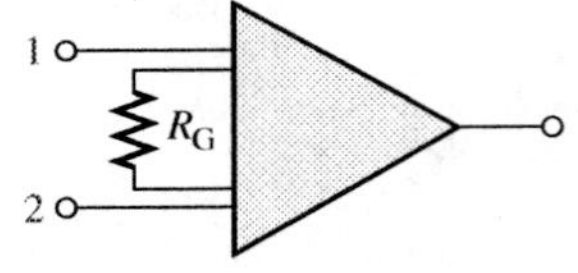

Instrumentation amplifier

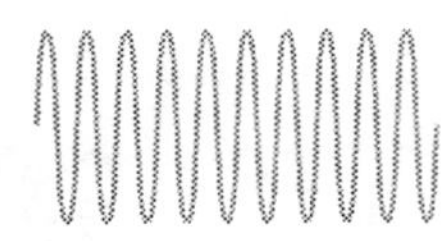

Amplified differential signal. No common-mode signal.

* *Answers are at the end of the chapter.*

A Specific Instrumentation Amplifier

Now that you have the basic idea of how an instrumentation amplifier works, let's look at a specific device. A representative device, the AD622, is shown in Figure 9–4. An IC package pin diagram is shown for reference. This instrumentation amplifier is based on the classic design using three op-amps as previously discussed.

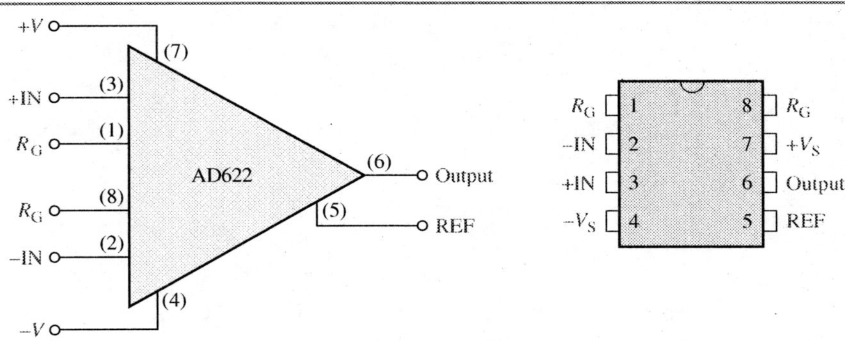

FIGURE 9–4

The AD622 instrumentation amplifier.

Some of the features of the AD622 are as follows. The voltage gain can be adjusted from 2 to 1000 with an external resistor R_G. There is unity gain with no external resistor. The input resistance is 10 GΩ. The common-mode rejection ratio (CMRR′) has a minimum value of 66 dB. Recall that a higher CMRR means better rejection of common-mode voltages. The AD622 has a bandwidth of 800 kHz at a gain of 10 and a slew rate of 1.2 V/μs.

Setting the Voltage Gain

For the AD622, an external resistor must be used to achieve a voltage gain greater than unity, as indicated in Figure 9–5. Resistor R_G is connected between the R_G terminals (pins 1 and 8). No resistor is required for unity gain. R_G is selected for the desired gain based on the following formula:

$$R_G = \frac{50.5\text{ k}\Omega}{A_v - 1} \qquad (9\text{–}3)$$

Notice that this formula is the same as Equation 9–2 for the three-op-amp configuration shown in Figure 9–2 where the internal resistors R_5 and R_6 are each 25.25 kΩ.

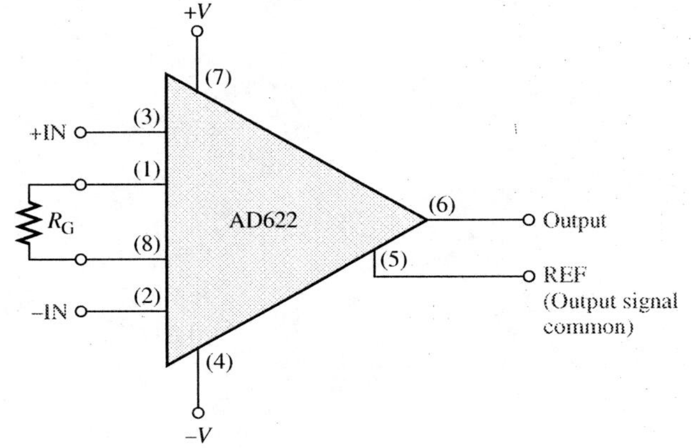

FIGURE 9–5

The AD622 with a gain-setting resistor.

Gain versus Frequency

The graph in Figure 9–6 shows how the gain varies with frequency for gains of 1, 10, 100, and 1000. As you can see, the bandwidth decreases as the gain increases. For example, for a gain of 1, the bandwidth is approximately 900 kHz, and for a gain of 1000, the bandwidth is approximately 3 kHz.

FIGURE 9–6

Gain versus frequency for the AD622 instrumentation amplifier.

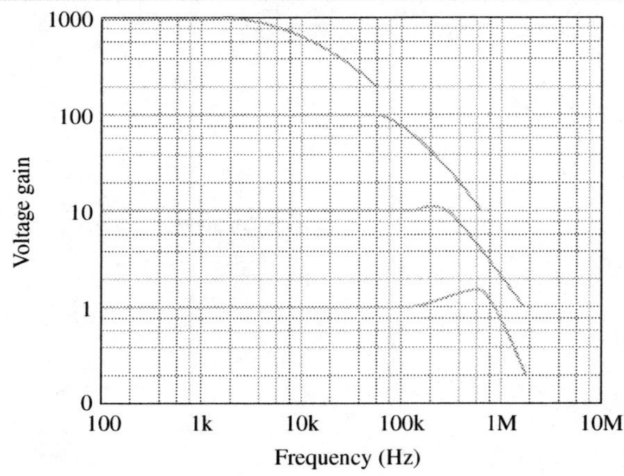

EXAMPLE 9–2

Problem

Calculate the gain and determine the approximate bandwidth using the graph in Figure 9–6 for the instrumentation amplifier in Figure 9–7.

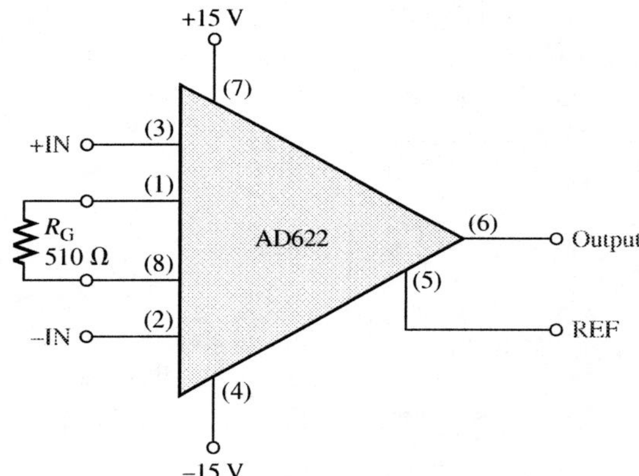

FIGURE 9–7

Solution

Determine the voltage gain as follows:

$$R_G = \frac{50.5\ \text{k}\Omega}{A_v - 1}$$

$$A_v - 1 = \frac{50.5\ \text{k}\Omega}{R_G}$$

$$A_v = \frac{50.5\ \text{k}\Omega}{510\ \Omega} + 1 = \mathbf{100}$$

The approximate bandwidth is determined from the graph.

$$BW \cong \mathbf{60\ kHz}$$

Question

How can you modify the circuit in Figure 9–7 for a gain of approximately 45?

Review Questions

Answers are at the end of the chapter.

1. What is the main purpose of an instrumentation amplifier and what are three of its key characteristics?
2. What components do you need to construct a basic instrumentation amplifier?
3. How is the gain determined in a basic instrumentation amplifier?
4. What value of R_G is required for an AD622 to achieve a voltage gain of 10?
5. In a certain AD622 configuration, $R_G = 10\ \text{k}\Omega$. What is the voltage gain?

ISOLATION AMPLIFIERS 9–2

The isolation amplifier is used for the protection of human life or sensitive equipment in those applications where hazardous power-line leakage or high-voltage transients are possible. The principal areas of application for isolation amplifiers are in medical instrumentation, power plant instrumentation, industrial processing, and automated testing.

In this section, you will learn how one type of isolation amplifier works and how it is applied.

The Basic Isolation Amplifier

An **isolation amplifier** is a device that provides dc isolation between the input and output. In some ways, the isolation amplifier can be viewed as an elaborate op-amp or instrumentation amplifier. The difference is that the isolation amplifier has an input stage, an output stage, and a power supply section that are all electrically isolated from each other. Many isolation amplifiers use optical coupling or capacitive coupling to achieve isolation. However, we will focus on a transformer-coupled device in this section which achieves the same results as other types of coupling. The circuit is in IC form, but the miniature, multiple-winding, toroid transformer is not fully integrated. This results in a package configuration that deviates somewhat from standard IC packages but is still designed for printed circuit board assembly.

A typical isolation amplifier is capable of operating with three independent grounds (indicated by three different symbols). As shown in Figure 9–8, the input and output stages

Diagram of a transformer-coupled isolation amplifier. FIGURE 9–8

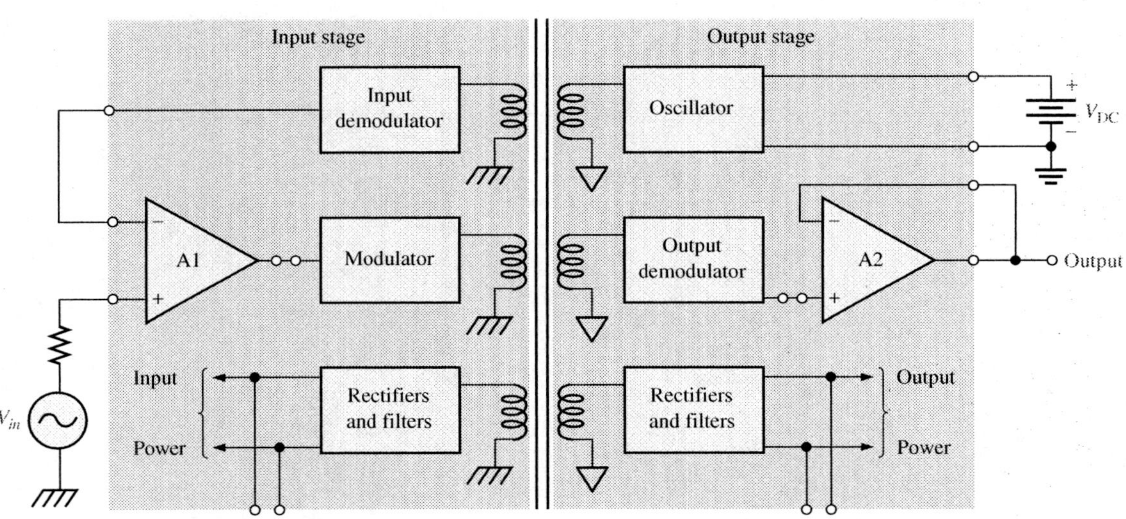

and the power supplies are transformer coupled to achieve isolation of the input and output signals and the power.

The input stage contains an op-amp, a demodulator, a modulator, and a dual-polarity power supply. The output stage contains an op-amp, an oscillator, a demodulator, and a dual-polarity power supply.

General Operation

The power to the input and output stages is produced as follows. An external dc supply voltage, V_{DC}, is applied to the oscillator (oscillators are covered in the next chapter). The oscillator converts the dc power to ac at a relatively high frequency. The ac output of the oscillator is coupled by the transformer to the input power supply (rectifiers and filters) and to the output power supply (rectifiers and filters) where it is rectified and filtered to produce dual-polarity (positive and negative) dc voltages for the input and output stages.

The oscillator output is also coupled to the modulator where it is combined with the input signal from the input op-amp A1. The modulator varies the amplitude of the relatively high oscillator frequency with the lower-frequency input signal. The higher modulated frequency allows a very small transformer to be used. To couple the lower-frequency input signal without modulation would require a prohibitively large transformer.

The modulated signal is coupled to the demodulator in the output stage. The demodulator recovers the original input signal from the higher oscillator frequency. The demodulated input signal is then applied to op-amp A2. The demodulator in the input stage is part of a feedback loop that forces the signal at the inverting input of A1 to equal the original input signal at the noninverting input.

Although the isolation amplifier is a fairly complex circuit, in terms of its overall function, it is still simply an amplifier. You apply a dc voltage, put a signal in, and you get an amplified signal out. The isolation function is an unseen process.

Applications

The isolation amplifier is used in applications requiring no common grounds between a transducer and the processing circuits where interfacing to sensitive equipment is required. In chemical, nuclear, and metal-processing industries, for example, millivolt signals typically exist in the presence of large common-mode voltages that can be in the kilovolt range. In this type of environment, the isolation amplifier can amplify small signals from noisy equipment and provide a safe output to sensitive equipment such as computers.

Another important application is in various types of medical equipment. In medical applications where body functions such as heart rate and blood pressure are monitored, the very small monitored signals are combined with large common-mode signals, such as 60 Hz power-line pickup from the skin. In these situations, without isolation, dc leakage or equipment failure could be fatal. Figure 9–9 shows a simplified diagram of an

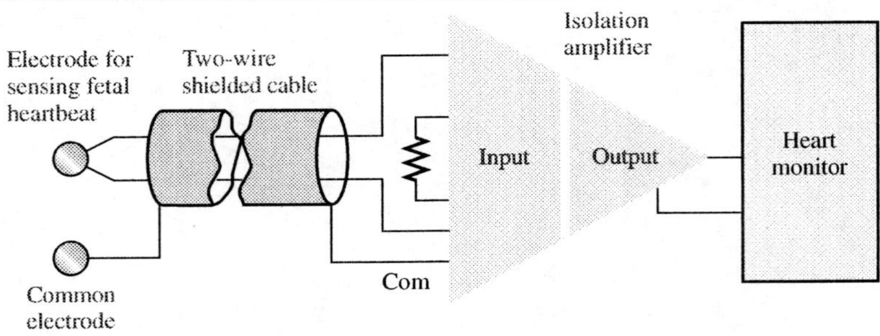

FIGURE 9–9

Fetal heartbeat monitoring using an isolation amplifier. The triangle with a "split" down the middle is one way of representing an isolation amplifier by indicating that the input and output stages are separated by transformer coupling.

isolation amplifier in a cardiac-monitoring application. In this situation, heart signals, which are very small, are combined with much larger common-mode signals caused by muscle noise, electrochemical noise, residual electrode voltage, and 60 Hz line pickup from the skin.

The monitoring of a fetal heartbeat, as illustrated, is the most demanding type of cardiac monitoring because in addition to the fetal heartbeat that typically generates 50 μV, there is also the mother's heartbeat that typically generates 1 mV. The common-mode voltages can run from about 1 mV to about 100 mV. The CMR (common-mode rejection) of the isolation amplifier separates the signal of the fetal heartbeat from that of the mother's heartbeat and from other common-mode signals. So, the signal from the fetal heartbeat is essentially all that the amplifier sends to the monitoring equipment.

SAFETY NOTE

Never wear rings or any type of metallic jewelry while working on a circuit. These items may accidentally come in contact with the circuit, causing shock and/or damage to the circuit.

A Specific Isolation Amplifier

Let's look at a representative device, the Burr-Brown 3656KG. The voltage gains of both the input stage and the output stage can be set with external resistors, as shown in Figure 9–10. The gain of the input stage is

$$A_{v1} = \frac{R_{f1}}{R_{i1}} + 1$$

The gain of the output stage is

$$A_{v2} = \frac{R_{f2}}{R_{i2}} + 1$$

The total amplifier gain is the product of the gains of the input and output stages.

$$A_{v(tot)} = A_{v1}A_{v2}$$

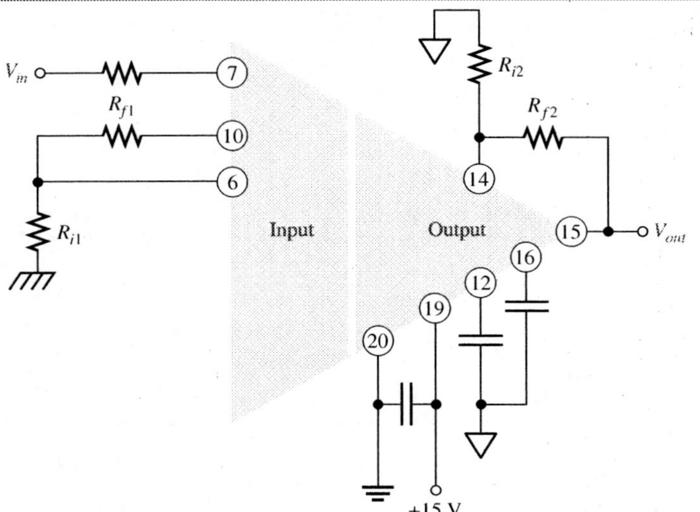

FIGURE 9–10

The 3656KG isolation amplifier.

EXAMPLE 9–3

Problem

Determine the total voltage gain of the 3656KG isolation amplifier in Figure 9–11.

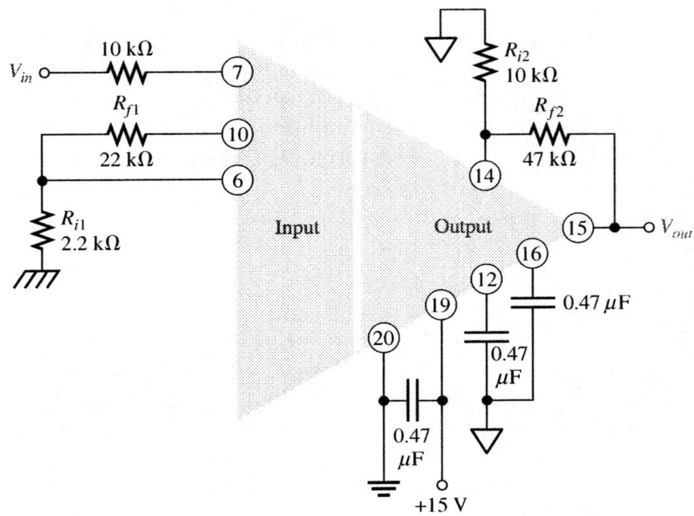

FIGURE 9–11

Solution

The gain of the input stage is

$$A_{v1} = \frac{R_{f1}}{R_{i1}} + 1 = \frac{22\text{ k}\Omega}{2.2\text{ k}\Omega} + 1 = 10 + 1 = 11$$

The gain of the output stage is

$$A_{v2} = \frac{R_{f2}}{R_{i2}} + 1 = \frac{47\text{ k}\Omega}{10\text{ k}\Omega} + 1 = 4.7 + 1 = 5.7$$

The total gain of the isolation amplifier is

$$A_{v(tot)} = A_{v1}A_{v2} = (11)(5.7) = \mathbf{62.7}$$

Question

What resistor values in Figure 9–11 will produce a total gain of approximately 100?

Review Questions

6. In what types of applications are isolation amplifiers used?
7. What are the two stages in a typical isolation amplifier?
8. How are the stages in an isolation amplifier connected?
9. What is the purpose of the oscillator in an isolation amplifier?
10. How is the total gain of an isolation amplifier set?

OPERATIONAL TRANSCONDUCTANCE AMPLIFIERS (OTAs) 9–3

Operational transconductance amplifiers (OTAs) are basically voltage-to-current amplifiers. Their applications include amplitude modulation and trigger circuits.

In this section, you will learn how OTAs work and how they are applied.

Conventional op-amps are, as you know, primarily voltage amplifiers in which the output voltage equals the gain times the input voltage. The **operational transconductance amplifier** (OTA) is a voltage-to-current amplifier in which the output current equals the gain times the input voltage.

Figure 9–12 shows the symbol for an OTA. The double circle symbol at the output represents an output current source that is dependent on a bias current. Like the conventional op-amp, the OTA has two differential input terminals, a high input resistance, and a high CMRR. Unlike the conventional op-amp, the OTA has a bias-current input terminal, a high output resistance, and no fixed open-loop voltage gain.

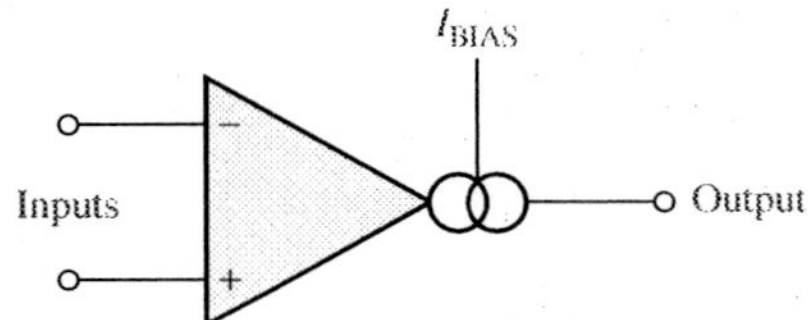

FIGURE 9–12

Symbol for an operational transconductance amplifier (OTA).

The Transconductance Is the Gain of an OTA

In general, the **transconductance** of an electronic device is the ratio of the output current to the input voltage. For an OTA, voltage is the input variable and current is the output variable; therefore, the ratio of output current to input voltage is its gain. Consequently, the voltage-to-current gain of an OTA is the transconductance, g_m.

$$g_m = \frac{I_{out}}{V_{in}}$$

In an OTA, the transconductance is dependent on a constant (K) times the bias current (I_{BIAS}) as indicated in Equation 9–4. The value of the constant is determined by the internal circuit design and is given for a particular device.

$$g_m = KI_{BIAS} \tag{9–4}$$

The output current is controlled by the input voltage and the bias current is

$$I_{out} = g_m V_{in} = KI_{BIAS}V_{in}$$

EXAMPLE 9–4

Problem

If an OTA has a $g_m = 1000\ \mu S$, what is the output current when the input voltage is 50 mV?

Solution

$$I_{out} = g_m V_{in} = (1000\ \mu S)(50\ mV) = \mathbf{50\ \mu A}$$

Question

Given that $K = 16\ \mu S/\mu A$, what is the bias current required to produce $g_m = 1000\ \mu S$?

Basic OTA Circuits

Figure 9–13 shows the OTA used as an inverting amplifier with fixed-voltage gain. The voltage gain is set by the transconductance and the load resistance as follows.

$$V_{out} = I_{out}R_L$$

Dividing both sides by V_{in},

$$\frac{V_{out}}{V_{in}} = \left(\frac{I_{out}}{V_{in}}\right)R_L$$

Since V_{out}/V_{in} is the voltage gain and $I_{out}/V_{in} = g_m$,

$$A_v = g_mR_L$$

The transconductance of the amplifier in Figure 9–13 is determined by the amount of bias current, which is set by the dc supply voltages and the bias resistor R_{BIAS}.

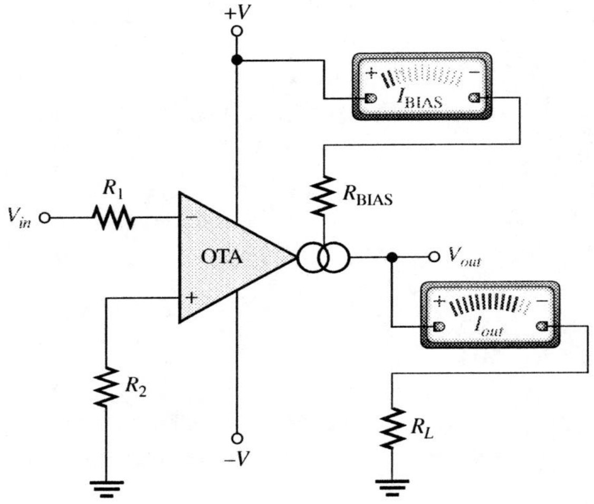

FIGURE 9–13

An OTA as an inverting amplifier with a fixed-voltage gain.

One of the most useful features of an OTA is that the voltage gain can be controlled by the amount of bias current. This can be done manually, as shown in Figure 9–14(a), by using a variable resistor in series with R_{BIAS} in the circuit of Figure 9–13. By changing the resistance, you can produce a change in I_{BIAS}, which changes the transconductance. A change in the transconductance changes the voltage gain. The voltage gain can also be controlled with an externally applied variable voltage as shown in Figure 9–14(b). A variation in the applied bias voltage causes a change in the bias current.

A Specific OTA

The LM13700 is a typical OTA and serves as a representative device. The LM13700 is a dual-device package containing two OTAs and buffer circuits. Figure 9–15 shows the pin configuration (numbers in parentheses) for a single OTA in the package. The maximum

FIGURE 9–14

An OTA as an inverting amplifier with a variable-voltage gain.

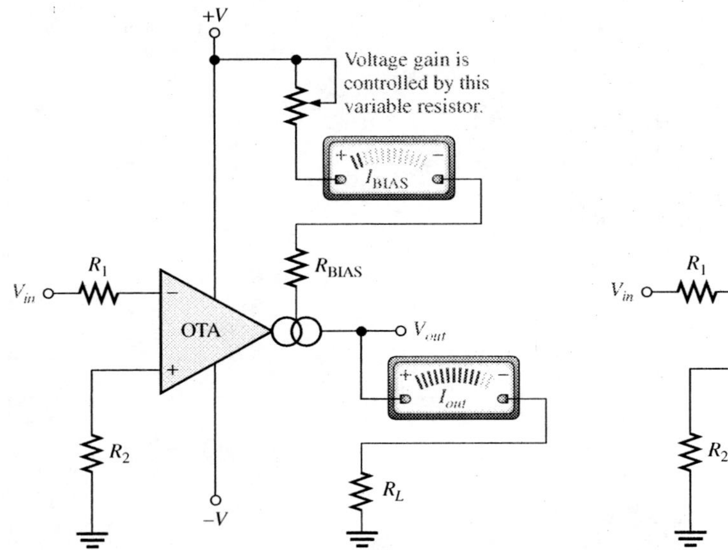

(a) Amplifier with resistance-controlled gain

(b) Amplifier with voltage-controlled gain

FIGURE 9–15

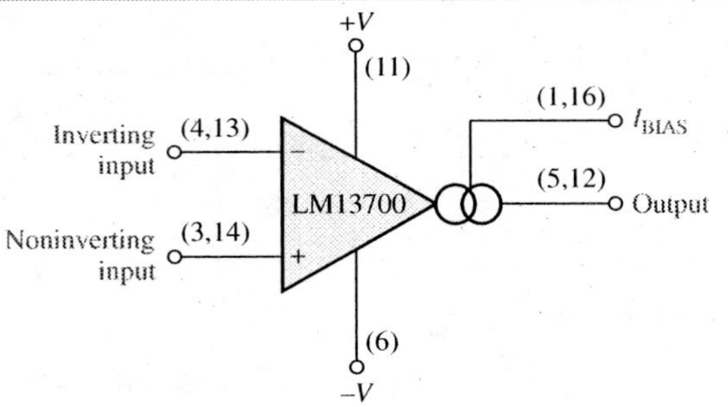

An LM13700 OTA. There are two in an IC package. The buffer transistors are not shown. Pin numbers for both OTAs are given in parentheses.

dc supply voltages are ±18 V. For an LM13700, the bias current is determined by the following formula:

$$I_{BIAS} = \frac{+V_{BIAS} - (-V) - 1.4\text{ V}}{R_{BIAS}} \tag{9–5}$$

The 1.4 V is due to the internal circuit where a base-emitter junction and a diode connect the external R_{BIAS} with the negative supply voltage ($-V$). The positive bias voltage may be obtained from the positive supply voltage.

Not only does the transconductance of an OTA vary with bias current, but so does the input and output resistances. Both the input and output resistances decrease as the bias current increases.

EXAMPLE 9–5

Problem

The OTA in Figure 9–16 is connected as an inverting fixed-gain amplifier. Determine the voltage gain. Assume $K = 16\ \mu S/\mu A$.

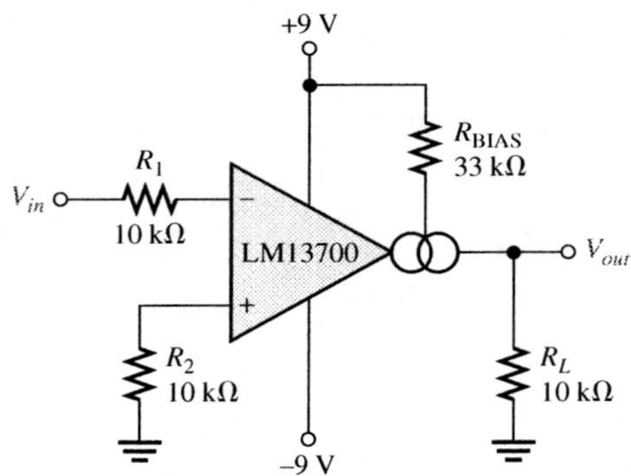

FIGURE 9–16

Solution

Calculate the bias current as follows:

$$I_{BIAS} = \frac{+V_{BIAS} - (-V) - 1.4\text{ V}}{R_{BIAS}} = \frac{9\text{ V} - (-9\text{ V}) - 1.4\text{ V}}{33\text{ k}\Omega} = 503\ \mu\text{A}$$

Since $K = 16\ \mu S/\mu A$, the transconductance corresponding to $I_{BIAS} = 503\ \mu A$ is

$$g_m = KI_{BIAS} = (16\ \mu\text{S}/\mu\text{A})(503\ \mu\text{A}) = 8.05 \times 10^3\ \mu\text{S}$$

Using this value of g_m, calculate the voltage gain.

$$A_v = g_m R_L = (8.05 \times 10^3\ \mu\text{S})(10\text{ k}\Omega) = \mathbf{80.5}$$

Question

If the OTA in Figure 9–16 is operated with dc supply voltages of ± 12 V, will this change the voltage gain and, if so, to what value?

Applications

Amplitude Modulator

Figure 9–17 illustrates an OTA connected as an amplitude modulator. The voltage gain is varied by applying a modulation voltage to the bias input. When a constant-amplitude input signal is applied, the amplitude of the output signal will vary according to the modulation voltage on the bias input. The gain is dependent on bias current, and bias current is related to the modulation voltage by the following formula:

$$I_{BIAS} = \frac{V_{mod} - (-V) - 1.4\text{ V}}{R_{BIAS}}$$

This modulating action is shown in Figure 9–17 for a higher frequency sinusoidal input voltage and a lower frequency sinusoidal modulating voltage.

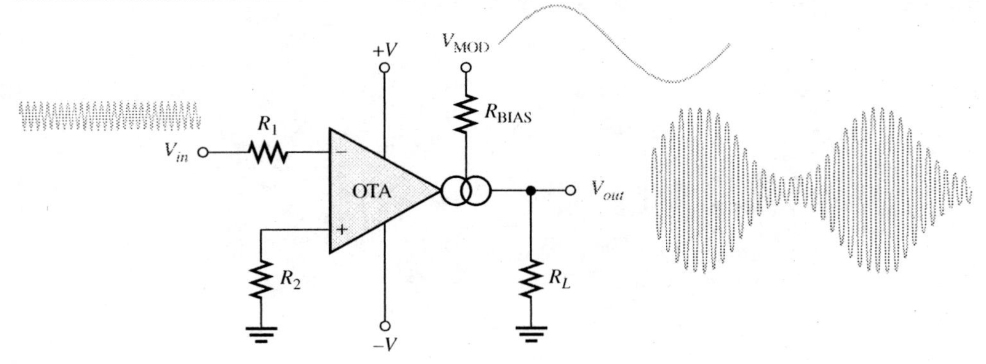

FIGURE 9–17

The OTA as an amplitude modulator.

EXAMPLE 9–6

Problem

The input to the OTA amplitude modulator in Figure 9–18 is a 50 mV peak-to-peak, 1 MHz sine wave. Determine the output signal, given the modulation voltage shown is applied to the bias input. Assume $K = 16\ \mu S/\mu A$.

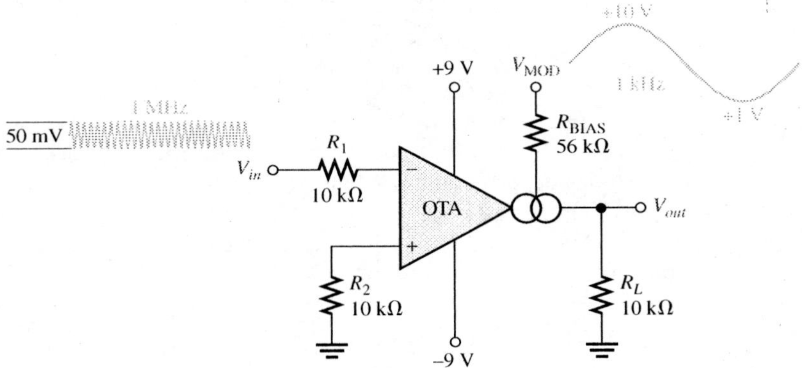

FIGURE 9–18

Solution

The maximum voltage gain is when I_{BIAS}, and thus g_m, is maximum. This occurs at the maximum peak of the modulating voltage, V_{mod}.

$$I_{BIAS(max)} = \frac{V_{mod(max)} - (-V) - 1.4\text{ V}}{R_{BIAS}} = \frac{10\text{ V} - (-9\text{ V}) - 1.4\text{ V}}{56\text{ k}\Omega} = 314\ \mu\text{A}$$

The constant K is given as $16\ \mu S/\mu A$.

$$g_m = KI_{BIAS(max)} = (16\ \mu\text{S}/\mu\text{A})(314\ \mu\text{A}) = 5.02\text{ mS}$$
$$A_{v(max)} = g_m R_L = (5.02\text{ mS})(10\text{ k}\Omega) = 50.2$$
$$V_{out(max)} = A_{v(min)} V_{in} = (50.2)(50\text{ mV}) = 2.51\text{ V}$$

The minimum bias current is

$$I_{BIAS(min)} = \frac{V_{mod(min)} - (-V) - 1.4\text{ V}}{R_{BIAS}} = \frac{1\text{ V} - (-9\text{ V}) - 1.4\text{ V}}{56\text{ k}\Omega} = 154\ \mu\text{A}$$
$$g_m = KI_{BIAS(min)} = (16\ \mu\text{S}/\mu\text{A})(154\ \mu\text{A}) = 2.46\text{ mS}$$
$$A_{v(min)} = g_m R_L = (2.46\text{ mS})(10\text{ k}\Omega) = 24.6$$
$$V_{out(min)} = A_{v(min)} V_{in} = (24.6)(50\text{ mV}) = 1.23\text{ V}$$

The resulting output voltage is shown in Figure 9–19.

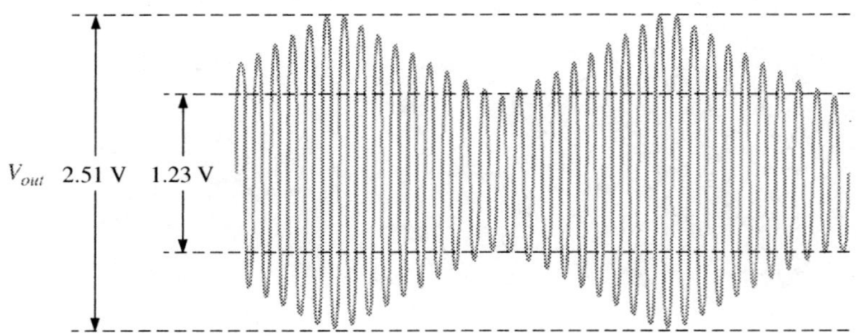

FIGURE 9–19

Question

What is the output signal if you repeat this example with the sinusoidal modulating signal replaced by a square wave with the same maximum and minimum levels and a bias resistor of 39 kΩ?

Schmitt Trigger

Figure 9–20 shows an OTA used in a Schmitt-trigger configuration. As described in Section 7–1, a Schmitt trigger is a comparator with hysteresis where the input voltage drives the device into either positive or negative saturation. When the input voltage exceeds a certain threshold value or trigger point, the device switches to one of its saturated output states. When the input falls back below another threshold value, the device switches back to its other saturated output state.

FIGURE 9–20

The OTA as a Schmitt trigger.

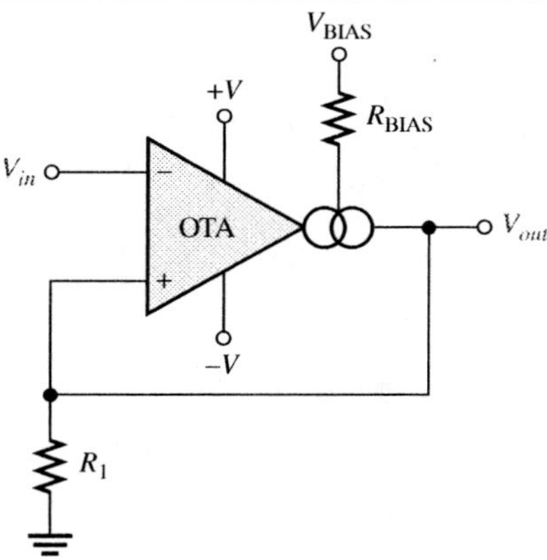

In the case of the OTA Schmitt trigger, the threshold levels are set by the current through resistor R_1. The maximum output current in an OTA equals the bias current. Therefore, in the saturated output states, $I_{out} = I_{BIAS}$. The maximum positive output voltage is $I_{out}R_1$, and this voltage is the positive threshold value or upper trigger point. When the input voltage exceeds this value, the output switches to its maximum negative voltage, which is $-I_{out}R_1$. Since $I_{out} = I_{BIAS}$, the trigger points can be controlled by the bias current. The trigger points are $\pm I_{BIAS}R_1$. Figure 9–21 illustrates this operation.

FIGURE 9–21
Basic operation of the OTA Schmitt trigger.

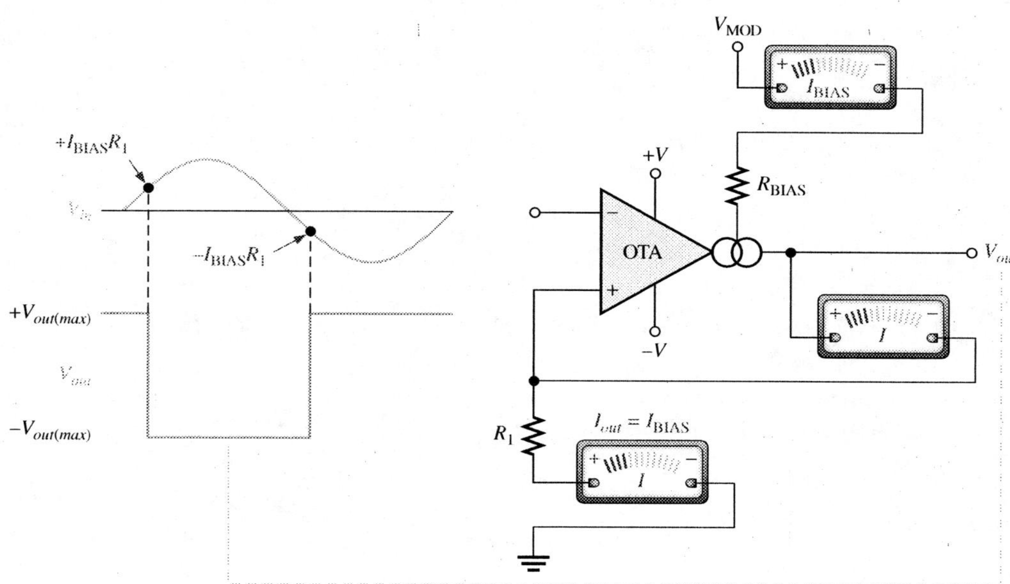

Review Questions

11. What does OTA stand for?
12. If the bias current in an OTA is increased, does the transconductance increase or decrease?
13. What happens to the voltage gain if the OTA is connected as a fixed-voltage amplifier and the supply voltages are increased?
14. What happens to the voltage gain if the OTA is connected as a variable-gain voltage amplifier and the voltage at the bias terminal is decreased?
15. What is meant by amplitude modulation?

CHAPTER REVIEW

Key Terms

Instrumentation amplifier A differential voltage-gain device that amplifies the differences between the voltages existing at its two input terminals.

Isolation amplifier A device that provides dc isolation between the input and output.

Operational transconductance amplifier An amplifier in which the output current is the gain times the input voltage.

Transconductance The ratio of output current to input voltage.

Important Facts

- ❑ A basic instrumentation amplifier is formed by three op-amps and seven resistors, including the gain-setting resistor, R_G.
- ❑ An instrumentation amplifier has high input impedance, high CMRR, low output offset, and low output impedance.
- ❑ The voltage gain of a basic instrumentation amplifier is set by a single external resistor.
- ❑ An instrumentation amplifier is useful in applications where small signals are embedded in large common-mode noise.
- ❑ A basic isolation amplifier has three electrically isolated parts: input, output, and power.
- ❑ Isolation amplifiers use capacitive coupling, optical coupling, or transformer coupling for isolation.
- ❑ Isolation amplifiers are used to interface sensitive equipment with high-voltage environments and to provide protection from electrical shock in certain medical applications.
- ❑ The operational transconductance amplifier (OTA) is a voltage-to-current amplifier.
- ❑ The output current of an OTA is the input voltage times the transconductance.
- ❑ In an OTA, transconductance varies with bias current; therefore, the gain of an OTA can be varied with a bias voltage or a variable resistor.

Formulas

Output voltage for the instrumentation amplifier:

$$V_{out} = \left(1 + \frac{2R}{R_G}\right)(V_{in2} - V_{in1}) \qquad \textbf{(9–1)}$$

Resistance of gain-setting resistor for instrumentation amplifier:

$$R_G = \frac{2R}{A_{cl} - 1} \qquad \textbf{(9–2)}$$

Resistance of gain-setting resistor for the AD622:

$$R_G = \frac{50.5\ \text{k}\Omega}{A_v - 1} \qquad \textbf{(9–3)}$$

Transconductance of an operational transconductance amplifier (OTA):

$$g_m = KI_{BIAS} \qquad \textbf{(9–4)}$$

Bias current for the LM13700:

$$I_{BIAS} = \frac{+V_{BIAS} - (-V) - 1.4\ \text{V}}{R_{BIAS}} \qquad \textbf{(9–5)}$$

Chapter Checkup

Answers are at the end of the chapter.

1. To make a basic instrumentation amplifier, it takes
   (a) one op-amp with a certain feedback arrangement
   (b) two op-amps and seven resistors
   (c) three op-amps and seven capacitors
   (d) three op-amps and seven resistors
2. Typically, an instrumentation amplifier has an external resistor used for
   (a) establishing the input resistance (b) setting the voltage gain
   (c) setting the current gain (d) interfacing with an instrument
3. Instrumentation amplifiers are used primarily in
   (a) high-noise environments (b) medical equipment
   (c) test instruments (d) filter circuits
4. Isolation amplifiers are used primarily in
   (a) remote, isolated locations
   (b) systems that isolate a single signal from many different signals
   (c) applications where there are high voltages and sensitive equipment
   (d) applications where human safety is a concern
   (e) answers (c) and (d)
5. The three parts of a basic isolation amplifier are
   (a) amplifier, filter, and power (b) input, output, and coupling
   (c) input, output, and power (d) gain, attenuation, and offset
6. The stages of most isolation amplifiers can be connected by
   (a) copper strips (b) transformers
   (c) microwave links (d) current loops
7. The characteristic that allows an isolation amplifier to amplify small signal voltages in the presence of much greater noise voltages is its
   (a) CMRR
   (b) high gain
   (c) high input resistance
   (d) magnetic coupling between input and output
8. The term *OTA* means
   (a) operational transistor amplifier
   (b) operational transformer amplifier
   (c) operational transconductance amplifier
   (d) output transducer amplifier
9. In an OTA, the transconductance is controlled by
   (a) the dc supply voltage (b) the input signal voltage
   (c) the manufacturing process (d) a bias current

10. The voltage gain of an OTA circuit is set by
    (a) a feedback resistor
    (b) the transconductance only
    (c) the transconductance and the load resistor
    (d) a gain-setting resistor
11. An OTA is basically a
    (a) voltage-to-current amplifier (b) current-to-voltage amplifier
    (c) current-to-current amplifier (d) voltage-to-voltage amplifier
12. If R_3 opens in the circuit of Figure 9–22, the output signal voltage will
    (a) increase (b) decrease
    (c) not change

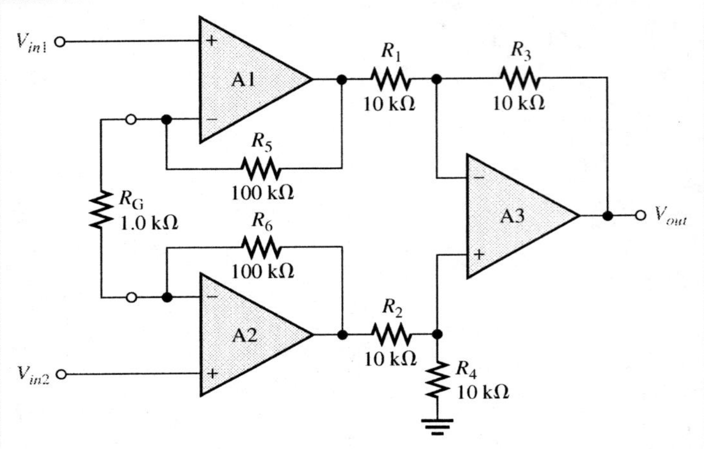

FIGURE 9–22

13. If R_G is shorted in the circuit of Figure 9–22, the output signal voltage will
    (a) increase (b) decrease
    (c) not change
14. For the IA in Figure 9–23, if R_G opens, the voltage gain will
    (a) increase (b) decrease
    (c) not change

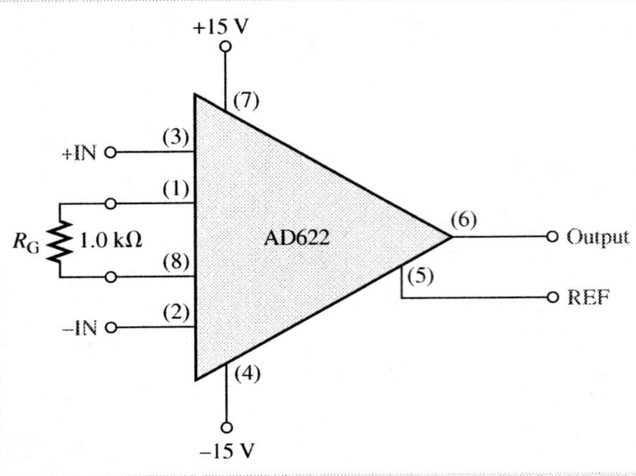

FIGURE 9–23

15. For the IA in Figure 9–23, if the value of R_G is larger than the specified value, the bandwidth will

(a) increase (b) decrease

(c) not change

Questions

Answers to odd-numbered questions are at the end of the book.

1. What is the closed-loop gain of the instrumentation amplifier in Figure 9–2 if $R_5 = R_6 = 20\ \text{k}\Omega$ and $R_G = 10\ \text{k}\Omega$?
2. If an instrumentation amplifier with a closed-loop gain of 10 has a 2 mV differential signal and a 50 mV common-mode signal on its inputs, what is on the output?
3. What is the range of the gain adjustment of the AD622 using an external resistor?
4. How does an isolation amplifier differ from an instrumentation amplifier?
5. In monitoring fetal heartbeat with an isolation amplifier, what separates the fetal heartbeat from the mother's heartbeat? Which signal is sent to the monitoring equipment?
6. If the gain of the input stage of an isolation amplifier is 5 and the gain of the output stage is 10, what is the total gain of the amplifier?
7. If you need a total gain of 25 in an isolation amplifier and you want to keep the gains of the input and output stages equal, what should each gain be?
8. For a certain OTA, the output current is 100 μA and the input voltage is 10 mV. What is the value of the transconductance?
9. If an OTA has a transconductance of 10 mS and a load resistance of 10 kΩ, what is the voltage gain?
10. How can the voltage gain of an OTA be controlled?
11. What are two OTA applications?
12. Basically, what is a Schmitt trigger circuit?

Basic Problems

PROBLEMS

Answers to odd-numbered problems are at the end of the book.

1. Find the overall differential voltage gain of the instrumentation amplifier in Figure 9–24.

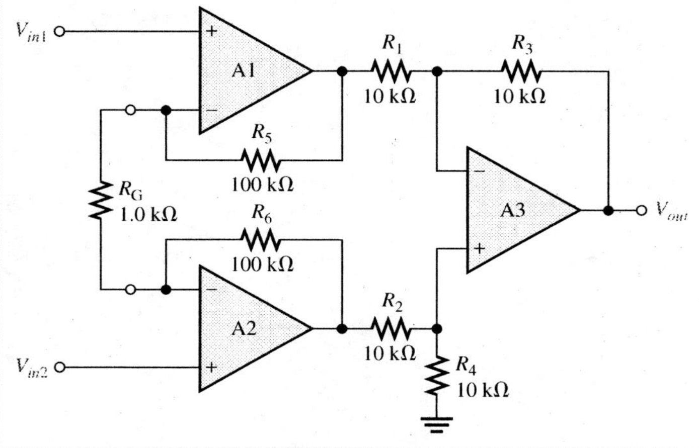

FIGURE 9–24

2. Determine the differential voltage gains of op-amps A1 and A2 for the instrumentation amplifier configuration in Figure 9–24.

3. The following dc voltages are applied to the instrumentation amplifier in Figure 9–24. $V_{in1} = 5$ mV, $V_{in2} = 10$ mV, and $V_{cm} = 225$ mV. Determine the final output voltage.

4. What value of R_G must be used to change the gain of the instrumentation amplifier in Figure 9–24 to 1000?

5. What is the voltage gain of the AD622 instrumentation amplifier in Figure 9–25?

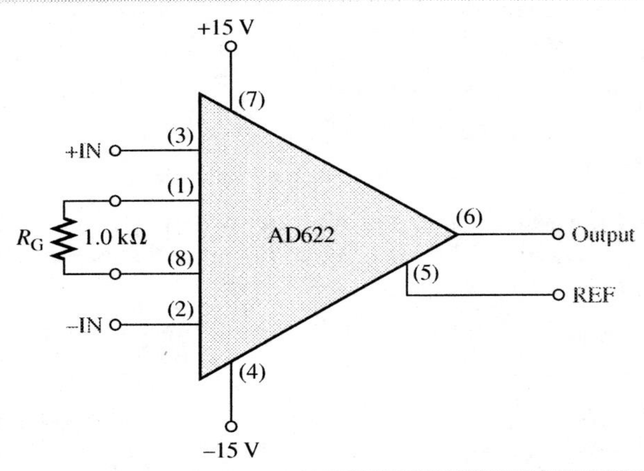

FIGURE 9–25

6. Determine the value of R_G in Figure 9–25 for a voltage gain of 20.

7. The op-amp in the input stage of a certain isolation amplifier has a voltage gain of 30. The output stage is set for a gain of 10. What is the overall voltage gain of this device?

8. Determine the overall voltage gain of each 3656KG in Figure 9–26.

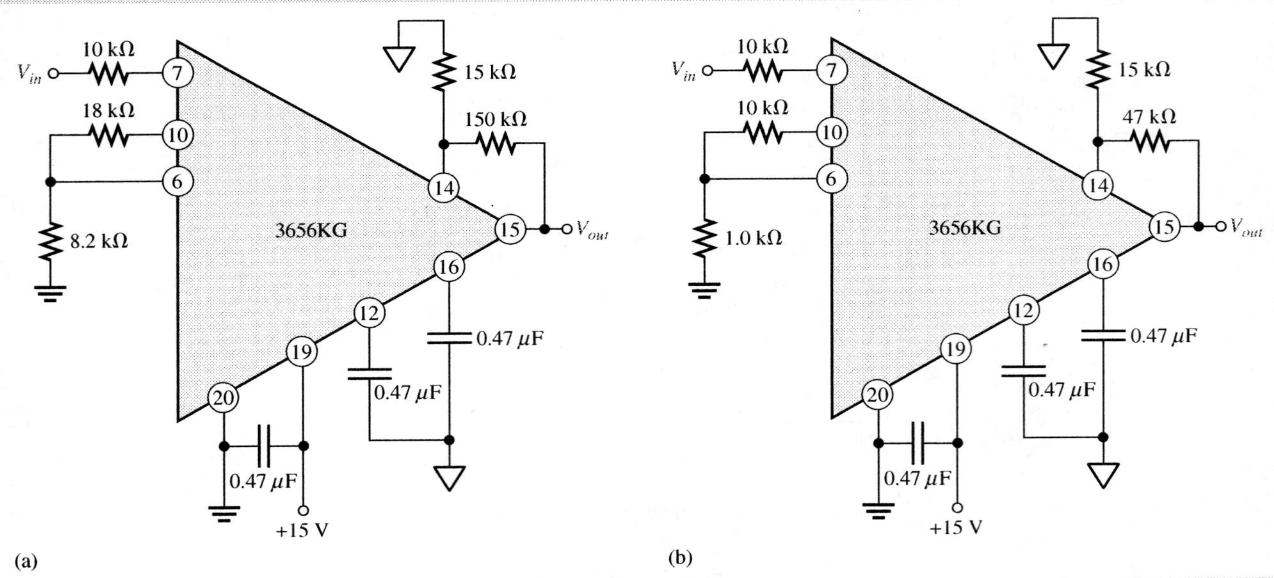

FIGURE 9–26

9. A certain OTA has an input voltage of 10 mV and an output current of 10 μA. What is the transconductance?

10. A certain OTA with a transconductance of 5000 μS has a load resistance of 10 kΩ. If the input voltage is 100 mV, what is the output current? What is the output voltage?

11. Determine the voltage gain of the OTA in Figure 9–27. Assume g_m = 2500 μS.

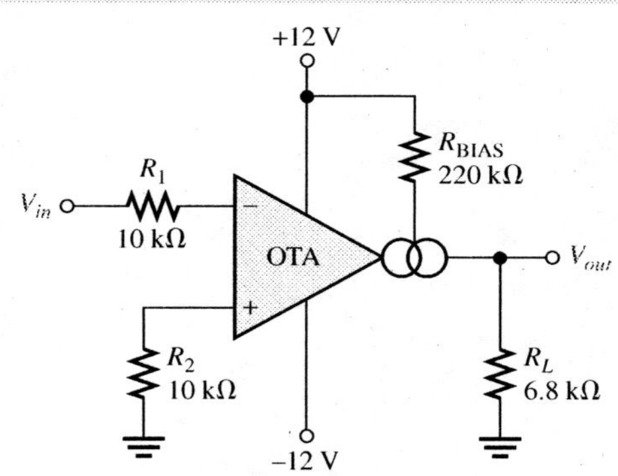

FIGURE 9–27

12. Determine the trigger points for the Schmitt-trigger circuit in Figure 9–28.

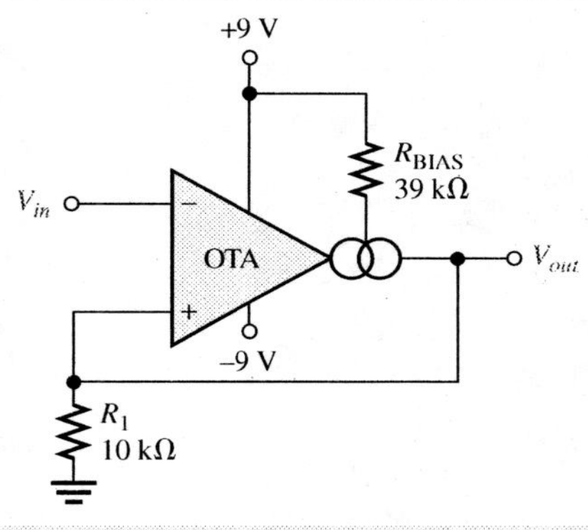

FIGURE 9–28

Basic-Plus Problems

13. Determine the approximate bandwidth of an AD622 instrumentation amplifier if the voltage gain is set to 10. Use the graph in Figure 9–6.

14. Specify what you must do to change the gain of the amplifier in Figure 9–25 to approximately 24.

15. Specify how you would change the overall gain of the amplifier in Figure 9–26(a) to approximately 100 by changing only the gain of the input stage.

16. Specify how you would change the overall gain in Figure 9–26(b) to approximately 440 by changing only the gain of the output stage.

17. Specify how you would connect each amplifier in Figure 9–26 for unity gain.

18. The output voltage of a certain OTA with a load resistance is determined to be 3.5 V. If its transconductance is 4000 μS and the input voltage is 100 mV, what is the value of the load resistance?

19. If a 10 kΩ rheostat is added in series with the bias resistor in Figure 9–27, what are the minimum and maximum bias current and minimum and maximum voltage gains? Assume $g_m = 2500\ \mu S$ at the maximum bias current and $2000\ \mu S$ at the minimum bias current.

20. The OTA in Figure 9–29 functions as an amplitude modulation circuit. Determine the output voltage waveform for the given input waveforms assuming $K = 16\ \mu S/\mu A$.

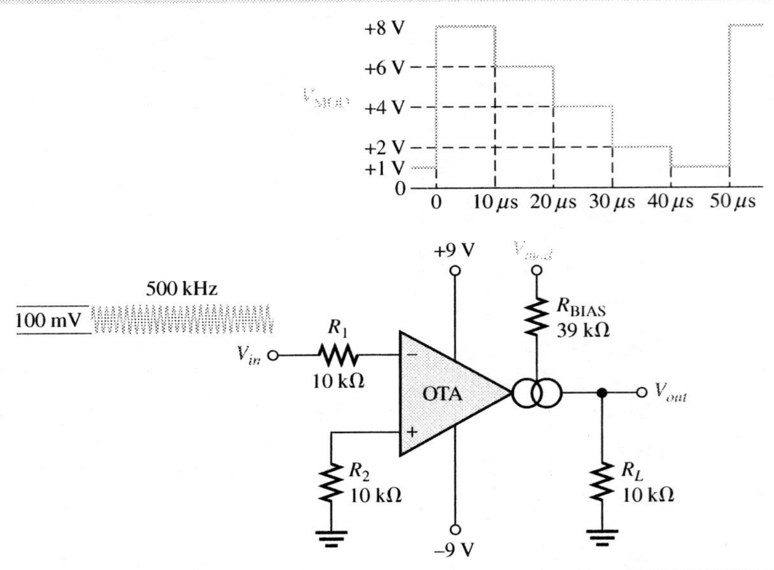

FIGURE 9–29

21. Determine the output voltage waveform for the Schmitt trigger in Figure 9–28 in relation to a 1 kHz sine wave with peak values of ±10 V.

22. In Section 9–2, a medical application of the isolation amplifier was described. Use the Internet as a resource to investigate other applications for isolation amplifiers. Summarize your findings in a short report.

ANSWERS

Example Questions

9–1: 240 Ω

9–2: Make $R_G = 1.1$ kΩ.

9–3: Many combinations are possible. Here is one: $R_{f1} = 10$ kΩ, $R_{i1} = 1.0$ kΩ, $R_{f2} = 10$ kΩ, $R_{i2} = 1.0$ kΩ

9–4: 62.5 μA.

9–5: Yes. Approximately 110.

9–6: $V_{out(max)} = 3.61$ V; $V_{out(min)} = 1.76$ V

Review Questions

1. The main purpose of an instrumentation amplifier is to amplify small signals that occur on large common-mode voltages. The key characteristics are high input impedance, high CMRR, low output impedance, and low output offset.

2. Three op-amps and seven resistors are required to construct a basic instrumentation amplifier (see Figure 9–2).

3. The gain is set by the external resistor R_G.
4. $R_G = 5.6\ k\Omega$
5. $A_v \cong 6$
6. Isolation amplifiers are used in medical equipment, power plant instrumentation, industrial processing, and automated testing.
7. The two stages of an isolation amplifier are input and output.
8. The stages are connected by transformer coupling.
9. The oscillator acts as a dc to ac converter so that the dc power can be ac coupled to the input and output stages.
10. By setting the gains of both the input and output stages
11. OTA stands for Operational Transconductance Amplifier.
12. Transconductance increases with bias current.
13. Assuming that the bias input is connected to the supply voltage, the voltage gain increases when the supply voltage is increased because this increases the bias current.
14. The gain decreases as the bias voltage decreases.
15. The amplitude of a higher frequency signal is varied according the amplitude of a lower frequency signal.

Answers to Chapter Checkup

1. (d)	2. (b)	3. (a)	4. (e)	5. (c)
6. (b)	7. (a)	8. (c)	9. (d)	10. (c)
11. (a)	12. (a)	13. (a)	14. (b)	15. (a)

OSCILLATORS AND TIMERS

CHAPTER OUTLINE

Study aids for this chapter are available at

http://www.prenhall.com/SOE

INTRODUCTION

Oscillators are circuits that generate a periodic waveform to perform timing, control, or communication functions. They are found in nearly all electronic systems, including analog and digital systems, and in most test instruments such as oscilloscopes and function generators.

Oscillators require a form of positive feedback, where a portion of the output signal is fed back to the input in a way that causes it to reinforce itself and thus sustain a continuous output signal. Although an external input is not strictly necessary, many oscillators use an external signal to control the frequency or to synchronize it with another source. Oscillators are designed to produce a controlled oscillation with one of two basic methods: the unity-gain method used with feedback oscillators and the timing method used with relaxation oscillators. Both will be discussed in this chapter.

Different types of oscillators produce various types of outputs including sine waves, square waves, triangular waves, and sawtooth waves. In this chapter, several types of basic oscillator circuits using an op-amp as the gain element are introduced. Also, a very popular integrated circuit, called the 555 timer, is discussed.

KEY OBJECTIVES

A section number is given for each objective. After completing this chapter, you should be able to

- **10–1** Describe the basic operating principles for all oscillators
- **10–2** Explain the operation of feedback oscillators
- **10–3** Describe and analyze the operation of basic *RC* sinusoidal feedback oscillators
- **10–4** Describe and analyze the operation of basic relaxation oscillators
- **10–5** Use a 555 timer in an oscillator application
- **10–6** Use a 555 timer as a one-shot

COMPUTER SIMULATIONS DIRECTORY

The following figures have Multisim circuit files associated with them.

- Figure 10–11 page 355
- Figure 10–18 page 360
- Figure 10–23 page 364

LABORATORY EXPERIMENTS DIRECTORY

The following exercises are for this chapter.

- **Experiment 19** The Wien-Bridge Oscillator
- **Experiment 20** A Triangular-Wave Oscillator

KEY TERMS

- Feedback oscillator
- Relaxation oscillator
- Positive feedback
- Wien-bridge oscillator
- Phase-shift oscillator
- Astable multivibrator
- One-shot

In all scientific work, it is vitally important that scientists have accurate standards to which to compare their measurements. The nation's primary standard for time and frequency is a cesium atomic clock. This clock contributes to the international group of atomic clocks that define the official world time. The uncertainty of the cesium atomic clock is less than 2×10^{-15}. This means that it does not gain or lose a second in 20 million years.

The cesium atomic clock uses a fountain-like movement of cesium atoms to measure frequency and time interval. Six infrared laser beams are directed at right angles to each other at the center of the clock's chamber. The lasers push the cesium atoms together to form a "ball." Also, the lasers slow down the movement of the atoms and cool them to near absolute zero temperature. Two vertically oriented lasers toss the ball of cesium atoms up through a microwave cavity. The ball then falls back through the microwave cavity under the influence of gravity. The microwave signal is tuned to a frequency that alters the states of the atoms to maximize their fluorescence. This frequency is used to define the second and equals 9,192,631,770 Hz, which is the natural resonant frequency of the cesium atom.

Research is now underway on an optical atomic clock that potentially is 1000 times more accurate than the cesium clock. This new clock is based on an energy transition in a single trapped mercury ion. The clock will monitor an optical frequency more than 100,000 times higher than the cesium clock's resonant frequency or about 1 quadrillion Hz.

10–1 THE OSCILLATOR

An **oscillator** is a circuit that produces a periodic waveform on its output with only the dc supply voltage as a required input. A repetitive input signal is not required but is sometimes used to synchronize oscillations. The output voltage can be either sinusoidal or nonsinusoidal, depending on the type of oscillator. Two major classifications for oscillators are feedback oscillators and relaxation oscillators.

In this section, you will learn the basic operating principles for all oscillators.

Types of Oscillators

Essentially, all oscillators convert electrical energy from the dc power supply to periodic waveforms that can be used for various timing, control, or signal-generating applications. A basic oscillator is illustrated in Figure 10–1. Oscillators are classified according to the technique for generating a signal.

FIGURE 10–1

The basic oscillator concept showing three common types of output waveforms.

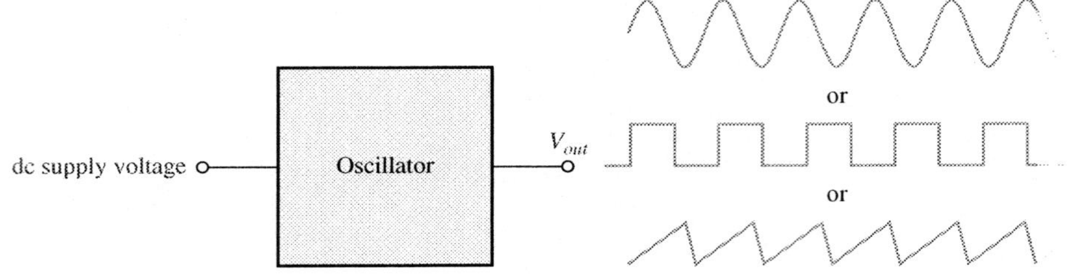

Feedback Oscillators

One type of oscillator is the feedback oscillator. A **feedback oscillator** returns a fraction of the output signal to the input with no net phase shift, resulting in a reinforcement of the output signal. After oscillations are started, the loop gain is maintained at 1.0 to maintain oscillations. A feedback oscillator consists of an amplifier for gain and a positive feedback circuit that produces phase shift and attenuates the signal, as shown in Figure 10–2. For an oscillator with an inverting amplifier (180° phase shift), the feedback circuit produces a phase shift of 180° to offset the phase shift in the amplifier. For an oscillator with a noninverting amplifier (0° phase shift), the feedback circuit produces a 0° or 360° phase shift so that, again, there is no net phase shift around the loop.

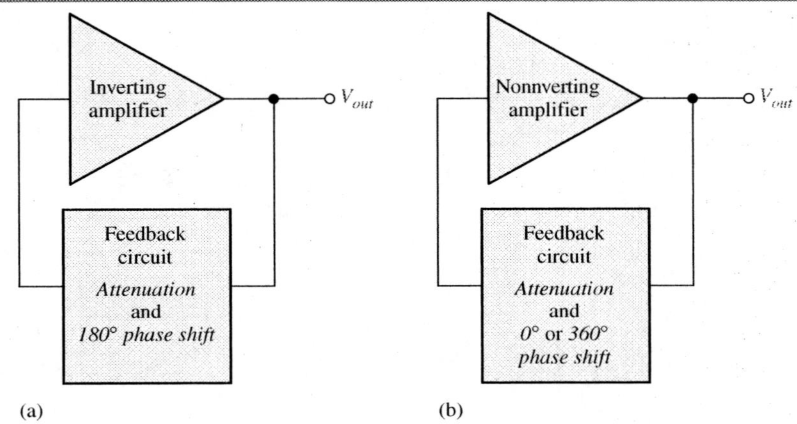

FIGURE 10–2

Basic elements of a feedback oscillator showing both inverting and noninverting amplifier configurations.

Relaxation Oscillators

A second type of oscillator is the relaxation oscillator. A **relaxation oscillator** uses an *RC* timing circuit instead of positive feedback to generate a waveform that is generally a square wave or other nonsinusoidal waveform. Typically, a relaxation oscillator uses a Schmitt trigger or other device that changes states to alternately charge and discharge a capacitor through a resistor. Relaxation oscillators are discussed in Section 10–4.

HISTORICAL NOTE

The first product from the Hewlett-Packard Co. was a two-stage *RC* oscillator using triode vacuum tubes that could operate over a five-decade frequency range. Introduced in 1939, it featured both positive and negative feedback loops, which resulted in a low-distortion output, primarily due to the negative feedback. Shortly after World War II, the oscillator was improved to extend the range to six decades (from 10 Hz to 10 MHz). The *RC* oscillator was described in the first volume of the *Hewlett-Packard Journal*, issues 3 and 4.

Review Questions

Answers are at the end of the chapter.

1. What is an oscillator?
2. What type of feedback does a feedback oscillator require?
3. What is the purpose of the feedback circuit?
4. How does a relaxation oscillator differ from a feedback oscillator?
5. Generally, which type of oscillator produces a square wave?

10–2 FEEDBACK OSCILLATOR PRINCIPLES

Feedback oscillator operation is based on the principle of positive feedback. Feedback oscillators are used to generate sinusoidal waveforms.

In this section, you will learn the operation of feedback oscillators.

Positive Feedback

Positive feedback is characterized by the condition wherein an in-phase portion of the output voltage of an amplifier is fed back to the input. This basic idea is illustrated with the sinusoidal oscillator shown in Figure 10–3. As you can see, the in-phase feedback voltage is amplified to produce the output voltage, which in turn produces the feedback voltage. That is, a loop is created in which the signal sustains itself and a continuous sinusoidal output is produced. This phenomenon is called *oscillation.*

FIGURE 10–3

Positive feedback produces oscillation. The feedback loop is indicated by the dashed arrows.

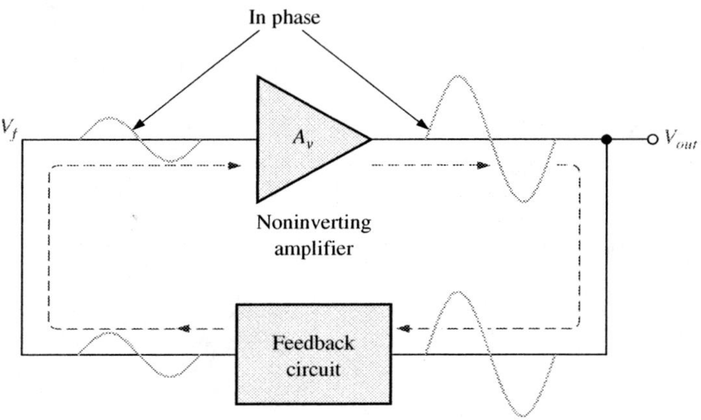

Conditions for Oscillation

Two conditions, illustrated in Figure 10–4, are required to sustain oscillations:

1. The phase shift around the feedback loop must be effectively 0°.
2. The voltage gain, A_{cl}, around the closed feedback loop (loop gain) must equal 1 (unity).

FIGURE 10–4

Conditions for oscillation.

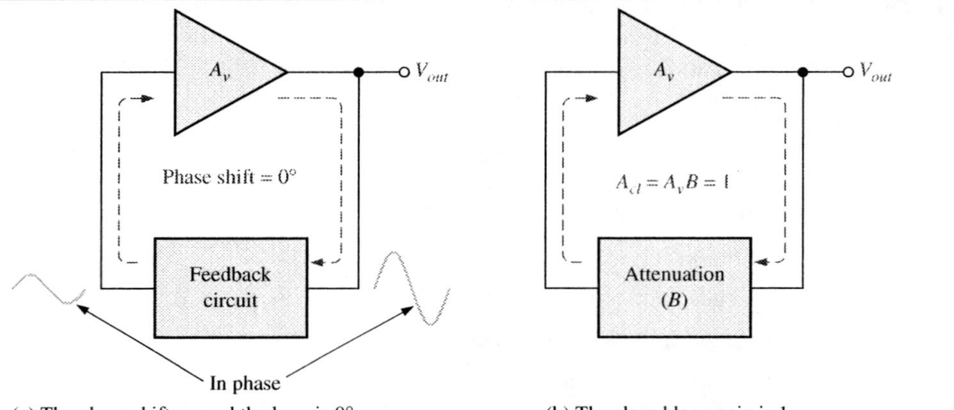

(a) The phase shift around the loop is 0°.

(b) The closed loop gain is 1.

The voltage gain around the closed feedback loop, A_{cl}, is the product of the amplifier gain, A_v, and the attenuation, B, of the feedback circuit.

$$A_{cl} = A_v B$$

If a sinusoidal wave is the desired output, a loop gain greater than 1 will rapidly cause the output to saturate at both peaks of the waveform, producing unacceptable distortion. To avoid this, some form of gain control must be used to keep the loop gain at exactly 1, once oscillations have started. For example, if the attenuation of the feedback circuit is 0.01, the amplifier must have a gain of exactly 100 to overcome this attenuation and not create unacceptable distortion ($0.01 \times 100 = 1.0$). An amplifier gain of greater than 100 will cause the oscillator to limit both peaks of the waveform.

Start-Up Conditions

So far, you have seen what it takes for an oscillator to produce a continuous sinusoidal output. Now let's examine the requirements for the oscillation to start when the dc supply voltage is turned on. As you know, the unity-gain condition must be met for oscillation to be sustained. For oscillation to begin, the voltage gain around the positive feedback loop must be greater than 1 so that the amplitude of the output can build up to a desired level. The gain must then decrease to 1 so that the output stays at the desired level and oscillation is sustained. (Several ways to achieve this reduction in gain after start-up are discussed in the next section.) The voltage-gain conditions for both starting and sustaining oscillation are illustrated in Figure 10–5.

A common question is this: If the oscillator is initially off and there is no output voltage, how does a feedback signal originate to start the positive feedback buildup process? Initially, a small positive feedback voltage develops from thermally produced broad-band noise in the resistors or other components or from power supply turn-on transients. The feedback circuit permits only a voltage with a frequency equal to the selected oscillation frequency to appear in phase on the amplifier's input. This initial feedback voltage is amplified and continually reinforced, resulting in a buildup of the output voltage as previously discussed.

FIGURE 10–5

When oscillation starts at t_0, the condition $A_{cl} > 1$ causes the sinusoidal output voltage amplitude to build up to a desired level. Then A_{cl} decreases to 1 and maintains the desired amplitude.

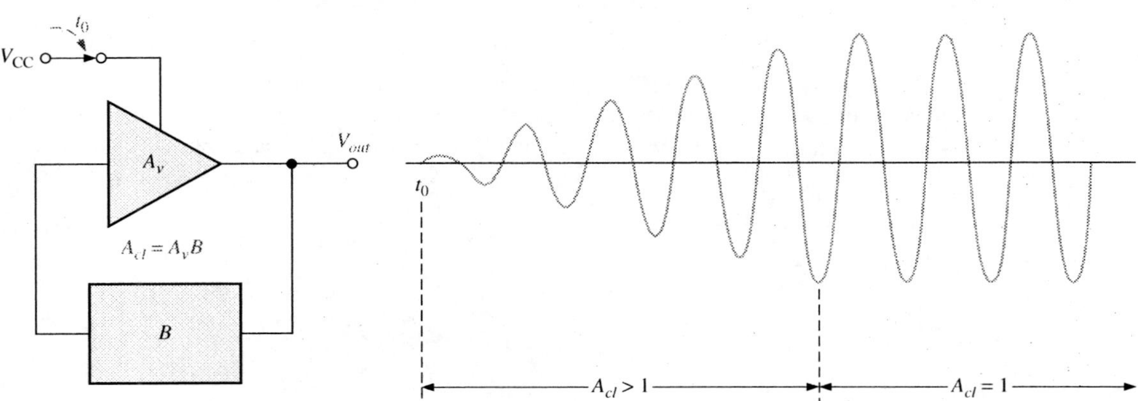

Review Questions

6. What are the conditions required for a circuit to oscillate?
7. Define positive feedback.

8. What is the voltage gain condition for oscillator start-up?
9. When oscillation starts, what gain condition causes the sinusoidal output to build up to a desired level?
10. What does the variable B represent?

10–3 SINUSOIDAL FEEDBACK OSCILLATORS

Three types of feedback oscillators that use *RC* circuits to produce sinusoidal outputs are the Wien-bridge oscillator, the phase-shift oscillator, and the twin-T oscillator. Generally, *RC* feedback oscillators are used for frequencies up to about 1 MHz. The Wien-bridge is by far the most widely used type of *RC* oscillator for this range of frequencies.

In this section, you will learn the operation of basic *RC* sinusoidal feedback oscillators.

The Wien-Bridge Oscillator

One type of sinusoidal feedback oscillator is the **Wien-bridge oscillator**. A fundamental part of the Wien-bridge oscillator is a lead-lag circuit like that shown in Figure 10–6(a). R_1 and C_1 together form the lag portion of the circuit; R_2 and C_2 form the lead portion. The operation of this lead-lag circuit is as follows. At lower frequencies, the lead circuit dominates due to the high reactance of C_2. As the frequency increases, X_{C2} decreases, thus allowing the output voltage to increase. At some specified frequency, the response of the lag circuit takes over, and the decreasing value of X_{C1} causes the output voltage to decrease.

FIGURE 10–6

A lead-lag circuit and its response curve.

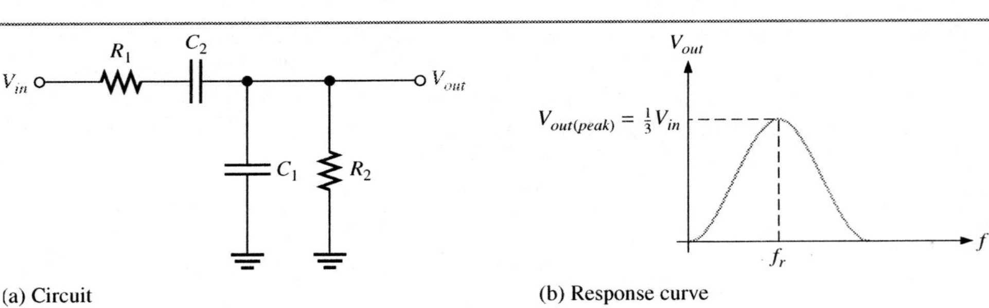

(a) Circuit

(b) Response curve

The response curve for the lead-lag circuit shown in Figure 10–6(b) indicates that the output voltage peaks at a frequency called the resonant frequency, f_r. At this point, the attenuation (V_{out}/V_{in}) of the network is 1/3 if $R_1 = R_2$ and $X_{C1} = X_{C2}$ as stated by the following equation.

$$\frac{V_{out}}{V_{in}} = \frac{1}{3} \quad (10\text{–}1)$$

The formula for the resonant frequency is

$$f_r = \frac{1}{2\pi RC} \quad (10\text{–}2)$$

To summarize, the lead-lag circuit in the Wien-bridge oscillator has a resonant frequency, f_r, at which the phase shift through the network is 0° and the attenuation is 1/3. Below f_r, the lead circuit dominates and the output leads the input. Above f_r, the lag circuit dominates and the output lags the input.

The Basic Circuit

The lead-lag circuit is used in the positive feedback loop of the op-amp, as shown in Figure 10–7(a). A voltage divider is used in the negative feedback loop. The Wien-bridge oscillator circuit can be viewed as a noninverting amplifier configuration with the input signal fed back from the output through the lead-lag circuit. Recall that the closed-loop gain of the amplifier is determined by the voltage divider.

FIGURE 10–7

Two ways to draw the schematic of a Wien-bridge oscillator.

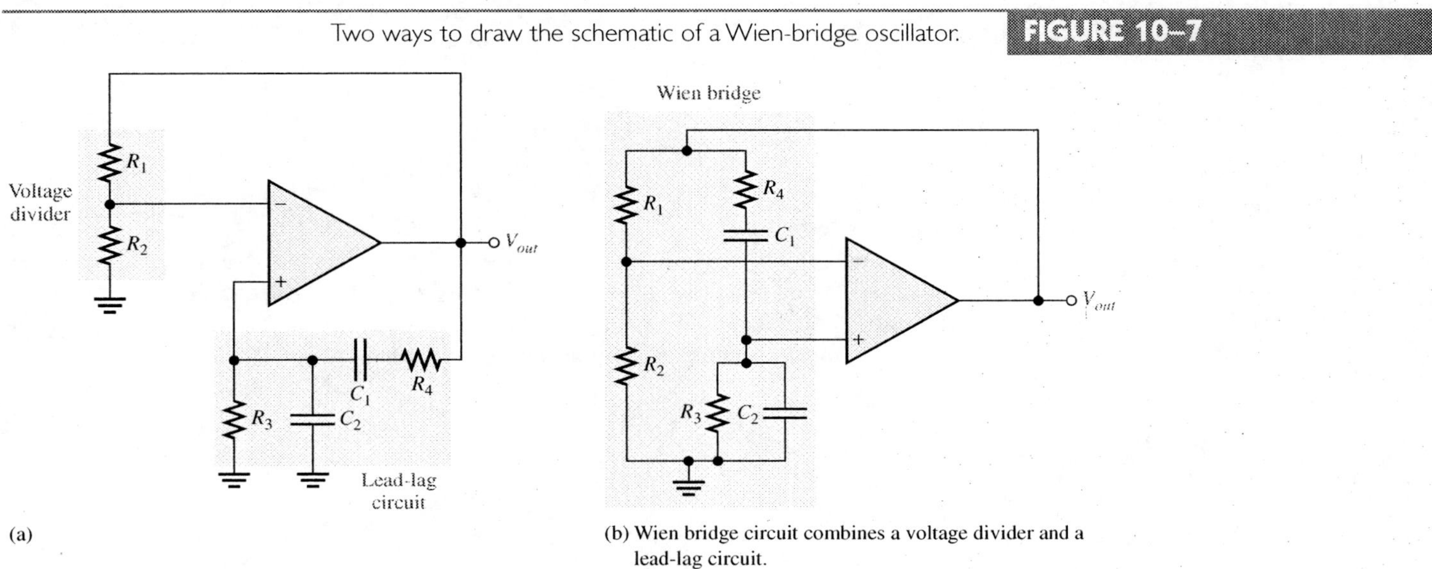

(a)

(b) Wien bridge circuit combines a voltage divider and a lead-lag circuit.

The circuit is redrawn in Figure 10–7(b) to show that the op-amp is connected across the bridge circuit. One leg of the bridge is the lead-lag circuit, and the other is the voltage divider.

Positive Feedback Conditions for Oscillation

As you know, for the circuit to produce a sustained sinusoidal output (oscillate), the phase shift around the positive feedback loop must be 0° and the gain around the loop must equal unity (1). The 0° phase-shift condition is met when the frequency is f_r because the phase shift through the lead-lag circuit is 0° and there is no inversion from the noninverting input (+) of the op-amp to the output. This is shown in Figure 10–8(a).

FIGURE 10–8

Conditions for oscillation.

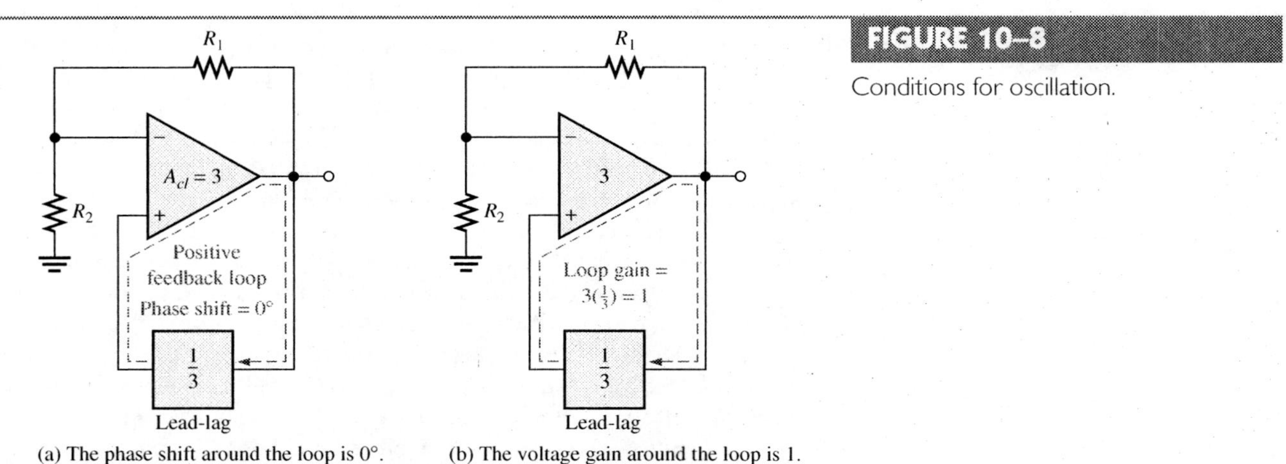

(a) The phase shift around the loop is 0°.

(b) The voltage gain around the loop is 1.

The unity-gain condition in the feedback loop is met when

$$A_{cl} = 3$$

This offsets the 1/3 attenuation of the lead-lag circuit, thus making the total gain around the positive feedback loop equal to 1, as depicted in Figure 10–8(b). To achieve a closed-loop gain of 3 for the amplifier,

$$R_1 = 2R_2$$

Then

$$A_{cl} = \frac{R_1 + R_2}{R_2} = \frac{2R_2 + R_2}{R_2} = \frac{3R_2}{R_2} = 3$$

Start-Up Conditions

Initially, the closed-loop gain of the amplifier itself must be more than 3 ($A_{cl} > 3$) until the output signal builds up to a desired level. Ideally, the gain of the amplifier must then decrease to 3 so that the total gain around the loop is 1 and the output signal stays at the desired level, thus sustaining oscillation. This is illustrated in Figure 10–9.

FIGURE 10–9 Oscillator start-up conditions.

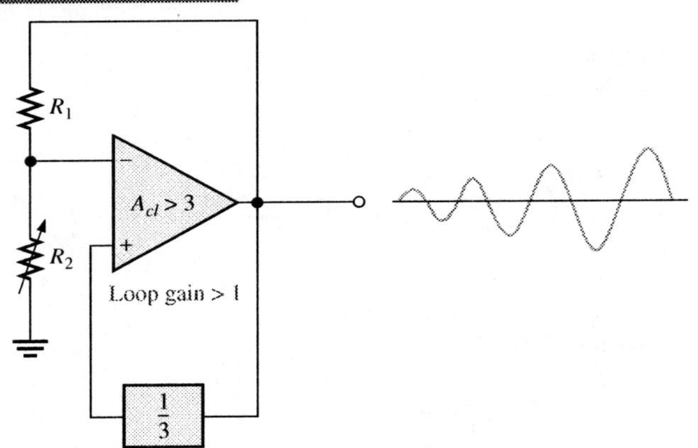

(a) Loop gain greater than 1 causes output to build up.

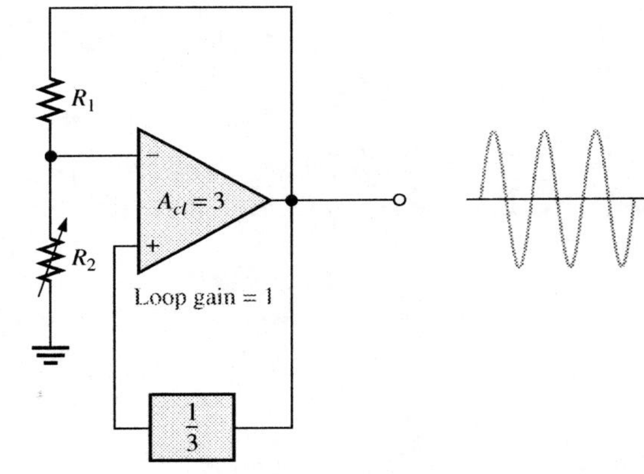

(b) Loop gain of 1 causes a sustained constant output.

One method to control the gain uses a JFET as a voltage-controlled resistor in a negative feedback path. This method can produce an excellent sinusoidal waveform that is stable. A JFET operating with a small or zero V_{DS} is operating in the ohmic region. As the gate voltage increases, the drain-source resistance increases. If the JFET is placed in the negative feedback path, automatic gain control can be achieved because of this voltage-controlled resistance.

A JFET stabilized Wien-bridge oscillator is shown in Figure 10–10. The gain of the op-amp is controlled by the components shown in the shaded box, which include the JFET. The JFET's drain-source resistance depends on the gate voltage. With no output signal, the gate is at zero volts, causing the drain-source resistance to be at the minimum. With this condition, the loop gain is greater than 1. Oscillations begin and rapidly build to a large output signal. Negative excursions of the output signal forward-bias D_1, causing capacitor C_3 to charge to a negative voltage. This voltage increases the drain-source resistance of the JFET and reduces the gain (and hence the output). This is classic negative feedback at work. With the proper selection of components, the gain can be stabilized at the required level. The following example illustrates a JFET stabilized oscillator.

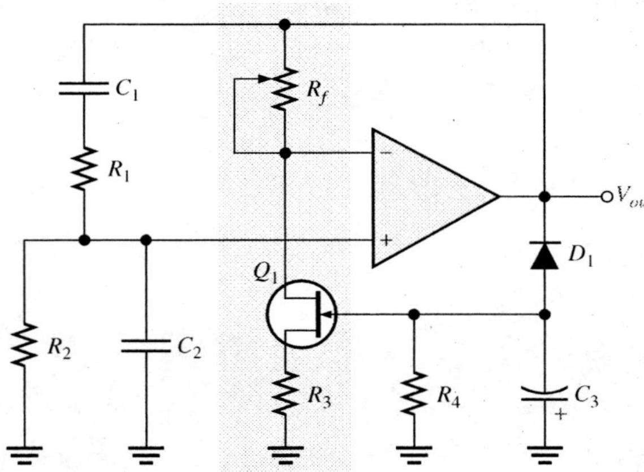

FIGURE 10–10

Self-starting Wien-bridge oscillator using a JFET in the negative feedback loop.

EXAMPLE 10–1

Problem

Determine the frequency for the Wien-bridge oscillator in Figure 10–11. Also, calculate the setting for R_f assuming the internal drain-source resistance, r'_{ds}, of the JFET is 500 Ω when oscillations are stable.

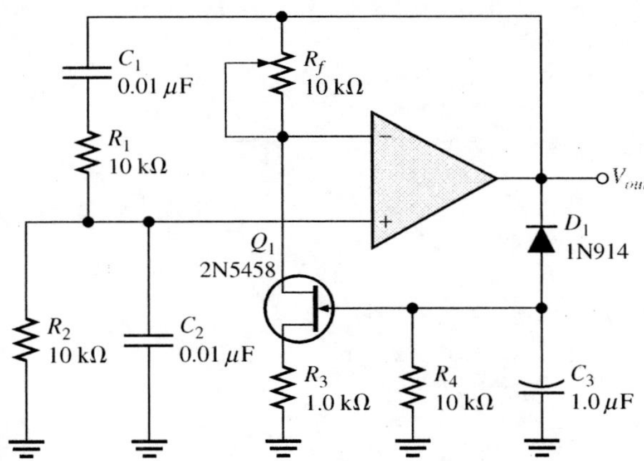

FIGURE 10–11

Solution

For the lead-lag circuit, $R_1 = R_2 = R = 10\ \text{k}\Omega$ and $C_1 = C_2 = C = 0.01\ \mu\text{F}$. The frequency is

$$f_r = \frac{1}{2\pi RC} = \frac{1}{2\pi(10\ \text{k}\Omega)(0.01\ \mu\text{F})} = \mathbf{1.59\ kHz}$$

The closed-loop gain must be 3.0 for oscillations to be sustained. For an inverting amplifier, the gain is that of a noninverting amplifier.

$$A_v = \frac{R_f}{R_i} + 1$$

R_i is composed of R_3 (the source resistor) and r'_{ds}. Substituting,

$$A_v = \frac{R_f}{R_3 + r'_{ds}} + 1$$

Rearranging and solving for R_f,

$$R_f = (A_v - 1)(R_3 + r'_{ds}) = (3 - 1)(1.0\text{ k}\Omega + 500\ \Omega) = \mathbf{3.0\ k\Omega}$$

Question*

What happens to the oscillations if the setting of R_f is too high? What happens if the setting is too low?

COMPUTER SIMULATION

Open the Multisim file F10-11DV on the website. Measure the frequency of the output.

The Phase-Shift Oscillator

A type of sinusoidal feedback oscillator called the **phase-shift oscillator** is shown in Figure 10–12. Each of the three *RC* circuits in the feedback loop can provide a maximum phase shift approaching 90°. Oscillation occurs at the frequency where the total phase shift through the three *RC* circuits is 180°. The inversion of the op-amp itself provides the additional 180° to meet the requirement for oscillation of a 360° (or 0°) phase shift around the feedback loop.

FIGURE 10–12

Op-amp phase-shift oscillator.

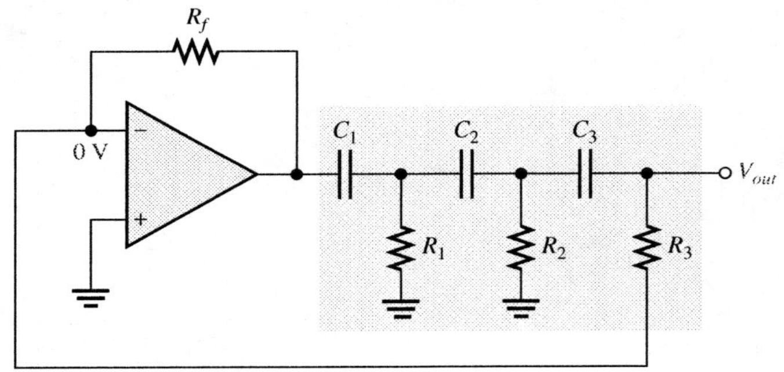

The attenuation, *B*, of the three-section *RC* feedback circuit is

$$B = \frac{1}{29} \tag{10–3}$$

where $B = R_3/R_f$. The derivation of this unusual result is beyond the scope of this book. To meet the greater-than-unity loop gain requirement, the closed-loop voltage gain of the op-amp must be greater than 29 (set by R_f and R_3). The frequency of oscillation is stated by the following formula, where $R_1 = R_2 = R_3 = R$ and $C_1 = C_2 = C_3 = C$.

$$f_r = \frac{1}{2\pi\sqrt{6}RC} \tag{10–4}$$

* *Answers are at the end of the chapter.*

Problem **EXAMPLE 10–2**

(a) Determine the value of R_f necessary for the circuit in Figure 10–13 to operate as an oscillator.

(b) Determine the frequency of oscillation.

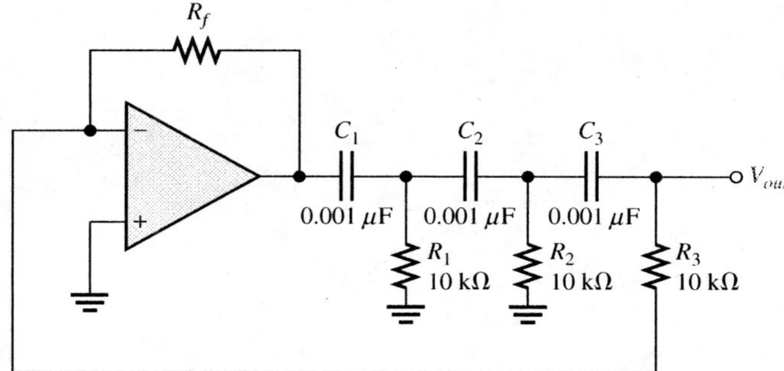

FIGURE 10–13

Solution

(a) $A_{cl} = 29$, and $B = \dfrac{1}{29} = \dfrac{R_3}{R_f}$.

$$\frac{R_f}{R_3} = 29$$

$$R_f = 29R_3 = 29(10\text{ k}\Omega) = \mathbf{290\text{ k}\Omega}$$

(b) $R_1 = R_2 = R_3 = R$ and $C_1 = C_2 = C_3 = C$.

$$f_r = \frac{1}{2\pi\sqrt{6}RC} = \frac{1}{2\pi\sqrt{6}(10\text{ k}\Omega)(0.001\ \mu\text{F})} \cong \mathbf{6.5\text{ kHz}}$$

Question

(a) If R_1, R_2, and R_3 in Figure 10–13 are changed to 8.2 kΩ, what value must R_f be for oscillation?

(b) What is the value of f_r?

Twin-T Oscillator

Another type of *RC* feedback oscillator is called the *twin-T* because of the two T-type *RC* filters used in the feedback loop, as shown in Figure 10–14(a). One of the twin-T filters has a low-pass response, and the other has a high-pass response. The combined parallel filters produce a band-stop or notch response with a center frequency equal to the desired frequency of oscillation, f_r, as shown in Figure 10–14(b).

Oscillation cannot occur at frequencies above or below f_r because of the negative feedback through the filters. At f_r, however, there is negligible negative feedback; thus, the positive feedback through the voltage divider (R_1 and R_2) allows the circuit to oscillate.

FIGURE 10–14 Twin-T oscillator and twin-T filter response.

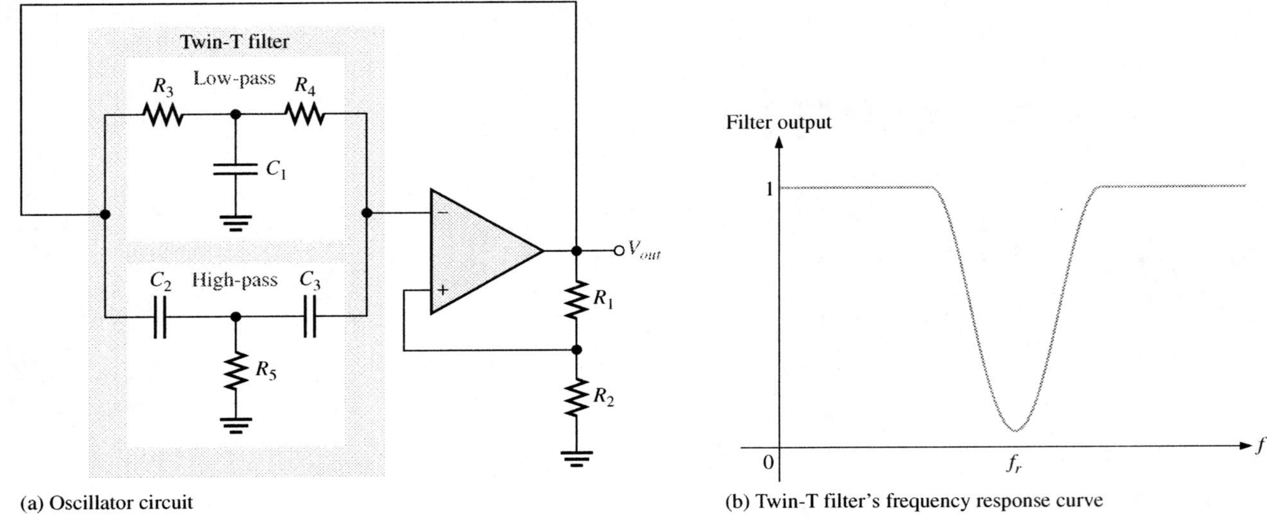

(a) Oscillator circuit

(b) Twin-T filter's frequency response curve

Review Questions

11. There are two feedback loops in the Wien-bridge oscillator. What is the purpose of each?
12. A certain lead-lag circuit has $R_1 = R_2$ and $C_1 = C_2$. An input voltage of 5 V rms is applied. The input frequency equals the resonant frequency of the circuit. What is the rms output voltage?
13. Why is the phase shift through the *RC* feedback circuit in a phase-shift oscillator equal to 180°?
14. What makes up the negative feedback circuit of a Twin-T oscillator?
15. What is the purpose of the JFET in the self-starting Wien-bridge oscillator?

10–4 RELAXATION OSCILLATORS

The second major category of oscillators is the relaxation oscillator. Relaxation oscillators use an *RC* timing circuit and a device that changes states to generate a periodic waveform.

In this section, you will learn the operation of basic relaxation oscillators.

A Triangular-Wave Oscillator

The op-amp integrator covered in Chapter 7 can be used as the basis for a triangular-wave generator. The basic idea is illustrated in Figure 10–15(a) where a dual-polarity, switched input is used. We use the switch only to introduce the concept; it is not a practical way to implement this circuit. When the switch is in position 1, the negative voltage is applied, and the output is a positive-going ramp. When the switch is thrown into position 2, a negative-going ramp is produced. If the switch is thrown back and forth at fixed intervals, the output is a triangular wave consisting of alternating positive-going and negative-going ramps, as shown in Figure 10–15(b).

FIGURE 10–15

Basic triangular-wave generator.

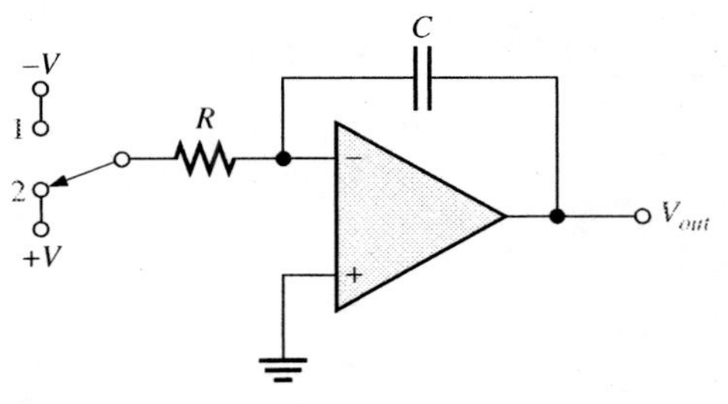

(a)

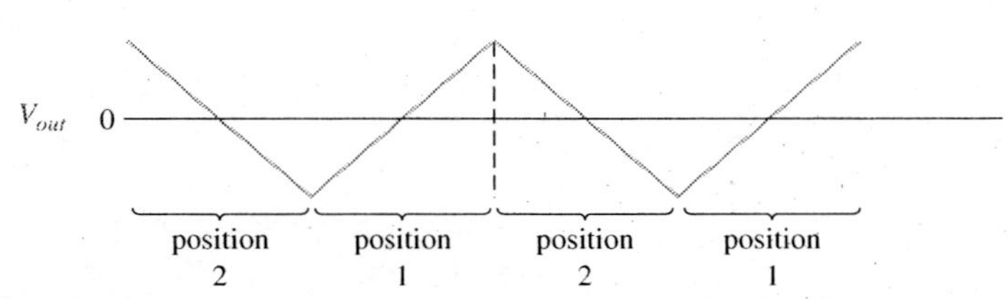

(b) Output voltage as the switch is thrown back and forth at regular intervals

A Practical Triangular-Wave Oscillator

One practical implementation of a triangular-wave generator utilizes an op-amp comparator set up as a Schmitt trigger to perform the switching function, as shown in Figure 10–16. The operation is as follows. To begin, assume that the output voltage of the comparator is at its maximum negative level. This output is connected to the inverting input of the integrator through R_1, producing a positive-going ramp on the output of the integrator. When the ramp voltage reaches the upper trigger point (UTP), the comparator switches to its maximum positive level. This positive level causes the integrator ramp to change to a negative-going direction. The ramp continues in this direction until the lower trigger point (LTP) of the comparator is reached. At this point, the comparator output switches back to the maximum negative level and the cycle repeats. This action is illustrated in Figure 10–17.

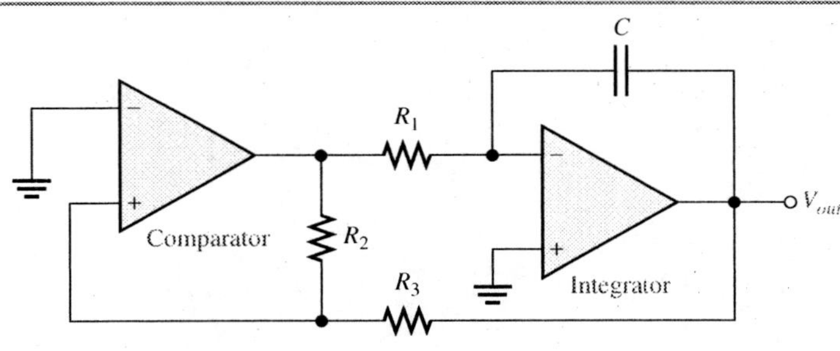

FIGURE 10–16

A triangular-wave generator using two op-amps.

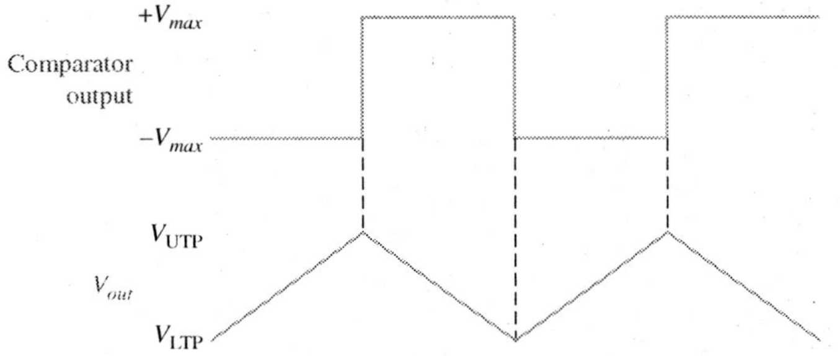

FIGURE 10–17

Waveforms for the circuit in Figure 10–16.

Since the comparator produces a square-wave output, the circuit in Figure 10–16 can be used as both a triangular-wave generator and a square-wave generator. Devices of this type are commonly known as *function generators* because they produce more than one output function. The output amplitude of the square wave is set by the output swing of the comparator, and resistors R_2 and R_3 set the amplitude of the triangular output by establishing the UTP and LTP voltages according to the following formulas:

$$V_{\text{UTP}} = +V_{max}\left(\frac{R_3}{R_2}\right)$$

$$V_{\text{LTP}} = -V_{max}\left(\frac{R_3}{R_2}\right)$$

where the comparator output levels, $+V_{max}$ and $-V_{max}$, are equal. The frequency of both waveforms depends on the R_1C time constant as well as the amplitude-setting resistors, R_2 and R_3. By varying R_1, the frequency of oscillation can be adjusted without changing the output amplitude. The approximate frequency of oscillation is

$$f \cong \frac{1}{4R_1C}\left(\frac{R_2}{R_3}\right) \qquad (10\text{–}5)$$

The circuit in Example 10–3 is the focus of Experiment 20 in the lab manual.

EXAMPLE 10–3

Problem

Determine the approximate frequency of oscillation of the circuit in Figure 10–18. To what value must R_1 be changed to make the frequency 20 kHz?

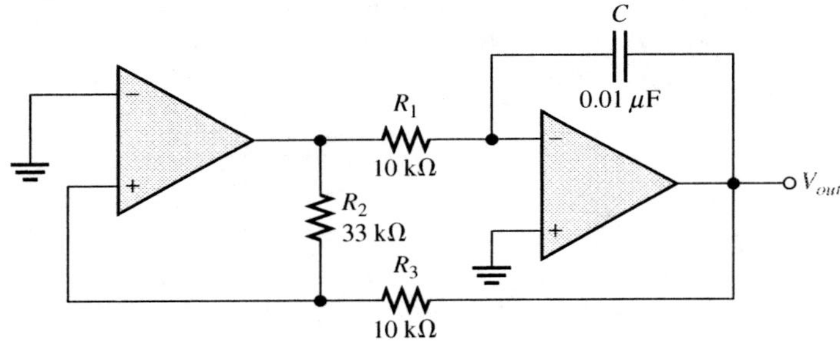

FIGURE 10–18

Solution

$$f \cong \frac{1}{4R_1C}\left(\frac{R_2}{R_3}\right) = \left(\frac{1}{4(10\text{ k}\Omega)(0.01\ \mu\text{F})}\right)\left(\frac{33\text{ k}\Omega}{10\text{ k}\Omega}\right) = \mathbf{8.25\ kHz}$$

To make $f = 20$ kHz,

$$R_1 = \frac{1}{4fC}\left(\frac{R_2}{R_3}\right) = \left(\frac{1}{4(20\text{ kHz})(0.01\ \mu\text{F})}\right)\left(\frac{33\text{ k}\Omega}{10\text{ k}\Omega}\right) = \mathbf{4.13\ k\Omega}$$

Question

What is the amplitude of the triangular wave in Figure 10–18 if the comparator output is ± 10 V?

COMPUTER SIMULATION

Open the Multisim file F10-18DV on the website. Measure the output frequency.

A Square-Wave Oscillator

The basic square-wave oscillator shown in Figure 10–19 is a type of relaxation oscillator because its operation is based on the charging and discharging of a capacitor. Notice that the op-amp's inverting input (−) is the capacitor voltage and the noninverting input (+) is a portion of the output fed back through resistors R_2 and R_3. When the circuit is first turned on, the capacitor is uncharged, and thus the inverting input is at 0 V. This makes the output a positive maximum, and the capacitor begins to charge toward V_{out} through R_1. When the capacitor voltage (V_C) reaches a value equal to the feedback voltage (V_f) on the noninverting input, the op-amp switches to the maximum negative state. At this point, the capacitor begins to discharge from $+V_f$ toward $-V_f$. When the capacitor voltage reaches $-V_f$, the op-amp switches back to the maximum positive state. This action continues to repeat, as shown in Figure 10–20, and a square-wave output voltage is obtained.

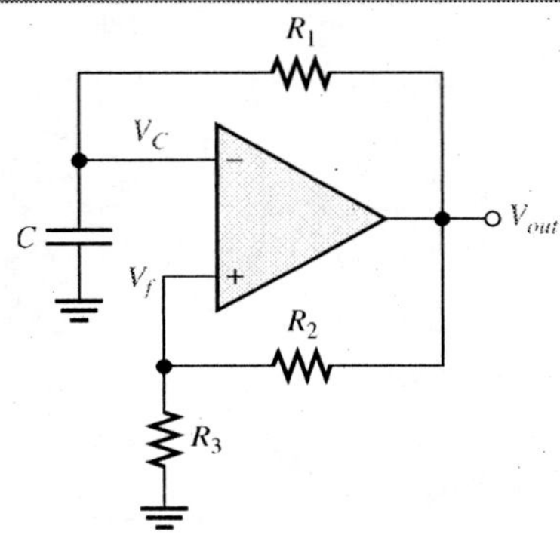

FIGURE 10–19

A square-wave relaxation oscillator.

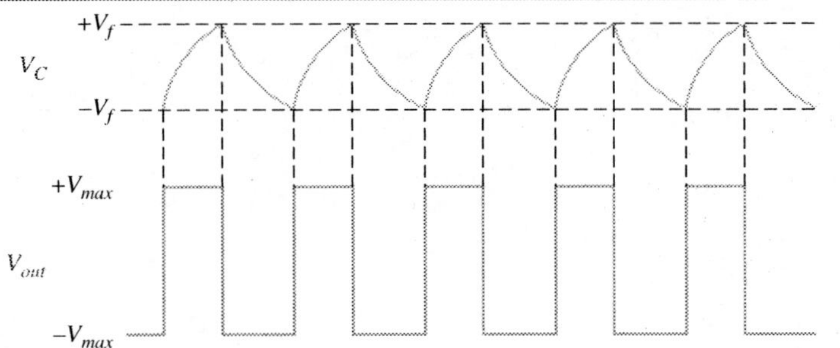

FIGURE 10–20

Waveforms for the square-wave relaxation oscillator.

Review Questions

16. Upon what principle does a relaxation oscillator operate?
17. What are the two general types of waveforms that are produced by relaxation oscillators?
18. What are V_{UTP} and V_{LTP}?
19. What type of relaxation oscillator uses a comparator and an integrator?
20. What is a function generator?

10–5 THE 555 TIMER AS AN OSCILLATOR

The 555 timer is a versatile integrated circuit with many applications. The term **astable** means no stable state.

In this section, you will learn how the 555 is configured as an astable multivibrator, which is essentially a square-wave oscillator.

Astable Operation

A 555 timer connected to operate as an **astable multivibrator**, which is a nonsinusoidal oscillator that produces a pulse waveform on its output, is shown in Figure 10–21. Notice that the threshold input (THRESH) is now connected to the trigger input (TRIG). The external components R_1, R_2, and C_{ext} form the timing circuit that sets the frequency of oscillation. The 0.01 μF capacitor connected to the control input (CONT) is strictly for decoupling and has no effect on the operation.

FIGURE 10–21

The 555 timer connected as an astable multivibrator.

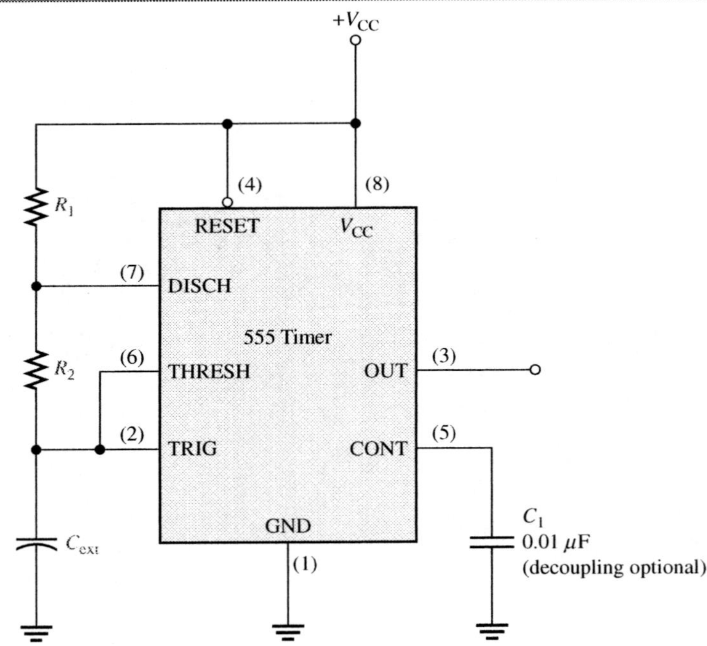

The frequency of oscillation is given by Equation 10–6.

$$f = \frac{1.44}{(R_1 + 2R_2)C_{ext}} \tag{10–6}$$

The duty cycle of a pulse waveform is the ratio of the pulse width (t_W) to the period (T) of the waveform expressed as a percentage.

$$\text{Duty cycle} = \left(\frac{t_W}{T}\right)100\%$$

By selecting R_1 and R_2, the duty cycle of the output of a 555 can be adjusted. Since C_{ext} charges through $R_1 + R_2$ and discharges only through R_2, duty cycles approaching a minimum of 50 percent can be achieved if $R_2 >> R_1$ so that the charging and discharging times are approximately equal. A formula to calculate the percent duty cycle of a 555 astable multivibrator is

$$\text{Duty cycle} = \left(\frac{R_1 + R_2}{R_1 + 2R_2}\right)100\% \qquad (10\text{–}7)$$

To achieve duty cycles of less than 50 percent, the circuit in Figure 10–21 can be modified so that C_{ext} charges through only R_1 and discharges through R_2. This is achieved with a diode, D_1, placed as shown in Figure 10–22. The duty cycle can be made less than 50 percent by making R_1 less than R_2. Under this condition, the formula for the percent duty cycle is

$$\text{Duty cycle} = \left(\frac{R_1}{R_1 + R_2}\right)100\% \qquad (10\text{–}8)$$

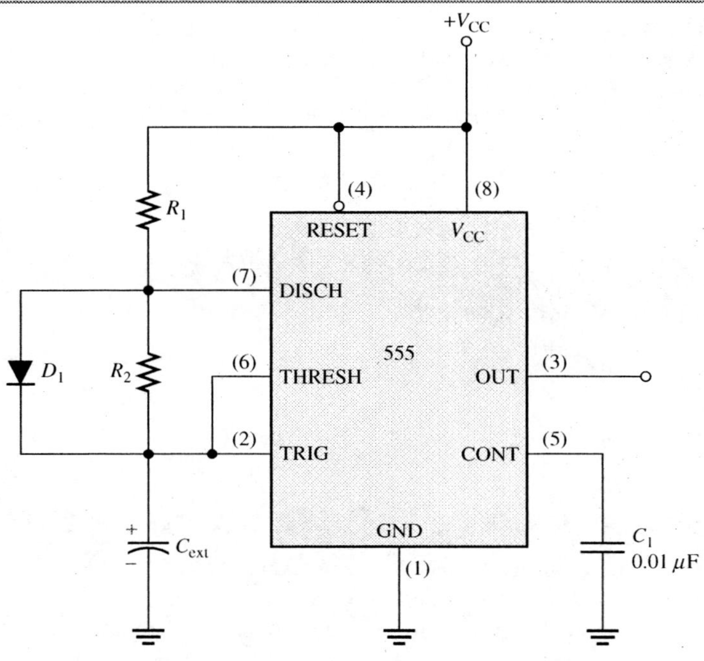

FIGURE 10–22

The addition of diode D_1 allows the duty cycle of the output to be adjusted to less than 50 percent by making $R_1 < R_2$.

EXAMPLE 10–4

Problem

A 555 timer configured to run in the astable mode (oscillator) is shown in Figure 10–23. Determine the frequency of the output and the duty cycle.

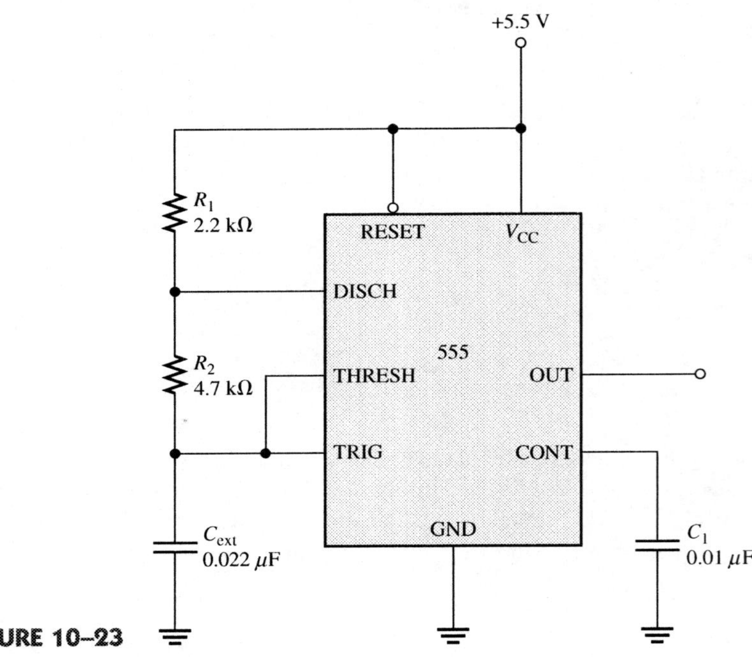

FIGURE 10–23

Solution

$$f = \frac{1.44}{(R_1 + 2R_2)C_{ext}} = \frac{1.44}{(2.2\ \text{k}\Omega + 9.4\ \text{k}\Omega)0.022\ \mu\text{F}} = \mathbf{5.64\ kHz}$$

$$\text{Duty cycle} = \left(\frac{R_1 + R_2}{R_1 + 2R_2}\right)100\% = \left(\frac{2.2\ \text{k}\Omega + 4.7\ \text{k}\Omega}{2.2\ \text{k}\Omega + 9.4\ \text{k}\Omega}\right)100\% = \mathbf{59.5\%}$$

Question

What is the duty cycle in Figure 10–23 if a diode is connected across R_2 as indicated in Figure 10–22?

COMPUTER SIMULATION

Open the Multisim file F10-23DV on the website. Measure the output frequency.

Operation as a Voltage-Controlled Oscillator (VCO)

The **voltage-controlled oscillator (VCO)** is a relaxation oscillator whose frequency can be changed by a variable dc control voltage. A 555 timer can be set up to operate as a VCO by using the same external connections as for astable operation, with the exception that a variable control voltage is applied to the CONT input (pin 5), as indicated in Figure 10–24.

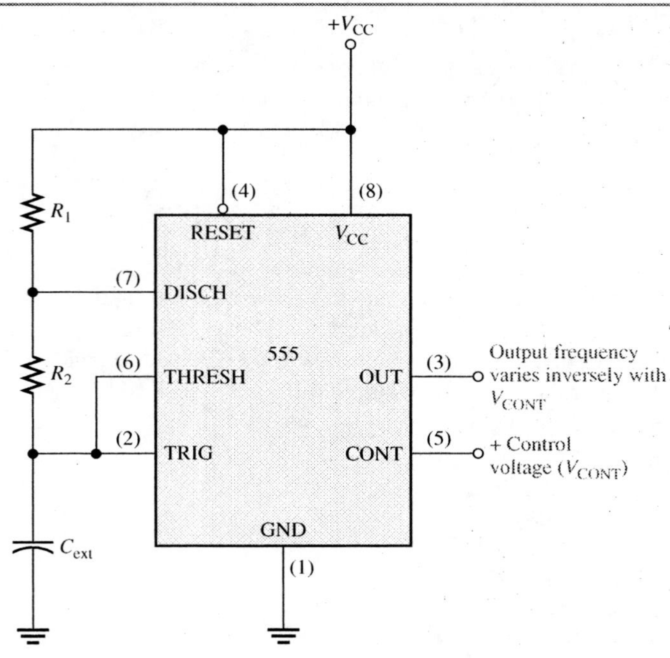

FIGURE 10–24

The 555 timer connected as a voltage-controlled oscillator (VCO). Note the variable control voltage input on pin 5.

For the capacitor voltage, as shown in Figure 10–25, the upper value is V_{CONT} and the lower value is $\frac{1}{2}V_{CONT}$. When the control voltage is varied, the output frequency also varies. An increase in V_{CONT} increases the charging and discharging time of the external capacitor and causes the frequency to decrease. A decrease in V_{CONT} decreases the charging and discharging time of the capacitor and causes the frequency to increase.

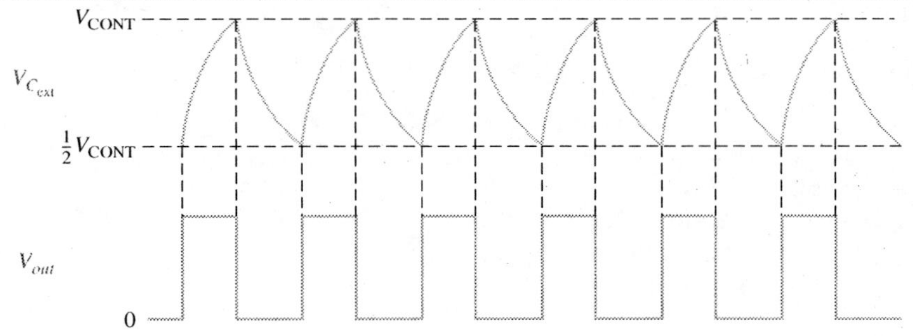

FIGURE 10–25

The VCO output frequency varies inversely with V_{CONT} because the charging and discharging time of C_{ext} is directly dependent on the control voltage.

An interesting application of the VCO is in phase-locked loops, which are used in various types of communications receivers to track variations in the frequency of incoming signals.

SAFETY NOTE

To avoid electrical shock, never touch a circuit while it is connected to a voltage source. If you need to handle a circuit, remove a component, or change a component, first make sure the voltage source is disconnected.

Review Questions

21. What the duty cycle of a pulse waveform?
22. When the 555 timer is configured as an astable multivibrator, how is the duty cycle determined?
23. When the 555 timer is used as a VCO, how is the frequency varied?
24. How do you reduce the duty cycle of a 555 oscillator to less than 50%?
25. What does VCO stand for?

10–6 THE 555 TIMER AS A ONE-SHOT

A **one-shot** is a monostable multivibrator that produces a single output pulse for each input trigger pulse. The term **monostable** means that the device has only one stable state.

In this section, you will learn how a 555 can be used as a one-shot.

A 555 timer connected for monostable or one-shot operation is shown in Figure 10–26. Compare this configuration to the one used for astable operation in Figure 10–21 and note the difference in the external circuit. When a one-shot is triggered, it temporarily goes to its unstable state but it always returns to its stable state. The time that it remains in its unstable state establishes the width of the output pulse and is set by the values of an external resistor and capacitor.

FIGURE 10–26

The 555 timer connected as a monostable multivibrator (one-shot).

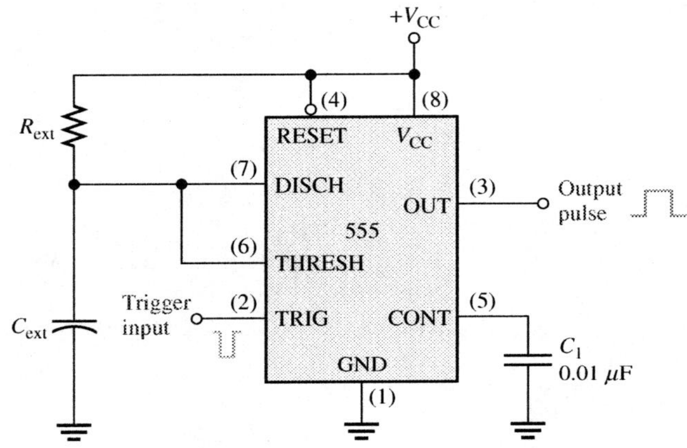

Monostable Operation

A negative-going input trigger pulse produces a single output pulse with a predetermined width. Once triggered, the one-shot cannot be retriggered until it completely times out; that is, it completes a full output pulse. Once it times out, the one-shot can then be triggered again to produce another output pulse. A low level on the reset input (RESET) can be used to prematurely terminate the output pulse. The width of the output pulse is determined by the following formula:

$$t_W = 1.1R_{ext}C_{ext} \tag{10–9}$$

EXAMPLE 10–5

Problem

A 555 timer is connected as a one-shot with $R_{ext} = 10\text{ k}\Omega$ and $C_{ext} = 0.1\ \mu\text{F}$. What is the pulse width of the output?

Solution

Use Equation 10–9.

$$t_W = 1.1R_{ext}C_{ext} = 1.1(10\text{ k}\Omega)(0.1\ \mu\text{F}) = \mathbf{1.1\text{ ms}}$$

Question

To what value must R_{ext} be changed to increase the one-shot's output pulse width to 5 ms?

Using One-Shots for Time Delay

In some applications, it is necessary to have a fixed time delay between certain events. Figure 10–27(a) shows two 555 timers connected as one-shots. The output of the first one-shot goes to the input of the second. When the first one-shot is triggered, it produces an output pulse whose width establishes a time delay. At the end of this pulse, the second one-shot is triggered. Therefore, there is an output pulse from the second one-shot that is delayed from the input trigger to the first one-shot by a time equal to the pulse width of the first one-shot, as indicated in the timing diagram in Figure 10–27(b).

Two one-shots produce a delayed output pulse. FIGURE 10–27

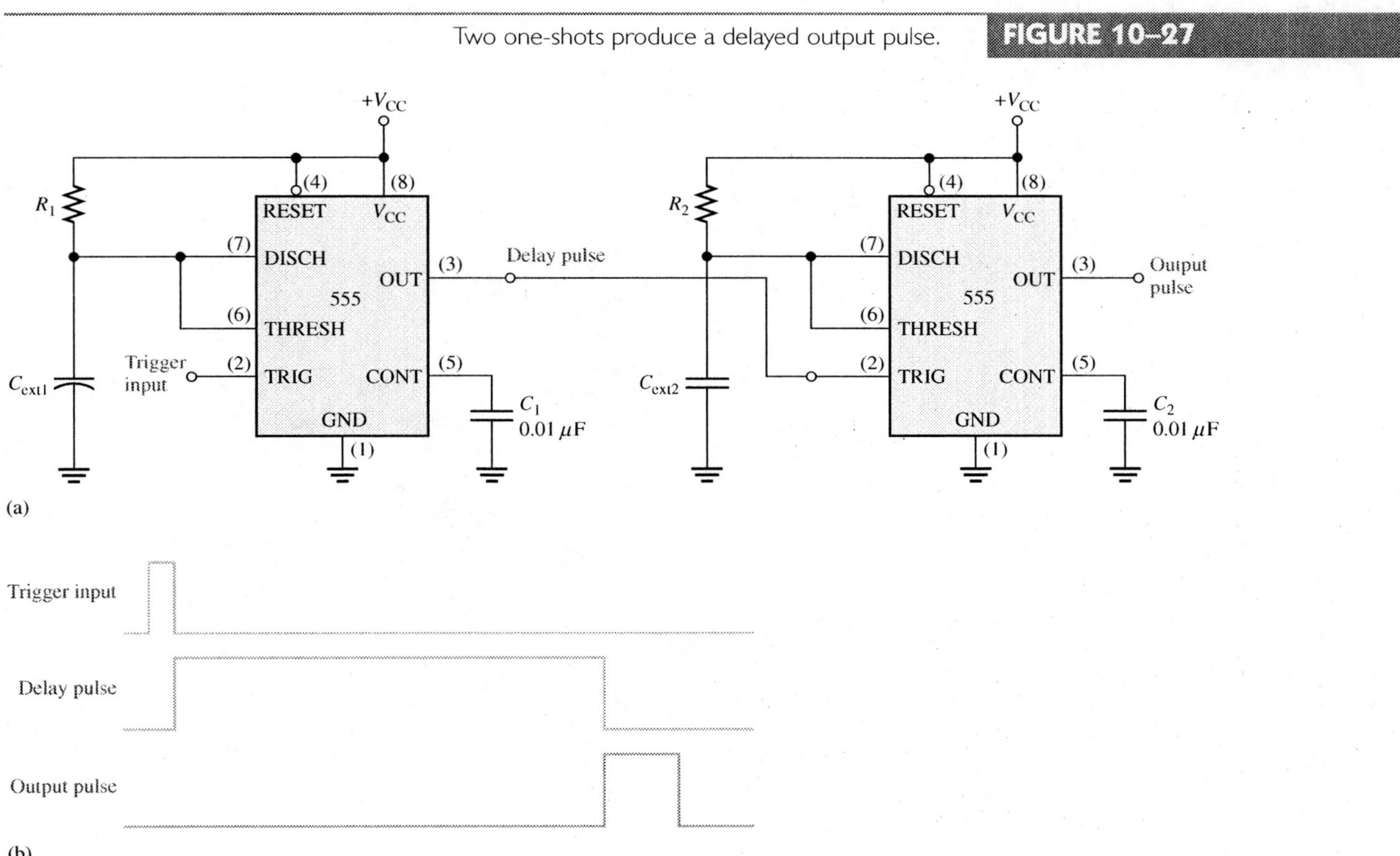

EXAMPLE 10–6

Problem

Determine the pulse widths and show the timing diagram (relationships of the input and output pulses) for the circuit in Figure 10–28.

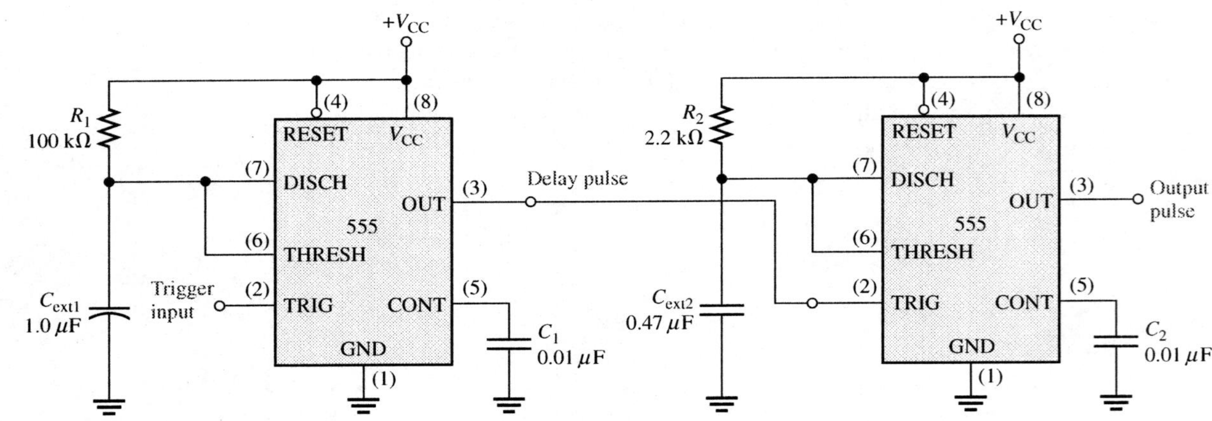

FIGURE 10–28

Solution

The time relationship of the inputs and outputs are shown in Figure 10–29. The pulse widths for the two one-shots are

$$t_{W1} = 1.1R_1C_{ext1} = 1.1(100\ \text{k}\Omega)(1.0\ \mu\text{F}) = \mathbf{110\ ms}$$
$$t_{W2} = 1.1R_2C_{ext2} = 1.1(2.2\ \text{k}\Omega)(0.47\ \mu\text{F}) = \mathbf{1.14\ ms}$$

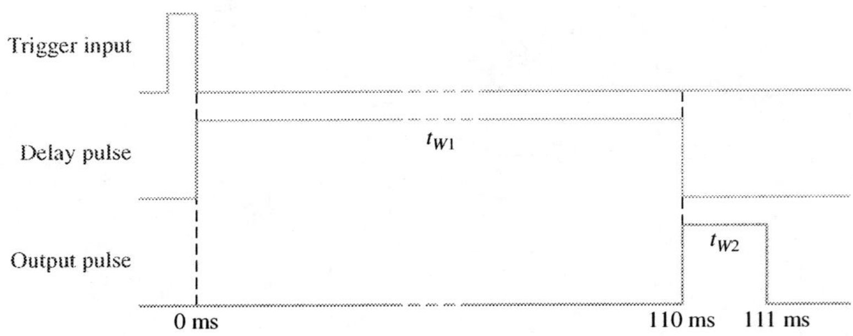

FIGURE 10–29

Question

What is a way that the circuit in Figure 10–28 can be modified so that the delay can be made adjustable from 10 ms to 200 ms?

Review Questions

26. What does *monostable* mean?
27. How many stable states does a one-shot have?
28. A certain 555 one-shot circuit has a time constant of 5 ms. What is the output pulse width?
29. How can you decrease the pulse width of a one-shot?
30. How many one-shots are used to produce a delayed output pulse?

Key Terms

Astable multivibrator A type of circuit that can operate as an oscillator and produces a pulse waveform output.

Feedback oscillator A type of oscillator that returns a fraction of output signal to the input with no net phase shift around the feedback loop resulting in a reinforcement of the output signal.

One-shot A monostable multivibrator that produces a single output pulse for each input trigger pulse.

Phase-shift oscillator A type of sinusoidal feedback oscillator that uses three *RC* circuits in the feedback loop.

Positive feedback A condition where an in-phase portion of the output voltage is fed back to the input.

Relaxation oscillator A type of oscillator that uses an *RC* timing circuit to generate a nonsinusoidal waveform.

Wien-bridge oscillator A type of sinusoidal feedback oscillator that uses an *RC* lead-lag circuit in the feedback loop.

Important Facts

- ❑ Feedback oscillators operate with positive feedback.
- ❑ The two conditions for positive feedback are the phase shift around the feedback loop must be 0° and the voltage gain around the feedback loop must equal 1.
- ❑ For initial start-up, the voltage gain around the feedback loop must be greater than 1.
- ❑ Sinusoidal feedback *RC* oscillators include the Wien-bridge, phase-shift, and twin-T.
- ❑ A relaxation oscillator uses an *RC* timing circuit and a device that changes states to generate a periodic waveform.
- ❑ The frequency in a voltage-controlled oscillator (VCO) can be varied with a dc control voltage.
- ❑ The 555 timer is an integrated circuit that can be used as an oscillator or as a one-shot by proper connection of external components.

Formulas

Wien-bridge positive feedback attenuation:

$$\frac{V_{out}}{V_{in}} = \frac{1}{3} \qquad \textbf{(10–1)}$$

Wien-bridge frequency:

$$f_r = \frac{1}{2\pi RC} \qquad \textbf{(10–2)}$$

Phase-shift feedback attenuation:

$$B = \frac{1}{29} \qquad \textbf{(10–3)}$$

Phase-shift oscillator frequency:

$$f_r = \frac{1}{2\pi\sqrt{6}RC} \quad \textbf{(10–4)}$$

Triangular wave generator frequency:

$$f \cong \frac{1}{4R_1C}\left(\frac{R_2}{R_3}\right) \quad \textbf{(10–5)}$$

555 astable frequency:

$$f = \frac{1.44}{(R_1 + 2R_2)C_{ext}} \quad \textbf{(10–6)}$$

555 astable (duty cycle ≥ 50%):

$$\text{Duty cycle} = \left(\frac{R_1 + R_2}{R_1 + 2R_2}\right)100\% \quad \textbf{(10–7)}$$

555 astable (duty cycle < 50% with diode):

$$\text{Duty cycle} = \left(\frac{R_1}{R_1 + R_2}\right)100\% \quad \textbf{(10–8)}$$

555 one-shot pulse width:

$$t_W = 1.1R_{ext}C_{ext} \quad \textbf{(10–9)}$$

Chapter Checkup

Answers are at the end of the chapter.

1. An oscillator differs from an amplifier because
   (a) it has more gain
   (b) it requires no input signal
   (c) it requires no dc supply
   (d) it always has the same output
2. Wien-bridge oscillators are based on
   (a) positive feedback
   (b) an *LC* circuit
   (c) the piezoelectric effect
   (d) high gain
3. One condition for oscillation is
   (a) a phase shift around the feedback loop of 180°
   (b) a gain around the feedback loop of one-third
   (c) a phase shift around the feedback loop of 0°
   (d) a gain around the feedback loop of less than one
4. A second condition for oscillation is
   (a) no gain around the feedback loop
   (b) a gain of one around the feedback loop
   (c) the attenuation of the feedback circuit must be one-third
   (d) the feedback circuit must be capacitive

5. In a certain oscillator, the attenuation of the feedback circuit is 0.02. A_v must be

(a) 1 (b) 3

(c) 10 (d) 50

6. For an oscillator to properly start, the gain around the feedback loop must initially be

(a) 1

(b) less than 1

(c) greater than 1

(d) equal to B

7. In a Wien-bridge oscillator, if the resistances in the feedback circuit are decreased, the frequency

(a) decreases

(b) increases

(c) remains the same

8. The Wien-bridge oscillator's positive feedback circuit is

(a) an *RL* circuit

(b) an *LC* circuit

(c) a voltage divider

(d) a lead-lag circuit

9. A phase-shift oscillator has

(a) three *RC* circuits

(b) three *LC* circuits

(c) a T-type circuit

(d) a π-type circuit

10. An oscillator whose frequency is changed by a variable dc voltage is known as

(a) a Wien-bridge oscillator

(b) a VCO

(c) a phase-shift oscillator

(d) an astable multivibrator

11. Which one of the following is not an input or output of the 555 timer?

(a) Threshold

(b) Control voltage

(c) Clock

(d) Trigger

(e) Discharge

(f) Reset

12. An astable multivibrator is

(a) an oscillator

(b) a one-shot

(c) a time-delay circuit

(d) characterized by having no stable states

(e) answers (a) and (d)

13. The output frequency of a 555 timer connected as an oscillator is determined by
    (a) the supply voltage
    (b) the frequency of the trigger pulses
    (c) the external *RC* time constant
    (d) the internal *RC* time constant
    (e) answers (a) and (d)
14. The term *monostable* means
    (a) one output
    (b) one frequency
    (c) one time constant
    (d) one stable state
15. A 555 timer connected as a one-shot has $R_{ext} = 2.0\ k\Omega$ and $C_{ext} = 2.0\ \mu F$. The output pulse has a width of
    (a) 1.1 ms (b) 4 ms
    (c) 4 μs (d) 4.4 ms
16. In Figure 10–30, if the op-amp dc supply voltage decreases, the frequency of oscillation will
    (a) increase (b) decrease
    (c) not change

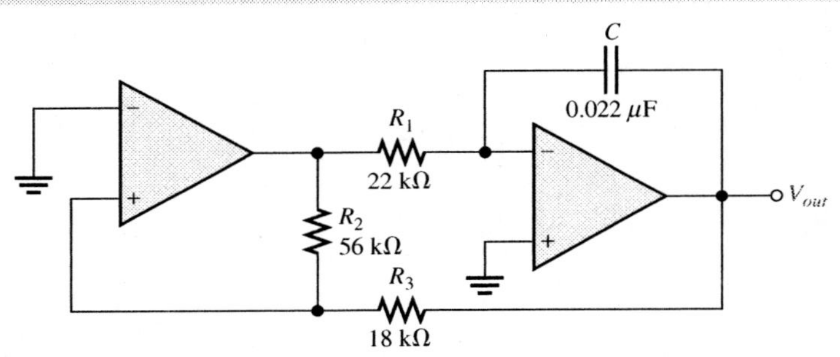

FIGURE 10–30

17. In Figure 10–30, if the op-amp dc supply voltage decreases, the amplitude of the triangular output will
    (a) increase
    (b) decrease
    (c) not change
18. In Figure 10–30, if the capacitor is larger, the frequency of oscillation will
    (a) increase
    (b) decrease
    (c) not change
19. In Figure 10–30, if the capacitor is larger, the output amplitude will
    (a) increase
    (b) decrease
    (c) not change

Questions

Answers to odd-numbered questions are at the end of the book.

1. Why does an oscillator with an inverting amplifier require a phase shift of 180° in the feedback circuit?
2. Why does an oscillator with a noninverting amplifier require a phase shift of either 0° (no phase shift) or 360° in the feedback circuit? Are 0° and 360° equivalent?
3. What is the phenomenon called in which a loop is created in an amplifier circuit so that a signal sustains itself and a continuous output waveform is produced?
4. Why must the loop gain of a feedback oscillator be equal to 1?
5. For self-starting, the initial loop gain of an oscillator must be what?
6. What type of circuit is used in the positive feedback loop of a Wien-bridge oscillator?
7. In a Wien-bridge oscillator, what is the attenuation of the lead-lag circuit at the frequency of oscillation?
8. What does the drain-to-source resistance of a JFET depend on?
9. If the R and C values in a phase shift oscillator are increased, what happens to the frequency of oscillation?
10. What is the difference between an astable multivibrator and a monostable multivibrator?
11. If the external capacitor value is halved in a 555 astable multivibrator, what happens to the frequency of oscillation? What happens to the duty cycle?
12. If the external resistor value is doubled in a 555 one-shot, what happens to the output pulse width?

Basic Problems

PROBLEMS

Answers to odd-numbered problems are at the end of the book.

1. Specify the type of input required for an oscillator.
2. Name the basic components of an oscillator circuit.
3. If the attenuation of the feedback circuit of a certain oscillator is 0.25, determine the voltage gain of the amplifier required to sustain oscillation.
4. Generally describe the change required in the oscillator of Problem 3 in order for oscillation to begin when the power is initially turned on.
5. Calculate the resonant frequency of a lead-lag circuit with the following values: $R_1 = R_2 = 6.2\text{ k}\Omega$, and $C_1 = C_2 = 0.02\ \mu\text{F}$.

6. Find the frequency of oscillation for the Wien-bridge oscillator in Figure 10–31.

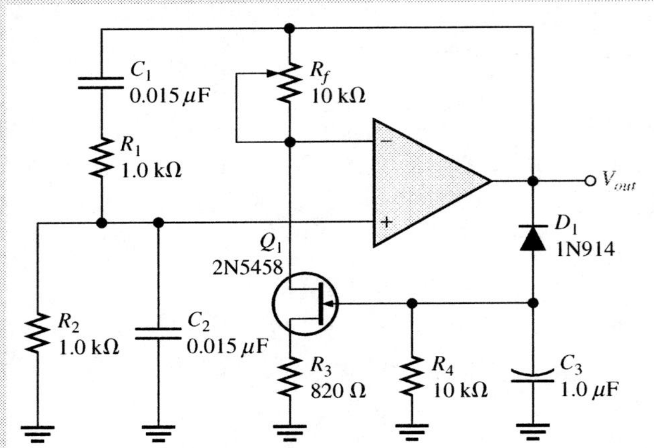

FIGURE 10–31

7. What value of R_f is required in Figure 10–32? What is f_r?

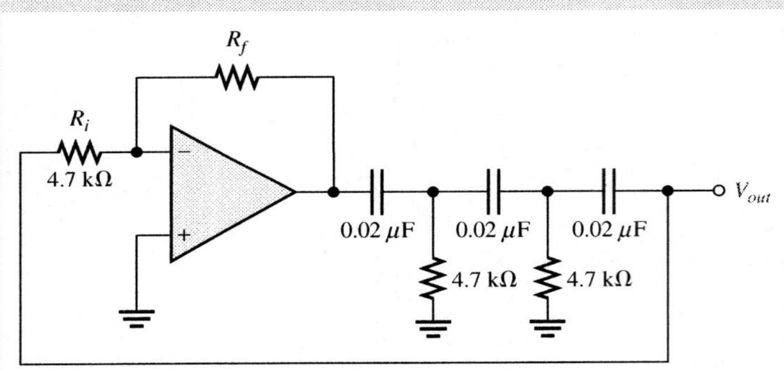

FIGURE 10–32

8. What type of signal does the circuit in Figure 10–33 produce? Determine the frequency of the output.

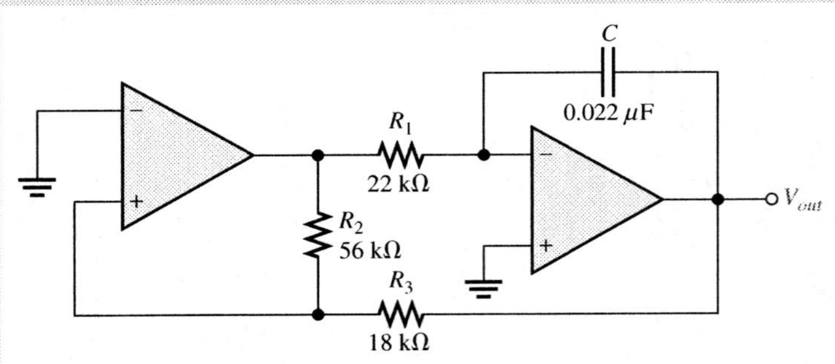

FIGURE 10–33

9. What are the two comparator reference voltages in a 555 timer when $V_{CC} = 10$ V?

10. Determine the frequency of oscillation for the 555 astable oscillator in Figure 10–34.

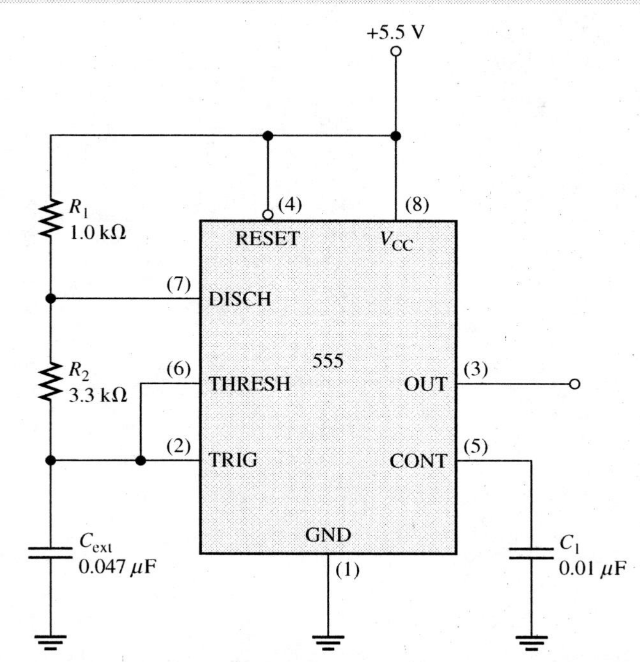

FIGURE 10–34

11. To what value must C_{ext} be changed in Figure 10–34 to achieve a frequency of 25 kHz?

12. A 555 timer connected in the monostable configuration has a 56 kΩ external resistor and a 0.22 μF external capacitor. What is the pulse width of the output?

Basic-Plus Problems

13. An equal-component lead-lag circuit has a resonant frequency of 3.5 kHz. What is the rms output voltage if an input signal with a frequency equal to f_r and with an rms value of 2.2 V is applied to the input?

14. For the Wien-bridge in Figure 10–31, calculate the setting for R_f, assuming the internal drain-source resistance, r'_{ds}, of the JFET is 350 Ω when oscillations are stable.

15. Show how to change the frequency of oscillation in Figure 10–33 to 10 kHz.

16. In an astable 555 configuration, the external resistor $R_1 = 3.3\ \text{k}\Omega$. What must R_2 equal to produce a duty cycle of 75 percent?

17. The output pulse width of a certain 555 one-shot is 12 ms. If $C_{ext} = 2.2\ \mu\text{F}$, what is R_{ext}?

18. Suppose that you need to hook up a 555 timer as a one-shot in the lab to produce an output pulse with a width of 100 μs. Select the appropriate values for the external components.

19. Devise a circuit to produce two sequential 50 μs pulses. The first pulse must occur 100 ms after an initial trigger and the second pulse must occur 300 ms after the first pulse ends.

20. Research and prepare a report on retriggerable monostable multivibrators. Cite the resources used in preparing your report.

ANSWERS

Example Questions

10–1: If R_f is too high, the output is distorted. If R_f is too small, oscillations cease.

10–2: (a) 238 kΩ

(b) 7.92 kHz

10–3: 6.06 V peak-to-peak

10–4: 31.9%

10–5: 45.5 kΩ

10–6: Replace R_1 with a potentiometer with a maximum resistance of at least 182 kΩ.

Review Questions

1. An oscillator is a circuit that produces a repetitive output waveform with only the dc supply voltage as an input.
2. Positive feedback
3. The feedback circuit provides attenuation and phase shift.
4. A relaxation oscillator does not use positive feedback.
5. The relaxation oscillator
6. Zero phase shift and unity voltage gain around the closed feedback
7. Positive feedback is when a portion of the output signal is fed back to the input of the amplifier such that it reinforces itself.
8. Loop gain greater than 1
9. $A_{cl} > 1$
10. B is feedback attenuation.
11. The negative feedback loop sets the closed-loop gain; the positive feedback loop sets the frequency of oscillation.
12. 1.67 V
13. The three RC circuits contribute a total of 180° and the inverting amplifier contributes 180° for a total of 360° around the loop.
14. High-pass and low-pass T-filters
15. The JFET stabilizes the voltage gain.
16. The basis of a relaxation oscillator is the charging and discharging of a capacitor.
17. Square waves and triangular
18. V_{UTP} is the upper trigger point voltage, and V_{LTP} is the lower trigger point voltage.
19. Triangular
20. An oscillator that produces more than one type of waveform
21. Duty cycle is the ratio of pulse width to period.
22. The duty cycle is set by the external resistors and the external capacitor.
23. The frequency of a VCO is varied by changing V_{CONT}.
24. Add a diode and make $R_1 < R_2$.
25. Voltage-controlled oscillator
26. One stable state
27. A one-shot has one stable state.

28. $t_W = 5.5$ ms

29. The pulse width can be decreased by decreasing the external resistance or capacitance.

30. Two

Chapter Checkup

1. (b)	2. (a)	3. (c)	4. (b)	5. (d)
6. (c)	7. (b)	8. (d)	9. (a)	10. (b)
11. (c)	12. (e)	13. (c)	14. (d)	15. (d)
16. (c)	17. (b)	18. (a)	19. (a)	

CHAPTER 11

VOLTAGE REGULATORS

CHAPTER OUTLINE

Study aids for this chapter are available at

http://www.prenhall.com/SOE

INTRODUCTION

A voltage **regulator** is an electronic circuit that provides a constant dc output voltage that is practically independent of the input voltage, output load current, and temperature. The voltage regulator is one part of a power supply. Its input voltage comes from the filtered output of a rectifier derived from an ac voltage or from a battery in the case of portable systems.

Most voltage regulators fall into two broad categories—linear regulators and switching regulators. In the linear regulator category, two general types are the linear series regulator and the linear shunt regulator. These are normally available for either positive or negative output voltages. A dual regulator provides both positive and negative outputs. In the switching regulator category, three general configurations are step-down, step-up, and inverting.

Switching regulators are also widely used for computers and applications where small size and high efficiency are needed. In this chapter, a specific IC switching regulator is introduced as representative of the wide range of available devices. In addition, the three-terminal fixed voltage regulators introduced in Section 2–6 are covered in further detail.

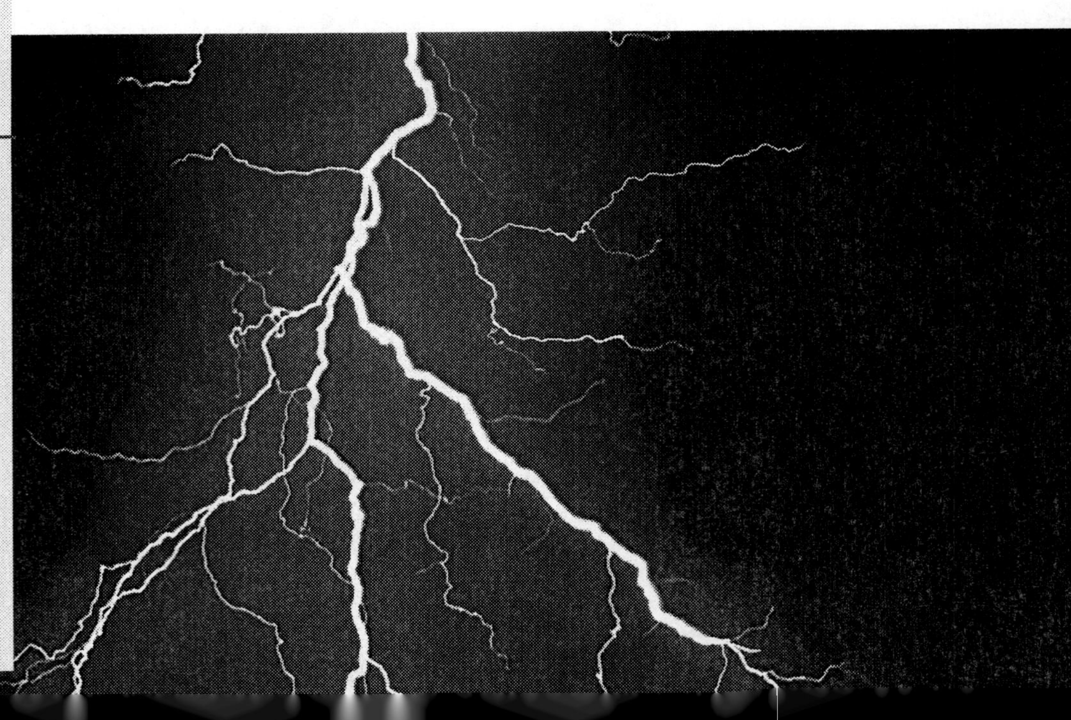

KEY OBJECTIVES

A section number is given for each objective. After completing this chapter, you should be able to

- **11–1** Given specific voltage parameters, calculate the line and load regulation of a regulator
- **11–2** Discuss the operation of a series load regulator
- **11–3** Discuss the operation of a shunt load regulator
- **11–4** Explain how switching regulators operate
- **11–5** Explain how to configure three-terminal regulators to obtain higher current and how to use them as a current source

COMPUTER SIMULATIONS DIRECTORY

The following figure has a Multisim circuit file associated with it.

- Figure 11–6 Page 386

LABORATORY EXPERIMENTS DIRECTORY

The following exercise is for this chapter.

- **Experiment 21** Voltage Regulators

KEY TERMS

- Line regulation
- Load regulation
- Linear regulator
- Switching regulator

ON THE JOB. . .

(Getty Images)

In many types of technical jobs, you will be required to write reports on projects. Reporting may be anything from simple logbook entries to formal technical reports. If you are involved in writing reports, find out the policy of the company. Some may require entries in ink and signed; others may have less formal requirements. If you are writing an explanation, keep in mind that an illustration of a problem may be a valuable addition to your report. If you are not a good speller, write out your report on a word processor and don't forget to use the spell checker!

As you have seen, there are two feedback mechanisms that work in electronic systems. Negative feedback tends to make a system stable, so it is used in amplifiers. Positive feedback tends to destabilize a system, so it is used in oscillators where it tends to reinforce the output.

In addition to electronic applications, many other systems in science have feedback mechanisms at work—both positive and negative feedback. Negative feedback tends to stabilize these systems, and positive feedback tends to destabilize them. One very large system for which many feedback mechanisms are at work has to do with the buildup of atmospheric carbon in the form of carbon dioxide, which contributes to global warming. Consider what might happen in your home if the signals from your thermostat were wired backwards. As the house warmed, the thermostat would call for even more heat. This is the kind of positive feedback mechanism that would cause a "runaway" temperature rise in your home.

This same thing that happens with a reversed thermostat appears to be happening globally with some of the feedback mechanisms known to be at work in the atmosphere. To cite one example, a large amount of carbon is "locked-up" in artic plants ("tundra"). The plants exchange carbon with the atmosphere. As temperatures warm, the tundra tends to emit more carbon dioxide, contributing to further warming. There may be other compensating feedback mechanisms. The full effects are not understood at this time, but show the need for further study in this complex issue.

11–1 VOLTAGE REGULATION

The requirement for a reliable source of constant voltage in virtually all electronic systems has led to many advances in power supply design. Designers have used feedback and operational amplifiers, as well as pulse circuit techniques, to develop reliable constant-voltage (and constant-current) power supplies. The heart of any regulated supply is the ability to establish a constant-voltage reference.

In this section, you will learn to calculate the line and load regulation of a regulator, given specific voltage parameters.

Line Regulation

Line regulation was introduced in Section 2–6. **Line regulation** is a measure of the ability of a power supply to maintain a constant output for changes in the input voltage. It is typically defined as a ratio of a change in output for a corresponding change in the input and expressed as a percentage.

$$\text{Line regulation} = \left(\frac{\Delta V_{\text{OUT}}}{\Delta V_{\text{IN}}}\right)100\% \qquad (11\text{–}1)$$

This equation was given earlier as Equation 2-3. Some specification sheets show line regulation differently. It can be specified as a percentage change in the output voltage per volt divided by change in the input voltage. In this case, line regulation is defined and expressed as a percentage as

$$\text{Line regulation} = \left(\frac{\Delta V_{\text{OUT}}/V_{\text{OUT}}}{\Delta V_{\text{IN}}}\right)100\% \qquad (11\text{–}2)$$

Because this definition is different, you need to be sure which definition is used when reading specifications. The key in a specification sheet is to look at the units. If the specification is a ratio of mV/V or other pure number, then Equation 11–1 is the defining equation. If the units are shown as %/mV or %/V, then Equation 11–2 is the defining equation.

EXAMPLE 11–1

Problem

When the input to a particular voltage regulator decreases by 5 V, the output decreases by 0.25 V. The nominal output is 15 V. Determine the line regulation expressed as a percentage and in units of %/V.

Solution

From Equation 11–1, the percent line regulation is

$$\text{Line regulation} = \left(\frac{\Delta V_{OUT}}{\Delta V_{IN}}\right)100\% = \left(\frac{0.25\ \text{V}}{5\ \text{V}}\right)100\% = \mathbf{5\%}$$

From Equation 11–2, the percent line regulation is

$$\text{Line regulation} = \left(\frac{\Delta V_{OUT}/V_{OUT}}{\Delta V_{IN}}\right)100\% = \left(\frac{0.25\ \text{V}/15\ \text{V}}{5\ \text{V}}\right)100\% = \mathbf{0.33\ \%/V}$$

Question*

The input of a certain regulator increases by 3.5 V. As a result, the output voltage increases by 0.42 V. The nominal output is 20 V. What is the regulation expressed as a percentage and in units of %/V?

Load Regulation

Load regulation was introduced in Section 2–6. When the amount of current through a load changes due to a varying load resistance, the voltage regulator must maintain a nearly constant output voltage across the load. The percent **load regulation** specifies how much change occurs in the output voltage over a certain range of load current values, usually from minimum current (no load, NL) to maximum current (full load, FL). Ideally, the percent load regulation is 0%. It can be calculated and expressed as a percentage with the following formula:

$$\text{Load regulation} = \left(\frac{V_{NL} - V_{FL}}{V_{FL}}\right)100\% \qquad (11\text{–}3)$$

where V_{NL} is the output voltage with no load and V_{FL} is the output voltage with full (maximum) load. This equation was given earlier as Equation 2–4. Equation 11–3 is expressed as a change due only to changes in load conditions; all other factors (such as input voltage and operating temperature) must remain constant. Normally, the operating temperature is specified as 25°C.

Sometimes power supply manufacturers specify the equivalent output resistance of a power supply (R_{OUT}) instead of its load regulation. An equivalent Thevenin circuit can be drawn for any two-terminal linear circuit. Figure 11–1 shows the equivalent Thevenin circuit for a power supply with a load resistor. The Thevenin voltage is the voltage from the supply with no load

* *Answers are at the end of the chapter.*

FIGURE 11–1

Thevenin equivalent circuit for a power supply with a load resistor.

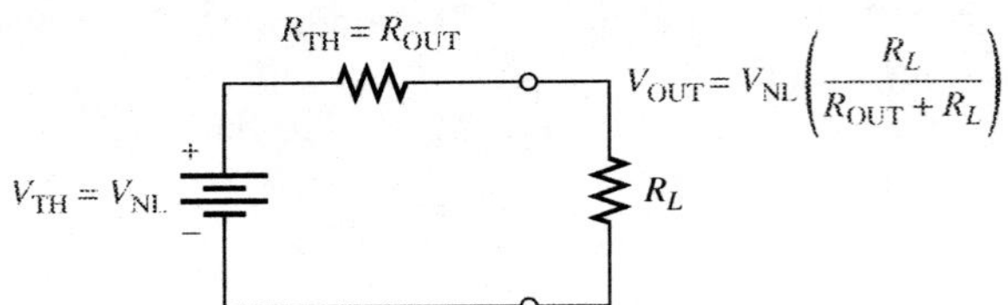

(V_{NL}), and the Thevenin resistance is the specified output resistance, R_{OUT}. Ideally, R_{OUT} is zero, corresponding to 0% load regulation, but in practical power supplies R_{OUT} is a small value. With the load resistor in place, the output voltage is found by applying the voltage-divider rule:

$$V_{OUT} = V_{NL}\left(\frac{R_L}{R_{OUT} + R_L}\right)$$

If we let R_{FL} equal the smallest-rated load resistance (largest-rated current), then the full-load output voltage (V_{FL}) is

$$V_{FL} = V_{NL}\left(\frac{R_{FL}}{R_{OUT} + R_{FL}}\right)$$

By rearranging and substituting into Equation 11–3, load regulation can be expressed in terms of the output resistance and smallest load resistor as

$$\text{Load regulation} = \left(\frac{R_{OUT}}{R_{FL}}\right)100\% \qquad (11\text{–}4)$$

Equation 11–4 is a useful way of finding the percent load regulation when the output resistance and minimum load resistance are specified.

Alternately, the load regulation can be expressed as a percentage change in output voltage for each mA change in load current. For example, a load regulation of 0.01 %/mA means that the output voltage changes 0.01% when the load current increases or decreases by 1 mA.

EXAMPLE 11–2

Problem

A certain voltage regulator has a +12.1 V output when there is no load ($I_L = 0$) and has a rated output current of 200 mA. With maximum current, the output voltage drops to +12.0 V. Determine the percentage load regulation and find the percent load regulation per mA change in load current.

Solution

The no-load output voltage is

$$V_{NL} = 12.1\text{ V}$$

The full-load output voltage is

$$V_{FL} = 12.0\text{ V}$$

The percent load regulation is

$$\text{Load regulation} = \left(\frac{V_{NL} - V_{FL}}{V_{FL}}\right)100\% = \left(\frac{12.1\text{ V} - 12.0\text{ V}}{12.0\text{ V}}\right)100\% = \mathbf{0.83\%}$$

The load regulation can also be expressed as

$$\text{Load regulation} = \frac{0.83\%}{200\text{ mA}} = \mathbf{0.0042\ \%/mA}$$

Question
What is the equivalent output resistance for this power supply?

Review Questions

Answers are at the end of the chapter.

1. What is line regulation?
2. What is load regulation?
3. The input of a certain regulator increases by 3.5 V. As a result, the output voltage increases by 0.042 V. The nominal output is 20 V. What is the line regulation expressed as a percent?
4. What is the line regulation for the regulator in question 3, expressed in %/V?
5. If a 5.0 V power supply has an output resistance of 80 mΩ and a specified maximum output current of 1.0 A, what is the load regulation expressed as a percent?

LINEAR SERIES REGULATORS 11–2

The fundamental classes of voltage regulators are linear regulators and switching regulators. Both of these are available in integrated circuit form. Two basic types of linear regulators are the series regulator and the shunt regulator.

In this section, you will learn the operation of a series regulator.

A **linear regulator** is a voltage regulator in which the control element operates in the linear region. A simple representation of a series type of linear regulator is shown in Figure 11–2(a), and the basic components are shown in the block diagram in Figure 11–2(b). Notice that the control element is in series with the load between input and output. The output

FIGURE 11–2

Simple series voltage regulator block diagram. The sample circuit "picks off" a part of the output, which is the feedback voltage, V_{FB}.

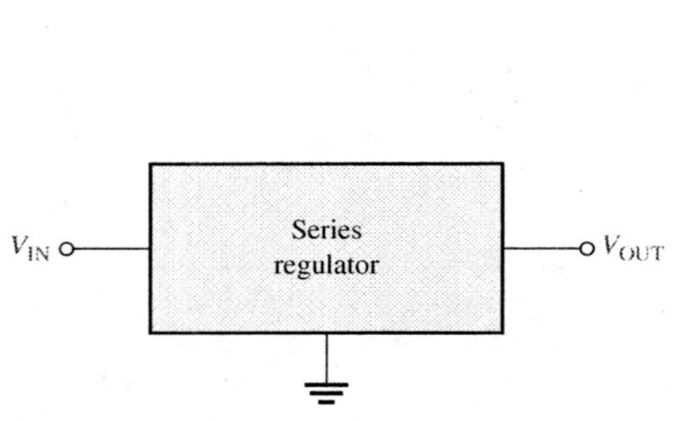

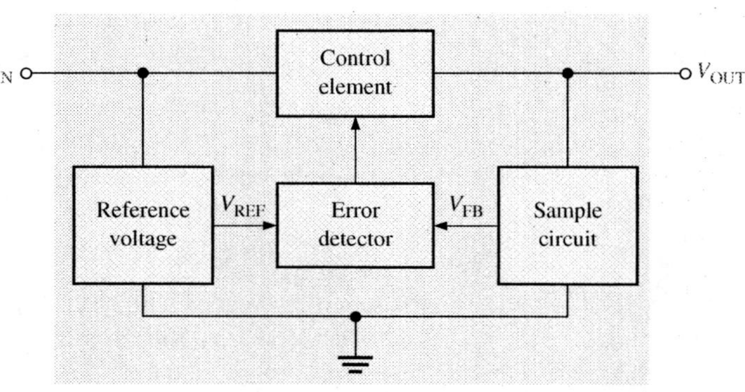

sample circuit senses a change in the output voltage and returns the sample as a feedback voltage, V_{FB}. The error detector compares the feedback voltage with a reference voltage and causes the control element to compensate in order to maintain a constant output voltage.

Voltage References

The ability of a voltage regulator to provide a constant output is dependent on the stability of a voltage reference to maintain a constant voltage for any change in temperature or other condition. Traditionally zener diodes (discussed in Section 2–6) were used as voltage references and are shown in many of the circuits in this chapter. Zeners are designed to break down at a specific voltage and maintain a fairly constant voltage if the current in the zener is constant and the temperature does not change. The drawback to zener diodes is they tend to be noisy and the zener voltage may change slightly as the zener ages (this is called *drift*). An even more serious effect is that the zener voltage is sensitive to temperature changes; the zener voltage can change hundreds of parts per million (ppm) for a change of just 1°C in temperature. This temperature effect varies widely among different types of zeners.

Special zener diode ICs have been designed to serve as voltage references with very low temperature drift (less than 10 ppm/°C). For low-voltage applications, zener-diode references are available that look and behave as diodes (but actually contain circuits to enhance their specifications). In the 8 V to 12 V range, two-terminal devices such as the LM329 and LM399 provide high stability and low noise, and have excellent temperature stability. The circuit symbol, which is the same as for a standard zener diode, and the internal construction of a representative IC voltage reference are shown in Figure 11–3. The circuit shown is called a *bandgap reference*. It is designed so that positive and negative temperature coefficients cancel, producing a reference voltage with almost no temperature coefficient. It uses a current mirror (Q_1) to set a particular current in Q_2. The output of the circuit is the sum of V_{BE} (from Q_3) and the voltage drop across R_2 (V_{R2}).

FIGURE 11–3

An IC voltage reference. The reference shown is a bandgap type that has a very small temperature coefficient.

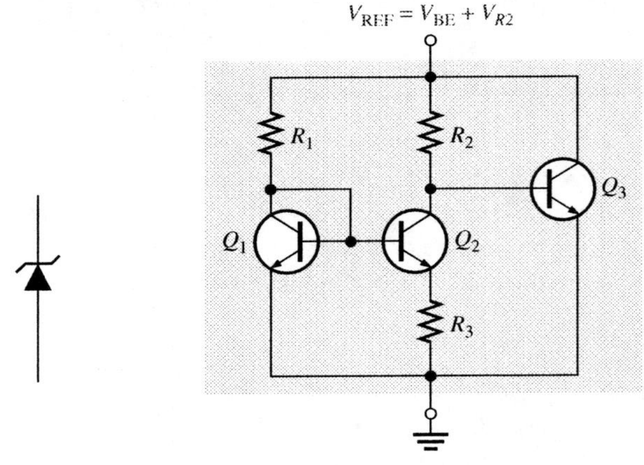

(a) Symbol (b) Internal construction

A more complicated voltage reference is the REF102 precision-voltage reference. The drift is laser trimmed to 2.5 ppm/°C maximum. It is a 10.00 V reference that is within 2.5 mV of this value. It uses a zener diode and op-amp in an 8-pin package.

Regulating Action

A basic op-amp series regulator circuit is shown in Figure 11–4. The operation of the series regulator is illustrated in Figure 11–5. The resistive voltage divider formed by R_2 and R_3 senses any change in the output voltage, and returns a feedback voltage, V_{FB}, to the error detector. Because of negative feedback, $V_{REF} \cong V_{FB}$.

FIGURE 11–4

Basic op-amp series regulator.

Control element
V_{IN}
V_{OUT}
R_1
Q_1
V_{REF}
Error detector
R_2
D_1
Sample circuit
V_{FB}
R_3

FIGURE 11–5

Illustration of series regulator action that keeps V_{OUT} constant when V_{IN} changes.

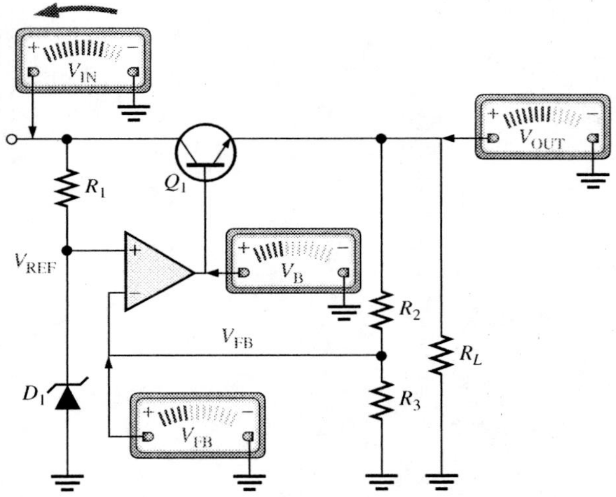

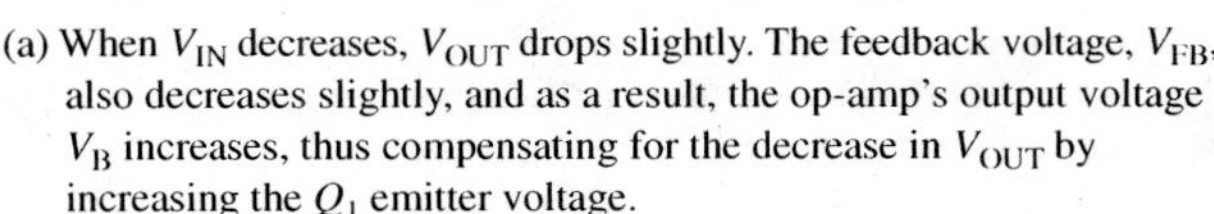
(a) When V_{IN} decreases, V_{OUT} drops slightly. The feedback voltage, V_{FB}, also decreases slightly, and as a result, the op-amp's output voltage V_B increases, thus compensating for the decrease in V_{OUT} by increasing the Q_1 emitter voltage.

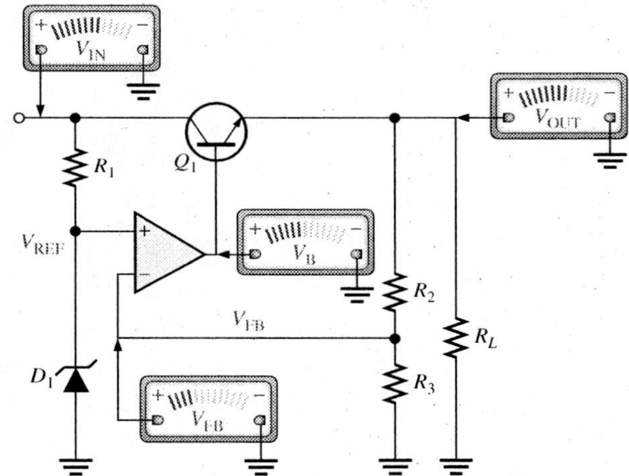

(b) When V_{IN} stabilizes at its new lower value, V_{OUT} will be nearly the same as before due to negative feedback.

Any attempt to change the output is "automatically" corrected. For example, assume the input voltage drops a small amount. Figure 11–5(a) illustrates what happens. A proportional voltage decrease is applied to the op-amp's inverting input by the voltage divider. Since the zener diode (D_1) holds the other op-amp input at a nearly fixed reference voltage, V_{REF}, a small difference voltage (error voltage) is developed across the op-amp's inputs. This difference voltage is amplified, and the op-amp's output voltage increases. This increase is applied to the base of Q_1, causing the emitter voltage V_{OUT} to increase until the voltage to the inverting input again equals the reference (zener) voltage. This action offsets the attempted decrease in output voltage, thus keeping it nearly constant, as shown in part (b).

The opposite action occurs when the output tries to increase for any reason. The op-amp in the series regulator is actually connected as a noninverting amplifier where the reference voltage V_{REF} is the input at the noninverting terminal, and the R_2/R_3 voltage divider forms the negative feedback network. The closed-loop voltage gain is

$$A_{cl} = 1 + \frac{R_2}{R_3}$$

SAFETY NOTE

Do not use any electrical equipment until you have familiarized yourself with its operation, know the proper operating procedures, and are aware of any potential hazards. Use equipment with three-wire power cords (three-prong plug). Make sure that the power cord is in good condition and that the grounding pin is not bent.

The base-emitter voltage (V_{BE}) of Q_1 is not included in this equation because it is inside the feedback loop. It acts like the active diode circuits discussed in Section 7–4. Therefore, the regulated output voltage of the series regulator is

$$V_{OUT} = \left(1 + \frac{R_2}{R_3}\right)V_{REF} \tag{11–5}$$

From this analysis, you can see that the output voltage is determined by the zener voltage (V_{REF}) and the feedback ratio of R_2/R_3. It is relatively independent of the input voltage, and therefore, regulation is achieved (as long as the input voltage and load current are within specified limits).

EXAMPLE 11–3

Problem

Determine the output voltage for the regulator in Figure 11–6 and the base voltage of Q_1.

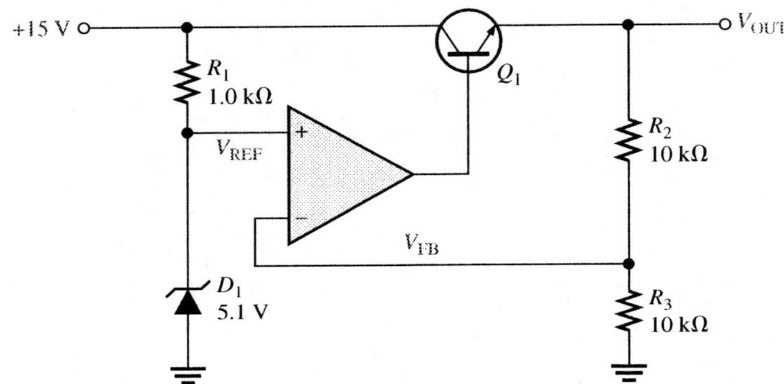

FIGURE 11–6

Solution

$V_{REF} = 5.1$ V, the zener voltage. The regulated output voltage is therefore

$$V_{OUT} = \left(1 + \frac{R_2}{R_3}\right)V_{REF} = \left(1 + \frac{10\text{ k}\Omega}{10\text{ k}\Omega}\right)5.1\text{ V} = (2)5.1\text{ V} = \mathbf{10.2\text{ V}}$$

The base voltage of Q_1 is

$$V_B = 10.2\text{ V} + V_{BE} = 10.2\text{ V} + 0.7\text{ V} = \mathbf{10.9\text{ V}}$$

Question

The following changes are made in the circuit in Figure 11–6: A 3.3 V zener replaces the 5.1 V zener, $R_1 = 1.8$ kΩ, $R_2 = 22$ kΩ, and $R_3 = 18$ kΩ. What is the output voltage?

COMPUTER SIMULATION

Open the Multisim file F11-06DV, on the website. Measure V_{IN}, V_{OUT}, V_{REF}, and V_{FB}.

Review Questions

6. What are the four basic components in a series regulator?
7. What are the two inputs to the error detector of a series regulator?
8. What are advantages of an integrated circuit reference over a zener diode reference?
9. A certain series regulator has an output voltage of 8 V. If the op-amp's closed loop gain is 4, what is the value of the reference voltage?
10. For the series regulator in Question 9, what is the value of the feedback voltage?

LINEAR SHUNT REGULATORS 11–3

The second basic type of linear voltage regulator is the shunt regulator. In the shunt regulator, the control element is a transistor in parallel (shunt) with the load.

In this section, you will learn the operation of a shunt regulator.

A simple representation of a shunt type of linear regulator is shown in part (a) of Figure 11–7, and the basic components are shown in the block diagram in part (b).

FIGURE 11–7 Simple shunt regulator and block diagram.

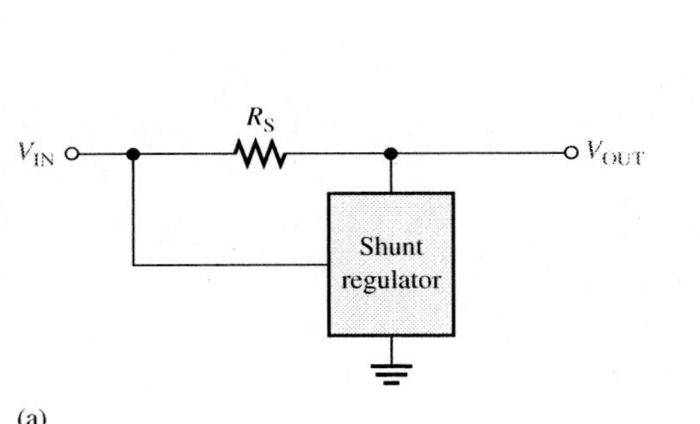

(a)

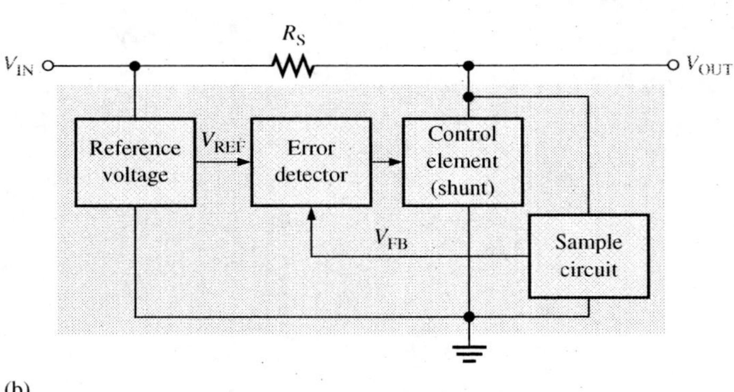

(b)

In the basic shunt regulator, the control element is a transistor, Q_1, in parallel with the load, as shown in Figure 11–8. A resistor, R_S is in series with the load. The operation of the circuit is similar to that of the series regulator, except that regulation is achieved by

FIGURE 11–8 Basic op-amp shunt regulator.

controlling the current through the parallel transistor Q_1. Notice that the feedback voltage, V_{FB}, is on the noninverting input of the error detector in this configuration.

When the output voltage tries to decrease due to a change in input voltage or load current caused by a change in load resistance, as shown in Figure 11–9, the attempted decrease is sensed by R_2 and R_3 and applied to the op-amp's noninverting input. The resulting difference voltage reduces the op-amp's output (V_B), driving Q_1 less, thus reducing its collector current (shunt current) and increasing its internal collector-to-emitter resistance r_{CE}. This action offsets the attempted decrease in V_{OUT} and maintains it at an almost constant level.

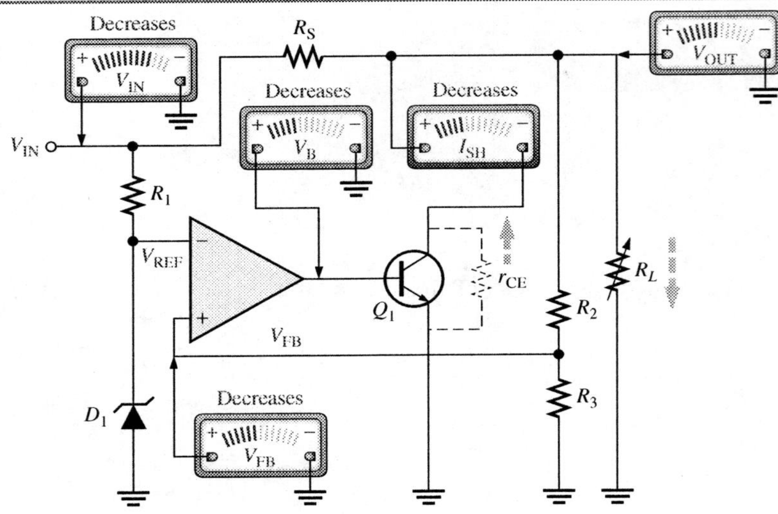

FIGURE 11–9

Sequence of responses when V_{OUT} tries to decrease as a result of a decrease in R_L or V_{IN}.

The opposite action occurs when the output tries to increase. With I_L and V_{OUT} constant, a change in the input voltage produces a change in shunt current (I_{SH}) as follows:

$$\Delta I_{SH} = \frac{\Delta V_{IN}}{R_S}$$

With a constant V_{IN} and V_{OUT}, a change in load current causes an opposite change in shunt current.

$$\Delta I_{SH} = -\Delta I_L$$

This formula says that if I_L increases, I_{SH} decreases, and vice versa. The shunt regulator is less efficient than the series type but offers inherent short-circuit protection. If the output is shorted ($V_{OUT} = 0$), the load current is limited by the series resistor R_S to a maximum value as follows ($I_{SH} = 0$).

$$I_{L(max)} = \frac{V_{IN}}{R_S} \qquad (11\text{–}6)$$

If a short is accidentally placed across the output, the resulting current is not enough to damage any components. This is an advantage of the shunt regulator.

Equation 11–5, developed for the series regulator, also applies to the shunt regulator. Resistors R_2 and R_3 determine the gain of the op-amp set up as a noninverting amplifier. Again, the base-emitter junction is inside the feedback loop, so it is not included in the equation for gain.

EXAMPLE 11–4

Problem
For the circuit in Figure 11–10, what is the output voltage?

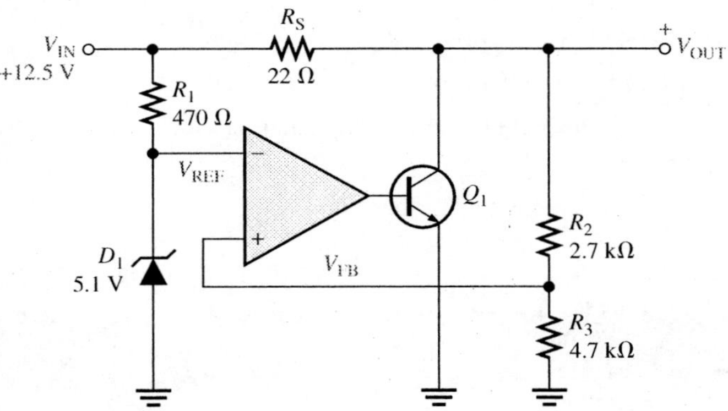

FIGURE 11–10

Solution
Apply Equation 11–5.

$$V_{OUT} = \left(1 + \frac{R_2}{R_3}\right)V_{REF} = \left(1 + \frac{2.7\text{ k}\Omega}{4.7\text{ k}\Omega}\right)5.1\text{ V} = \mathbf{8.03\ V}$$

Question
How could you change the circuit to enable the output to be varied between 5.1 V and 8.0 V?

EXAMPLE 11–5

Problem
In Figure 11–10, what power rating must R_S have if the maximum input voltage is 12.5 V?

Solution
The worst-case power dissipation in R_S occurs when the output is short-circuited. $V_{OUT} = 0$, and when $V_{IN} = 12.5$ V, the voltage dropped across R_S is $V_{IN} - V_{OUT} = 12.5$ V. The power dissipation in R_S is

$$P_{R_S} = \frac{V_{R_S}^2}{R_S} = \frac{(12.5\text{ V})^2}{22\ \Omega} = 7.1\text{ W}$$

Therefore, a resistor with at least a **10 W** rating should be used.

Question
In Figure 11–10, R_S is changed to 33 Ω. What must be the power rating of R_S if the maximum input voltage is 24 V?

Example 11–5 illustrates the major drawback to the shunt regulator. The series resistor dissipates a lot of power, so the efficiency is poor. For higher current requirements the series regulator is a better choice.

Review Questions

11. How does the control element in a shunt regulator differ from that in a series regulator?
12. What is one advantage of a shunt regulator over a series type?
13. What is a disadvantage of a shunt regulator over a series regulator?
14. Why isn't the V_{BE} of the transistor included in the gain calculation for either the series or shunt regulator?
15. For the circuit in Figure 11–10, what is the current in R_1?

11–4 SWITCHING REGULATORS

The switching regulator is different from the linear regulator; the control element operates as a switch rather than in the linear region. A greater efficiency can be realized with this type of voltage regulator than with the linear types because the transistor is not always conducting.

In this section, you will learn how switching regulators operate.

A **switching regulator** is a very efficient dc to dc converter that can change an unregulated dc input to another voltage, and produce a regulated output. There are three basic configurations of switching regulators: step-down, step-up, and inverting. All switching regulators work by turning on and off the input very rapidly (from 10 kHz to 100 kHz) creating a pulse train, which is then filtered. The voltage level of the output is controlled by changing the on and off times of the pulses. The rapid switching makes the output easier to filter, but has a drawback in that the output is prone to radiate radio frequency interference (RFI). To minimize interference, switching regulators need to be well shielded.

Step-Down Configuration

In the step-down configuration, the output voltage is always less than the input voltage. A basic step-down switching regulator is shown in Figure 11–11(a), and its simplified equivalent is shown in Figure 11–11(b). Transistor Q_1 is used to switch the input voltage at a duty cycle

FIGURE 11–11 Basic step-down switching regulator.

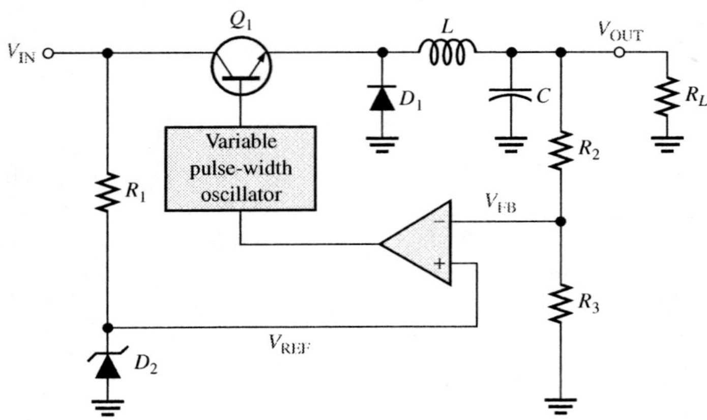

(a) Typical circuit

(b) Simplified equivalent circuit

that is based on the regulator's load requirement. The LC filter is then used to average the switched voltage. Since Q_1 is either *on* (saturated) or *off*, the power lost in the control element is relatively small. Therefore, the switching regulator is useful primarily in higher power applications or in applications such as computers where efficiency is of utmost concern.

The on and off intervals of Q_1 are shown in the waveform of Figure 11–12(a). The capacitor charges during the on-time (t_{on}) and discharges during the off-time (t_{off}). When the on-time is increased relative to the off-time, the capacitor charges more, thus increasing the output voltage, as indicated in Figure 11–12(b). When the on-time is decreased relative to the off-time, the capacitor discharges more, thus decreasing the output voltage, as in Figure 11–12(c). Therefore, by adjusting the duty cycle, $t_{on}/(t_{on} + t_{off})$, of Q_1, the output voltage can be varied. The inductor further smooths the fluctuations of the output voltage caused by the charging and discharging action.

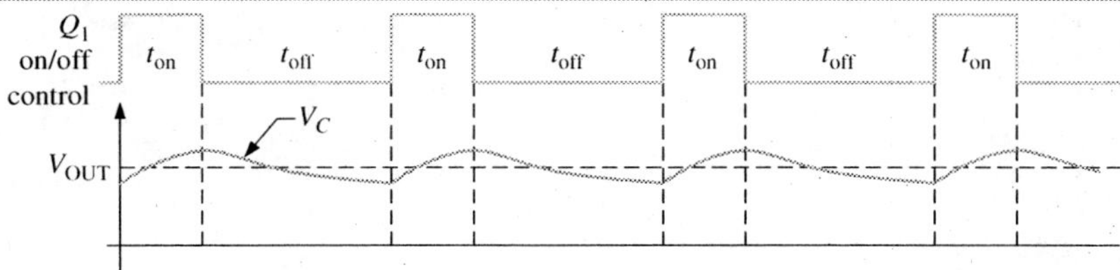

(a) V_{OUT} depends on the duty cycle.

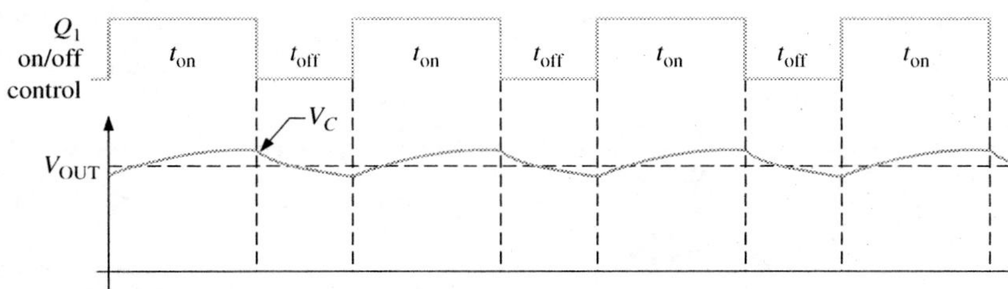

(b) Increase the duty cycle and V_{OUT} increases.

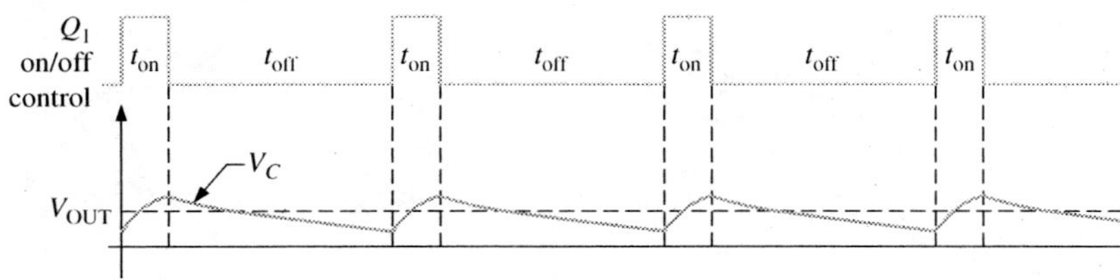

(c) Decrease the duty cycle and V_{OUT} decreases.

FIGURE 11–12

Switching regulator waveforms. The V_C waveform is shown for no inductive filtering to illustrate the charge and discharge action (ripple). L and C smooth V_C to a nearly constant level, as indicated by the dashed line for V_{OUT}.

HISTORICAL NOTE

In 1965, Arno A. Penzias and Robert W. Wilson of the Bell Telephone Laboratories were making a careful study of microwave radio noise capable of interfering with satellite communication systems. They discovered that wherever they pointed their precisely calibrated horn antenna, a background radiation was present corresponding to that of a perfect "blackbody" radiation source at a temperature of 3 K, which permeates the universe. Penzias and Wilson's experiments led theoreticians to develop models of the very early history of the universe—within seconds of its formation. Penzias and Wilson were awarded a Nobel Prize in 1978 for their momentous discovery.

Ideally, the output voltage is expressed as

$$V_{OUT} = \left(\frac{t_{on}}{T}\right)V_{IN} \quad (11\text{–}7)$$

T is the period of the on-off cycle of Q_1 and is related to the frequency by $T = 1/f$. The period is the sum of the on-time and the off-time.

$$T = t_{on} + t_{off}$$

The ratio t_{on}/T is the duty cycle. As you can see from Equation 11–7, the output voltage is just the duty cycle multiplied by the input voltage.

The regulating action is as follows and is illustrated in Figure 11–13. When V_{OUT} tries to decrease, the on-time of Q_1 is increased, causing an additional charge on the capacitor, C, to offset the attempted decrease. When V_{OUT} tries to increase, the on-time of Q_1 is decreased, causing C to discharge enough to offset the attempted increase.

FIGURE 11–13 Basic regulating action of a step-down switching regulator.

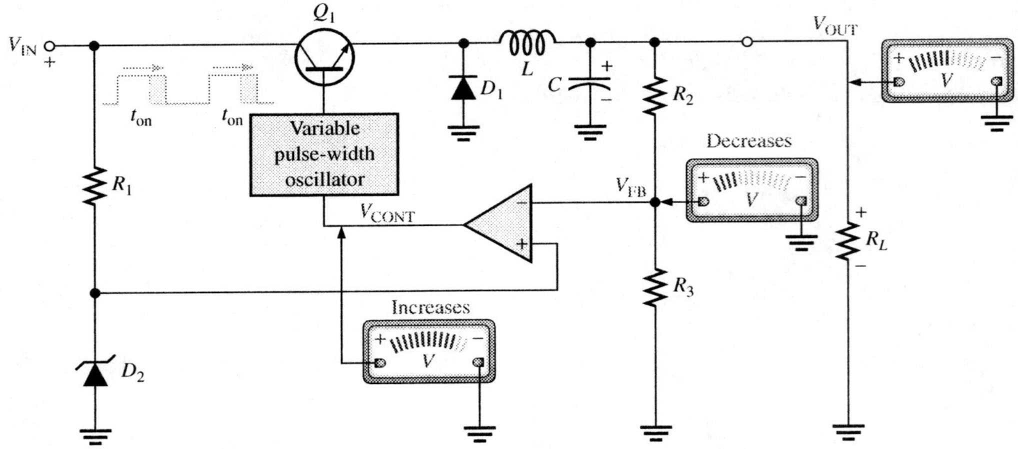

(a) When V_{OUT} attempts to decrease, the on-time of Q_1 increases.

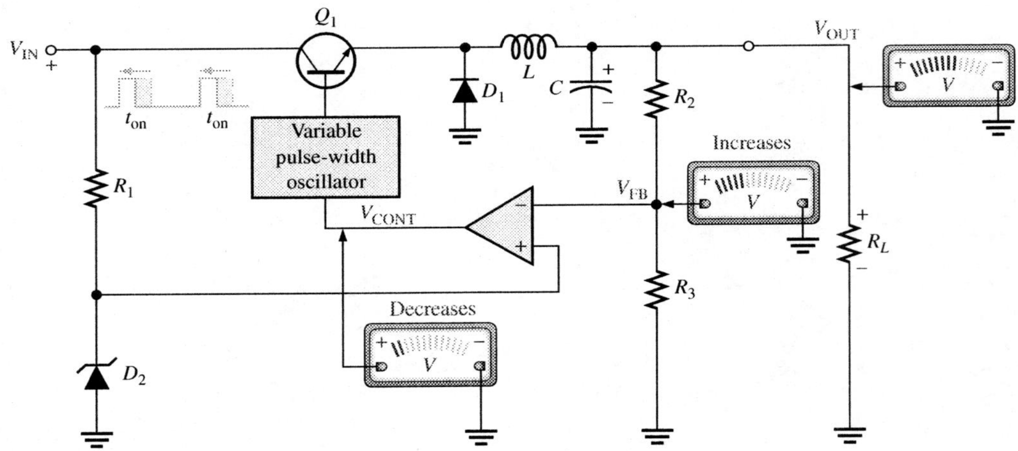

(b) When V_{OUT} attempts to increase, the on-time of Q_1 decreases.

Step-Up Configuration

A basic step-up type of switching regulator is shown in Figure 11–14, where transistor Q_1 operates as a switch to ground.

The switching action is illustrated in Figures 11–15 and 11–16. When Q_1 turns on, a voltage equal to approximately V_{IN} is induced across the inductor with a polarity as indicated in Figure 11–15. During the on-time (t_{on}) of Q_1, the inductor voltage, V_L, decreases from its initial maximum and diode D_1 is reverse-biased. The longer Q_1 is on, the smaller V_L becomes. During the on-time, the capacitor only discharges an extremely small amount through the load.

When Q_1 turns off, as indicated in Figure 11–16, the inductor voltage suddenly reverses polarity and adds to V_{IN}, forward-biasing diode D_1 and allowing the capacitor to charge. The

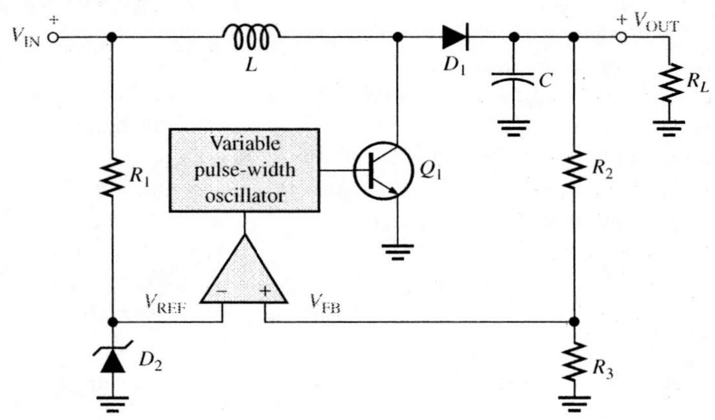

FIGURE 11–14

Basic step-up switching regulator.

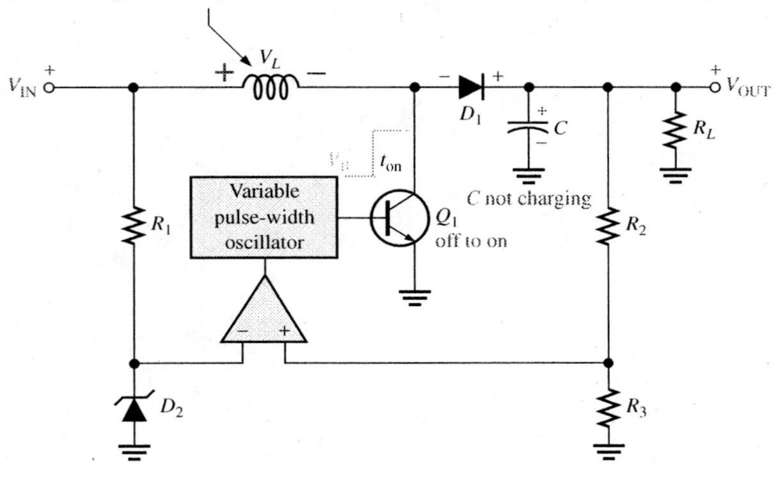

FIGURE 11–15

Basic action of a step-up regulator when Q_1 is on.

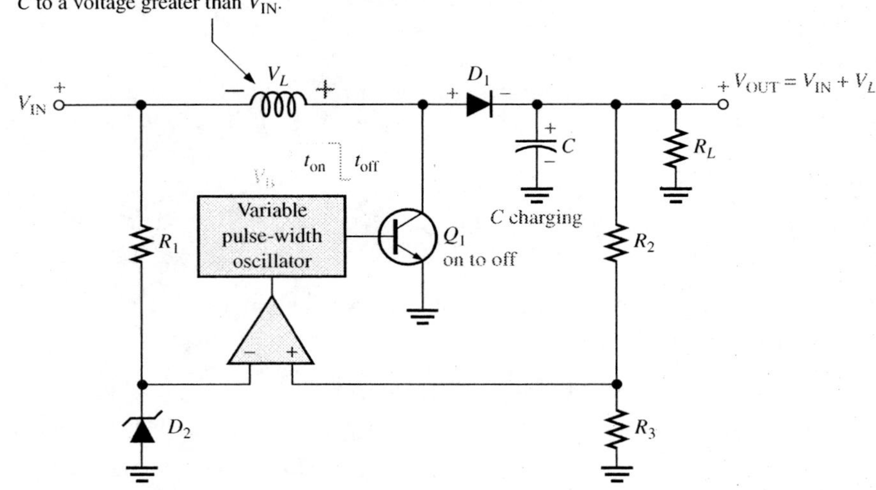

FIGURE 11–16

Basic switching action of a step-up regulator when Q_1 turns off.

output voltage is equal to the capacitor voltage and can be larger than V_{IN} because the capacitor is charged to V_{IN} plus the voltage induced across the inductor during the off-time of Q_1.

The longer the on-time of Q_1, the more the inductor voltage will decrease and the greater the magnitude of the voltage when the inductor reverses polarity at the instant Q_1 turns off. As you have seen, this reverse polarity voltage is what charges the capacitor above V_{IN}. The output voltage is dependent on both the inductor's magnetic field action (determined by t_{on}) and the charging of the capacitor (determined by t_{off}).

Voltage regulation is achieved by the variation of the on-time of Q_1 (within certain limits) as related to changes in V_{OUT} due to changing load or input voltage. If V_{OUT} tries to increase, the on-time of Q_1 will decrease, which results in a decrease in the amount that C will charge. If V_{OUT} tries to decrease, the on-time of Q_1 will increase, which results in an increase in the amount that C will charge. This regulating action maintains V_{OUT} at an essentially constant level.

Voltage-Inverter Configuration

A third type of switching regulator produces an output voltage that is opposite in polarity to the input. A basic diagram is shown in Figure 11–17.

FIGURE 11–17

Basic inverting switching regulator.

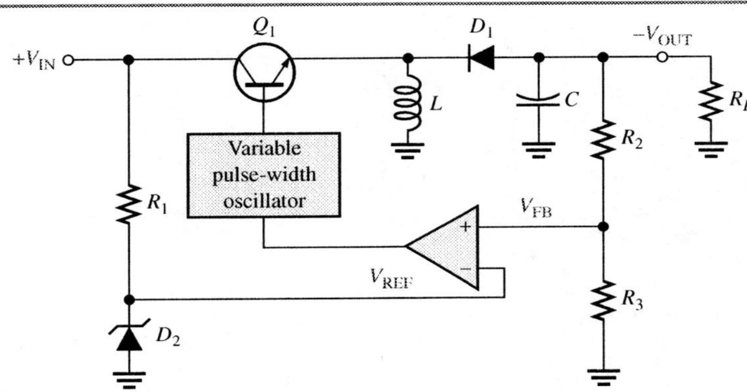

When Q_1 turns on, the inductor voltage jumps to approximately V_{IN} and the magnetic field rapidly expands, as shown in Figure 11–18(a). While Q_1 is on, the diode is reverse-biased and the inductor voltage decreases from its initial maximum. When Q_1 turns off, the magnetic field collapses and the inductor's polarity reverses, as shown in Figure 11–18(b). This forward-biases the diode, charges C, and produces a negative output voltage, as indicated. The repetitive on-off action of Q_1 produces a repetitive charging and discharging that is smoothed by the LC filter action.

In the inverting switching regulator, the output voltage varies inversely with the time that Q_1 is on. This is the same as for the step-up regulator. Switching regulator efficiencies can be greater than 90 percent.

Review Questions

16. Why is the switching frequency high in a switching regulator?
17. What is RFI?
18. What are three types of switching regulators?
19. What is the primary advantage of switching regulators over linear regulators?
20. How are changes in output voltage compensated for in the switching regulator?

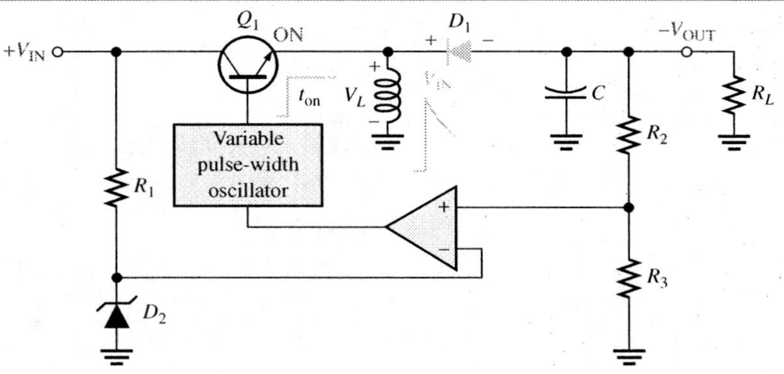

(a) When Q_1 is on, D_1 is reverse-biased.

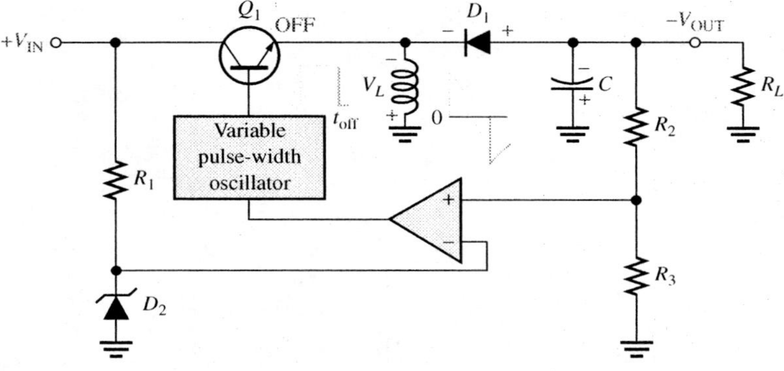

(b) When Q_1 turns off, D_1 is forward-biased.

FIGURE 11–18

Basic inverting action of an inverting switching regulator.

INTEGRATED CIRCUIT VOLTAGE REGULATORS 11–5

Several types of both linear and switching regulators are available in integrated circuit (IC) form. Generally, the linear regulators are three-terminal devices that provide either positive or negative output voltages that can be either fixed or adjustable.

In this section, you will learn how to configure three-terminal regulators to obtain higher current and how to use them as a current source.

Fixed Voltage Regulators

Although many types of IC regulators are available, the 7800 series of IC regulators is representative of three-terminal devices that provide a fixed positive output voltage. The 7900 series is typical of three-terminal IC regulators that provide a fixed negative output voltage. Both of these types were discussed in Section 2–6. Recall that the fixed three-terminal regulators give outstanding regulation with only an input and output capacitor as external components for filtering.

Most IC regulators require that the input voltage is at least 2 V above the output voltage in order to maintain regulation. If the voltage is too high, the regulator will become hot, so heat sinking is required. The 7800 series regulators have internal thermal overload protection and short-circuit current-limiting features. Thermal overload occurs when the internal power dissipation becomes excessive. A regulator that is too hot may show symptoms of drift, excess ripple, or the output may fall out of regulation.

Adjustable Voltage Regulators

Recall that an adjustable voltage regulator is designed to have an output voltage controlled by the user. The LM317 is an example of a basic three-terminal adjustable regulator with a positive output of 1.25 V between its output terminal and the adjustment terminal. The LM337 is the negative counterpart to the LM317 as discussed in Section 2–6. Figure 11–19 shows a typical LM317 regulator circuit, and Figure 11–20 shows a typical LM337 circuit.

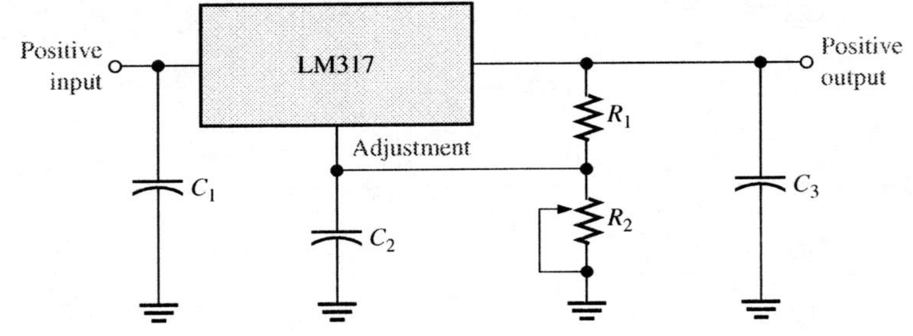

FIGURE 11–19

The LM317 three-terminal adjustable positive voltage regulator.

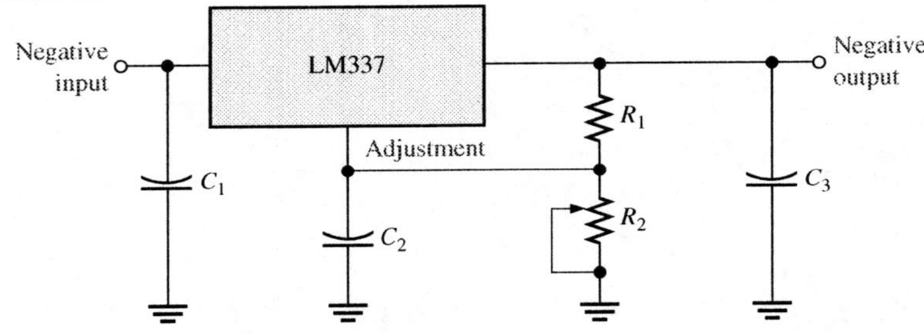

FIGURE 11–20

The LM337 three-terminal adjustable negative voltage regulator.

Troubleshooting Three-Terminal Regulators

Three-terminal regulators are very reliable devices. When problems occur, the indication is usually an incorrect voltage, high ripple, noisy or oscillating output, or drift. Troubleshooting a regulator circuit is best done with an oscilloscope because problems such as excessive ripple or noise won't show up using a DMM. Before starting, it is useful to review the possible causes of a failure (analysis) and plan measurements that will point to the failure.

If the output voltage is too low, the input voltage should be checked; the problem may be in the circuit preceding the regulator. Also check the load resistor: Does the problem go away when the load is removed? If so, it may be that the load draws too much current. A high output can occur with adjustable regulators if the feedback resistors are the wrong value or open.

If there is ripple or noise on the output, check the capacitors for an open, a wrong value, or that they are installed with the proper polarity. A useful quick check of a capacitor is to place another capacitor of the same or larger size in parallel with the capacitor to be tested.

If the output is oscillating, has high ripple, or is drifting, check that the regulator is not too hot or supplying more than its rated current. If heat is a problem, make sure the regulator is firmly secured to the heat sink with special heat sink paste. (Heat sink paste is a special heat conductive material that helps move heat from the IC to the heat sink.)

Switching Voltage Regulators

As an example of an IC switching voltage regulator, let's look at the 78S40. This is a universal device that can be used with external components to provide step-up, step-down, and inverting operation.

The internal circuitry of the 78S40 is shown in Figure 11–21. This circuit can be compared to the basic switching regulators that were covered in Section 11–4. For example, look back at Figure 11–11(a). The oscillator and comparator functions are directly comparable. The gate and flip-flop, which are digital devices, were not included in the basic circuit of Figure 11–11(a), but they provide additional regulating action. Transistors Q_1 and Q_2 effectively perform the same function as Q_1 in the basic circuit. The 1.25 V reference block in the 78S40 has the same purpose as the zener diode in the basic circuit, and diode D_1 in the 78S40 corresponds to D_1 in the basic circuit.

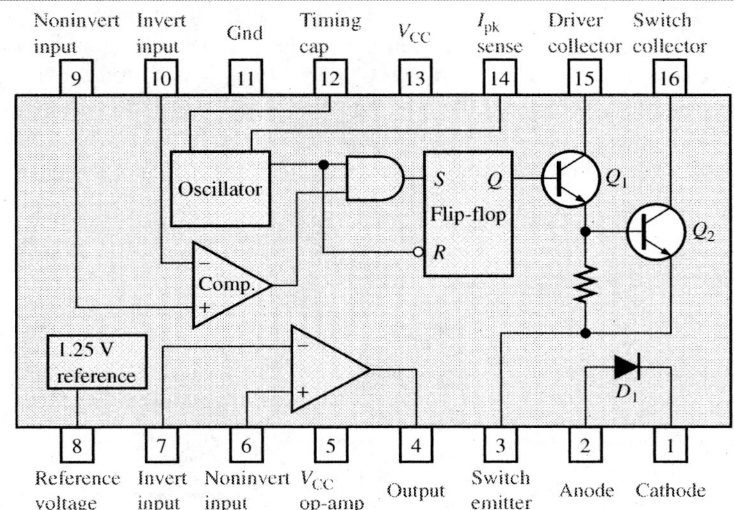

FIGURE 11–21

The 78S40 switching regulator.

The 78S40 also has an "uncommitted" op-amp thrown in for good measure. It is not used in any of the regulator configurations. External circuitry is required to make this device operate as a regulator.

Increasing the Current from an IC Regulator

As you know, an IC voltage regulator is capable of delivering only a certain amount of output current to a load. For example, the 7800 series regulators can handle a peak output current of 1.3 A (more under certain conditions). If the load current exceeds the maximum allowable value, there will be thermal overload and the regulator will shut down. A thermal overload condition means that there is excessive power dissipation inside the device.

If an application requires more than the maximum current that the regulator can deliver, an external pass transistor can be used. Figure 11–22 illustrates a three-terminal regulator

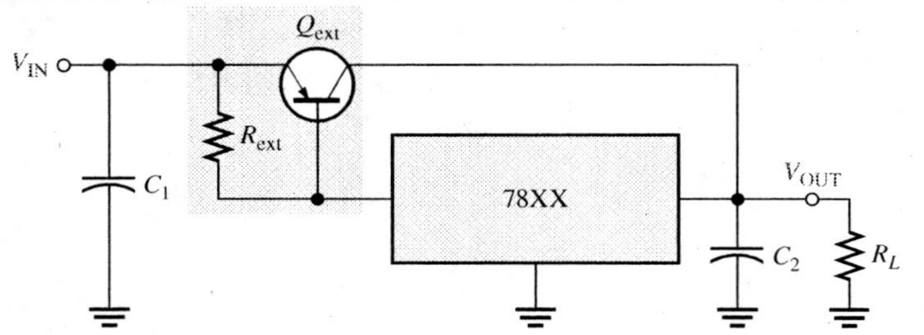

FIGURE 11–22

A 7800-series three-terminal regulator with an external pass transistor.

with an external pass transistor for handling currents in excess of the output current capability of the basic regulator.

The value of the external current-sensing resistor R_{ext} determines the value of current at which Q_{ext} begins to conduct because it sets the base-to-emitter voltage of the transistor. As long as the current is less than the value set by R_{ext}, the transistor Q_{ext} is off, and the regulator operates normally as shown in Figure 11–23(a). This is because the voltage drop across R_{ext} is less than the 0.7 V base-to-emitter voltage required to turn Q_{ext} on. R_{ext} is determined by the following formula, where I_{max} is the highest current that the voltage regulator is to handle internally.

$$R_{ext} = \frac{0.7\text{ V}}{I_{max}}$$

When the current is sufficient to produce at least a 0.7 V drop across R_{ext}, the external pass transistor Q_{ext} turns on and conducts any current in excess of I_{max}, as indicated in Figure 11–23(b). Q_{ext} will conduct more or less, depending on the load requirements. For example, if the total load current is 3 A and I_{max} was selected to be 1 A, the external pass transistor will conduct 2 A, which is the excess over the internal regulator current I_{max}.

The external pass transistor is typically a power transistor with heat sink that must be capable of handling a maximum power of

$$P_{ext} = I_{ext}(V_{IN} - V_{OUT})$$

FIGURE 11–23

Operation of the regulator with an external pass transistor.

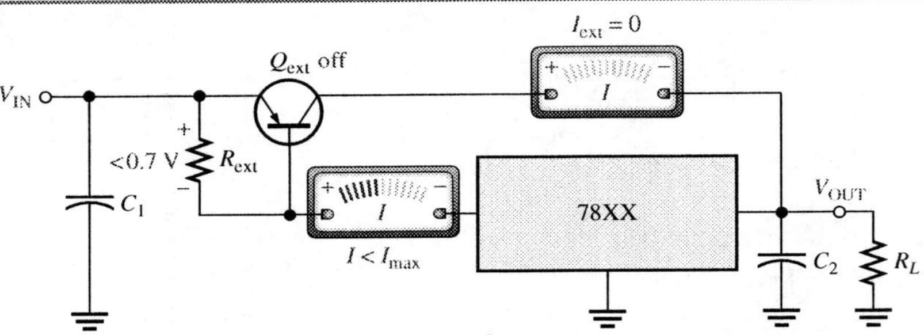

(a) When the regulator current is less than I_{max}, the external pass transistor is off and the regulator is handling all of the current.

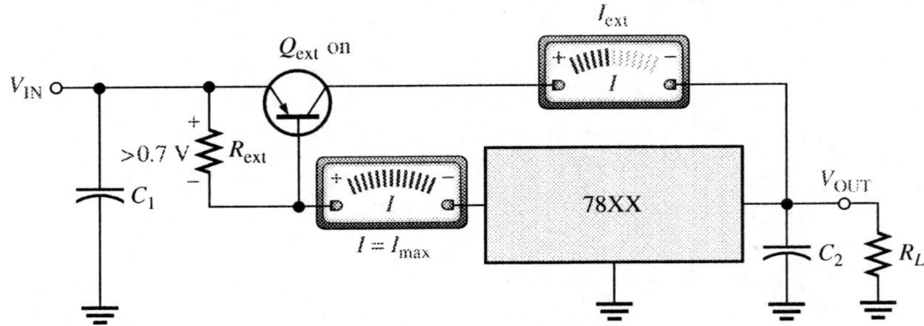

(b) When the load current exceeds I_{max}, the drop across R_{ext} turns Q_{ext} on and the transistor conducts the excess current.

EXAMPLE 11–6

Problem
What value is R_{ext} if the maximum current to be handled internally by the voltage regulator in Figure 11–22 is set at 700 mA?

Solution

$$R_{ext} = \frac{0.7\text{ V}}{I_{max}} = \frac{0.7\text{ V}}{0.7\text{ A}} = \mathbf{1\ \Omega}$$

Question
If R_{ext} is changed to 1.5 Ω, at what current value will Q_{ext} turn on?

A Current Regulator

The three-terminal regulator can be used as a current source when an application requires that a constant current be supplied to a variable load. The basic circuit is shown in Figure 11–24 where R_1 is the current-setting resistor. The regulator provides a fixed constant voltage, V_{OUT}, between the ground terminal (not connected to ground in this case) and the output terminal. This determines the constant current supplied to the load.

$$I_L = \frac{V_{OUT}}{R_1} + I_G$$

The current, I_G, from the ground terminal is very small compared to the output current and can often be neglected.

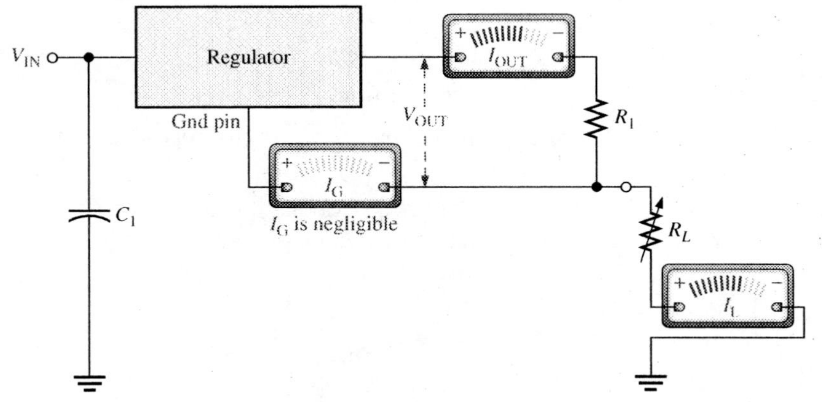

FIGURE 11–24
The three-terminal regulator as a current source.

EXAMPLE 11–7

Problem
What value of R_1 is necessary in a 7805 regulator to provide a constant current of 1 A to a variable load that can be adjusted from 0–10 Ω?

Solution
First, 1 A is within the limits of the 7805's capability (it can handle at least 1.3 A without an external pass transistor).

The 7805 produces 5 V between its ground terminal and its output terminal. Therefore, if you want 1 A of current, the current-setting resistor must be (neglecting I_G)

$$R_1 = \frac{V_{OUT}}{I_L} = \frac{5\text{ V}}{1\text{ A}} = \mathbf{5.0\ \Omega}$$

The circuit is shown in Figure 11–25.

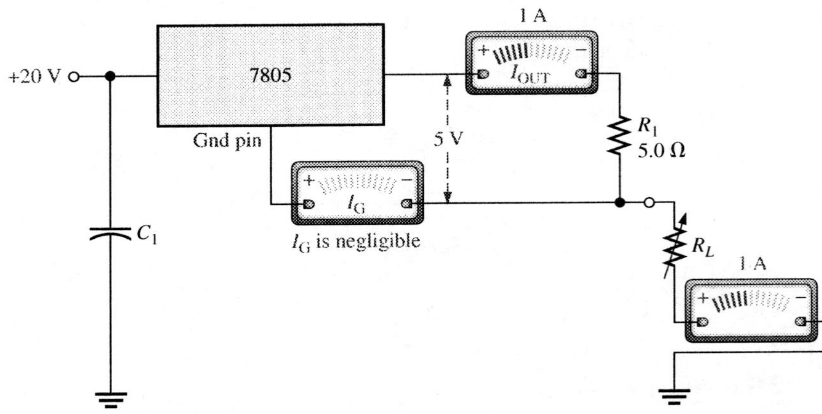

FIGURE 11–25
A 1 A constant-current source.

Question
If a 7808 regulator (+8 V output) is used instead of the 7805, to what value would you change R_1 to maintain a constant current of 1 A?

Review Questions

21. If the output voltage of a three-terminal regulator is low, what is the next logical check you should make?
22. What external components are required for a basic LM317 configuration?
23. How can you increase the current from a three-terminal regulator?
24. How is the maximum current to the three-terminal regulator set in the circuit in Figure 11–22?
25. If the circuit in Figure 11–22 has an external resistance (R_{ext}) of 0.82 Ω, what is the maximum current in the three-terminal regulator?

CHAPTER REVIEW

Key Terms

Linear regulator A voltage regulator in which the control element operates in the linear region.

Line regulation The percentage change in output voltage for a given change in line (input) voltage.

Load regulation The percentage change in output voltage for a given change in load current.

Switching regulator A voltage regulator in which the control element is a switching device.

Important Facts

- ❑ Voltage regulators keep a constant dc output voltage when the input or load varies within limits.
- ❑ A basic voltage regulator consists of a reference voltage source, an error detector, a sampling element, and a control device. Protection circuitry is also found in most regulators.
- ❑ Two basic categories of voltage regulators are linear and switching.
- ❑ Two basic types of linear regulators are series and shunt.
- ❑ In a series linear regulator, the control element is a transistor in series with the load.
- ❑ In a shunt linear regulator, the control element is a transistor in parallel with the load.
- ❑ Three configurations for switching regulators are step-down, step-up, and inverting.
- ❑ Switching regulators are more efficient than linear regulators and are particularly useful in low-voltage, high-current applications.
- ❑ Three-terminal linear IC regulators are available for either fixed output or variable output voltages of positive or negative polarities.
- ❑ An external pass transistor increases the current capability of a regulator.
- ❑ The 7800 series are three-terminal IC regulators with fixed positive output voltage.
- ❑ The 7900 series are three-terminal IC regulators with fixed negative output voltage.
- ❑ The LM317 is a three-terminal IC regulator with a positive variable output voltage.
- ❑ The LM337 is a three-terminal IC regulator with a negative variable output voltage.
- ❑ The 78S40 is a switching voltage regulator.

Formulas

Percent line regulation:

$$\text{Line regulation} = \left(\frac{\Delta V_{\text{OUT}}}{\Delta V_{\text{IN}}}\right)100\% \qquad \textbf{(11–1)}$$

Percent line regulation per volt:

$$\text{Line regulation} = \left(\frac{\Delta V_{\text{OUT}}/V_{\text{OUT}}}{\Delta V_{\text{IN}}}\right)100\% \qquad \textbf{(11–2)}$$

Percent load regulation:

$$\text{Load regulation} = \left(\frac{V_{\text{NL}} - V_{\text{FL}}}{V_{\text{FL}}}\right)100\% \qquad \textbf{(11–3)}$$

Percent load regulation given output resistance and minimum load resistance:

$$\text{Load regulation} = \left(\frac{R_{\text{OUT}}}{R_{\text{FL}}}\right)100\% \qquad \textbf{(11–4)}$$

Series or shunt regulator output:

$$V_{\text{OUT}} = \left(1 + \frac{R_2}{R_3}\right)V_{\text{REF}} \qquad \textbf{(11–5)}$$

Maximum load current for a shunt regulator:

$$I_{\text{L(max)}} = \frac{V_{\text{IN}}}{R_{\text{S}}} \qquad \textbf{(11–6)}$$

Output voltage for step-down switching regulator:

$$V_{OUT} = \left(\frac{t_{on}}{T}\right)V_{IN} \quad \textbf{(11–7)}$$

Chapter Checkup

Answers are at the end of the chapter.

1. In the case of line regulation,
   (a) when the temperature varies, the output voltage stays constant
   (b) when the output voltage changes, the load current stays constant
   (c) when the input voltage changes, the output voltage stays constant
   (d) when the load changes, the output voltage stays constant
2. In the case of load regulation,
   (a) when the temperature varies, the output voltage stays constant
   (b) when the input voltage changes, the load current stays constant
   (c) when the load changes, the load current stays constant
   (d) when the load changes, the output voltage stays constant
3. All of the following are parts of a basic voltage regulator *except*
   (a) control element (b) sampling circuit
   (c) voltage follower (d) error detector
   (e) reference voltage
4. The basic difference between a series regulator and a shunt regulator is
   (a) the amount of current that can be handled
   (b) the position of the control element
   (c) the type of sample circuit
   (d) the type of error detector
5. In a basic series regulator, V_{OUT} is determined by
   (a) the control element (b) the sample circuit
   (c) the reference voltage (d) answers (b) and (c)
6. In a linear regulator, the control transistor is conducting
   (a) a small part of the time (b) half the time
   (c) all of the time (d) only when the load current is excessive
7. In a switching regulator, the control transistor is conducting
   (a) part of the time
   (b) all of the time
   (c) only when the input voltage exceeds a set limit
   (d) only when there is an overload
8. The LM317 is an example of an IC
   (a) three-terminal negative voltage regulator
   (b) fixed positive voltage regulator
   (c) switching regulator
   (d) variable positive voltage regulator

9. An external pass transistor is used for
   (a) increasing the output voltage
   (b) improving the regulation
   (c) increasing the current that the regulator can handle
   (d) short-circuit protection
10. If D_1 in the circuit in Figure 11–26 is mistakenly replaced with a 4.7 V zener, the output voltage will
   (a) increase (b) decrease
   (c) not change

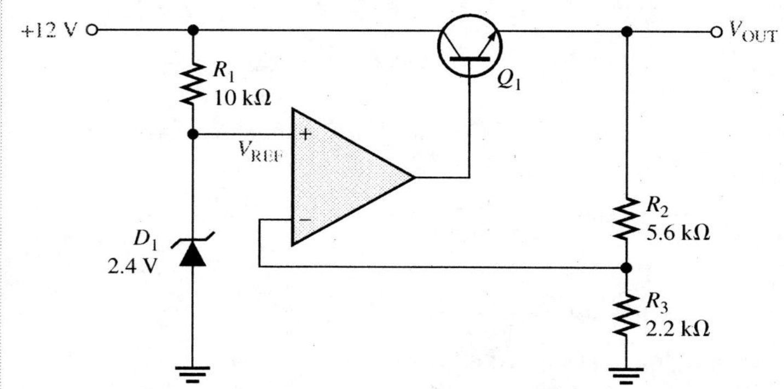

FIGURE 11–26

11. If D_1 in the circuit in Figure 11–26 is mistakenly replaced with a 4.7 V zener, the voltage across Q_1 from collector to emitter will
   (a) increase (b) decrease
   (c) not change
12. If R_3 in the circuit in Figure 11–26 opens, the output voltage will
   (a) increase (b) decrease
   (c) not change
13. If the output voltage in the circuit in Figure 11–26 increases from +12 V to +12.5 V, V_{OUT} will
   (a) increase (b) decrease
   (c) not change
14. If R_1 in the circuit in Figure 11–26 is replaced with a 15 kΩ resistor, V_{REF} will
   (a) increase (b) decrease
   (c) not change

Questions

Answers to odd-numbered questions are at the end of the book.

1. Why should the equivalent output resistance of a power supply be very small?
2. What is the difference between a linear regulator and a switching regulator?
3. In a series regulator, if the feedback voltage increases, what happens to the voltage from the error detector?

4. In which type of linear regulator is the feedback applied to the noninverting terminal of the error detector?
5. What type of regulator is the most efficient? (series, shunt, or switching)
6. Which type of regulator is least efficient? (series, shunt, or switching)
7. What type of regulator is prone to producing radio frequency interference?
8. What is the purpose of the inductor in a switching regulator?
9. What type of regulator has an output that is of the opposite polarity of the input?
10. What is the typical minimum voltage difference requirement between the input and output on a three-terminal regulator?
11. What is the difference between an LM317 regulator and an LM337 regulator?
12. What is the purpose of heat sink paste?
13. Where is the load resistor connected when a 78XX series three-terminal regulator is used as a current source?
14. What determines the output current when a 78XX series three-terminal regulator is used as a current source?

PROBLEMS

Basic Problems

Answers to odd-numbered problems are at the end of the book.

1. The nominal output voltage of a certain regulator is 8 V. The output changes 2 mV when the input voltage goes from 12 V to 18 V. Determine the line regulation and express it as a percentage change.
2. Express the line regulation found in Problem 1 in units of %/V.
3. A certain regulator has a no-load output voltage of 10 V and a full-load output voltage of 9.90 V. What is the percent load regulation?
4. In Problem 3, if the full-load current is 250 mA, express the load regulation in %/mA.
5. Label the functional blocks for the voltage regulator in Figure 11–27.

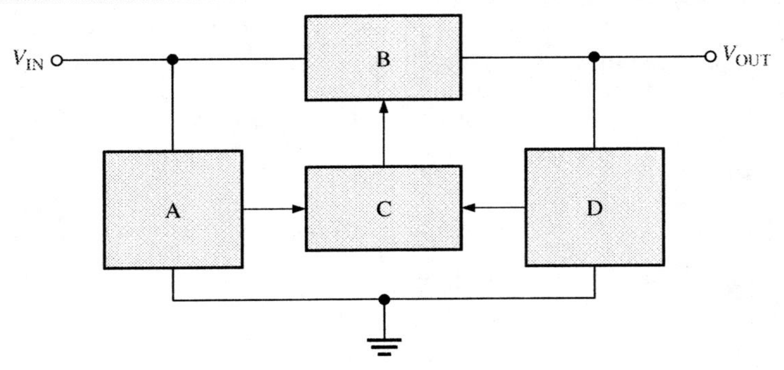

FIGURE 11–27

6. Determine the output voltage for the regulator in Figure 11–28.

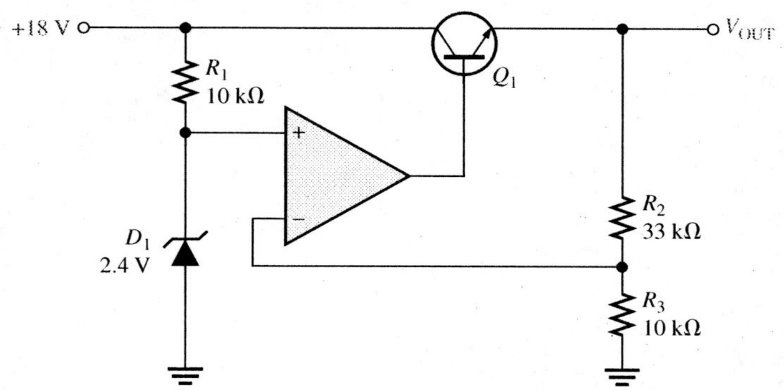

FIGURE 11–28

7. If the zener voltage is 3.3 V instead of 2.4 V in Figure 11–28, what is the output voltage?
8. In the shunt regulator of Figure 11–29, when the current through R_L increases, does Q_1 conduct more or less? Why?

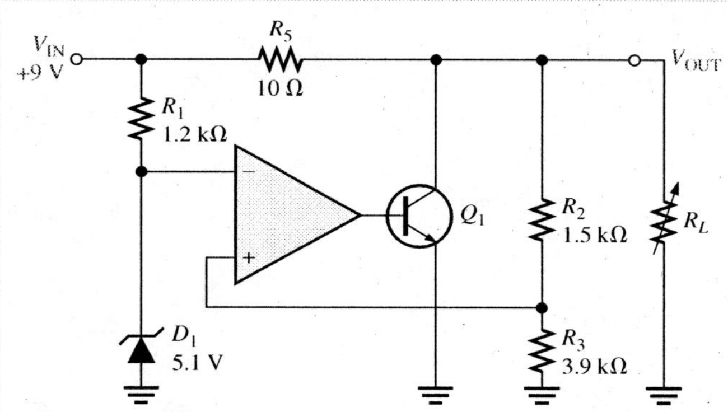

FIGURE 11–29

9. A basic switching regulator is shown in Figure 11–30. If the switching frequency of the transistor is 10 kHz with an off-time of 60 μs, what is the output voltage?

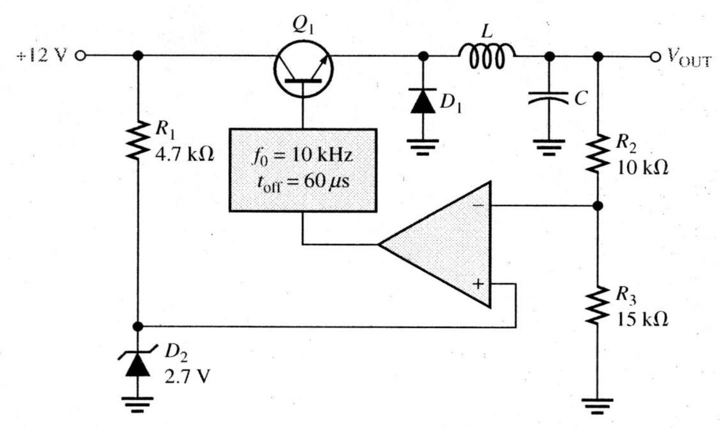

FIGURE 11–30

10. What is the duty cycle of the transistor in Problem 9?
11. When does the diode D_1 in Figure 11–31 become forward-biased?

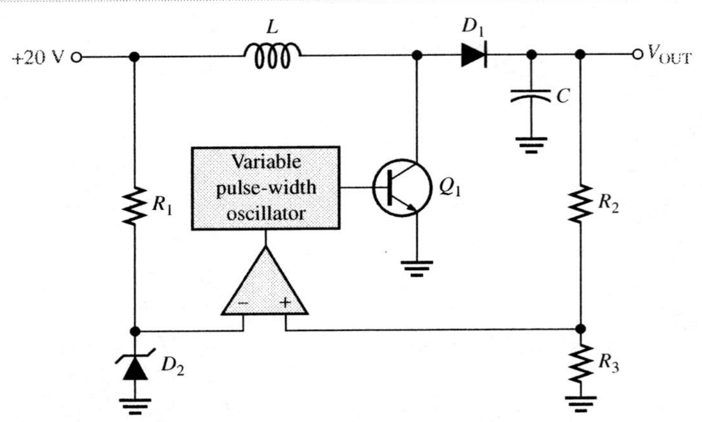

FIGURE 11–31

12. If the on-time of Q_1 in Figure 11–31 is decreased, does the output voltage increase or decrease?
13. Determine the output voltage of the regulator in Figure 11–32. $I_{ADJ} = 50\ \mu A$.

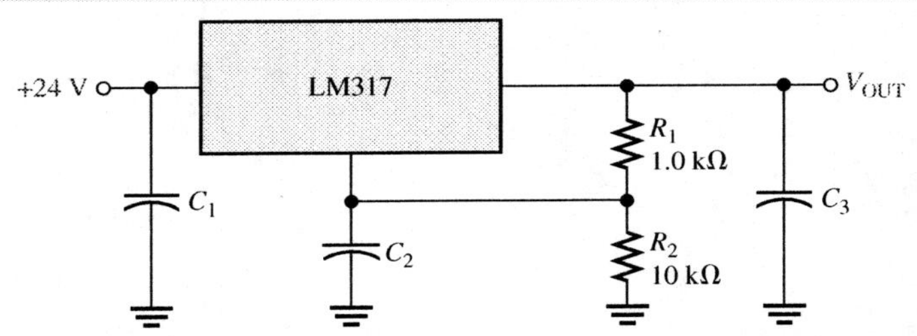

FIGURE 11–32

14. Determine the minimum and maximum output voltages for the circuit in Figure 11–33. $I_{ADJ} = 50\ \mu A$.

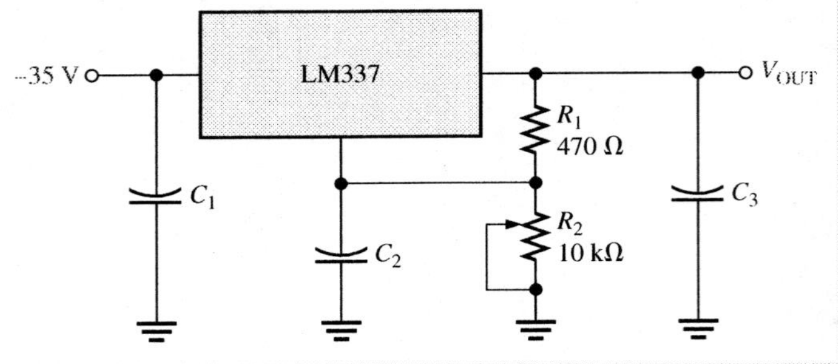

FIGURE 11–33

15. With no load connected, how much current is there through the regulator in Figure 11–32? Neglect the adjustment terminal current.

Basic-Plus Problems

16. Assume the current through R_L remains constant and V_{IN} changes by 1 V in Figure 11–29. What is the change in the collector current of Q_1?
17. With a constant input voltage of 9 V, the load resistance in Figure 11–29 is varied from 1.0 kΩ to 1.2 kΩ. Neglecting any change in output voltage, how much does the shunt current through Q_1 change?
18. If the maximum allowable input voltage in Figure 11–29 is 10 V, what is the maximum possible output current when the output is short-circuited? What power rating should R_5 have?
19. Select the values for the external resistors to be used in an LM317 circuit that is required to produce an output voltage of 12 V with an input of 18 V. The maximum regulator current with no load is to be 2 mA. There is no external pass transistor.
20. In the regulator circuit of Figure 11–34, determine R_{ext} if the maximum internal regulator current is to be 250 mA.

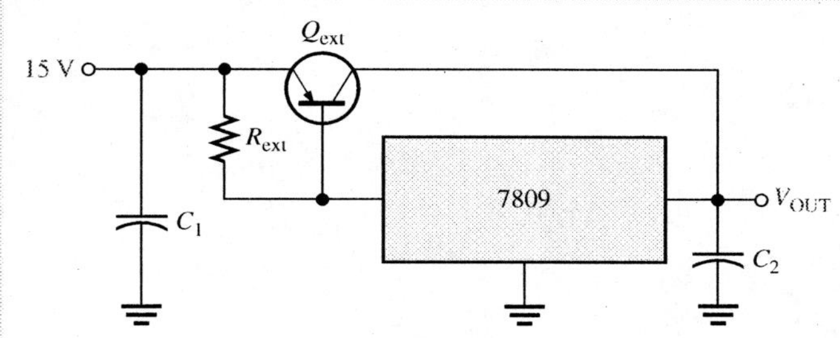

FIGURE 11–34

21. Using a 7812 voltage regulator and a 10 Ω load in Figure 11–34, how much power will the external pass transistor have to dissipate? The maximum internal regulator current is set at 500 mA by R_{ext}.
22. Using an LM317, design a circuit that will provide a constant current of 500 mA to a load.

Example Questions

ANSWERS

11–1: 12%, 0.6 %/V

11–2: 0.5 Ω

11–3: 7.33 V

11–4: Change R_3 to a rheostat.

11–5: 17.5 W

11–6: 467 mA

11–7: 8 Ω

Review Questions

1. The percentage change in the output voltage for a given change in input voltage
2. The percentage change in output voltage for a given change in load current
3. 1.2%
4. 0.06 %/V
5. 1.6%
6. Control element, error detector, sampling element, reference source
7. V_{REF} and V_{FB}
8. Very low temperature drift, lower noise
9. 2 V
10. 2 V
11. In a shunt regulator, the control element is in parallel with the load rather than in series.
12. A shunt regulator has inherent current limiting.
13. A disadvantage is that a shunt regulator is less efficient than a series regulator.
14. The V_{BE} of the transistor is compensated for by the op-amp because it is in the feedback loop.
15. 15.7 mA
16. It makes filtering the output easier.
17. Radio Frequency Interference
18. Step-down, step-up, inverting
19. Switching regulators operate at a higher efficiency.
20. The duty cycle varies to regulate the output.
21. Check the input voltage to see that it is at least 2 V above the output and is filtered.
22. A two-resistor voltage divider and capacitors across the input, output, and adjustment terminals.
23. A pass transistor can be used to increase the current.
24. The current is determined by R_{ext}, which turns on Q_{ext} when $V_{BE} \geq 0.7$ V.
25. 853 mA

Chapter Checkup

1. (c)	2. (d)	3. (c)	4. (b)	5. (d)
6. (c)	7. (a)	8. (d)	9. (c)	10. (a)
11. (b)	12. (b)	13. (c)	14. (c)	

MEASUREMENT AND CONTROL CIRCUITS

CHAPTER OUTLINE

Study aids for this chapter are available at

http://www.prenhall.com/SOE

INTRODUCTION

Physical quantities, such as temperature, force, motion, or light are converted to electrical quantities for processing. The components that perform these conversions are known as electrical transducers. Electrical transducers make it possible to measure and control the many nonelectrical quantities in our environment.

Most transducers use passive devices that apply principles you have studied in basic dc and ac circuits. One type of widely used microphone responds to sound by varying the distance (and hence the capacitance) between two plates of a capacitor. Another important transducer (the strain gauge) operates on the principle that wire resistance is a function of length and diameter. These are but two of many ways that passive components find direct application for sensing quantities before processing them with electronic circuits.

The chapter begins by examining some important principles of transducers, then common attributes and selection criteria of transducers in general are discussed. Specific transducers for particular physical quantities are considered with an emphasis on passive devices. In particular, temperature and force measurements are considered in more detail because these are the most widely used applications of transducers. Force measurements also apply an important circuit, the Wheatstone bridge. In addition, pressure, motion, and light sensors are discussed. Thyristors are also introduced and their application in power control circuits is discussed.

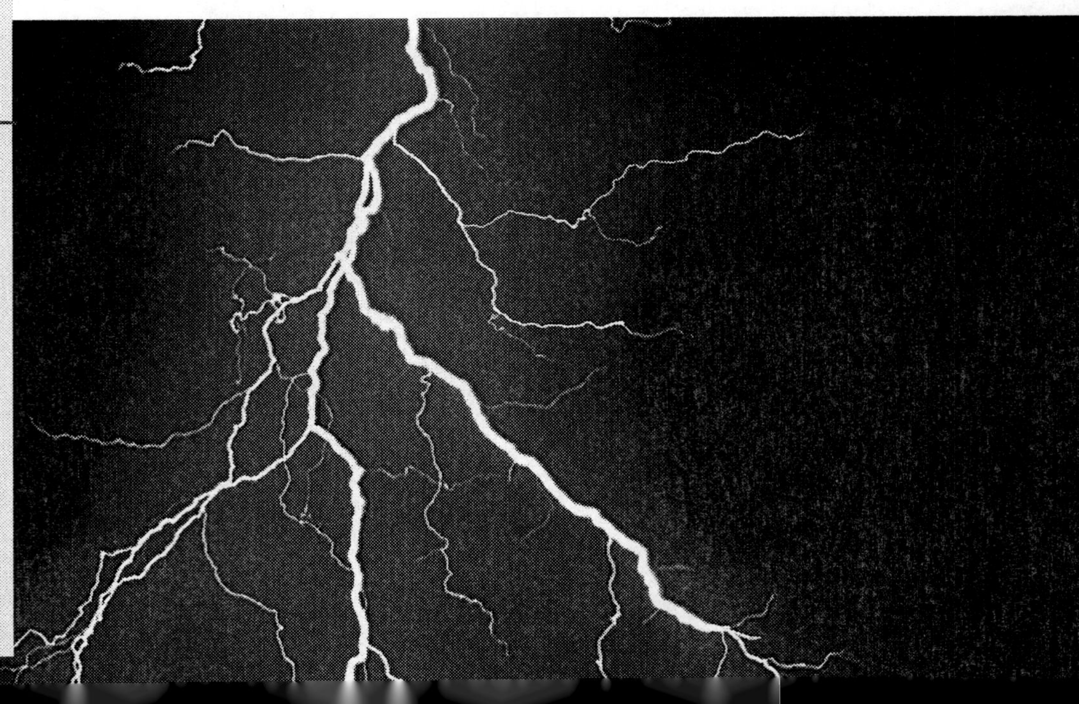

KEY TERMS

- Transducer
- Passive transducer
- Active transducer
- Excitation
- Strain gauge
- Resolution
- Thermocouple
- Resistance temperature detector
- Thermistor
- Elastic limit
- Strain
- Young's modulus
- Stress
- Gauge factor
- Load cell
- Pressure
- Thyristor
- SCR
- Triac

KEY OBJECTIVES

A section number is given for each objective. After completing this chapter, you should be able to

12–1 Describe common methods to convert a physical parameter into an electrical quantity and list examples of some electrical transducers

12–2 Describe important parameters to consider when selecting a transducer

12–3 List five types of temperature transducers and describe advantages and applications for each type

12–4 Explain how strain gauges and load cells are used for measuring forces

12–5 Describe electrical transducers for measuring pressure, motion, and light

12–6 Describe the operation of thyristors and how they can be used to control power to a load

LABORATORY EXPERIMENTS DIRECTORY

The following exercises are for this chapter.

- **Experiment 22**
  Measuring Rotational Speed
- **Experiment 23**
  The SCR

Since the time of Galileo, science has advanced as measurements became more precise. Health science is no exception. Health professionals depend on many types of sensors that are used for detecting the abnormal conditions of the body. One type of biosensor that is currently in the development phase shows promise for detecting lung cancer. This device, known as Cyranose, could offer a fast, noninvasive diagnostic method by picking up the scent of certain compounds exhaled in the breath of lung cancer patients. Cyranose is based on the fact that a lung cancer patient's breath contains a mixture of chemicals that is high in alkanes and benzene derivatives.

In tests, the Cyranose picks up chemicals that are present in the breath and produces a pattern on a video monitor. The patterns produced by lung cancer patients are generally distinguishable from patients with other types of lung diseases or from healthy people. However, further research is needed to improve the sensitivity of the Cyranose in order to produce patterns with even more distinct differences to make it suitable for widespread practical diagnostic applications.

12–1 TRANSDUCER CHARACTERISTICS

Transducers receive energy in some form, such as light or heat, and convert it into some other form, such as a resistance change. Although transducers can convert energy to many forms, for electronic systems the output needs to be an electrical parameter.

In this section, you will learn about common methods to convert a physical parameter into an electrical quantity and be able to list examples of some electrical transducers.

Active and Passive Transducers

Broadly speaking, a **transducer** is a device that converts energy from one form to another. Many useful transducers have an output that is not electrical in nature—the common mercury thermometer is an example. In the case of the thermometer, temperature change is converted to height of mercury in a glass tube. For electronic systems, transducers such as the mercury thermometer are not particularly useful because the output is not electrical in nature. For electronic systems, a transducer is often restricted to include only those devices that convert energy to an electrical quantity. This type of transducer is the input device for many systems.

Many transducers applied to electronic systems are considered to be passive. A **passive transducer** is one that produces an output without any direct interaction with a source of power. Passive transducers are considered to be self-generating. Other transducers are active. An **active transducer** is one whose output power is derived from a source other than the quantity being measured. Most active transducers employ passive components.

The common mercury thermometer is an example of a passive transducer. The mercury inside the glass tube responds to temperature by expanding. The energy source is the temperature to be measured.

A potentiometer used as a voltage divider is an example of an active electrical transducer. If you turn the knob in one direction, the output voltage increases; if you turn it in the other direction, the output voltage decreases. The potentiometer is "automatically" converting the angular position of the shaft to an output voltage that is proportional to the angle. The dc voltage supplied to the potentiometer is necessary to provide an output, fitting the definition for an active transducer. This outside source of energy to operate a transducer is given the name **excitation**. Excitation can be any form of energy, but for electrical transducers, it is usually provided by a dc power supply or an ac source.

When an electronic system uses a passive circuit element as a basic sensing element, an electrical quantity (resistance, capacitance, or inductance) must change its value in response to a physical stimulus. The passive circuit element often modifies a dc or ac excitation voltage to produce a usable output.

For electronic systems, the fact that a given transducer is active or passive usually isn't the most important consideration; rather, the output must be easily related to the quantity to be measured and it must be in electrical form. It can be a voltage, current, frequency, capacitance, resistance, pulse width, or some other electrical variable that is related to the quantity you are measuring. In addition, it needs to be able to sense the range of values expected and meet environmental requirements such as shock or heat.

Electrical Transducers

Measurements always involve some source of energy. Energy can be in one of six fundamental forms: mechanical, thermal, nuclear radiation, electromagnetic radiation, magnetic, and chemical. Each of these basic forms can easily be converted to an electrical parameter. For example, nuclear radiation can be converted directly into electrical pulses with a Geiger counter. Another conversion is the resistance change that occurs when a wire is mechanically stretched. The stretching increases the length and decreases the diameter of the wire; therefore, it will have greater electrical resistance. Table 12–1 lists some examples of other methods for converting the six forms of energy into an electrical parameter.

TABLE 12–1

Representative transducers that convert input energy into an electrical parameter.

Energy source	Representative transducer	Comment
Mechanical	Capacitive displacement	Capacitance changes when plate moves.
Thermal	Thermistor	Resistance is a function of temperature.
Nuclear radiation	Geiger counter	Radiation causes electrical pulses.
Electromagnetic	Antenna	Converts radio waves directly to voltage
Magnetic	Hall-effect sensor	Voltage is produced in a current carrying conductor.
Chemical	pH sensor	Converts hydrogen ion concentration into a voltage

As mentioned, there are many types of transducers and different principles of operation. One classification scheme is to consider the operating principle and show different measurements that can be made using that principle. For example, capacitive changes are used in a variety of measurements including displacement, proximity, pressure, and even in low-temperature thermometers. In the remainder of this section, you will be introduced to several important methods for converting a physical quantity to an electrical quantity. Keep in mind there are many more.

Resistive Transducers

The variation of resistance is one of the most common principles of transducers. The resistance of the transducer can be varied by several methods, including sliding a wiper along a resistive material, varying light intensity to a photosensitive material, or changing the temperature. A widely used resistive transducer is the **strain gauge**, which is a thin conductor made in a back and forth pattern. It responds to force by changing its resistance as it is stretched or compressed. Stretching increases the length of the wire and decreases its cross-sectional area, resulting in greater resistance; compression does the opposite. This is clear from the wire resistance formula:

$$R = \frac{\rho l}{A}$$

The change in resistance of a strain gauge is measured in a Wheatstone bridge circuit. Strain gauges are used in many types of transducers, most noticeably in load cells. Load cells are a force-measuring system found in scales ranging from postal scales to those for weighing trucks. Strain gauges and load cells are covered further in Section 12–4.

Resistive transducers can be used to measure other quantities such as temperature. Temperature transducers include RTDs (resistance temperature detectors) and thermistors (*therm*al res*istors*). RTDs are very stable thermometers that are usually constructed from platinum and exhibit a direct relationship between resistance and temperature. RTDs are some of the most accurate thermometers made and exploit the fact that the resistivity (ρ) of metals is directly proportional to the temperature.

Thermistors are another type of resistive thermometer but are formed from semiconductor materials rather than wire. The resistivity of thermistors is opposite to the change in a metal. Heat frees electrons in the semiconductor material. This produces a change in resistance that is inversely proportional to temperature. They are not as accurate as RTDs, but they are inexpensive and small, so they are widely used in noncritical applications, such as thermostats.

Another resistive transducer is the CdS (for Cadmium Sulfide) cell. The CdS cell is a light-sensitive resistor that can be used to measure light or trigger a digital circuit when light level changes. When light strikes the CdS cell, charge-carriers are increased in the material, lowering the resistance.

HISTORICAL NOTE

The first transducer in space was a solar cell used by American scientists on the second satellite launched by the United States—the *Vanguard I*, on March 14, 1958. The U.S. physicist John A. O'Keefe used the satellite to detect slight variations in gravitational intensity over the earth's surface; he deduced that the earth deviated from the oblate spheroid theory predicted. The satellite's source of power for radio transmission was solar cells; they converted the sunlight that struck them into electrical energy sufficient to power a radio transmitter.

Capacitive Transducers

The capacitance of a parallel plate capacitor is given by

$$C = 8.85 \times 10^{-12}\ \text{F/m}\left(\varepsilon_r \frac{A}{d}\right)$$

Capacitive transducers change the capacitance by varying either the material between the plates (ε_r), the size of the plates (A), or the spacing between the plates (d), as illustrated in Figure 12–1. For example, a capacitive microphone uses acoustical pressure to vary the spacing between the plates, which changes an audio signal into a variation of capacitance. Another capacitive transducer is a fluid-level indicator. As a nonconducting liquid rises between the plates, the dielectric constant of the capacitor changes, which in turn affects the capacitance. This method is common in aircraft fuel measurements. In addition, the effective plate area can be varied by displacement of one of the plates, as in a capacitive displacement transducer. Capacitance transducers have a wide range of applications including measuring displacement, velocity, force, pressure, flow, and relative humidity.

To get a usable electrical signal from a capacitive transducer, the change in capacitance needs to relate to the original stimulus. One way is to use the capacitive change to modify the

FIGURE 12–1

Capacitive transducers vary capacitance in some way.

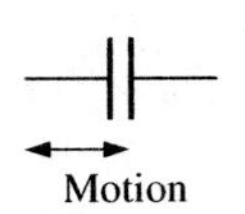
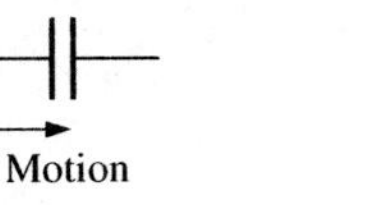

(a) Distance between plates

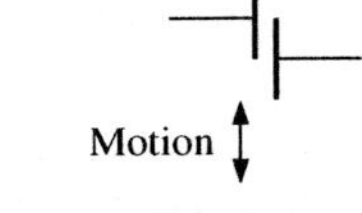

(b) Effective plate area

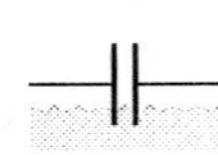

(c) Dielectric

amplitude of an ac signal; this is frequently done with an ac bridge circuit (reactive components are in the arms of the bridge). The capacitance can also change the frequency of a resonant circuit. This later method has been applied to shock wave measurements, where the shock wave crushes a cable, changing the capacitance and in turn the frequency of a resonant circuit. Data is analyzed by comparing the recorded frequency to the known distance of crushed cable.

Inductive Transducers

Inductive transducers convert the quantity to be measured into a change in inductance of a coil. One common method for changing the inductance is to move a magnetic core, linked to a sensing element, within the windings of the coil. A device called a linear variable differential transformer (LVDT) uses this principle and can be used in displacement, velocity, acceleration, pressure, and force sensors. An advantage is that there is no wear due to sliding contacts in this type of transducer. The LVDT is described in Section 12–5.

Electromagnetic Transducers

A voltage is produced in a conductor by the relative motion between the conductor and a perpendicular magnetic field. The relative motion induces a voltage across the conductor. It doesn't matter whether the coil moves or the magnetic field moves, only that there is relative motion. Figure 12–2 shows a two-part transducer that is a sensor for rotational speed. This is the principle used in many bicycle speedometers. The same principle can be applied to flow meters, where a conductive fluid has a voltage induced across it as it crosses under a magnet.

FIGURE 12–2

A transducer that includes a magnet and coil. When the magnet moves under the coil, a voltage is induced.

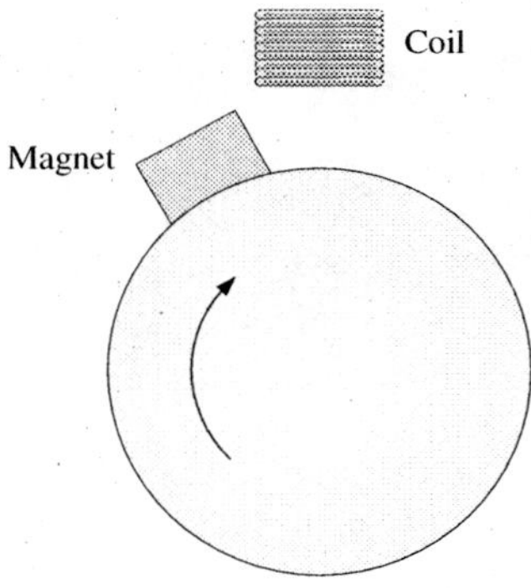

Photoelectric Transducers

The photoelectric effect is exploited in various types of light sensors. The photoelectric effect is the direct conversion of light (or other radiant energy) into electricity. One of the most widely used sensors is solar cells or photovoltaic cells. In addition to their role as a light sensor, solar cells are used to power electronic equipment in space (see Historical Note) and in remote locations as well as other power-generating applications.

Review Questions

Answers are at the end of the chapter.

1. What is a transducer?
2. With respect to transducers, what does the term *excitation* refer to?
3. Assuming the plates of a capacitive pressure transducer are moved together with an increase of pressure, what happens to the capacitance?
4. What are three ways to vary the capacitance of a capacitive transducer?
5. What is an LVDT?

TRANSDUCER SELECTION 12–2

The selection of a transducer begins with the specifications of the physical quantity to be measured. The range of measurement, accuracy requirements, the environment and calibration procedure must all be considered. Each of these considerations is part of the selection criteria for transducers.

In this section, you will learn about important parameters to consider when selecting a transducer.

Measurement Parameters

Several parameters must be considered in selecting a transducer for a particular job.

Range

The **range** is the set of values a transducer is designed to measure. The minimum and maximum values of the transducer's range are called the endpoints. Some transducers can be adjusted to cover a different range by attenuating the input; for example, a sensitive light transducer can be used if the light is attenuated using a filter. It is not always possible to find a single transducer to cover the entire range of input values; sometimes transducers with overlapping ranges must be selected.

Threshold

The **threshold** is the smallest detectable value of the measured quantity starting near the zero value. For a particular input level to be discerned, it must be possible to assign a unique number to it. The selection of a transducer requires that it respond in some discernable manner to the threshold.

Dynamic Behavior

The **dynamic behavior** of a transducer specifies how the transducer responds to a changing input. The transducer's dynamic performance is usually specified as a frequency response or as a response time, depending on the type of transducer and the measurement. The response time is the time required to reach a specified percentage (typically 90% to 99%) of the final value for a given change of the input quantity. For example, the response time of a temperature sensor can be given in the amount of time that is required for the sensor to respond if placed in a different thermal environment. The response time is measured in much the same way as the time constant for an *RC* or *RL* circuit. (The time constant is the time required for the output to reach 63% of its final value.)

Accuracy

Accuracy is the difference between the measured value and accepted value of a measurement. The accepted value is normally a standard quantity that can be traced to a national standard. The accuracy requirements for a particular measurement can greatly affect the total cost of the measurement system, including the transducer. In addition, certain transducers, such as strain gauges and pressure transducers, have a fatigue life that can change the accuracy, particularly if they have been subjected to cyclic behavior. In some cases, the accuracy isn't as important as the ability to detect a small change (resolution) as when quantities are being compared. For example, in underground tank testing, the interface between the liquid and air can be located by observing the small temperature difference between the air and liquid. In other cases, resolution and consistency are more important than accuracy.

Resolution

Resolution is the magnitude of the smallest detectable change in a quantity that is being measured. Typically, the resolution is specified in terms of a percentage of the full-scale output. Noise can affect many transducers because of their small output voltages, which in turn affects the resolution.

Repeatability

Repeatability is a measure of how well the transducer responds to the same input multiple times. It is usually expressed as a percentage difference of the full-scale output. To measure repeatability, the measured quantity must be approached from the same direction.

Hysteresis Error

Hysteresis error is the maximum difference between consecutive measurements for the same quantity when the measured point is approached each time from a different direction for the full range of the transducer. An example is when backlash in gearing causes the readings of a dial to be different, depending on whether the gearing was turned in one direction or the other. Another example is to test if a temperature reading of a substance is the same when the thermometer is cooling as when the thermometer is warming up.

EXAMPLE 12–1

Problem

Assume you need to detect weights up to 20 tons to within 25 pounds. What is the minimum resolution required of the scale?

Solution

$$\text{Number of pounds in 20 tons} = 20\,\cancel{t}\left(\frac{2000\text{ lbs}}{\cancel{t}}\right) = 40{,}000\text{ lbs}$$

$$\text{Resolution} = \frac{25\text{ lb}}{40{,}000\text{ lb}} \times 100\% = \mathbf{0.063\%}$$

Question*

Is it possible to meet this resolution specification with an inaccurate scale?

Operational and Environmental Considerations

In addition to measurement parameters, a transducer obviously must survive in the environment that it is placed. The transducer, wiring, and connectors must all be able to withstand the effects of exposure to the environment. Natural hazards need to be considered. These include the effects of dust and dirt, high or low temperatures, water (including salt water), and humid conditions. Wire insulation is available that is resistant to natural effects as well as to solvents, acids, and bases. In contrast, the transducer should not present a hazard to the environment in which it is placed, including causing electrical problems such as explosion hazard or shock hazard.

Human-caused hazards include high-radiation environments, corrosive or dangerous chemicals, immersion, abrasion, vibration, and explosive environments, to name a few. The electrical signal from the transducer may be interfered with if the signal cables are routed in an electrically noisy environment, another possible hazard for proper operation, particularly with transducers that have a low output signal.

Power requirements depend on the type of transducer. Active transducers require a source of excitation, which can be a dc or ac source. If the transducer is being operated in a remote or noisy environment, the power leads become a source of potential problems.

If the transducer produces a very small signal, or it is located in an electrically noisy environment (motors, arc welders, etc.), amplification or other signal conditioning may be required before the signal is sent to the instrumentation system. The transducer output may need to be converted into a different format (such as a digital signal). Also, physical limitations such as space available for the installation may need to be considered.

Loading effects are another important consideration in selecting a transducer. Loading can also occur with other types of measurements. All measurements in some way modify the quantity to be measured. For example, a rotating-vane flowmeter extracts a small amount of the energy from the fluid to turn the vane and thus changes the speed of the fluid from its undisturbed value. Loading can also occur in the electrical-measuring circuit that is connected to the transducer. Electrical loading occurs when a transducer with a high Thevenin resistance is connected to an amplifier with low input resistance. In this case, the transducer signal is reduced as soon as it is connected to the amplifier. Many transducers have a very high equivalent Thevenin resistance (pH electrodes, for example), so it is important to be aware of the need for compatible amplifiers.

---

* *Answers are at the end of the chapter.*

Calibration Requirements

One final consideration for selecting a transducer is the calibration requirements. **Calibration** refers to the comparison of a measurement device or instrument to a standard of known accuracy. The process of calibration is to adjust the device or instrument to substantially agree with the standard. Some transducers require regular calibration. The calibration can consist of comparing the transducer to a known reference instrument or a physical standard (such as a known mass for a load cell) or of using a physical reference (such as the triple point of water for a temperature transducer). The calibration interval is determined by the operating life of the transducer and other factors such as long-term sensitivity shift, zero-shift, and the accuracy requirements of the application.

Frequently, the physical parameter is varied and the response of the transducer is observed and compared to a known standard. Generally, the specific calibration points should extend over the full range of the quantity being measured to avoid the need for extrapolation. For example, a pressure transducer should be subjected to the same range of pressures in the calibration process that it will experience in the application to reveal any nonlinearities or other potential problem. The data should be taken in both an increasing and decreasing direction to reveal any hysteresis present. Data taken during calibration is called a *calibration record*. The calibration record should list the standard that was used for the accepted value to allow for tracing. A line connecting measured data points during calibration is called a calibration curve for the particular transducer and becomes part of the record.

Review Questions

6. When selecting a transducer, why is it important to consider its range?
7. How is dynamic behavior of a transducer specified?
8. What is the difference between *accuracy* and *resolution*?
9. How does loading affect a measurement?
10. What is meant by calibration?

12–3 TEMPERATURE MEASUREMENTS

In industrial process-control, temperature is the most frequently controlled and measured variable. It is necessary to monitor temperature in a wide variety of industries because the proper operation of most industrial plants—and frequently the safety of their operation—is related to the temperature of a process.

In this section, you will learn about five types of temperature transducers, including the advantages and applications for each type.

All materials contain molecules, which are in constant motion. The point where molecular motion totally ceases, and there is no thermal energy, is called absolute zero.* In a solid, molecules are confined to a particular location and molecular motion is confined to molecules vibrating in place. In liquids, the molecules have sufficient energy to move within the material. As more thermal energy is added to a body, the velocity of molecules increases to the point that the molecules are free to roam independently of each other, forming a gas. Theoretically, there is no upper limit to temperature; as temperature is increased, molecules lose electrons and the molecules break down into atoms, forming plasma. This is the condition found in the stars.

---

* According to quantum theory, molecules still have energy at absolute zero.

Temperature (from the Greek words for "heat measure") is related to how vigorously the atoms of a substance are moving and colliding with other atoms. Temperature difference is one of the factors that determine if heat can be transferred between two bodies. If the temperature of two bodies is identical, no heat will transfer between them no matter how large or small they are.

Although the word *temperature* is derived from heat measure, in physics heat is an entirely different concept than temperature. **Heat** is a measure of the total internal energy of a body, measured in joules. A substance with a higher temperature contains less heat if it has a smaller total internal kinetic energy. For example, a hot cup of coffee has a higher temperature than an iceberg because its atoms are moving rapidly, but the coffee has less heat than the iceberg because the iceberg has much greater total internal energy because of its mass.

Temperature Scales

Although he did not invent the thermometer, Gabriel Fahrenheit, a Dutch instrument maker, was recognized for producing the first mercury thermometers; they were the first ones accurate enough for scientific work. His scale was calibrated on the basis of the lowest temperature he could obtain (a mixture of ice water and ammonium chloride at 0°F) and the temperature of the human body as 96° (although 98.6° was later found to be more accurate). The scale, used primarily in the United States, indicates the freezing point of water as 32° and the boiling point of water as 212° at standard pressure. Absolute zero is at a temperature of −459.6° on the Fahrenheit scale.

The Celsius scale is more widely used worldwide than the Fahrenheit scale. The Celsius scale defines the freezing point of water as 0° and the boiling point as 100° at standard pressure (although Celsius originally defined these points in the reverse). The Celsius scale has absolute zero at −273.2°.

The Kelvin scale is an absolute scale in which all temperatures are positive; it is the temperature scale used in most scientific work. The Kelvin scale simply defines the lowest temperature on the Celsius scale as 0 K. (The degree symbol is not used with the Kelvin scale.) The magnitude of a degree on the Celsius and Kelvin scales is identical. Thus, the number of degrees between the freezing point of water and the boiling point of water is 100° on both the Celsius and the Kelvin scales.

Conversion of temperatures between the scales can be done with the following equations:

$$F = \frac{9}{5}C + 32 \quad (12\text{–}1)$$

$$C = \frac{5}{9}(F - 32) \quad (12\text{–}2)$$

$$K = C + 273.2 \quad (12\text{–}3)$$

where F is temperature (°F), C is temperature (°C), and K is temperature (K).

EXAMPLE 12–2

Problem

A temperature of 65°F is what temperature in Celsius?

Solution

Substitute into Equation 12–2.

$$C = \frac{5}{9}(F - 32) = \frac{5}{9}(65 - 32) = \mathbf{18.3°C}$$

Question

What is the same temperature in K?

The Thermocouple

When two dissimilar metal wires are joined at one end and heated, a small thermionic voltage appears between the wires that is proportional to the temperature. This effect was discovered by Thomas Seebeck in 1821 and is named the Seebeck effect. The small voltage is frequently referred to as an emf (for electromotive force). The junction between the two metals is called a **thermocouple junction**.

If both ends of the wires are joined together, two thermocouple junctions are produced. If one junction is at a different temperature than the other, there is current in the circuit, as illustrated in Figure 12–3. The amount of current is a function of the temperature difference between the two junctions and the type of metals used in the wires. To be useful as a temperature measurement, one junction is the sensing, or "hot," junction whereas the other junction is the reference, or "cold," junction. If the cold junction is at a known temperature, such as that of melting ice, the current in the circuit can be calibrated in terms of the temperature of the sensing junction. A **thermocouple** is a temperature-sensing transducer that produces a current proportional to the difference in temperature between a measuring junction and a reference junction.

FIGURE 12–3

A basic two-wire thermocouple. Current in the circuit depends on the type of wires and the temperature difference between the junctions.

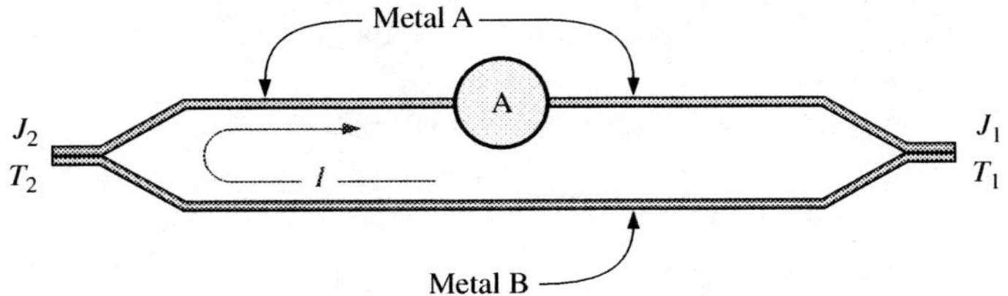

If you break the circuit and try to measure the thermionic voltage created at a junction with a voltmeter, you encounter a problem. This is because when you connect the leads of a voltmeter to the dissimilar metals of the junction, you create two new junctions (called parasitic junctions) that are themselves thermocouples, as shown in Figure 12–4. As long as both meter leads are at the same temperature, the voltmeter responds only to the difference between the temperatures of the meter leads and the original junction that you are attempting to measure.

FIGURE 12–4

Connection of a voltmeter creates two new thermocouple junctions. The voltmeter will read the algebraic sum of the three junction voltages.

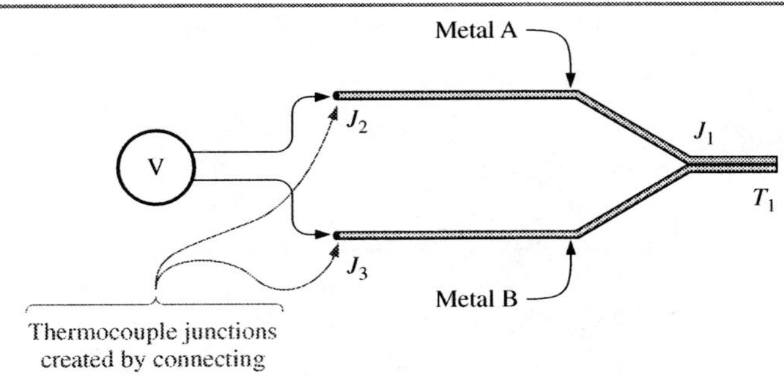

The solution to the dilemma is to move both junctions connecting the voltmeter onto an isothermal (same temperature) block and place the block at a known reference temperature. The voltage from the unknown junction will now be proportional to the type of materials and the temperature difference between the unknown and the isothermal reference block. Thus, the circuit is equivalent to the basic thermocouple but with the addition of a meter. This idea is illustrated in Figure 12–5.

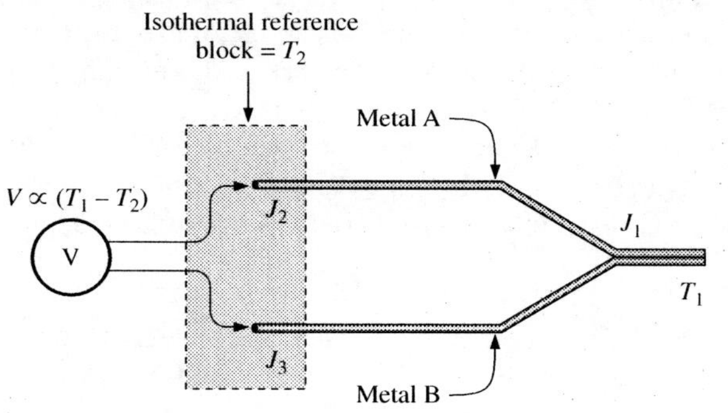

FIGURE 12–5

By connecting the voltmeter to the thermocouple on an isothermal block, the thermocouple is equivalent to the basic one shown in Figure 12–3. The block represents the reference.

The method of adding the isothermal block is commonly used, but the block temperature needs to be known. Usually, this is done with another temperature sensor that cannot stand the temperature of the sensing junction but can measure the reference junction with high accuracy. With the addition of some electronic circuitry, the output of the thermocouple can be calibrated for measuring temperatures directly.

Standard thermocouples cover different ranges of temperature and have different sensitivity, linearity, stability, and cost. Figure 12–6 shows the relationship between the temperature and the thermoelectric voltage for several common types of thermocouples. The output voltage is shown for a reference temperature (T1) of 0°C, which is at the origin. One of the more linear types is type K; it is linear over the range of 0°C to 1250°C and is widely used for this reason.

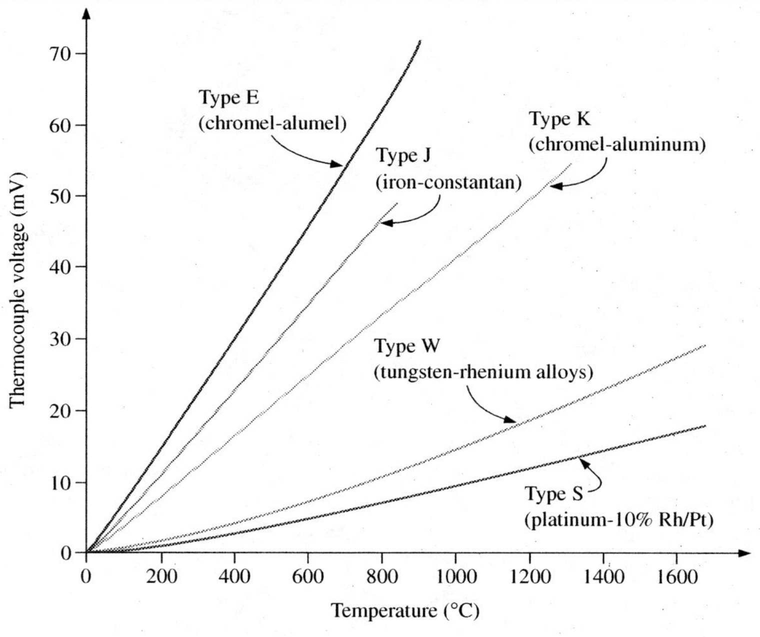

FIGURE 12–6

Comparison of some common thermocouples (0° reference temperature).

Precautions with Thermocouples

Because thermocouples produce a relatively small output voltage (typically a few millivolts) special precautions must be observed to prevent interference because the wire can act as an antenna for pickup. To avoid this, the thermocouple wires should be as short as possible; twisting the lead wires and shielding may be necessary in some cases. When longer runs are required, the thermocouple wires should be extended with special extension wire designed for the particular thermocouple.

In harsh chemical environments, thermocouples may deteriorate; water can cause problems because of dissolved substances. At extreme temperatures, the metal of the thermocouple can boil off, changing the alloy and affecting the reading. These types of deteriorations require the thermocouple to be replaced periodically. As a check on deterioration, the thermocouple's electrical resistance can be logged. To measure the resistance, the ohmmeter should be used on the same range every time; readings are taken with the leads on one set of contacts and then reversed. The average of the readings is used.

The RTD

A **resistance temperature detector (RTD)** is a type of temperature transducer in which resistance is directly proportional to temperature. They are the most accurate type of thermometers for temperatures from about −50°C to 450°C. (Higher ranges are possible with less accuracy.) Nearly all RTDs are constructed from fine wire of platinum, although wires made from nickel, germanium, and carbon-glass are occasionally used for specialized applications. Two examples of RTDs are shown in Figure 12–7. The highest quality RTDs are made from platinum wire (PRTDs) that is mounted in a manner to avoid strain-induced change in the resistance.

FIGURE 12–7 Platinum resistance thermometers.

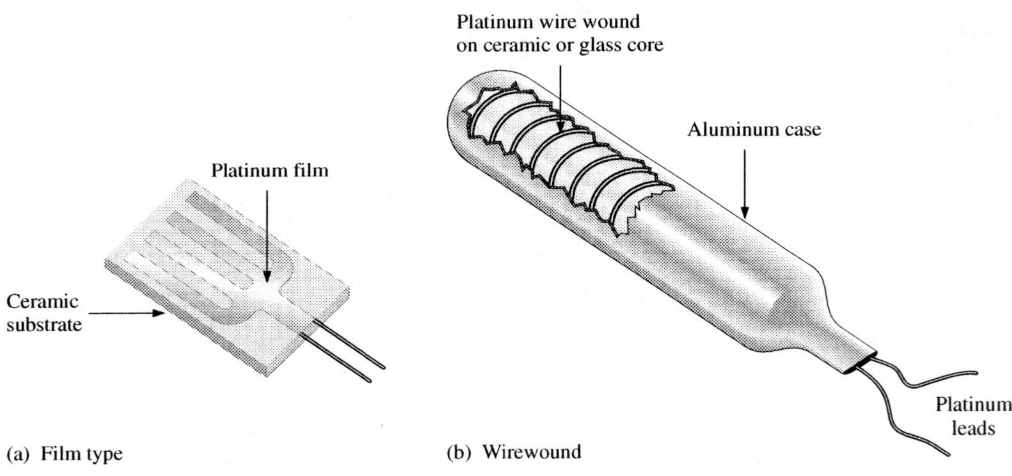

A Wheatstone bridge or a 4-wire resistance measurement generally is used to measure the resistance of RTDs. In the bridge method, the RTD is placed in one leg of the bridge and the output voltage is sensed. The output voltage is a function of the temperature. The resistance of RTDs is typically 100 Ω, so care must be taken to avoid problems with lead resistance. The 4-wire resistance method requires a special ohmmeter that uses two leads to source a constant current into the RTD and two leads to sense the voltage developed across the RTD. This method is extremely accurate for low resistance measurements of wires because it cancels the wire resistance of the meter leads.

Thermistors

Like RTDs, **thermistors** change resistance as a function of temperature, but they have a negative coefficient (resistance decreases with increasing temperature). Thermistors are available in a variety of packages, including glass beads, probes, discs, washers, and rods. Typically, they are manufactured as a small, encapsulated bead made of a sintered mixture of transition metal oxides (nickel, manganese, iron, or cobalt). These oxides exhibit a very large resistance change for a change in temperature, which means they are very sensitive. At room temperature, the resistance is about 2000 Ω. In addition to the advantage of great sensitivity, thermistors are chemically stable, have fast response times, and are physically small. The small physical size and fast response of thermistors make them ideal for monitoring temperatures in limited space, such as the case temperature of a power transistor or the internal body temperature of animals. The major limiting factors for thermistors are the limited range—from −50°C to a maximum of 300°C—and a nonlinear response. Figure 12–8 illustrates a typical thermistor response.

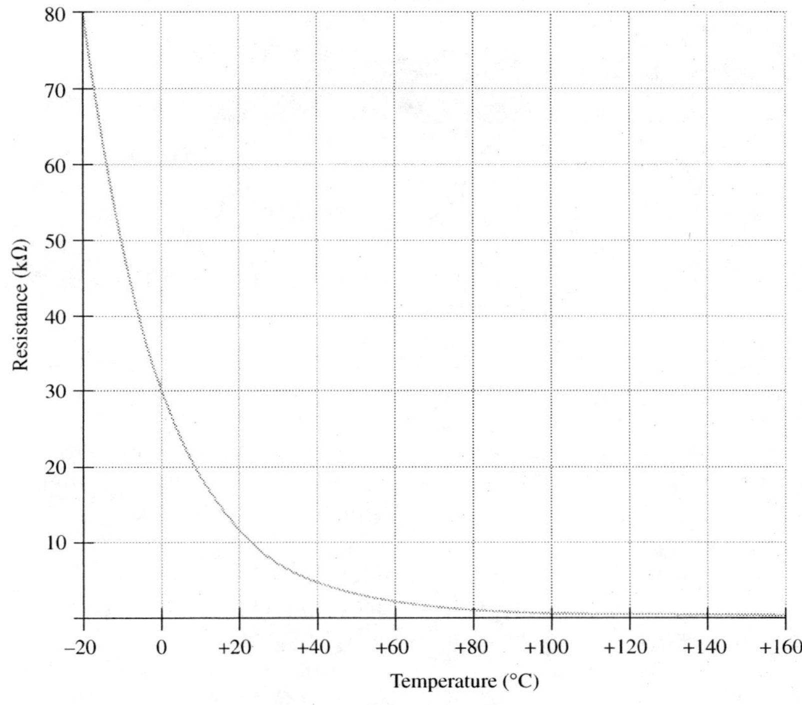

FIGURE 12–8

Typical thermistor response.

Integrated Circuit Temperature Sensors

Integrated circuit (IC) temperature sensors are convenient and low-cost alternatives to thermistors and thermocouples. IC sensors are available in conventional transistor and IC packages with either a voltage or a current output that is proportional to temperature. An example of a sensor with a voltage output is National Semiconductor's LM135. The LM135 operates over the range from −55°C to 150°C. A device with a current output is Analog Devices AD590. This two-terminal temperature transducer is connected in series with a low-voltage power supply, and the series current is regulated to be equal to 1 mA/K.

IC sensors have limited range and they are fragile; however, they can be easily calibrated, are low cost, and have an output that can be read directly in degrees when connected to a DMM.

Radiation Pyrometers

The radiation **pyrometer** is a noncontacting temperature sensor that detects infrared radiation from a source and converts the radiation to a voltage or current that is proportional to temperature. It is possible to measure temperatures from a remote location. It is usually used to observe high temperatures, such as hot ovens, but with recent developments is capable of measuring temperatures to as low as −50°C. Pyrometers are primarily used for measuring high temperatures in inaccessible locations or in environments that a thermocouple cannot operate.

Review Questions

11. What is the freezing point of water in the Fahrenheit, Celsius, and Kelvin scales?
12. What is the Seebeck effect?
13. What is the range of temperatures a type K thermocouple can measure?
14. What are three advantages of thermistors over RTDs?
15. What is the advantage of a radiation pyrometer to sense temperature?

12–4 FORCE MEASUREMENTS

Many products ranging from small food items to truckloads of gravel need to be weighed before being sold. Force transducers are used to convert weight (force exerted) to an electrical quantity. The most common force transducer is the strain gauge.

In this section, you will learn how strain gauges and load cells are used for measuring force.

Hooke's Law

If you apply a force to an elastic material, it will deform to some extent. **Elasticity** is the property of a material to recover to its original size and shape after a deforming force has been removed. A spring is an example of an elastic material that deforms under a force. If a small force is applied to the spring, it changes its length (longer or shorter, depending on the force). When the force is removed, the spring returns to its original size, provided the force is not too large.

Although the effect is smaller, a weight (force) added to any solid metal block will compress the block by an amount that is proportional to the weight. The applied force can be either a positive force (such as the weight) or a negative force that tends to stretch the block, as shown in Figure 12–9. As long as the material remains in the elastic region, the change in length is proportional to the applied force. This relationship is known as Hooke's law.

$$F \propto \Delta l$$

FIGURE 12–9

The change in length, Δl, is proportional to the applied force until the elastic limit is exceeded.

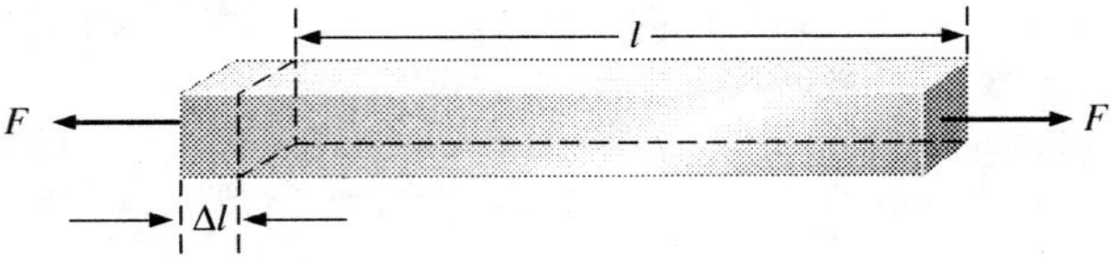

(a) Tensile forces (positive).

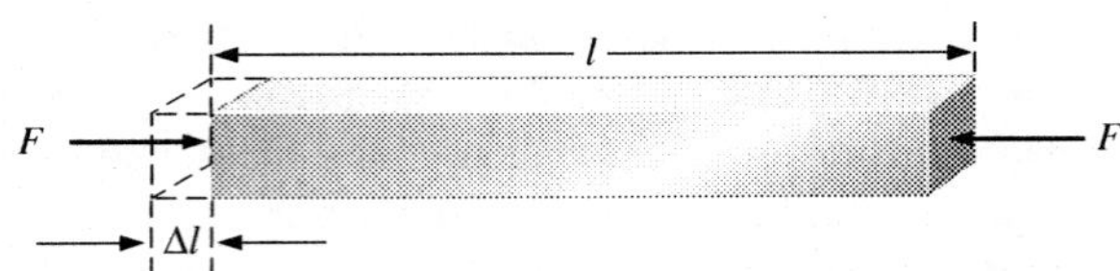

(b) Compressive forces (negative).

where F is applied force (newtons) and Δl is change in length due to the applied force (meters). The symbol $\propto$ means, "is proportional to." By inserting a constant, k, Hooke's law can be written as an equation

$$F = k\Delta l \tag{12-4}$$

where k is a constant of proportionality, n/m.

The tendency of an elastic body to return to its original shape is limited. As more and more force is applied to an elastic body, it reaches the **elastic limit**, which is the point where permanent deformation results; Hooke's law is no longer valid after this point. For materials such as steel, the elastic limit is reached if the change in length is more than a few percent of its initial length. Once this point is reached, additional force will cause plastic flow and fracture. Unlike solid metals, some materials, such as modeling clay, have no elasticity and will not return to the original dimensions regardless of how small a force is applied. Clay is said to be a plastic material.

Strain and Stress

When a block is subjected to a force, it will change length. For a given force, the *change* in length depends on three factors: the original length of the block, its cross-sectional area, and its composition. Let's examine the effect of length first. Assume you have two identical blocks except for their length. If the same compression force is applied to each block, the longer block will compress an amount that is proportionally greater than the shorter block. Interestingly, the *change in length* divided by the original *length* of the block will be the same for both blocks.

$$\frac{\Delta l_1}{l_1} = \frac{\Delta l_2}{l_2}$$

where $\Delta l_1/l_1$ is change in length divided by length of block 1 and $\Delta l_2/l_2$ is change in length divided by length of block 2.

The quantity $\Delta l/l$ is called **strain**, which is abbreviated with the symbol $\in$. It is the elastic deformation of the length of a body due to an applied force. As a defining equation, strain is

$$\in = \frac{\Delta l}{l} \tag{12-5}$$

where a is strain, a dimensionless number (often expressed as in./in.).

In practice, the magnitude of strain is a tiny number for metals; therefore, it is common practice to express strain in units of microstrain. Microstrain is $\in \times 10^{-6}$ and it is written as $\mu\in$.

The second factor that affects the block's change in length is its cross-sectional area. Imagine a block that is supporting a load. The load causes the block to be compressed by some amount. If a second identical block is added to share the load, the load force is distributed equally to the two blocks; the change in length is half as much as before. Evidently, an increase in area reduces the effect of the force. That is,

$$\frac{F}{A} \propto \frac{\Delta l}{l}$$

where A is cross-sectional area, m^2.

The third factor that affects the change in length is the material itself. Some materials, like steel, are very stiff and do not deform easily. Others, like wood, deform much more easily.

For example, a wooden block the same size as a steel block will deform more than ten times the amount that the steel would deform when subjected to a force.

A constant called **Young's modulus**, *E*, is introduced in order to account for the properties of the material. Young's modulus is specified for either stretching or bending. It is a measure of the stiffness of a material and, for a given cross-sectional area, of the material's ability to resist a change in length (stretching) when a load is added. The proportional relationship is changed into an equation by inserting Young's modulus as

$$\frac{F}{A} = E\frac{\Delta l}{l} \tag{12–6}$$

where *E* is Young's modulus (N/m^2).

The quantity *F/A* is called *stress*. **Stress** is the force per unit area on a given plane within a solid body; it can be either due to a tension force or a compression force. The unit for stress is the N/m^2.

Equation 12–6 shows that stress (*F/A*) and strain ($\Delta l/l$) are proportional to each other and related by Young's modulus. A stress-strain diagram is shown in Figure 12–10 for low-carbon steel. Notice that it begins with a proportional region where Hooke's law applies (the applied force is proportional to the change in length). If the force is removed along this region, the steel will return to its original shape. After the elastic limit is reached, a point is reached called the *yield point* occurs, where permanent deformation of the material occurs. This region is called the plastic range. Additional force will lead to failure of the material. Force transducers normally never operate beyond the yield point.

FIGURE 12–10

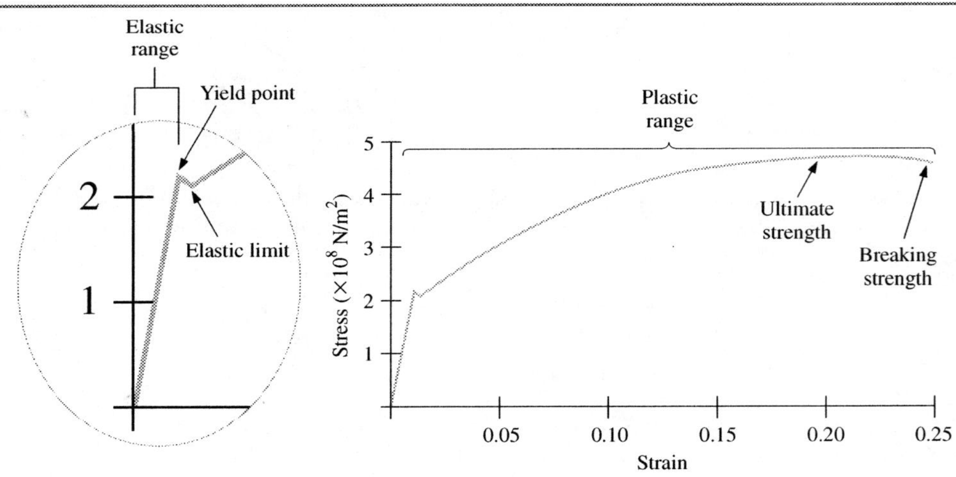

Stress-strain diagram for low-carbon steel. By conventional engineering practice, strain is plotted as the independent variable.

You have seen how a force that compresses a block along its length produces strain that is proportional to the force as long as it remains in the elastic region. When the force is applied perpendicular to the plane of the material, it is called normal strain (in physics, the word *normal* means perpendicular to). Normal strain is illustrated in Figure 12–11.

There are other types of strain that are caused by forces that are not perpendicular to the surface. This type of force produces a torque about the pivot point. The strain produced by this torque is illustrated in Figure 12–12. With the force applied as shown, the top of the beam will be stretched and the bottom will be compressed. In this case, Δl will be positive along the top and negative along the bottom. The strain will still be proportional to the applied force.

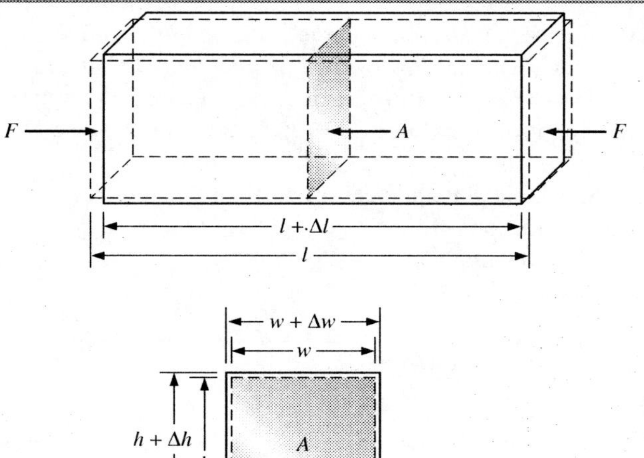

FIGURE 12–11

Normal strain. The length, width and height all change.

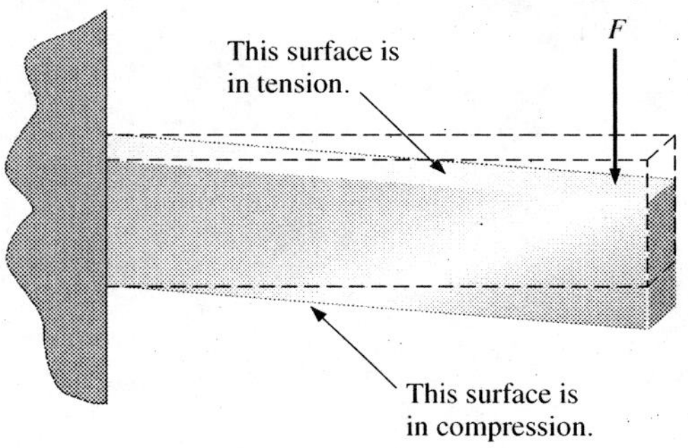

FIGURE 12–12

Bending strain on a cantilevered beam due to torque.

The Strain Gauge

The strain gauge is a transducer that is made of a very thin metallic conductor, which is firmly bonded to a solid object to detect strain in the object. When a force causes the object to be deformed, the strain gauge undergoes the same deformation, causing the resistance of the gauge to change. Recall that wire resistance is proportional to the length of the wire. For this reason, the resistance change, ΔR, divided by the unstrained resistance, R, is proportional to the strain.

$$\frac{\Delta R}{R} \propto \frac{\Delta l}{l}$$

By introducing a constant, called the **gauge factor**, G, the proportion can be changed to an equality. The gauge factor depends on the particular strain gauge, but is typically a number between 2 and 5 (2 is typical for metallic strain gauges). Thus,

$$\frac{\Delta R}{R} = G\frac{\Delta l}{l} \tag{12–7}$$

where G is gauge factor, which is dimensionless.

Metallic strain gauges are made from very thin conductive foils with a conductor folded back and forth to allow a long path for the conductive elements while retaining a short gauge length. A typical strain gauge, along with strain gauge nomenclature is illustrated in

Figure 12–13. Gauge lengths vary from as small as 0.2 mm to over 10 cm. The back and forth pattern is designed to make the gauge sensitive only to strain in the direction parallel to the wire and insensitive in the direction perpendicular to the wire. Typically, the foils are made from special alloys such as constantan, a combination of 60% copper and 40% nickel. Standard foil resistances are 120 Ω and 350 Ω; some gauges are available with resistances of up to 5000 Ω.

Strain is a directly measurable quantity, but stress, usually the quantity of interest, is not. For this reason, strain gauges are used to determine stress. Stress can be directly related to weight, so strain gauges are widely used in scales.

FIGURE 12–13

A typical foil-type strain gauge.

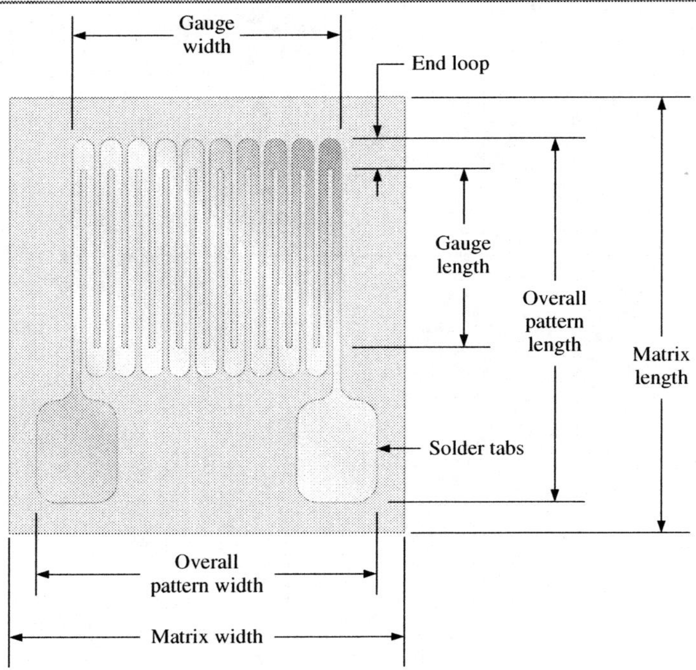

EXAMPLE 12–3

Problem

A strain gauge is bonded to a beam and changes resistance under load by 2.5 mΩ. If the unstrained resistance of the strain gauge is 350 Ω, and the gauge factor is 2.2, what is the strain?

Solution

Rearrange Equation 12–7.

$$\frac{\Delta l}{l} = \frac{\Delta R/R}{G}$$

Substituting,

$$\frac{\Delta l}{l} = \frac{2.5\ \text{m}\Omega/350\ \Omega}{2.2} = \mathbf{3.25\ \mu\in}$$

Question

If the same strain is measured with a strain gauge that has a nominal resistance of 350 Ω but a gauge factor of 3, what is the change in resistance?

The Load Cell

The basis of almost all weighing systems is the **load cell**. Load cells come in a wide variety of sizes, shapes, and ranges. A typical load cell is illustrated in Figure 12–14. Load cells are force-sensing transducers with a metal body that is deformed by the applied force (weight) and designed to operate within the elastic region of its body material. The heart of a load cell is the strain gauge. Most load cells will have four strain gauges bonded to it. With four gauges, two are in tension and two in compression when a force is applied. The gauges are connected in a Wheatstone bridge arrangement as shown in Figure 12–15. Typically, an excitation voltage of 10 V to 15 V is applied to the bridge. A force on the load cell decreases the resistance of the strain gauges in compression and increases the resistance of the two in tension, but only by a very small amount. This small change in resistance unbalances the bridge, changing the output voltage by a small amount. The output voltage is related to the force that was applied.

Load cells are generally constructed with 350 Ω strain gauges mounted in a full-bridge arrangement such as the one shown in Figure 12–15. Notice that the resistance, measured from excitation leads to output leads will also measure a nominal 350 Ω in the unstrained condition. When the strain gauges are mounted in a load cell in the full-bridge configuration, the full-scale output of the load cell is normally designated to be 1, 2 or 3 millivolts of signal per volt of excitation. For example, a load cell that is specified as 3 mV/V will have a 30 mV full-scale output if the excitation voltage is 10 V. The output voltage is nearly a linear relationship of the applied force.

SAFETY NOTE

Five rules should be observed when using hand or power tools:

- Keep all tools in good condition with regular maintenance.
- Use the right tool for the job.
- Examine each tool for damage before use and do not use damaged tools.
- Operate tools according to the manufacturers' instructions.
- Use the right personal protective equipment.

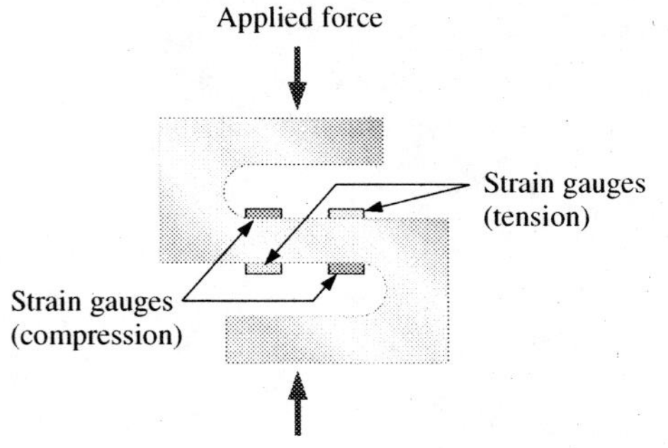

FIGURE 12–14

A typical load cell with four strain gauges. Strain gauges are bonded to the metal body to measure the deformation when a force is applied.

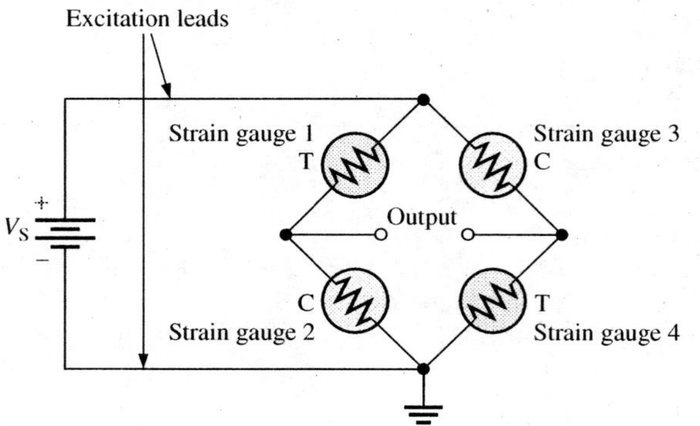

FIGURE 12–15

A Wheatstone bridge for measuring the output of a load cell. Two strain gauges (on opposite diagonals) are in tension; two are in compression.

EXAMPLE 12–4

Problem

A load cell with a 2 mV/V specification has four 350 Ω strain gauges (two in compression, two in tension). The excitation voltage is 15 V. If the Wheatstone bridge is balanced with no force applied, what is the resistance, R_2, of strain gauge 2 at full load?

Solution

The normal balanced bridge is illustrated in Figure 12–16(a). When balanced, each side of the bridge has one-half of the excitation voltage. When fully loaded, the bridge is unbalanced, as shown in Figure 12–16(b). The output voltage is then

$$V_{OUT} = (2\text{ mV/V})\,(15\text{ V}) = 30\text{ mV}$$

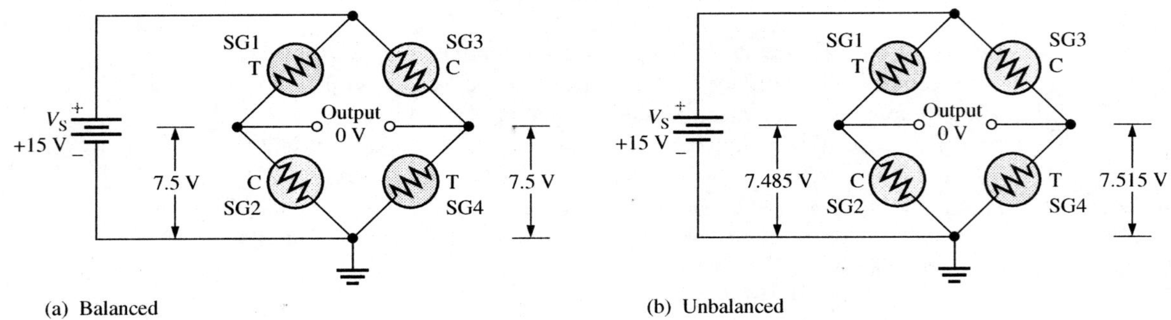

FIGURE 12–16

Each half of the bridge will be unbalanced by 15 mV. Thus, one side of the bridge will have an output that is 15 mV less than 7.5 V; the other half will have an output that is 15 mV greater than 7.5 V as shown.

The total resistance of each half of the bridge is 700 Ω, which is independent of the load. The voltage-divider equation can be written for the left side of the bridge.

$$V_{OUT} = V_S\left(\frac{R_2}{R_T}\right)$$

where R_2 represents the resistance of strain gauge 2.

Rearranging and substituting,

$$R_2 = \left(\frac{V_{OUT}}{V_S}\right)R_T = \left(\frac{7.485\text{ V}}{15\text{ V}}\right)700\ \Omega = \mathbf{349.3\ \Omega}$$

This result shows that even at the maximum output, the change in resistance of the strain gauge is less than 1 Ω, which represents a 0.2% change.

Question

What is the resistance of SG1 at full-scale output?

Sometimes only two strain gauges are used for a measurement. For bending measurements, they are installed so that one is in tension and the other is in compression by installing one directly on top of the other and on opposite sides of a beam such as that shown in Figure 12–17. The gauges are typically connected in adjacent arms of a Wheatstone

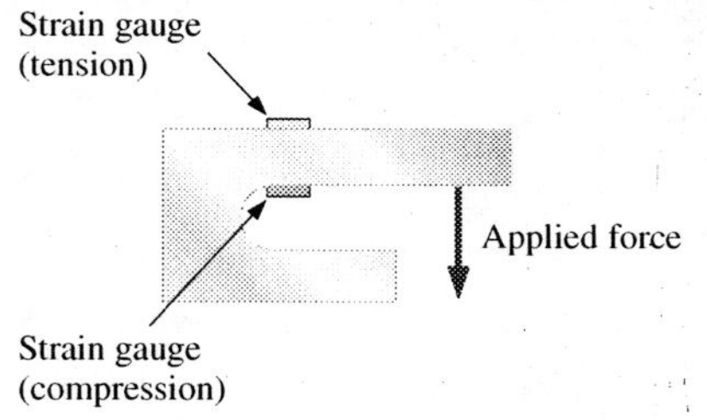

FIGURE 12–17

Cantilevered device.

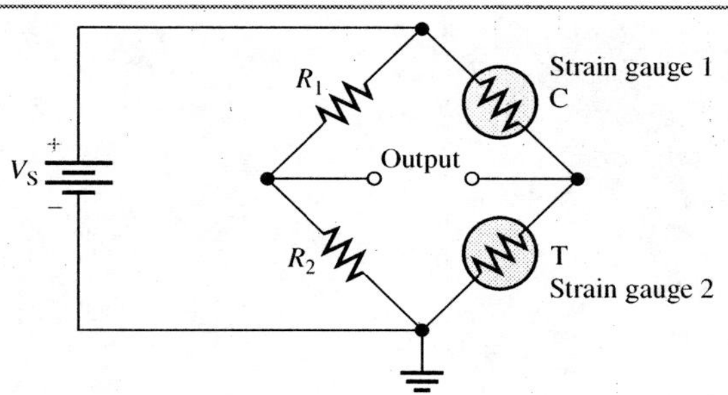

FIGURE 12–18

Half-bridge arrangement. One strain gauge is in tension; the other is in compression.

bridge, as illustrated in Figure 12–18. This configuration is called a half-bridge. This arrangement tends to cancel the adverse effect of temperature changes on the resistance of the gauges because both gauges will be affected in the same way.

EXAMPLE 12–5

Problem

A 2 mV/V load cell is excited with a 10 V source and monitored with a digital voltmeter. The load cell is used in a scale designed for a full-scale output with a 100 lb load. If the required resolution of the measurement is 0.1 lb, what is the minimum required resolution of the voltmeter, expressed as a voltage?

Solution

The required resolution of the measurement, expressed as a percentage is

$$\text{Resolution} = \left(\frac{0.1\ \text{lb}}{100\ \text{lb}}\right) \times 100\% = 0.1\%$$

The full-scale output from the bridge is

$$\text{FSO} = (2\ \text{mV/V})(10\ \text{V}) = 20\ \text{mV}$$

The resolution of the voltmeter, expressed as a voltage is

$$\text{Voltmeter resolution} = (0.1\%)(20\ \text{mV}) = \mathbf{20\ \mu V}$$

As you can see from this result, the measurement requires a sensitive meter and very low noise to avoid masking the small voltages involved.

Question

How is the required resolution affected if the excitation voltage is 15 V instead of 10 V?

Review Questions

16. What is meant by the elastic limit of a metal?
17. What is the difference between strain and stress?
18. What is Young's modulus?
19. How does a strain gauge respond to strain?
20. What is a load cell?

12–5 PRESSURE, MOTION, AND LIGHT MEASUREMENTS

Three additional measurements are pressure measurements, motion measurements (displacement, velocity, acceleration), and light measurements. There are common electrical transducers for measuring each of these quantities.

In this section, you will learn about electrical transducers for measuring pressure, motion, and light.

Pressure Measurements

Consider a fluid such as water in a flat-bottomed container. The weight of the water exerts a force on the bottom of the container that is distributed over the entire bottom surface. Each square meter of the bottom area carries the same weight as every other square meter. The force per unit area is defined as **pressure**; the area must be measured perpendicular to the force.

$$P = \frac{F}{A}$$

where P is pressure, n/m^2 (pascal), F is force, N, and A is area, m^2 (perpendicular to the force).

Pressure is measured in newtons/meter^2, which is given the special name pascal (abbreviated Pa). This unit represents a very small pressure, so it is common to see the unit written with the metric prefix of *kilo* or *mega*. In the English system, if the force is measured in pounds and the area is measured in inches, the pressure is in pounds per inch^2 (abbreviated psi). One psi is approximately 6.895 kPa. Pressure can also be measured in terms of the height of a column of mercury it can support, a common procedure for atmospheric pressure. This unit is somewhat misleading; instead of force per area, it is given as a height (mm of mercury). Atmospheric pressure is the force on a unit area due to the weight of atmosphere on that area; it could be written as either 760 mm of mercury, 29.92 inches of mercury, 14.7 psi, or 101 kPa.

As mentioned, it is common practice to express pressure measurements in terms of the equivalent pressure at the bottom of a column of a liquid of a stated height. Liquids used in pressure measurements are generally either mercury (because of its very high density) or water. Thus, 29.9 inches of mercury is the pressure at the bottom of a mercury column 29.9 inches high. Interestingly, because mercury is 13.6 times denser than water, the atmosphere can support a column of water approximately 406 inches high (about 33 feet). This is the maximum limit that a siphon can work because it is operated by atmospheric pressure.

With gases, pressure is exerted equally in all directions. If a gas is inside a closed container, the pressure of the gas is the force per area that the gas exerts on the walls of the container. In the atmosphere, as we move above sea level, the pressure decreases because of the decreased weight of the air that is supported. At the top of Mt. Everest, the air pressure is only one-third that of sea level.

Gauge Pressure and Absolute Pressure

If you experience a flat tire, and check the tire pressure, the gauge reads zero. Although you might remark that there is no air in the tire, the fact is that the tire has air in it that is at the same pressure as the atmosphere! The gauge is simply reading the *difference* between the atmospheric pressure and the pressure inside the tire. This difference is known as gauge pressure (shown in the English system as psig). In other words, the atmospheric pressure is the reference when the term *gauge pressure* is used.

Pressure readings that include the atmospheric contribution are called absolute pressure (shown in the English system as psia). Absolute pressure is referenced to a vacuum. Most pressure gauges are designed to read gauge pressure; it is important to keep in mind which pressure you are using. Another pressure measurement is called differential pressure. As the name implies, differential pressure is the difference between two input pressures (shown in the English system as psid).

Pressure Transducers

All pressure transducers operate on the principle of balancing an unknown pressure against a known load. A common technique is to use a diaphragm to balance the unknown pressure against the mechanical restraining force keeping the diaphragm in place. A diaphragm is a flexible disk fastened on its periphery that changes shape under pressure. A spring may be used to push against the diaphragm and provide a load. The amount of movement of the diaphragm is proportional to the pressure. Diaphragms can be used on a wide range of pressures, from about 15 psi to 6000 psi. In simple pressure gauges, the displacement of the diaphragm is mechanically linked to an indicator.

In an electronic-measuring system, it is necessary to convert the mechanical motion of the pressure-sensing element into an electrical signal. There are many conversion techniques possible including the capacitive pressure gauge and a resistive gauge.

In capacitive transducers, one plate of the capacitor is attached to a movable diaphragm. The other plate is stationary. An increase in pressure causes the plates to move together, increasing the capacitance, as shown in Figure 12–19. Another type of capacitive transducer uses a movable center plate that is mounted between two fixed plates. As the center plate moves in response to pressure changes, the capacitance of one of the capacitors increases while the other decreases. The capacitors are electrically connected into a bridge circuit. Capacitive pressure transducers have a high frequency response (due to low mass), implying that they can respond quickly to changes in pressure. They are available for a wide range of pressures, from very sensitive sensors for blood pressure to sensors that can respond to as much as 5000 psi.

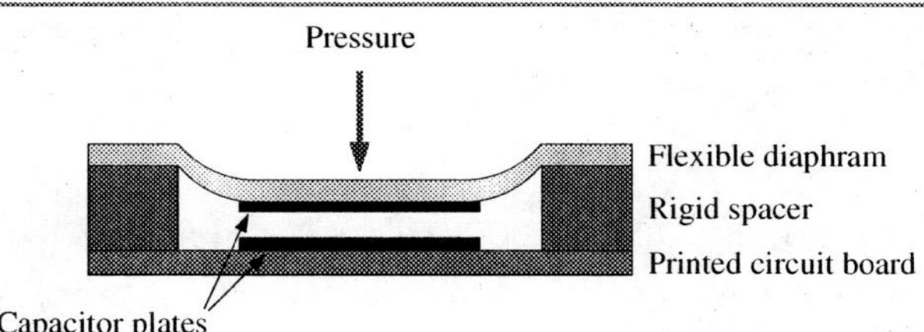

FIGURE 12–19
A basic capacitive pressure sensor.

A strain gauge can be used as the sensing element of a pressure transducer by bonding it to the diaphragm. Pressure on the diaphragm introduces strain, which is sensed by the gauges and converted to an electrical resistance change. Typically, gauges are bonded on both sides of the diaphragm and connected into a full- or half-bridge arrangement, described in the last section.

Measurement of Motion

Motion can be along a straight line or it can be circular. The measurement of motion includes displacement, velocity, and acceleration. Displacement is a vector quantity that indicates the change in position of a body or point. Velocity is the rate of change of displacement, and acceleration is a measure of how fast velocity changes. Angular displacement is measured in degrees or radians.

Displacement Transducers

Displacement transducers can be either contacting or noncontacting. Contacting transducers typically use a sensing shaft with a coupling device to follow the position of an object. The sensing shaft can be connected to the wiper arm of a potentiometer. The electrical output signal can be either a voltage or a current.

Displacement can also be converted into an electrical quantity using a variable inductor and monitoring the change in inductance. The inductance can be changed by moving the core material, varying the coil dimensions, or by moving a sliding contact along the coil.

An important displacement transducer is the linear variable differential transformer (LVDT). The sensing shaft is connected to a moving magnetic core inside a specially wound transformer. A typical LVDT is shown in Figure 12–20. The primary of the transformer is in line and located between two identical secondaries. The primary winding is excited with ac (usually in the range of 1 to 5 kHz). When the core is centered, the voltage induced in each secondary is equal. As the core moves off center, the voltage in one secondary will be greater than the other. A circuit called a *demodulator* causes the polarity of the output to change as the core passes the center position. The transducer has excellent sensitivity, linearity, and repeatability.

FIGURE 12–20

An LVDT displacement transducer.

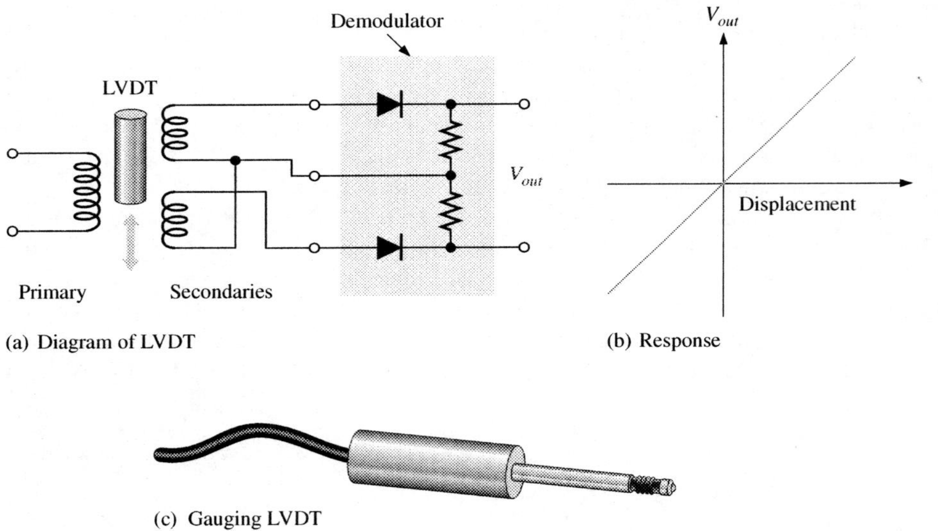

(a) Diagram of LVDT

(b) Response

(c) Gauging LVDT

Noncontacting displacement transducers include optical and capacitive transducers. Photocells can be arranged to observe light through holes in an encoding disk or to count fringes painted on the surface to be measured. Optical systems are fast, but noise can be a problem with optical sensors. To avoid problems with ambient light, the light detector should be directed only to receive the light that is part of the measuring system; shielding may be necessary to block other light.

Capacitive displacement sensors are also used as sensitive proximity transducers. The capacitance is varied by moving one of the plates of a capacitor with respect to the second plate. The moving plate can be any metallic surface, such as the diaphragm of a capacitive microphone or a rotating camshaft.

Velocity Transducers

Since velocity is the rate of change of displacement, velocity can be determined using a displacement sensor and measuring the time between two points. A direct measurement of velocity is possible with certain transducers that have an output proportional to the velocity to be measured. They sense either linear or angular velocity. Linear velocity transducers can be constructed using a permanent magnet inside a concentric coil forming a simple motor by generating an emf proportional to the velocity. Either the coil or the magnet can be fixed and the other moved with respect to the fixed component. The output is taken from the coil.

There are various transducers that are designed to measure angular velocity. Tachometers are a class of angular velocity transducers that provide a dc or ac voltage output. DC tachometers are small generators with a coil that rotates in a constant magnetic field. A voltage is induced in the coil as it rotates in the magnetic field. The average value of the induced voltage is proportional to the speed of rotation, and the polarity is indicative of the direction of rotation, an advantage with dc tachometers. AC tachometers can be designed as generators that provide an output frequency that is proportional to the rotational speed.

Another technique for measuring angular velocity is to rotate a shutter over a photosensitive element. The shutter interrupts a light source from reaching the photocells and causing the output of the photocells to vary at a rate proportional to the rotational speed.

Acceleration Transducers

Acceleration is usually measured by use of a spring-supported seismic mass, mounted in a suitable enclosure as shown in Figure 12–21. Damping is provided by a dashpot. The relative motion between the case and the mass is proportional to the acceleration. A secondary transducer such as a resistive displacement transducer is used to convert the relative motion to an electrical output. In the ideal world, the mass does not move when the case accelerates because of its inertia; in practice it does because of forces applied to it through the spring. The accelerometer has a natural frequency, the period of which should be shorter than the time required for the measured acceleration to change. Accelerometers used to measure vibration should also be used at frequencies less than the natural frequency.

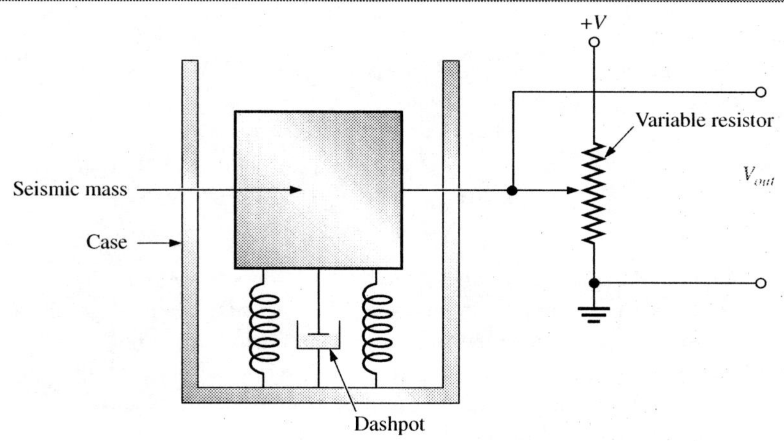

FIGURE 12–21

A basic accelerometer. Motion is converted to a varying voltage.

Measurement of Light

The Electromagnetic Spectrum

The light we see is only a small portion of the vast electromagnetic spectrum. The **electromagnetic spectrum** is the whole range of frequencies that comprise electromagnetic radiation. The theory of electromagnetic radiation was developed in the 1860s when James C. Maxwell combined the known laws of electricity with those of magnetism into a set of laws that describes the behavior of light. Light has similar characteristics to radio waves but at a very much shorter wavelength. Electromagnetic waves travel at a velocity of 3.00×10^8 m/s in a vacuum. The frequency is related to the wavelength by the equation

$$f = \frac{c}{\lambda}$$

where c is velocity of electromagnetic radiation, m/s, f is frequency, Hz, and l is wavelength, m.

Exploring the electromagnetic spectrum (see Figure 12–22), we find radio waves that extend from wavelengths hundreds of meters long to wavelengths that are millimeters long (microwaves). At shorter wavelengths is the infrared region that extends to the region of visible light. The wavelength of visible light is a thousand times shorter than the shortest radio waves; the light that our eyes respond to has a wavelength between approximately 390 nanometers (violet) and 760 nanometers (red). Immediately above visible light is the ultraviolet region and beyond this is the x-ray and gamma ray portion of the electromagnetic spectrum.

FIGURE 12–22 The electromagnetic spectrum.

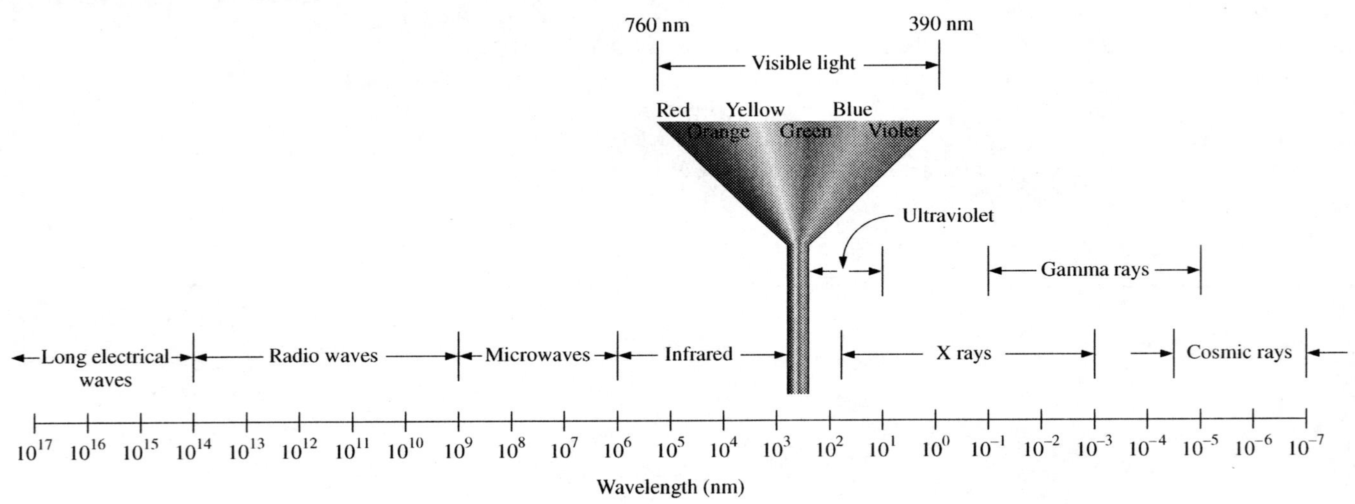

The boundaries between these regions are not well defined; they are simply names associated with parts of a continuous spectrum of which visible light is a tiny fraction of the total. Our eyes do not respond to infrared and ultraviolet radiation, but this radiation is are often referred to loosely as infrared and ultraviolet "light." Many detectors are sensitive across the boundaries of these regions. For example, photoelectric cells are sensitive in both the ultraviolet and visible portions of the spectrum.

The spectrum of common light sources is dependent on how the light was generated. Incandescent sources such as ordinary light bulbs produce a continuous spectrum that is dependent on the temperature of the filament. At higher temperatures, the spectral content shifts toward the blue end and away from the red end. By varying the operating voltage of a tungsten light, the specific color temperature can be shifted. Sunlight also produces a continuous spectrum that is temperature dependent, but it includes certain absorption lines that identify the elements in the solar atmosphere.

In electronics, LEDs play an important role as light emitters, including as a source for optical fiber communication. They can be turned on and off very rapidly. The light from an LED has a narrow spectrum, which depends on the color of the LED. Light distribution is dependent on the geometry, with most light being emitted in the forward direction.

Optical Transducers

Nearly all optical transducers can be classed into one of three basic forms of photodetection: photovoltaic sensors that are self-generating semiconducting devices that convert light directly into an emf, photoconductive sensors that act as light-sensitive resistors, and photoemissive sensors that contain a light-sensitive cathode that emits electrons when struck by light.

Photodiodes are active transducers that are constructed from a diode, a device that conducts in only one direction. When a photon of light passes through a transparent layer, it can be absorbed. If the energy of the photon is high enough, an electron is released, creating a source for current in an external circuit.

Another active light sensor is the phototransistor. Phototransistors are more sensitive than photodiodes because they have internal gain. Phototransistors are excellent light detectors where a sensitive sensor is needed; however, because of nonlinear response and poor temperature characteristics, they are not useful for light measurement applications.

Photoconductive sensors include cadmium sulfide (CdS), cadmium selenide (CdSe), lead sulfide (PbS), and other cells. CdS and CdSe cells are popular when it is necessary for the cell to respond to visible light. Photoconductive sensors consist of a photosensitive crystalline material sandwiched between two electrodes. Absorption of photons in the crystalline material causes the resistance of the crystalline material to decrease. They are inexpensive, sensitive, and can withstand high voltages, which makes them suited for control applications such as outdoor lighting control. Their principal disadvantage is that they are slow.

Review Questions

21. What is the difference between gauge pressure and absolute pressure?
22. How does a capacitive pressure sensor work?
23. How does a tachometer work?
24. What is the wavelength range of visible light?
25. What is a CdS cell?

POWER-CONTROL CIRCUITS 12–6

A useful application of electronic circuits is to control power to a load. Two devices that are widely used in power-control applications are the SCR and the triac. These devices are members of a class of devices known as thyristors.

In this section, you will learn the operation of thyristors and how they can be used to control power to a load.

The Silicon-Controlled Rectifier

A **thyristor** is a semiconductor switch composed of four or more layers of alternating *pnpn* material. A thyristor can be thought of as an electronic switch that can rapidly turn on or off a large current to a load. There are various types of thyristors; the type principally depends on the number of layers and the particular connections to the layers. When a connection is made to the first, second, and fourth layer of a four-layer thyristor, a form of gated diode known as an **SCR** (silicon-controlled rectifier) is formed. This is one of the most important devices in the thyristor family because it acts like a diode that can be turned on when required. The basic structure and schematic symbol for an SCR is shown in Figure 12–23. For an SCR, the three connections are labeled the anode (A), cathode (K), and gate (G) as shown.

FIGURE 12–23

The silicon-controlled rectifier (SCR).

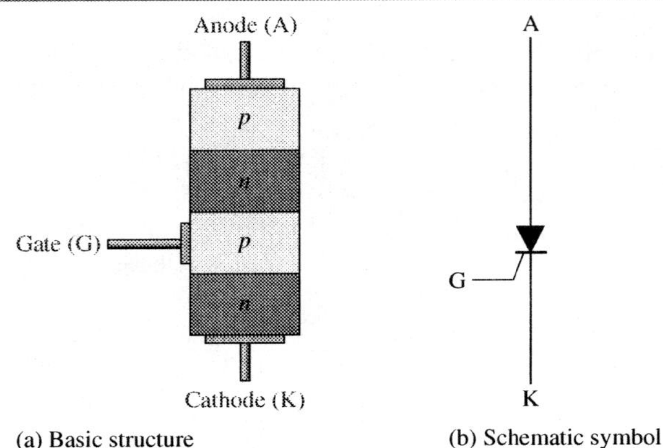

(a) Basic structure (b) Schematic symbol

The characteristic curve for an SCR is shown in Figure 12–24(a) for a gate current of zero. There are a total of four regions of the characteristic curve of interest. The reverse characteristic (plotted in quadrant 3) is the same as a normal diode with regions called the reverse-blocking region and the reverse-avalanche region. The reverse-blocking region is equivalent to an open switch. The reverse voltage that must be applied to an SCR to drive it into the avalanche region is typically several hundred volts or more. SCRs are normally not operated in the reverse-avalanche region.

The forward characteristic (plotted in quadrant 1) is divided into two regions. There is a forward-blocking region, where the SCR is basically off and the very high resistance between the anode and cathode can be approximated by an open switch. The second region is the forward-conduction region, where anode current occurs as in a normal diode. To move an SCR into this region, the forward-breakover voltage, $V_{BR(F)}$ must be exceeded. When an SCR is operated in the forward-conduction region, it approximates a closed switch between anode and cathode. Notice the similarity to a normal diode characteristic except for the forward-blocking region.

Turning the SCR On

There are two ways to move an SCR into the forward-conduction region. In both cases, the anode to cathode must be forward-biased; that is, the anode must be positive with respect to the cathode. The first method has already been mentioned and requires the application of forward voltage that exceeds the forward breakover voltage, $V_{BR(F)}$. Breakover voltage triggering is not normally used as a triggering method. The second method requires a positive pulse of current (trigger) on the gate. This pulse reduces the forward-breakover voltage, as shown in Figure 12–24(b) and the SCR conducts. The greater the gate current, the lower the value of $V_{BR(F)}$. This is the normal method for turning on an SCR.

Once the SCR is turned on, the gate loses control. In effect, the SCR is latched and will continue to approximate a closed switch as long as anode current is maintained. When the anode current drops below a value of current called the holding current, the SCR will drop out of conduction. The holding current is indicated in Figure 12–24.

Turning the SCR Off

There are two basic methods for turning off an SCR: anode current interruption and forced commutation. The anode current can be interrupted by opening the path in the anode circuit, causing the anode current to drop to zero, turning off the SCR. One common "automatic" method to interrupt the anode current is to connect the SCR in an ac circuit. The negative cycle of the ac waveform will turn off the SCR.

The forced commutation method requires momentarily forcing current through the SCR in the direction opposite to the forward conduction so that forward current is reduced below the holding value. This can be implemented by various circuits. Probably the simplest is to electronically switch a charged capacitor across the SCR in the reverse direction.

FIGURE 12–24

SCR characteristic curves.

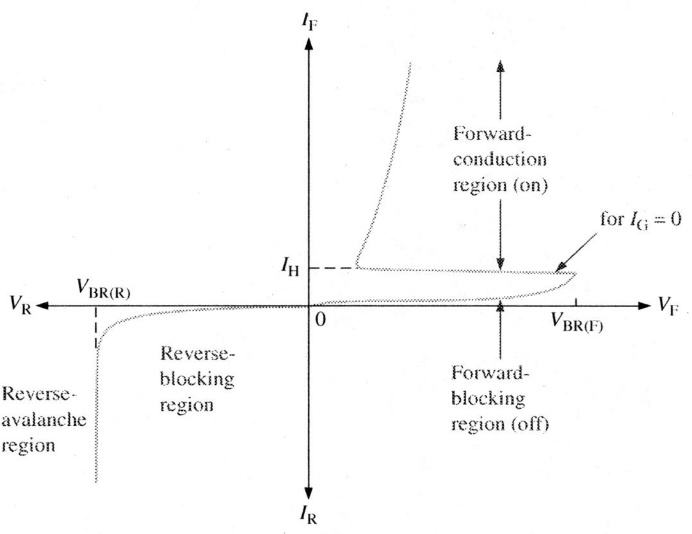

(a) When $I_G = 0$, $V_{BR(F)}$ must be exceeded to move into the conduction region.

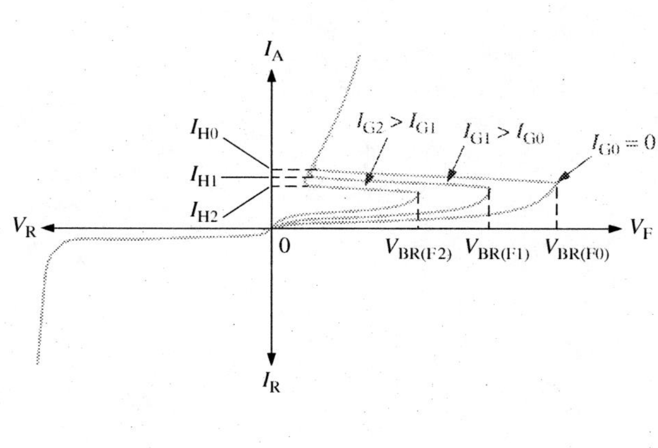

(b) I_G controls the value of $V_{BR(F)}$ required for turn on.

The Triac

The **triac** is a thyristor with the ability to pass current bidirectionally and is therefore an ac power control device. Although it is one device, its performance is equivalent to two SCRs connected in parallel in opposite directions but with a common gate terminal. The basic characteristic curves for a triac are illustrated in Figure 12–25. Because a triac is like two back-to-back SCRs, there is no reverse characteristic.

FIGURE 12–25

Triac characteristic curves.

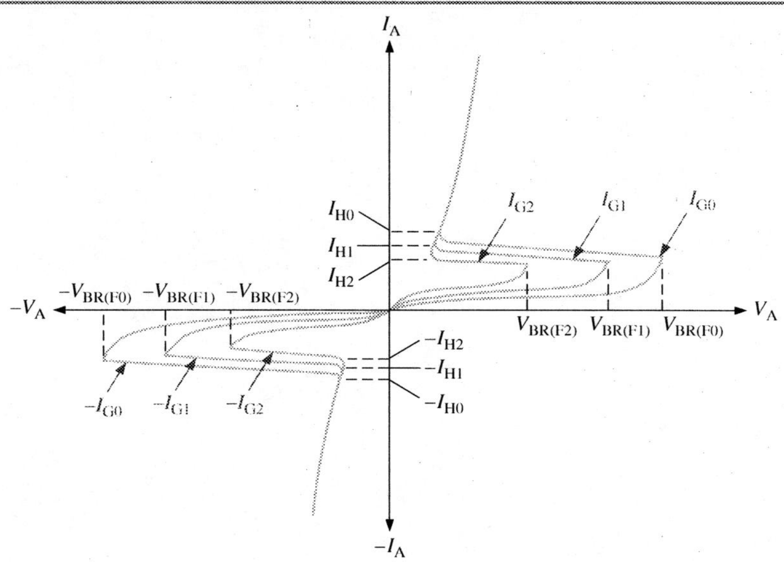

As in the case of the SCR, gate triggering is the usual method for turning on a triac. Application of current to the triac gate initiates the latching mechanism discussed in the previous section. Once conduction has been initiated, the triac will conduct on with either polarity, hence it is useful as an ac controller. A triac can be triggerted such that ac power is supplied to the load for a portion of the ac cycle. This enables the triac to provide more or less power to the load depending on the trigger point. This basic operation is illustrated with the circuit in Figure 12–26.

FIGURE 12–26

Basic triac phase control. The timing of the gate trigger determines the portion of the ac cycle passed to the load.

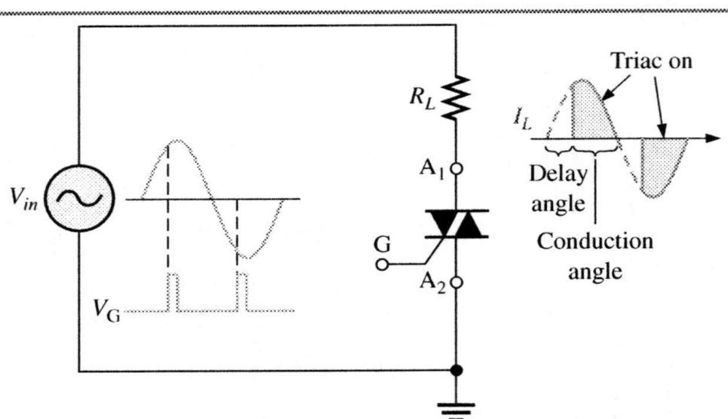

The Zero-Voltage Switch

One problem that arises with triggering an SCR or triac when it is switched on during the ac cycle is generation of **RFI** (radio frequency interference) due to switching transients. If the SCR or triac is suddenly switched on near the peak of the ac cycle, for example, there would be a sudden inrush of current to the load. When there is a sudden transition of voltage or current, many high-frequency components are generated. These high-frequency components can radiate into sensitive electronic circuits, creating serious disturbances, even catastrophic failures. By switching the SCR or triac on when the voltage across it is zero, the sudden increase in current is prevented because the current will increase sinusoidally with the ac voltage. **Zero-voltage switching** also prevents thermal shock to the load which, depending on the type of load, may shorten its life.

Not all applications can use zero-voltage switching, but when it is possible, noise problems are greatly reduced. For example, the load might be a resistive heating element, and the power is typically turned on for several cycles of the ac and then turned off for several cycles to maintain a certain temperature. The zero-voltage switch uses a sensing circuit to determine when to turn power on. The idea of zero-voltage switching is illustrated in Figure 12–27.

FIGURE 12–27

Comparison of zero-voltage switching to nonzero switching of power to a load.

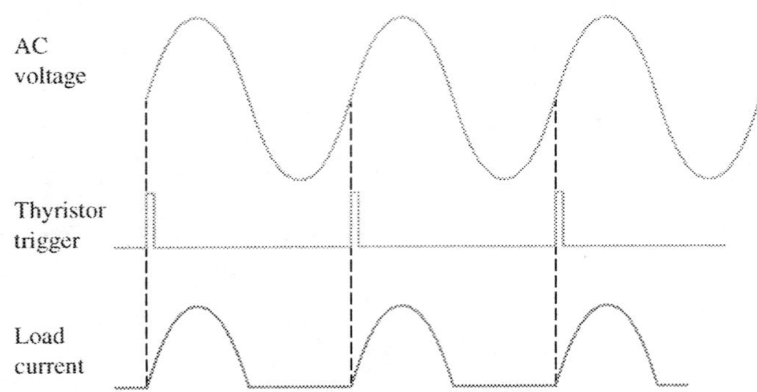

(a) Zero-voltage switching of load current

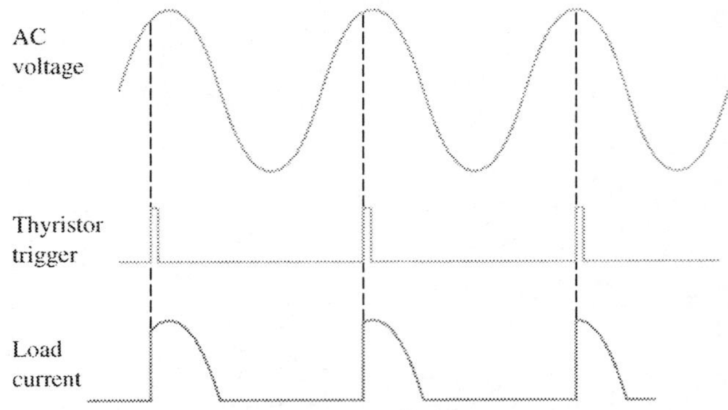

(b) Nonzero switching of load current produces current transients that cause RFI.

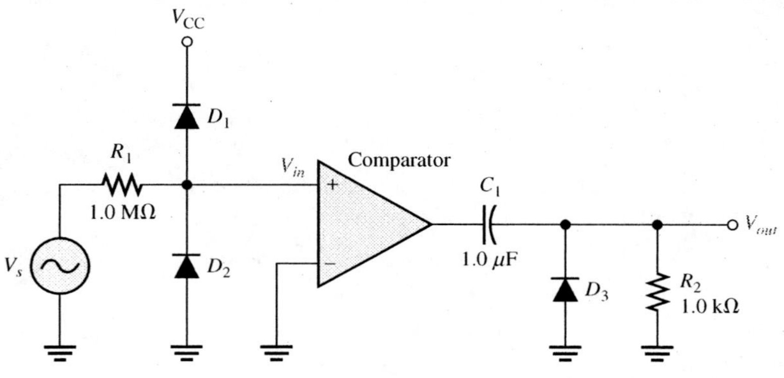

FIGURE 12–28

A circuit that can provide triggers when the ac waveform crosses the zero axis in the positive direction.

A basic circuit that can provide a trigger pulse as the ac waveform crosses the zero axis in the positive direction is shown in Figure 12–28. Resistor R_1 and diodes D_1 and D_2 protect the input of the comparator from excessive voltage swings. The output voltage level of the comparator is a square wave. C_1 and R_2 form a differentiating circuit to convert the square wave output to trigger pulses. Diode D_3 limits the output to positive triggers only.

Microcontrollers

SCRs and triacs are often used in systems that have many additional requirements. For instance, a system as basic as a washing machine requires timing functions, speed or torque regulation, motor protection, sequence generation, display control and so on. Systems like this can be controlled by a special class of computers called **microcontrollers**. A microcontroller is constructed as a single integrated circuit with all of the basic features found in a microprocessor with special input/output (I/O) circuits, ADCs (analog-to-digital converters), counters, timers, oscillators, memory, and other features. Microcontrollers can be configured for a specific system and offer an inexpensive alternative to older methods for providing a trigger to an SCR or triac.

Review Questions

26. What is a thyristor?
27. What does SCR stand for?
28. How does an SCR differ from a triac in terms of delivering power to a load?
29. Explain the basic purpose in zero-voltage switching.
30. For an SCR, what do the letters *A*, *K*, and *G* stand for?

CHAPTER REVIEW

Key Terms

Active transducer A type of transducer whose output power is derived from a source other than the quantity being measured.

Elastic limit The point where permanent deformation results when force is applied to an elastic body.

Excitation An outside source of energy to operate a transducer.

Gauge factor A dimensionless number that is the constant of proportionality that represents the fractional change in resistance for a given strain.

Load cell A metal body that is deformed within its elastic range by an applied force. The metal body is instrumented with strain gauges.

Passive transducer A type of transducer that produces an output without any direct interaction with a source of power.

Pressure The force per unit area; the area is measured perpendicular to the force.

Resistance temperature detector (RTD) A type of temperature transducer in which resistance is directly proportional to temperature.

Resolution The magnitude of the smallest detectable change in a quantity that is being measured.

SCR Silicon controlled rectifier; a type of three-terminal thyristor.

Strain The elastic deformation of a body due to an applied force; it is the ratio of a change in length divided by the length of an object.

Strain gauge A thin conductor made in a back and forth pattern. It responds to force by changing its resistance as it is stretched or compressed.

Stress The force per area in a solid body; it can be either due to a tension force or a compression force.

Thermistor A sensitive-resistive temperature sensor that is made of a sintered mixture of transition metal oxides; the resistance is inversely proportional to the temperature.

Thermocouple A temperature transducer that produces a current proportional to the difference in temperature between a measuring junction and a reference junction.

Thyristor A class of four-layer (*pnpn*) semiconductor switching devices.

Transducer A device that converts energy from one form to another. For electronic systems, the output is an electrical parameter.

Triac A 3-terminal thyristor that can conduct current in either direction.

Young's modulus A constant for an elastic material that depends on the geometry and material. It is the ratio of stress divided by strain; it is specified differently for stretching and bending.

Important Facts

- ❑ A classification scheme for transducers is to group them by the conversion principle.
- ❑ Common electrical transducers convert a physical quantity into a change in resistance, capacitance, or inductance. Others convert a physical quantity into a voltage or current directly.
- ❑ Some important criteria to consider for selecting a transducer for a particular application are the range, threshold, dynamic behavior, accuracy, resolution, repeatability, and hysteresis error. Other criteria are the environment, power requirements, loading effects, and calibration requirements.
- ❑ Temperature is specified in one of three scales: Fahrenheit, Celsius, and Kelvin.
- ❑ The thermocouple is a temperature transducer that produces a current proportional to the difference in temperature between a measuring junction and a reference junction.
- ❑ Thermocouples are widely used for measuring high temperatures. The K-type can measure up to 1250°C.

- ❑ The RTD (resistance temperature detector) is the most accurate type of thermometer for temperatures from about −50°C to 450°C.
- ❑ Thermistors are temperature sensors used in noncritical measurements such as thermostats where a small sensitive sensor is desirable.
- ❑ Integrated circuit temperature sensors are small and accurate temperature transducers for temperatures from −55°C to 150°C.
- ❑ The change in resistance of a strain gauge is very small, so it is normally detected with a Wheatstone bridge.
- ❑ A load cell consists of a metal body that is deformed within its elastic range by an applied force. The metal body is instrumented with strain gauges.
- ❑ A common transducer for sensing pressure has a flexible diaphragm that responds to pressure. The diaphragm movement may be sensed by a strain gauge or may be detected by a capacitive change.
- ❑ An important motion detector is the linear variable differential transformer (LVDT). The output of an LVDT is determined by the position of a movable core.
- ❑ A tachometer measures angular velocity by generating a voltage that is proportional to the velocity.
- ❑ Photovoltaic sensors, photoconductive sensors, and photoemissive sensors can sense light.
- ❑ A CdS cell responds to light by changing its resistance.
- ❑ The thyristor is a category of electronic switches.
- ❑ The SCR (silicon-controlled rectifier) and triac are two types of thyristors used in power control.
- ❑ A zero-voltage switch generates pulses at the zero crossings of an ac voltage for triggering a thyristor.

Formulas

Temperature conversion formulas:

$$F = \frac{9}{5}C + 32 \qquad \textbf{(12–1)}$$

$$C = \frac{5}{9}(F - 32) \qquad \textbf{(12–2)}$$

$$K = C + 273.2 \qquad \textbf{(12–3)}$$

Hooke's law:

$$F = k\Delta l \qquad \textbf{(12–4)}$$

Definition of strain:

$$\in \, = \frac{\Delta l}{l} \qquad \textbf{(12–5)}$$

Definition of stress:

$$\frac{F}{A} = E\frac{\Delta l}{l} \qquad \textbf{(12–6)}$$

Resistance change per nominal resistance of a strain gauge as a function of strain:

$$\frac{\Delta R}{R} = G\frac{\Delta l}{l} \qquad \textbf{(12–7)}$$

Chapter Checkup

Answers are at the end of the chapter.

1. An RTD measures
(a) resistance (b) force
(c) temperature (d) velocity

2. A sensor that varies its resistance in response to light is a
(a) strain gauge (b) CdS cell
(c) phototransistor (d) thermistor

3. The capacitance of a capacitive transducer will increase if the
(a) area of the plates is increased
(b) spacing between plates is increased
(c) air replaces a nonconducting fluid between the plates
(d) all of these answers

4. Capacitive transducers can be used to measure
(a) displacement (b) pressure
(c) relative humidity (d) all of these answers

5. The direct conversion of light to electricity occurs in a
(a) CdS cell (b) LVDT
(c) phototransistor (d) solar cell

6. The smallest detectable value that a transducer can respond to is called the
(a) range (b) threshold
(c) resolution (d) repeatability

7. The difference between the measured value and accepted value is a measure of
(a) resolution (b) repeatability
(c) accuracy (d) hysteresis error

8. Temperature is a measure of the
(a) total heat contained in a body
(b) average kinetic energy of molecules in a body
(c) mass of a body
(d) all of these answers

9. The boiling point of water on the Kelvin scale is approximately
(a) 100 K (b) 212 K
(c) 273 K (d) 373 K

10. The Seebeck effect has to do with
(a) thermocouples (b) RTDs
(c) strain gauges (d) CdS cells

11. For temperatures from −50°C to 450°C, the most accurate type of thermometer is a
(a) thermocouple (b) RTD
(c) thermistor (d) IC temperature sensor

12. Strain that is measured by a strain gauge can be expressed as

    (a) $\in \ = \dfrac{\Delta R/R}{G}$

    (b) $\in \ = \dfrac{\Delta l}{l}$

    (c) $\in \ = \dfrac{\sigma}{E}$

    (d) all of the above

13. Young's modulus is the constant that relates

    (a) resistance to strain (b) heat to temperature

    (c) capacitance to pressure (d) none of these answers

14. What type of measurement is made by a load cell?

    (a) acceleration (b) temperature

    (c) force (d) power

15. Pressure is defined as force divided by

    (a) volume (b) length

    (c) area (d) time

16. Zero-voltage switching is commonly used in

    (a) determining thermocouple voltage

    (b) SCR and triac power-control circuits

    (c) in balanced bridge circuits

    (d) RFI generation

17. A major disadvantage of nonzero switching of power to a load is

    (a) lack of efficiency

    (b) possible damage to the thyristor

    (c) RF noise generation

Questions

Answers to odd-numbered questions are at the end of the book.

1. What is the difference between an active and a passive transducer?
2. Give examples of transducers that use resistive, inductive, and capacitive principles of operation.
3. How does a capacitive microphone work?
4. Describe how to covert linear motion to an electrical parameter.
5. What are four applications for capacitive sensors?
6. Explain how an electromagnetic transducer can be used to measure the speed of a rotating shaft.
7. How is the dynamic behavior of a transducer specified?
8. What is hysteresis error?
9. What does the range of a transducer mean?
10. Under what conditions is it necessary to add an amplifier very close to the transducer?
11. What is included in a calibration record?

12. What criteria would you use to specify a pressure transducer for an underground gasoline tank? The pressure will be converted into a liquid level indicator.
13. What is 0 K called?
14. Why are special wires necessary if you wish to extend the length of a thermocouple?
15. When you connect a voltmeter to a thermionic junction, the meter does not read the emf of the junction. Why not?
16. What are three advantages of thermistors for measuring temperatures?
17. Compare a thermistor to an RTD. Which is more sensitive? Which is more accurate?
18. How does the range of an IC sensor compare to the range of a thermocouple? Which sensor would you choose to measure temperatures from −40°C to 0°C?
19. What is Hooke's law?
20. When is Hooke's law not valid?
21. What are the measurement units for stress?
22. What is the ratio of stress divided by strain called?
23. What is the reason most strain gauge measurements involve using a Wheatstone bridge?
24. What is meant by a half-bridge?
25. What is the definition of a pascal?
26. Explain how the height of a column of mercury can be used as a pressure measurement.
27. What is meant by differential pressure?
28. What is the difference between displacement and velocity?
29. What are the similarities and differences between a radio wave and a light wave?
30. How does a photodiode detect light?
31. Why are phototransistors not good for measuring the *amount* of light?
32. Once an SCR is triggered on, what is the minimum current called that is required to keep it on?
33. What are the two ways to turn an SCR off?

PROBLEMS

Basic Problems

Answers to odd-numbered problems are at the end of the book.

1. Assume you need to detect weights up to 25 pounds to within 1 oz. What is the minimum resolution required of the scale?
2. Convert 90°F to Celsius and Kelvin.
3. Convert −80°F to Celsius and Kelvin.
4. Convert −160°C to Fahrenheit and Kelvin.
5. Convert 370°C to Fahrenheit and Kelvin.
6. Convert 400 K to Celsius and Fahrenheit.
7. Convert 200 K to Celsius and Fahrenheit.
8. Assume a strain gauge changes resistance under load by 1.8 mΩ. If the unstrained resistance of the strain gauge is 500 Ω, and the gauge factor is 2.0, what is the strain?

9. A 6.00 cm long steel cylinder is loaded with a force that compresses it by 0.3 μm.
   (a) What is the strain?
   (b) If a nominal 350 Ω strain gauge is used to measure the strain, and the gauge factor is 2, what is the change in resistance of the gauge?
10. The load cell shown in Figure 12–29 has a 2 mV/V specification and four 350 Ω strain gauges (two in compression, two in tension). The excitation voltage is +12 V. If the Wheatstone bridge is balanced (0 V output) with no force applied, what is the resistance, R_1, of strain gauge 1 at full load?

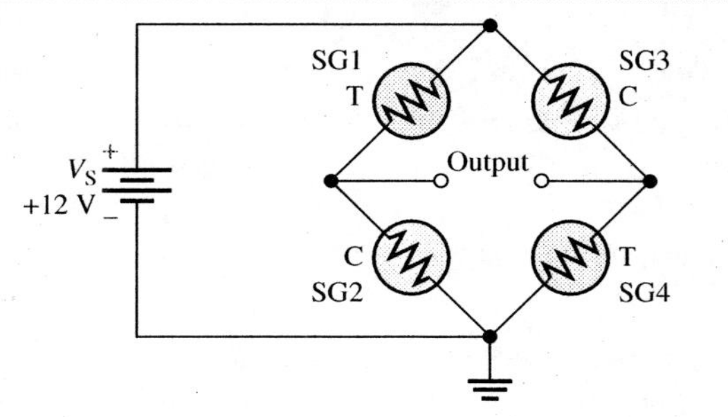

FIGURE 12–29

11. A 3 mV/V load cell is excited with a 15 V source and monitored with a digital voltmeter. The meter has a resolution of 200 μV. What is the smallest detectable change in weight the system can respond to if the full-scale weight is 10,000 pounds?
12. Name two ways an SCR can be placed in the forward-conduction region.
13. Sketch the V_R waveform for the circuit in Figure 12–30, given the indicated relationship of the input waveforms.

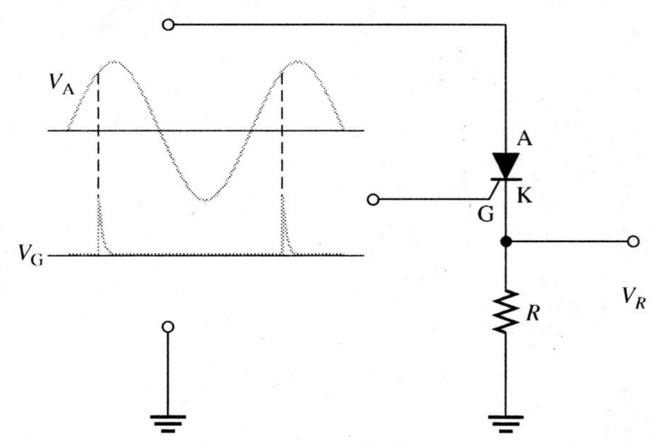

FIGURE 12–30

Basic-Plus Problems

14. A capacitive level-sensing transducer has two equal circular plates that are 2.5 cm in diameter and are separated by 1 mm.
   (a) Calculate the capacitance in air.
   (b) Calculate the capacitance if oil, with a relative permittivity of 4000, fills the space between the plates.

15. A capacitive pressure sensor uses two plates that have an area of 4×10^{-3} m^2 and a separation of 1 mm when the pressure is 0. The dielectric is air. When pressure is applied, the separation is reduced to 0.4 mm.

    (a) What is the capacitance with no pressure?

    (b) What is the capacitance under pressure?

16. At what temperature is the reading on the Fahrenheit and Celsius scale the same?

17. A full-bridge arrangement is used with four equal strain gauges to measure a strain of 200 $\mu\in$. The gauge factor is 2.06. What is the output voltage for each volt of excitation voltage?

18. Two strain gauges with nominal resistances of 350 Ω and each with a gauge factor of 2.0 are connected into a half-bridge circuit. The fixed bridge resistors are also 350 Ω. The excitation voltage is set to 10 V. Assume initially the bridge is balanced with no load and has an output of 1.5 mV under load. What strain is applied? (*Hint*: Start by finding the change in resistance.)

19. The frequency of light from a certain LED is 5.6×10^{14} Hz. What is the color of the LED?

20. For the circuit in Figure 12–31, describe the waveform at the output of the comparator and at the output of the circuit in relation to the input. Assume the input is a 115 V rms sine wave and the comparator and the power supply voltages for the comparator are ±10 V.

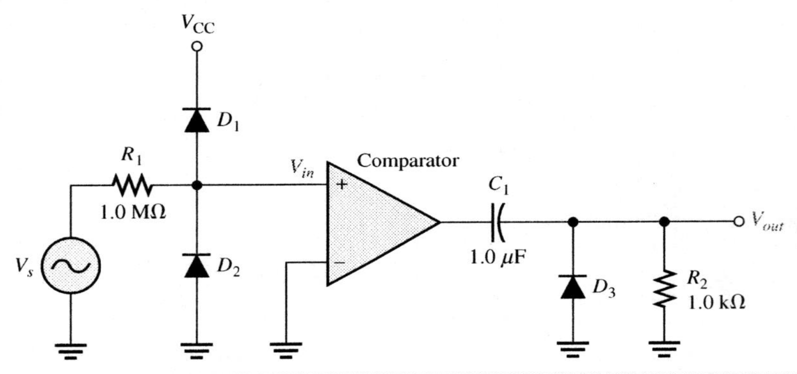

FIGURE 12–31

21. What change to the circuit in Figure 12–31 would you make if you wanted to have positive triggers on the negative slope of the input waveform?

ANSWERS

Example Questions

12–1: Yes. Accuracy and resolution are different; just because the scale has high resolution does not mean it is accurate.

12–2: 291.5 K

12–3: 3 mΩ

12–4: 350.7 Ω

12–5: The required resolution is 30 μV, instead of 20 μV.

Review Questions

1. A transducer is a device that converts energy from one form to another.
2. Excitation refers to the outside source of energy to operate a transducer.
3. Capacitance increases.

4. (1) Vary the material between the plates, (2) change the size of the plates, or (3) change the spacing between the plates.
5. An LVDT is a linear variable differential transformer.
6. The range specifies the values the transducer is intended to measure.
7. Dynamic behavior is specified as either a frequency response or as a response time.
8. Accuracy is the difference between the measured value and accepted value, whereas resolution is the magnitude of the smallest detectable change in a quantity.
9. Loading can change the quantity being measured and thus affect the measurement.
10. Calibration is the process of comparing a measurement device to a known standard.
11. 32°F, 0°C, 273.2 K
12. The Seebeck effect is the small voltage that appears when two unlike wires are joined and heated.
13. From approximately 0°C to 1250°C
14. They are sensitive, small, and inexpensive.
15. It can detect the temperature in inaccessible or remote locations.
16. The elastic limit of a metal is the point where permanent deformation results when force is applied to an elastic body.
17. Strain is the elastic deformation of a body due to a force; stress is force per area on a body.
18. Young's modulus is a constant that when multiplied by the strain gives the stress.
19. The strain gauge responds to strain by changing its resistance by a very small amount.
20. A load cell is a force-sensing transducer with a metal body that is deformed by the applied force (weight); it is instrumented with strain gauges.
21. Gauge pressure is measured with respect to atmospheric pressure. Absolute pressure includes atmospheric pressure.
22. Most capacitive pressure sensors change the spacing between a movable diaphragm and a fixed plate that form a capacitor.
23. A tachometer generates a pulse across a coil as it rotates in a constant magnetic field. The average voltage is proportional to the speed.
24. From about 390 nm to about 760 nm
25. A CdS cell is a photoconductive cell that responds to light by changing its resistance.
26. A 4-layer semiconductor switch
27. Silicon-controlled rectifier
28. An SCR is unidirectional and therefore allows current through the load only during half of the ac cycle. A triac is bidirectional and allows current during the complete cycle.
29. Zero-voltage switching eliminates fast transitions in the current to a load, thus reducing RFI emissions and thermal shock to the load element.
30. A—anode, K—cathode, G—gate

Chapter Checkup

1. (c)	2. (b)	3. (a)	4. (d)	5. (d)
6. (b)	7. (c)	8. (b)	9. (d)	10. (a)
11. (b)	12. (d)	13. (d)	14. (c)	15. (c)
16. (b)	17. (c)			

APPENDIX

Logarithms and Decibels

Logarithms

A widely used unit in electronics is the *decibel,* which is based on logarithms. Before defining the decibel, let's quickly review logarithms (sometimes called *logs*). A logarithm is simply an exponent. Consider the equation

$$y = b^x$$

The value of y is determined by the exponent of the base (b). The exponent, x, is said to be the logarithm of the number represented by the letter y.

Two bases are in common use—base ten and base e (discussed in mathematics courses). To distinguish the two, the abbreviation "log" is written to mean base ten, and the letters "ln" are written to mean base e. Base ten is standard for work with decibels. Thus, for base ten,

$$y = 10^x$$

Solving for x,

$$x = \log_{10} y$$

The subscript 10 can be omitted because it is implied by the abbreviation "log."

Logarithms are useful when you multiply or divide very large or small numbers. When two numbers written with exponents are multiplied, the exponents are simply added. That is,

$$10^x \times 10^y = 10^{x+y}$$

This is equivalent to writing

$$\log xy = \log x + \log y$$

This concept is applied to problems involving multiple stages of amplification or attenuation.

Decibel Power Ratios

Power ratios are often very large numbers. Early in the development of telephone communication systems, engineers devised the decibel as a means of describing large ratios of gain or attenuation (a signal reduction). The decibel (dB) is defined as 10 multiplied by the logarithmic ratio of the power gain.

$$\text{dB} = 10 \log\left(\frac{P_2}{P_1}\right)$$

where P_1 and P_2 are the two power levels being compared.

Power gain is defined as the ratio of power delivered from an amplifier to the power supplied to the amplifier. To show power gain, A_p, as a decibel ratio, we use a prime in the abbreviation.

$$A'_p = 10 \log\left(\frac{P_{out}}{P_{in}}\right)$$

where A'_p is power gain expressed as a decibel ratio, P_{out} is power delivered to a load, and P_{in} is power delivered to the amplifier.

The decibel (dB) is a dimensionless quantity because it is a ratio. Any two power measurements with the same ratio are the same number of decibels. For example, the power ratio between 500 W and 1 W is 500:1, and the number of decibels this ratio represents is 27 dB. There is exactly the same number of decibels between 100 mW and 0.2 mW (500:1) or 27 dB. When the power ratio is less than 1, there is a power loss or attenuation. The decibel ratio is *positive* for power gain and *negative* for power loss.

One important power ratio is 2:1. This ratio is the defining power ratio for specifying the cutoff frequency of instruments, amplifiers, filters, and the like. By substituting into the decibel power ratio equation, the dB equivalent of a 2:1 power ratio is

$$\text{dB} = 10 \log\left(\frac{P_2}{P_1}\right) = 10 \log\left(\frac{2}{1}\right) = 3.01 \text{ dB}$$

This result is usually rounded to 3 dB.

Since 3 dB represents a doubling of power, 6 dB represents another doubling of the original power (a power ratio of 4:1). Nine decibels represents an 8:1 ratio of power and so forth. If the ratio is the same, but P_2 is smaller than P_1, the decibel result remains the same except for the sign.

$$\text{dB} = 10 \log\left(\frac{P_2}{P_1}\right) = 10 \log\left(\frac{1}{2}\right) = -3.01 \text{ dB}$$

The negative result indicates that P_2 is less than P_1.

Another useful ratio is 10:1. Since the log of 10 is 1, 10 dB equals a power ratio of 10:1. With this in mind, you can quickly estimate the overall gain (or attenuation) in certain situations. For example, if a signal is attenuated by 23 dB, it can be represented by two 10 dB attenuators and a 3 dB attenuator. Two 10 dB attenuators are a factor of 100 and another 3 dB represents another factor of 2 for an overall attenuation ratio of 1:200.

It is common in certain applications of electronics (microwave transmitters, for example) to combine several stages of gain or attenuation. When working with several stages of gain or attenuation, the total voltage gain is the product of the gains in absolute form.

$$A_{v(tot)} = A_{v1} \times A_{v2} \times \cdots \times A_{vn}$$

Decibel units are useful when combining these gains or losses because they involve just addition or subtraction. The algebraic addition of decibel quantities is equivalent to multiplication of the gains in absolute form.

$$A'_{v(tot)} = A'_{v1} = A'_{v2} + \cdots + A'_{vn}$$

Although decibel power ratios are generally used to compare two power levels, they are occasionally used for absolute measurements when the reference power level is understood. Although different standard references are used depending on the application, the most common absolute measurement is the dBm. A dBm is the power level when the reference is understood to be 1 mW developed in some assumed load impedance. For radio frequency systems, this is commonly 50 Ω; for audio systems, it is generally 600 Ω. The dBm is defined as

$$\text{dBm} = 10 \log\left(\frac{P_2}{1 \text{ mW}}\right)$$

The dBm is commonly used to specify the output level of signal generators and is used in telecommunications to simplify the computation of power levels.

Decibel Voltage Ratios

Since power is given by the ratio of V^2/R, the decibel power ratio can be written as

$$\text{dB} = 10 \log\left(\frac{V_2^2/R_2}{V_1^2/R_1}\right)$$

where R_1, R_2 are resistances in which P_1 and P_2 are developed and V_1, V_2 are voltages across the resistances R_1 and R_2. If the resistances are equal, they cancel.

$$\text{dB} = 10 \log\left(\frac{V_2^2}{V_1^2}\right)$$

A property of logarithms is

$$\log x^2 = 2 \log x$$

Thus, the decimal voltage ratio is

$$\text{dB} = 20 \log\left(\frac{V_2}{V_1}\right)$$

When V_2 is the output voltage (V_{out}) and V_1 is the input voltage (V_{in}) for an amplifier, the equation defines the decibel voltage gain. By substitution,

$$A_v' = 20 \log\left(\frac{V_{out}}{V_{in}}\right)$$

where A_v' is voltage gain expressed as a decibel ratio, V_{out} is voltage delivered to a load, and V_{in} is voltage delivered to the amplifier. This equation gives the decibel voltage gain, a logarithmic ratio of amplitudes. It was originally derived from the decibel power equation when both the input and load resistances are the same (as in telephone systems).

Both the decibel voltage gain equation and decibel power gain equation give the same ratio if the input and load resistances are the same. However, it has become common practice to apply the decibel voltage equation to cases where the resistances are *not* the same. When the resistances are not equal, the two equations do not give the same result.*

In the case of decibel voltage gain, note that if the amplitudes have a ratio of 2:1, the decibel voltage ratio is very close to 6 dB (since 20 log 2 = 6). If the signal is attenuated by a factor of 2 (ratio = 1:2), the decibel voltage ratio is −6 (since 20 log 1/2 = −6). Another useful ratio is when the amplitudes have a 10:1 ratio; in this case, the decibel voltage ratio is 20 dB (since 20 log 10 = 20).

*The *IEEE Standard Dictionary of Electrical and Electronic Terms* recommends that a specific statement accompany this application of decibels to avoid confusion.

ANSWERS TO ODD-NUMBERED QUESTIONS

Chapter 1

1. Two classifications of integrated circuits are digital and analog.
3. $y = mx + b$, where y is the dependent variable, x is the independent variable, m is the slope, and b is the y-axis intercept.
5. A transducer is a device that converts energy from one form to another.
7. The 5th harmonic of 500 Hz is 2500 Hz.
9. When a square wave is applied to an oscilloscope, selective attenuation of high or low frequencies can be determined from the observed waveshape. High frequency attenuation produces a slower rise time; low frequency attenuation causes "sagging" in the normally flat parts of the pulse.
11. (1) Digital scopes can store waveforms of known good circuits for comparison of a circuit under test. (2) Displays can be pre- and post-triggered to catch intermittent signals and events leading up to or following a trigger event.
13. A DMM measures voltage, current, and resistance.

Chapter 2

1. An ion is an atom that has acquired a charge due to an imbalance in the number of protons and electrons.
3. A valence electron is in the outermost shell of the atom. A conduction (free) electron is one that has broken free of the atom.
5. Electron current is produced by the movement of free electrons. Hole current occurs at the valence level when valence electrons move into a hole created by a free electron. This effectively is the same as the hole moving from one atom to another.
7. Electrons near the junction drift across and recombine with holes leaving positive ions on one side of the junction and negative ions on the other.
9. Peak inverse voltage (PIV)
11. Holes are the majority carriers in a p material.
13. Half-wave rectification produces an output during half of the sine wave cycle. Full-wave rectification produces an output during the entire sine wave cycle.
15. The charging and discharging of the filter capacitor on the output of a rectifier produces a variation in the dc voltage which is called the ripple voltage.
17. A zener diode is operated in the reverse-breakdown region.
19. A photodiode is operated in reverse bias.
21. There would be a half-wave output voltage instead of a full-wave voltage if there is an open diode.

Chapter 3

1. The emitter current is the largest.
3. The base-emitter junction is normally forward-biased. The base-collector junction is normally reverse-biased.
5. The dc load line touches the x-axis at the cutoff point.
7. Base bias is dependent on the β_{DC} of the transistor.
9. Stiff bias means that the circuit is essentially independent of the β_{DC}.

11. The parameter h_{fe} is ac current gain, β_{ac}.
13. V_{CC} is at ground potential for the ac signal.
15. The voltage gain of a CE amplifier is the ratio of the ac collector resistance to the ac emitter resistance.
17. A low output resistance results in less loading on the amplifier.
19. The output of a CB amplifier is taken at the collector.

Chapter 4

1. A JFET has a *pn* junction but a MOSFET does not.
3. The MOSFET has an insulated gate.
5. The source and drain are connected to the channel.
7. The arrow points out in a *p*-channel JFET.
9. The JFET is on when there is no gate voltage.
11. I_{DSS} is the maximum drain current and is specified at $V_{GS} = 0$ V.
13. $V_{GS(off)}$ is the cutoff voltage that makes I_D approximately zero.
15. A resistor (R_G) from gate to ground makes the gate voltage 0 V and creates a self-biased condition.
17. The source resistor is necessary in a JFET with voltage-divider bias in order to reverse bias the gate-to-source *pn* junction.
19. The vertical channel line in a D-MOSFET symbol is solid whereas an E-MOSFET has a broken vertical channel line indicating the absence of a physical channel.
21. The D-MOSFET can work in either depletion or enhancement mode.
23. The transconductance, g_m, is a change in drain current divided by a change in gate-to-source voltage.
25. A CD amplifier is also called a *source-follower*.
27. The $r_{DS(on)}$ is the channel-on resistance and its value determines the amount of attenuation of the input signal.
29. A small voltage, $V_{DS(on)}$, appears between the source and drain when a MOSFET is on.

Chapter 5

1. A stage in an amplifier is generally one transistor biased to operate as an amplifier.
3. The Thevenin equivalent voltage is found by drawing an amplifier as an equivalent circuit and determining the Thevenin equivalent voltage and resistance.
5. The ac resistance (r_c') is approximately 25 mV divided by I_E.
7. A capacitor from V_{CC} to ground is called a decoupling capacitor and shorts noise voltage on the power supply line to ground.
9. The advantage of IF is that it is a fixed frequency and requires no changes in the tuned circuit for any RF frequency (within limits).
11. A decoupling network is an *RC* circuit that helps isolate one circuit from another and prevents unwanted oscillations.
13. Open-loop gain is without feedback. Closed loop gain is with negative feedback.
15. A Darlington pair is two cascaded transistors connected so that the emitter of the first drives the base of the second.

17. A class B amplifier is biased at cutoff and operates in the linear region for 180° of the input cycle. A class AB amplifier is biased into slight conduction and operates in the linear region for slightly more than 180°.
19. The dc load line crosses the x-axis at V_{CEQ}.
21. The current in the emitter resistor of a differential amplifier is approximately twice the collector current in each transistor.

Chapter 6

1. The inverting input of an op-amp is indicated by a negative sign (−) and the noninverting input is indicated by a plus sign (+).
3. The peak-to-peak output of an op-amp is limited to slightly less than the supply voltages.
5. The input offset voltage forces the practical op-amp output to 0 V.
7. Differential input resistance is the resistance between the inverting and noninverting inputs. Common-mode input resistance is the resistance from each input to ground.
9. Op-amp B has the highest CMRR because it has less output due to the common-mode signal.
11. If R_f is increased, the voltage gain increases.
13. The voltage gain A_v is R_f/R_i = 100 kΩ/1 kΩ = 100.
15. The gain-bandwidth product equals 15 kHz.

Chapter 7

1. The reference voltage is $(R_2/(R_1 + R_2))V$ = (10 kΩ/20 kΩ)9 V = 4.5 V.
3. The thermistor resistance decreases as the temperature increases.
5. The output voltage is (11.5 V)0.25 = 2.875 V.
7. The output will change by (−10 mV/μs)(50 μs) = −500 mV.
9. The output is a square wave at a frequency of 10 kHz.
11. In addition to the op-amp there is a capacitor, resistor, and diode.

Chapter 8

1. The three regions of a low-pass filter response are the passband, the transition region, and the stop-band region.
3. The maximum output voltage is (1.414)(1 V) = 1.414 V.
5. BW = 12 kHz − 10 kHz = 2 kHz
7. For a band-stop filter, frequencies within the passband are blocked. For a band-pass filter, frequencies within the passband are passed.
9. The damping factor is determined by the negative feedback circuit.

Chapter 9

1. The voltage gain, A_{cl}, is $1 + 2R/R_G$ = 1 + 2(20 kΩ)/10 kΩ = 5.
3. The voltage gain can be adjusted from 2 to 1000 with an external resistor.

5. The common-mode rejection separates the fetal's and mother's heartbeats. The signal from the fetal heartbeat is sent to the monitoring equipment.
7. Each stage gain should be 5 to produce a total gain of 25.
9. The voltage gain is $g_m R_L = (10 \text{ mS})(10 \text{ k}\Omega) = 100$.
11. An OTA can be used as an amplitude modulator and a Schmitt trigger circuit.

Chapter 10

1. The inverting amplifier produces a phase shift of 180°, so an additional 180° phase shift is required because the total phase shift around the feedback loop must be effectively 0°.
3. Oscillation
5. For self-starting, the initial loop gain must be greater than 1.
7. The attenuation of the lead-lag circuit is 1/3 at the frequency of oscillation.
9. Increasing the R and C values causes the frequency of oscillation to decrease.
11. Halving the external capacitor causes the frequency of oscillation to double. The duty cycle is not affected by a change in the external capacitor value.

Chapter 11

1. The equivalent output resistance of a power supply should be small in order for the value of percent load regulation to be low.
3. If the feedback voltage increases, the output of the error detector decreases to cause a decrease in output voltage.
5. The switching regulator is the most efficient.
7. The switching regulator is prone to produce rf interference.
9. The voltage-inverter configuration of the switching regulator produces an output with polarity opposite to the input.
11. The LM337 is the negative-output counterpart of the LM317.
13. The load resistor is connected to the ground pin for a 78XX used as a current source.

Chapter 12

1. An active transducer requires operating power (called excitation) from a source other than the quantity being measured; a passive transducer derives power from the quantity being measured and produces an output with no external power source.
3. Sound pressure waves cause the spacing between two plates to vary at a rate proportional to the sound frequency, which changes the audio signal into a variation of capacitance.
5. Examples of capacitive sensors include types used for measuring displacement, velocity, force, pressure, and flow.
7. Dynamic behavior is specified as either the frequency response or as the response time to a change in the input.
9. Range is the set of values a transducer is designed to measure.
11. The calibration record should include the standards used to allow tracing. In addition, a table or graph with measured data points taken during calibration is part of the record.

13. Absolute zero
15. It will not read the emf of its own junctions as long as the temperatures of both leads of the voltmeter are the same.
17. Within their ranges, the thermistor is more sensitive; the RTD is more accurate.
19. The change in length of a material is proportional to the applied force.
21. The mks unit is the newton/meter^2.
23. The Wheatstone bridge is very sensitive for measuring change in resistance and can be connected with gauges in both tension and compression.
25. The pascal is a small pressure unit defined as 1 newton per square meter.
27. It is the difference between two input pressures.
29. Both are electromagnetic waves, which are transverse waves that travel 3.00×10^8 m/s in a vacuum. They differ in their frequency; radio waves have much longer wavelengths (lower frequency) than light waves.
31. Because of nonlinear response and poor temperature characteristics
33. Drop the conduction below the holding current level or use capacitor commutation, which temporarily connects a reverse-biased capacitor between the anode and cathode.

ANSWERS TO ODD-NUMBERED PROBLEMS

Chapter 1

1. 45.5 μS
3. See Figure ANS–1.

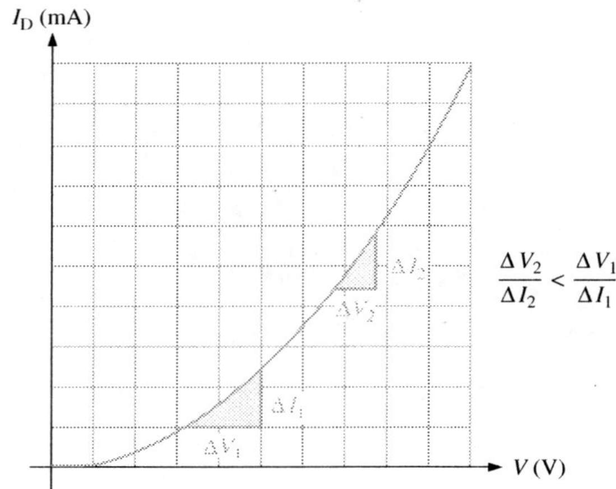

FIGURE ANS–1

5. $f = 31.8$ Hz; $T = 31.4$ ms
7. 0.1 ms
9. 0.652 A
11. 1.11
13. 14.5 Ω
15. (a) 318 Hz
    (b) 3.14 ms
17. A basic test plan for the system in Figure 1–17 (text) is
    1. Check for a signal at the input of the amplifier. If present, go to step 2; if not, go to step 3.
    2. Check for a signal at the output of the amplifier. If present, problem is with the speakers; if not, go to step 4.
    3. Check for a signal at the output of the microphones. If present, switch is bad; if not, go to step 5.
    4. Check for power to the amplifier. If present, amplifier is bad; if not, problem is with the power supply.
    5. Check for power to the microphones. If present, microphones are bad; if not, batteries need to be replaced.

Chapter 2

1. See Figure ANS–2.

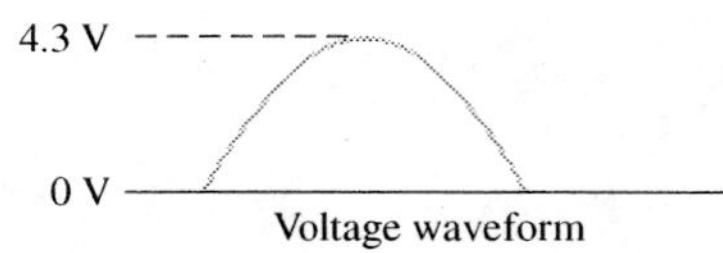

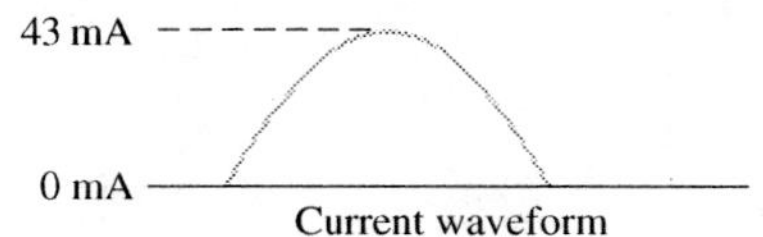

FIGURE ANS–2

3. (a) Full-wave rectifier
   (b) 28.3 V
   (c) 14.2 V (reference is center tap)
   (d) See Figure ANS–3.
   (e) 13.5 mA
   (f) 28.3 V (ideal)

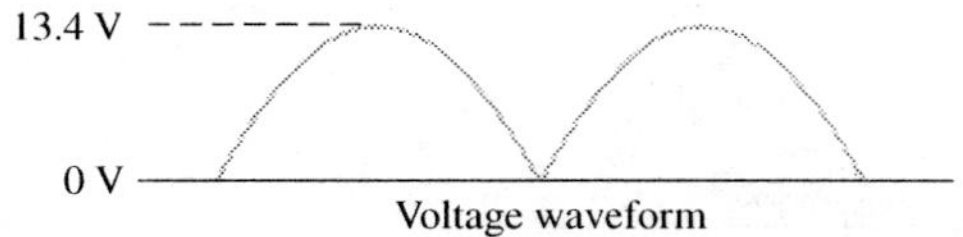

FIGURE ANS–3

5. 78.5 V
7. 4 %
9. $V_{IN(MIN)} = 6.12$ V; $V_{IN(MAX)} = 21.8$ V
11. DMM1 is correct but DMM2 is reading the rectified average voltage rather than the peak voltage that it would show if the capacitor was in the circuit. DMM3, indicating no voltage, implies an open circuit between the bridge and the output. The most likely cause is an open path along the output line between the bridge and the filter capacitor.
13. 720 Ω
15. (a) Readings are correct.
    (b) Open zener diode
    (c) Open switch or fuse blown
    (d) Open capacitor
    (e) Open transformer winding (less likely: more than one diode open)

Chapter 3

1. 5.29 mA
3. 29.4 mA
5. $I_B = 0.276$ mA; $I_C = 20.7$ mA; $V_C = 15.1$ V
7. $I_B = 13.6\ \mu$A; $I_C = 3.41$ mA; $V_C = 6.59$ V
9. $V_{CE(sat)} = 0.1$ V; $I_{C(sat)} = 3.67$ mA
11. (a) Decrease (to zero)
    (b) Remain the same
    (c) Increase
    (d) Increase
13. $I_C = 2.03$ mA; $V_{CE} = -9.51$ V
15. $I_C = 36.2$ mA; $V_{CE} = 9.23$ V
17. 199
19. Figure 3–57 in text shows a common-base amplifier. $I_C = 2.55$ mA
21. 34.7 kΩ
23. (a) small-signal
    (b) power
    (c) power
    (d) small-signal
    (e) rf
25. $A_{v(min)} = 2.93$; $A_{v(max)} = 123$
27. (a) 0.383 V
    (b) $\leqq 1.083$ V

Chapter 4

1. (a) Depletion region widens, creating narrower channel.
   (b) Increase
   (c) Less
3. (a) +2 V
   (b) −6 V
5. (a) +4 V
   (b) 2.5 mA
   (c) 15.8 V
7. (a) +2.1 V
   (b) 2.1 mA
   (c) 5.97 V
9. (a) $V_{DS} = 7.29$ V; $V_{GS} = -0.3$ V
   (b) $V_{DS} = -1.65$ V; $V_{GS} = +2.35$ V
11. +3 V
13. (a) The device is on.
    (b) The device is off.

15. -21.9

17. (a) $I_D = 4.85$ mA; $V_{DS} = 9.30$ V
    (b) -5.4
    (c) 3.38 MΩ
    (d) Enhancement mode

19. (a) 10 mA
    (b) 4 GΩ

21. $A_{v(min)} = 0.64$; $A_{v(max)} = 0.90$

23. Q_1 or Q_2 open, R_E open, no negative supply voltage, or open path between transistors

25. 0.953 mA

Chapter 5

1. 812

3. (a) See Figure ANS–4.
   (b) 6000
   (c) 3600

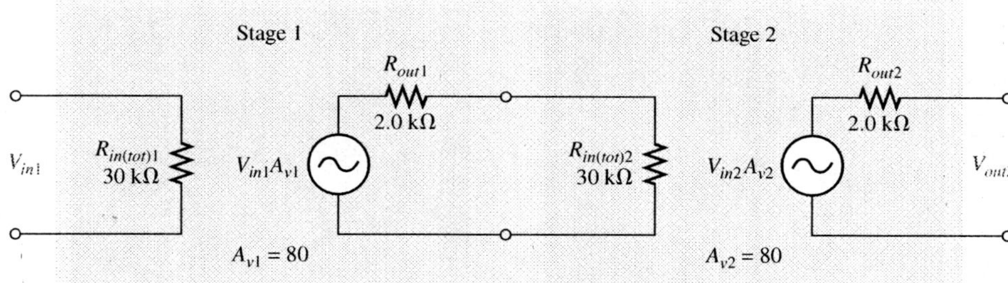

FIGURE ANS–4

5. $f_r = 356$ kHz; $Q = 47.1$; $BW = 7.56$ kHz

7. 8

9. (a) $I_{C(Q2)} = I_{E(Q1)} = 5.3$ mA; $V_{B(Q2)} = 0.7$ V; $V_{E(Q3)} = 0$ V; $I_{C(Q3)} = 120$ mA
   (b) 0.25 W

11. If R_4 opens, $V_{BE(Q3)} = 0.7$ V when $V_{IN} = 0$ V. Q_3 is forward-biased due to R_L. Since $V_L = 0$, $I_C = 0$. The amplifier is operating as a class B follower and will clip the output below 0 V.

13. (a) $I_{CQ} = 68.4$ mA; $V_{CEQ} = 5.14$ V
    (b) $A_v = 11.7$; $A_p = 263$

15. The changes are shown in Figure ANS–5.

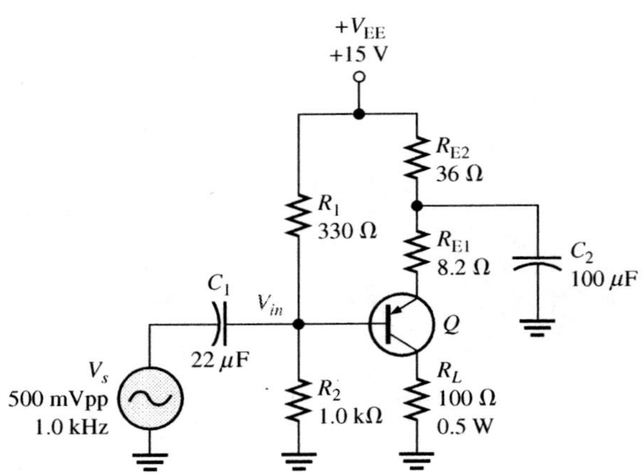

FIGURE ANS–5

17. (a) $V_{B(Q1)} = 0.7$ V; $V_{B(Q2)} = -0.7$ V; $V_E = 0$ V; $I_{CQ} = 8.3$ mA; $V_{CEQ(Q1)} = 9$ V; $V_{CEQ(Q2)} = -9$ V
    (b) 0.5 W
19. (a) $V_{B(Q1)} = 8.2$ V; $V_{B(Q2)} = 6.8$ V; $V_E = 7.5$ V; $I_{CQ} = 6.8$ mA; $V_{CEQ(Q1)} = 7.5$ V: $V_{CEQ(Q2)} = -7.5$ V
    (b) 167 mW
21. (a) C_2 open or Q_2 open
    (b) Power supply off, open R_1, Q_1 base shorted to ground
    (c) Q_1 has collector-to-emitter short
    (d) One or both diodes shorted
23. (a) 6.5 mA
    (b) 3.25 mA
25. See Figure ANS–6.

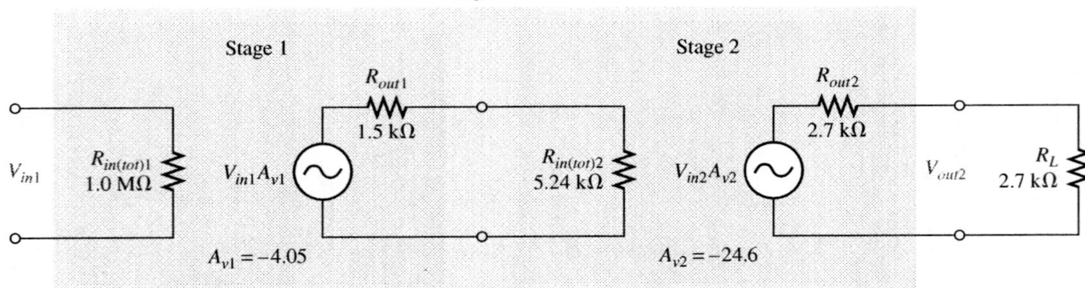

FIGURE ANS–6

27. R_3 and C_3 from a low-pass filter to block high frequencies from the power supply.

Chapter 6

1. Practical op-amp: High open-loop gain, very high input resistance and very low output resistance.

   Ideal op-amp: Infinite open-loop gain, infinite input resistance and zero output resistance.

3. 8.1 μA
5. 972,222
7. (a) Voltage-follower

   (b) Noninverting amplifier

   (c) Inverting amplifier
9. (a) 1

   (b) −1

   (c) 22.3

   (d) −10
11. 1.6 V/μs
13. $V_f = 49.5$ mV; $V_{in} = 49.5$ mV
15. (a) 49 kΩ

    (b) 3 MΩ

    (c) 84 kΩ

    (d) 165 kΩ
17. (a) 0.455 mA

    (b) 0.455 mA

    (c) −10 V

    (d) −10
19. 750 kHz
21. (a) Output is 0 V.

    (b) Output will saturate.

    (c) No affect on ac; may add or subtract a small dc voltage to the output.

    (d) Instead of a gain of −10, the gain is −0.1.

Chapter 7

1. 24 V pp with distortion due to clipping.
3. $V_{UTP} = 2.77$ V; $V_{LTP} = -2.77$ V

5. See Figure ANS–7.

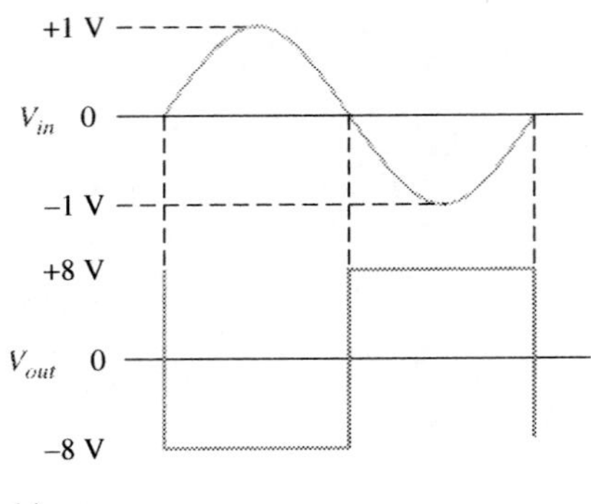

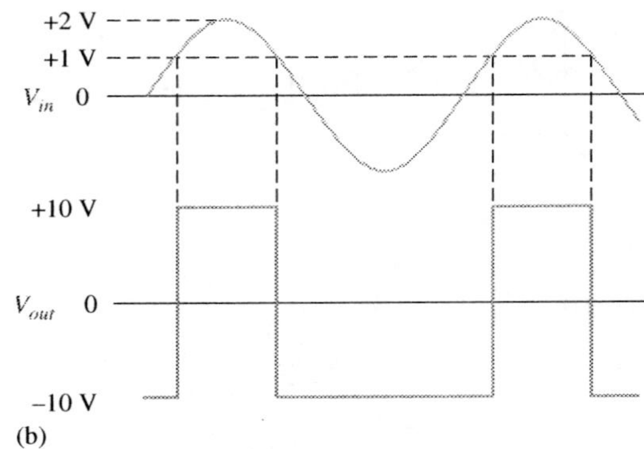

FIGURE ANS–7 (a) (b)

7. −0.357 mA

9. (a) A sine wave with a positive peak at +0.7 V, a negative peak at −7.3 V, and a dc value of −3.3 V.

   (b) A sine wave with a positive peak at +29.3 V, a negative peak at −0.7 V, and a dc value of +14.3 V.

11. See Figure ANS–8.

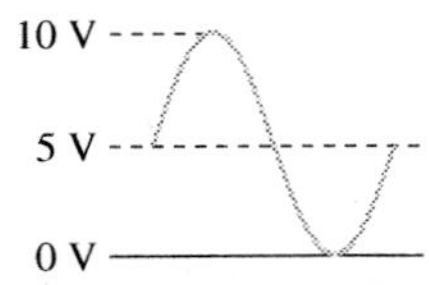

FIGURE ANS–8

13. (a) 7.76 V

    (b) 6.86 V

15. 110 kΩ

17. R_f = 100 kΩ, R_1 = 100 kΩ, R_2 = 50 kΩ, R_3 = 25 kΩ, R_4 = 12.5 kΩ, R_5 = 6.25 kΩ, R_6 = 3.125 kΩ

19. 1.0 mA

21. See Figure ANS–9.

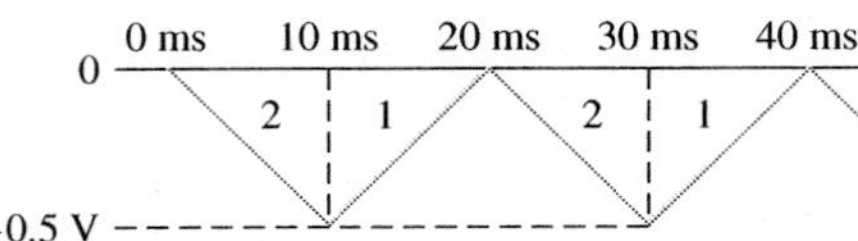

FIGURE ANS–9

23. See Figure ANS–10.

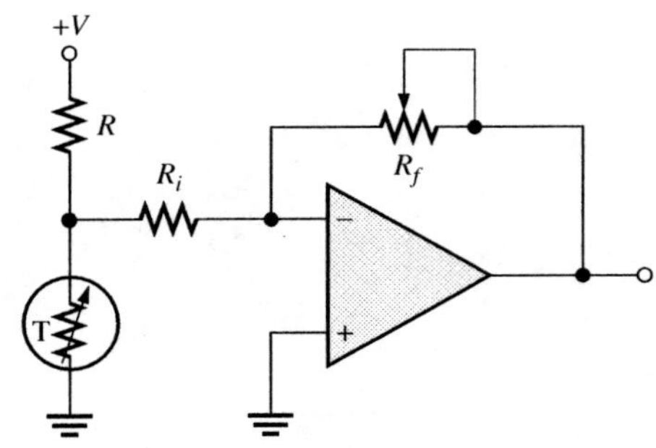

FIGURE ANS–10

25. R_2 is open.

Chapter 8

1. (a) Band-pass
   (b) High-pass
   (c) Low-pass
   (d) Band-stop
3. 48.2 kHz
5. $BW = 700$ Hz; $Q = 5.05$
7. (a) 1.43
   (b) 1.44
9. Close to ideal Butterworth; −80 dB/decade
11. See Figure ANS–11.

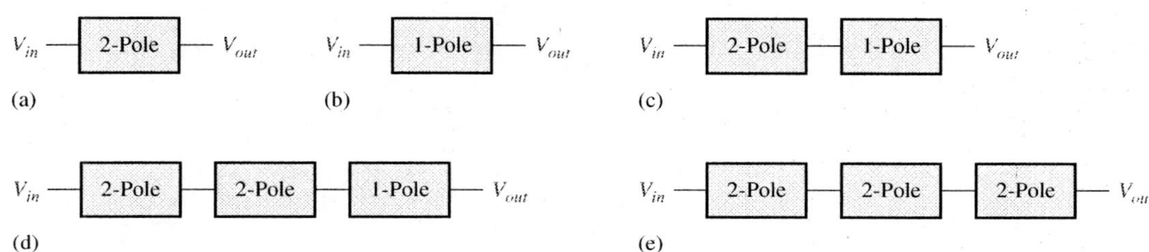

FIGURE ANS–11

13. (a) Multiple feedback
    (b) State-variable
15. Response is already Butterworth.
17. See Figure ANS–12.

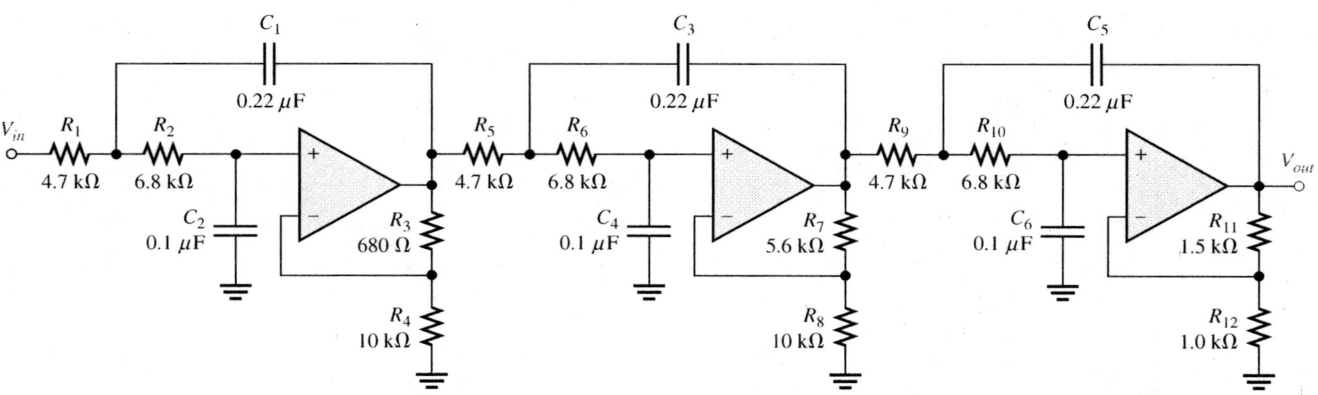

FIGURE ANS–12

19. Answers will vary. To obtain one-half the critical frequency, *either* the resistors or the capacitors (but not both) in the frequency determining network can be doubled.

Chapter 9

1. 201
3. 1.005 V
5. 51.5

7. 300
9. 1 mS
11. 17.0
13. 300 kHz
15. 66.4 kΩ (nearest standard value is 68 kΩ)
17. To achieve unity gain in the 3656KG, connect pin 15 and pin 14 together to produce a gain in the output section of 1.

    Connect pin 6 directly to pin 10 to produce a gain for the input section of 1. The overall gain is

    $A_{v(total)} = A_{v(input)}A_{v(output)} = (1)(1) = 1$
19. $A_{v(max)} = 17.0$; $A_{v(min)} = 13.6$
21. See Figure ANS–13.

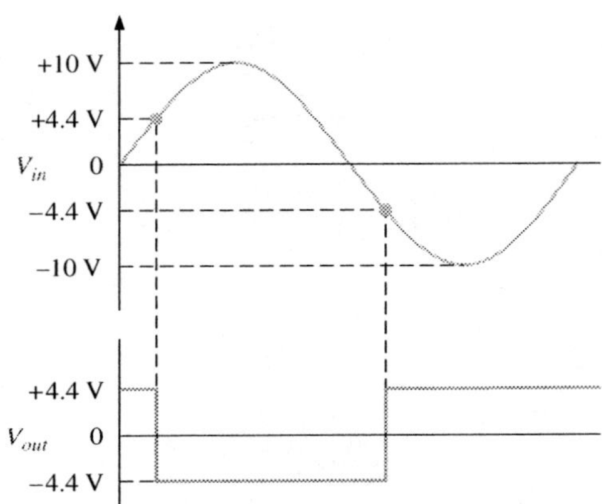

FIGURE ANS–13

Chapter 10

1. An oscillator requires no input other than the dc power supply voltage.
3. 4
5. 1.28 kHz
7. $R_f = 136$ kΩ; $f_r = 1.69$ kHz
9. $V_{REF1} = 3.33$ V; $V_{REF2} = 6.67$ V
11. 0.0076 μF
13. 733 mV
15. Change R_1 to 3.54 kΩ (nearest standard value is 3.6 kΩ).
17. 4.96 kΩ
19. See Figure ANS–14.

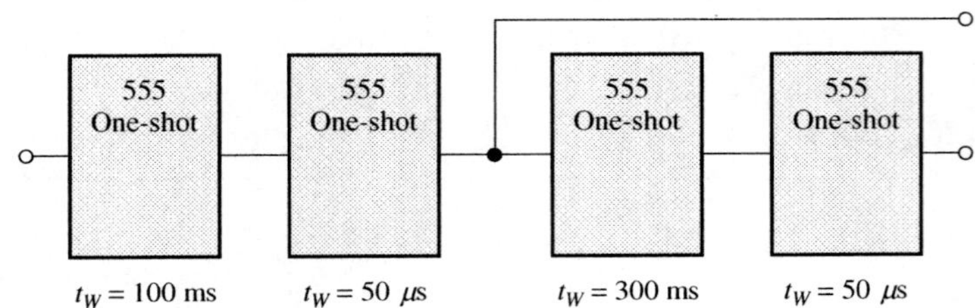

FIGURE ANS–14

Chapter 11

1. 0.0333%
3. 1.01%
5. See Figure ANS–15.

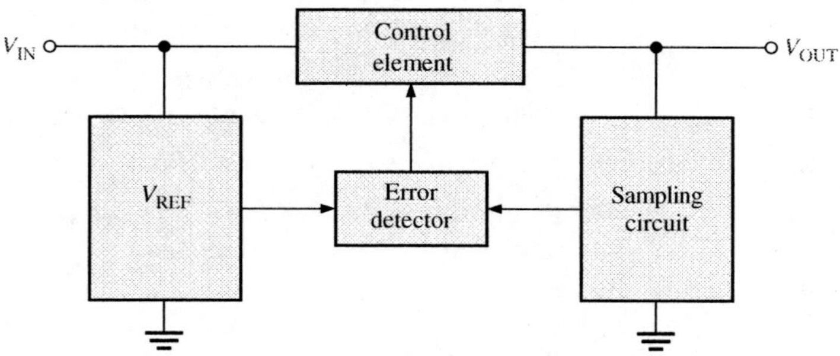

FIGURE ANS–15

7. 14.2 V
9. 4.8 V
11. The diode becomes forward biased when Q_1 turns off.
13. 14.3 V
15. 1.25 mA
17. +1.18 mA
19. $R_1 = 625\ \Omega$ (nearest standard value is 620 Ω)
    $R_2 = 5.33\ k\Omega$ (nearest standard value is 5.6 kΩ)
21. 2.1 W

Chapter 12

1. 0.25 %
3. $C = -62.2°C$; $K = 211$ K
5. $F = 698.0°F$; $K = 643.2$ K
7. $C = -73.2°C$; $F = -99.8°F$
9. (a) 5.0 $\mu\in$
   (b) 3.5 mΩ
11. 44.4 lbs.
13. See Figure ANS–16.
15. (a) 35.4 pF
    (b) 88.5 pF
17. 0.412 mV/V
19. Green
21. Reverse the comparator inputs.

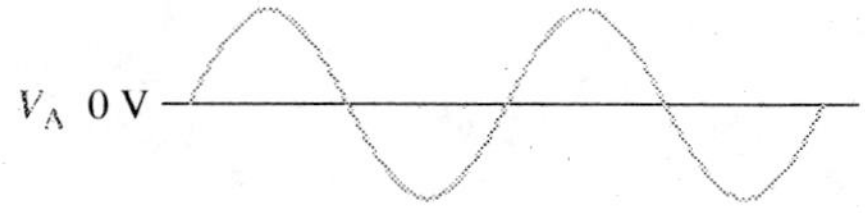

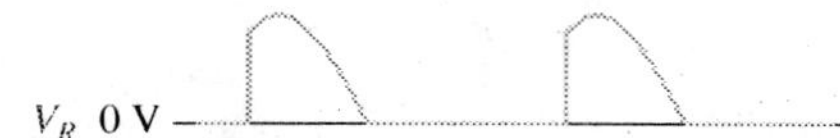

FIGURE ANS–16

GLOSSARY

ac beta (β_{ac}) The ratio of a change in collector current to a corresponding change in base current in a bipolar junction transistor.

Accuracy The difference between the measured value and the accepted value of a measurement.

ac resistance The ratio of a small change in voltage divided by a corresponding change in current for a given device; also called *dynamic, small-signal,* or *bulk resistance.*

Active filter A frequency-selective circuit consisting of active devices such as transistors or op-amps coupled with reactive components.

Active transducer A type of transducer whose output power is derived from a source other than the quantity being measured.

A/D conversion A process whereby information in analog form is converted into digital form.

Amplification The process of producing a larger voltage, current, or power using a smaller input signal as a "pattern."

Amplifier An electronic circuit having the capability of amplification and designed specifically for that purpose.

Analog signal A signal that can take on a continuous range of values within certain limits.

Analog switch A type of semiconductor switch that connects an analog signal from input to output with a control input.

Analog-to-digital converter (ADC) A device used to convert an analog signal to a sequence of digital codes.

Anode (semiconductor diode definition) The terminal of a semiconductor diode that is more positive with respect to the other terminal when it is biased in the forward direction.

Astable Characterized by having no stable states; a type of oscillator.

Astable multivibrator A type of circuit that can operate as an oscillator and produces a pulse waveform output.

Attenuation The reduction in the level of power, current, or voltage.

Audio Related to the range of frequencies that can be heard by the human ear and generally considered to be in the 20 Hz to 20 kHz range.

Band-pass filter A type of filter that passes a range of frequencies lying between a certain lower frequency and a certain higher frequency.

Band-stop filter A type of filter that blocks or rejects a range of frequencies lying between a certain lower frequency and a certain higher frequency.

Bandwidth A measure of a filter's passband; the difference between the upper and lower cutoff (critical) frequencies of the passband.

Barrier potential The inherent voltage across the depletion region of a *pn* junction.

Base One of the semiconductor regions in a BJT.

Base bias A form of bias in which a single resistor is connected between a BJT's base and V_{CC}.

Bessel A type of filter response having a linear phase characteristic and less than −20 dB/decade/pole roll-off.

Bias The application of dc voltage to a diode or other electronic device to produce a desired mode of operation.

Bipolar Characterized by two *pn* junctions.

Bipolar junction transistor (BJT) A transistor constructed with three doped semiconductor regions separated by two *pn* junctions.

Butterworth A type of filter response characterized by flatness in the passband and a −20 dB/decade/pole roll-off.

Bypass capacitor A capacitor connected in parallel with a resistor to provide the ac signal with a low impedance path.

Calibration The process of comparing a measurement device or instrument to a standard of known accuracy.

Cathode (semiconductor diode definition) The terminal of a diode that is more negative with respect to the other terminal when it is biased in the forward direction.

Center tap A connection at the midpoint of the secondary of a transformer.

Characteristic curve A plot which shows the relationship between two variable properties of a device. For most electronic devices, a characteristic curve refers to a plot of the current, *I,* plotted as a function of voltage, *V.*

Chebyshev A type of filter response characterized by ripples in the passband and a greater than −20 dB/decade/pole roll-off.

Clamper A circuit that adds a dc level to an ac signal; also called a *dc restorer.*

Class A An amplifier that operates in the active region at all times.

Class AB An amplifier that is biased into slight conduction. The Q-point is slightly above cutoff.

Class B An amplifier that has the Q-point located at cutoff, causing the output current to vary only during one-half of the input cycle.

Closed-loop An op-amp configuration in which the output is connected back to the input through a feedback circuit.

Closed-loop voltage gain The net voltage gain of an amplifier when negative feedback is included.

Cold junction A reference thermocouple held at a fixed temperature and used for compensation in thermocouple circuits.

Collector One of the semiconductor regions in a BJT.

Common-base (CB) A BJT amplifier configuration in which the base is the common terminal to an ac signal.

Common-collector (CC) A BJT amplifier configuration in which the collector is the common terminal to an ac signal.

Common-drain (CD) A FET amplifier configuration in which the drain is the ac ground terminal.

Common-emitter (CE) A BJT amplifier configuration in which the emitter is the common terminal to an ac signal.

Common-gate (CG) A FET amplifier configuration in which the gate is the ac ground terminal.

Common-mode The input condition where two identical signals are applied to the inputs of a differential amplifier.

Common-mode input resistance The ac resistance between each input and ground.

Common-mode input voltage range The range of input voltage, which when applied to both inputs, will not cause clipping or other output distortion.

Common-mode rejection ratio (CMRR) A measure of the ability of an amplifier to reject common-mode signals; it is the ratio of the differential gain to the common-mode gain.

Common-source (CS) A FET amplifier configuration in which the source is the ac ground terminal.

Comparator A circuit which compares two input voltages and produces an output in either of two states indicating the greater than or less than relationship of the inputs.

Complementary symmetry transistors These are a matching pair of *npn/pnp* BJTs or a matching pair of *n*-channel/*p*-channel FETs.

Conduction electron An electron that has broken away from the valance band of the parent atom and is free to move from atom to atom within the atomic structure of a material: also called a *free electron*.

Constant-current region The region on the drain characteristic of a FET in which the drain current is independent of the drain-to-source voltage.

Coupling capacitor A capacitor connected in series with the ac signal and used to block dc voltages.

Covalent bond A type of chemical bond in which atoms share electron pairs.

Critical frequency The frequency that defines the end of the passband of a filter; also called *cutoff frequency.*

Crossover distortion Distortion in the output of a class B push-pull amplifier at the point where each transistor changes from the cutoff state to the on state.

Crystal A solid in which the particles form a regular, repeating pattern.

Current mirror A circuit that uses matching diode junctions to form a current source. The current in a diode junction is reflected as a matching current in the other junction (which is typically the base-emitter junction of a transistor). Current mirrors are commonly used to bias a push-pull amplifier.

Cutoff The nonconducting state of a transistor.

Cycle The complete sequence of values that a waveform exhibits before another identical pattern occurs.

D/A conversion The process of converting a sequence of digital codes to an analog form.

Damping factor (DF) A filter characteristic that determines the type of response.

dc beta (β_{DC}) The ratio of collector current to base current in a bipolar junction transistor.

Darlington pair A configuration of two transistors in which the collectors are connected and the emitter of the first drives the base of the second to achieve beta multiplication.

Decibel A dimensionless quantity that is 10 times the logarithm of a power ratio or 20 times the logarithm of a voltage ratio.

Decoupling network A low-pass filter that provides a low-impedance path to ground for high-frequency signals.

Depletion mode A class of FETs that is on with zero-gate voltage and is turned off by increasing gate voltage, which has a polarity such that it decreases channel conductivity. All JFETs and some MOSFETS are depletion-mode devices.

Depletion region The area near a *pn* junction on both sides that has no majority carriers.

Differential amplifier (diff-amp) An amplifier that produces an output voltage proportional to the difference of the two input voltages.

Differential input resistance The total resistance between the inverting and the noninverting inputs.

Differential-mode The input condition where two opposite-polarity signals are applied to the inputs of a differential amplifier.

Differentiator A circuit that produces an inverted output which approximates the rate of change of the input function.

Digital signal A noncontinuous signal that has discrete numerical values assigned to the specific steps.

Digital-to-analog converter (DAC) A device in which information in digital form is converted to an analog form.

Diode An electronic device that permits current in only one direction.

Discrete device An individual electrical or electronic component that must be used in combination with other components to form a complete functional circuit.

Domain The values assigned to the independent variable. Frequency or time is typically used as the independent variable for plotting signals.

Doping The process of imparting impurities to an intrinsic semiconductive material in order to control its conduction characteristics.

Drain One of the three terminals of a field-effect transistor; it is one end of the channel.

Dynamic behavior A transducer parameter that specifies how the transducer responds to a changing input.

Dynamic emitter resistance (r_e') The ac resistance of the emitter; it is determined by the dc emitter current.

Efficiency (power) The ratio of the signal power supplied to the load to the power from the dc supply.

Elasticity The property of a material to recover to its original size and shape after a deforming force has been removed.

Elastic limit The point where permanent deformation results when force is applied to an elastic body.

Electromagnetic spectrum The whole range of frequencies that comprise electromagnetic radiation.

Electron The basic particle of negative electrical charge in matter.

Electrostatic discharge (ESD) The discharge of a high voltage through an insulating path that frequently destroys a device.

Emitter One of the three semiconductor regions in a BJT.

Emitter bias A very stable form of bias requiring two power supplies. The emitter is connected through a resistor to one supply; another resistor is connected between a BJT's base and ground.

Emitter-follower Another name for a common-collector amplifier.

Energy The ability to do work.

Enhancement mode A MOSFET in which the channel is formed (or enhanced) by the application of a gate voltage, thus increasing channel conductivity.

Excitation An outside source of energy to operate a transducer.

Feedback oscillator A type of oscillator that returns a fraction of output signal to the input with no net phase shift around the feedback loop, resulting in a reinforcement of the output signal.

Field-effect transistor (FET) A voltage-controlled device in which the voltage at the gate terminal controls the amount of current through the device.

Filter A type of electrical circuit that passes certain frequencies and rejects all others.

Flash A method of A/D conversion.

Forward bias The condition in which a *pn* junction conducts current.

Frequency The number of repetitions per unit of time for a periodic waveform.

Full-wave rectifier A circuit that converts an alternating sine wave into a pulsating dc voltage consisting of both halves of a sine wave for each input cycle.

Gain The amount of amplification. Gain is a ratio of an output quantity to an input quantity (e.g., voltage gain is the ratio of the output voltage to the input voltage).

Gain-bandwidth product A constant which is the product of the closed-loop gain and the closed-loop critical frequency; the frequency at which the op-amp's open-loop gain is unity (1).

Gate One of the three terminals of a field-effect transistor. A voltage applied to the gate controls drain current.

Gauge factor A dimensionless number that is the constant of proportionality that represents the fractional change in resistance for a given strain.

Germanium A semiconductive material.

Half-wave rectifier A circuit that converts an alternating sine wave into a pulsating dc voltage consisting of one-half of a sine wave for each input cycle.

Harmonics Higher-frequency sinusoidal waves that are integer multiples of a fundamental frequency.

Heat A measure of the total internal energy of a body, measured in joules.

High-pass filter A type of filter that passes frequencies above a certain frequency while rejecting lower frequencies.

Hole A mobile vacancy in the electronic valence structure of a semiconductor. A hole acts like a positively charged particle.

Hysteresis The property that permits a circuit to switch from one state to the other at one voltage level and switch back to the original state at another lower voltage level.

Hysteresis error The maximum difference between consecutive measurements for the same quantity when the measured point is approached each time from a different direction for the full range of the transducer.

Input bias current The average dc current required by the inputs of an op-amp to properly operate the device.

Input offset voltage (V_{OS}) The differential dc voltage required between the op-amp inputs to force the differential output to zero volts.

Instrumentation amplifier A differential voltage-gain device that amplifies the difference between the voltage existing at its two input terminals.

Intregrated circuit (IC) A type of circuit in which all the components are constructed on a single chip of silicon.

Integrator A circuit that produces an inverted output which approximates the area under the curve of the input function.

Intermediate frequency A fixed frequency that is lower than the RF, produced by beating an RF signal with an oscillator frequency.

Intrinsic (pure) An intrinsic semiconductor is one in which the charge concentration is essentially the same as a pure crystal with relatively few free electrons.

Inverting amplifier An op-amp closed-loop circuit in which the input signal is applied to the inverting input.

Ion An atom or group of atoms that has gained or lost one or more valence electrons, resulting in a net positive or negative charge.

Isolation amplifier A device that provides dc isolation between the input and output.

Junction field-effect transistor (JFET) A type of FET that operates with a reverse-biased *pn* junction to control current in a channel. It is a depletion-mode device.

Large-signal A signal that operates an amplifier over a significant portion of its load line.

Light-emitting diode (LED) A type of diode that emits light when there is forward current.

Limiter A circuit that removes part of a waveform above or below a specified level; also called a *clipper.*

Linear component A component in which an increase in current is proportional to the applied voltage.

Linear regulator A voltage regulator in which the control element operates in the linear region.

Line regulation The change in output voltage for a given change in line (input) voltage, normally expressed as a percentage.

Load cell A metal body that is deformed within its elastic range by an applied force. The metal body is instrumented with strain gauges.

Loading effect A change in circuit parameters when a load is connected.

Load line A straight line plotted on a current versus voltage plot that represents all possible operating points for an external circuit.

Load regulation The change in output voltage for a given change in load current, normally expressed as a percentage.

Logarithm An exponent; the logarithm of a quantity is the exponent or power to which a given number called the base must be raised in order to equal the quantity.

Low-pass filter A type of filter that passes frequencies below a certain frequency while rejecting higher frequencies.

Microcontroller A specialized microprocessor designed for control functions.

Mixer A nonlinear circuit that combines two signals and produces the sum and difference frequencies; a device for down-converting frequencies in a receiver system.

Monostable Characterized by having one stable state.

MOSFET Metal-oxide semiconductor field-effect transistor; one of two major types of FET. It uses a SiO_2 layer to insulate the gate lead from the channel. MOSFETs can be either depletion mode or enhancement mode.

Negative feedback The process of returning a portion of the output back to the input in a manner to cancel a fraction of the input.

Negative ion An atom or group of atoms that has a negative charge because it has gained one or more electrons.

Noise An unwanted voltage or current fluctuation.

Noninverting amplifier An op-amp closed-loop circuit in which the input signal is applied to the noninverting input.

Ohmic region The region on the drain characteristic of a FET with low values of V_{DS} in which the channel resistance can be changed by the gate voltage; in this region the FET can be operated as a voltage-controlled resistor.

One-shot A monostable multivibrator that produces a single output pulse for each input trigger pulse.

Open-loop A condition in which an op-amp has no feedback.

Open-loop voltage gain The internal voltage gain of an amplifier without external feedback.

Operational amplifier (op-amp) An electronic device that amplifies the difference voltage between the two inputs. It has very high voltage gain, very high input resistance, very low output resistance, and good rejection of common-mode signals.

Order The number of poles in a filter.

Oscillator An electronic circuit that operates with positive feedback and produces a time-varying output signal without an external input signal.

Output resistance The ac resistance viewed from the output terminal of an op-amp.

Passband The region of frequencies that are allowed to pass through a filter with minimum attenuation.

Passive transducer A type of transducer that produces an output without any direct interaction with a source of power.

Period (T) The time for one cycle of a repeating wave.

Periodic A waveform that repeats at regular intervals.

Phase angle (in radians) The fraction of a cycle that a waveform is shifted from a reference waveform of the same frequency.

Phase shift The relative angular displacement of a time-varying function relative to a reference.

Phase-shift oscillator A type of sinusoidal feedback oscillator that uses three *RC* circuits in the feedback loop.

Photodiode A diode whose reverse resistance changes with incident light.

Pinch-off voltage The value of the drain-to-source voltage of a FET at which the drain current becomes constant when the gate-to-source voltage is zero.

***PN* junction** The boundary between *n*-type and *p*-type materials.

Pole A network containing one resistor and one capacitor that contributes −20 dB/decade to a filter's roll-off rate.

Positive feedback A condition where an in-phase portion of the output voltage is fed back to the input.

Positive ion An atom or group of atoms that has a positive charge because it has lost one or more electrons.

Power gain The ratio of the power delivered to the load to the input power of an amplifier.

Power supply A device that converts ac or dc voltage into a voltage or current suitable for use in various applications to power electronic equipment. The most common form is to convert ac from the utility line to a constant dc voltage.

Pressure The force per unit area; the area is measured perpendicular to the force.

Push-pull A type of class B amplifier with two transistors in which one transistor conducts for one half-cycle and the other conducts for the other half-cycle.

Pyrometer A noncontacting temperature sensor that detects infrared radiation from a source and converts the radiation to a voltage or current that is proportional to temperature.

Quality factor (Q) A dimensionless number that is the ratio of the maximum energy stored in a cycle to the energy lost in a cycle. The ratio of a band-pass filter's center frequency to its bandwidth.

Quantizing The process of assigning numbers to sampled data.

Quiescent point The point on a load line that represents the current and voltage conditions for a circuit with no signal (also called operating or Q-point). It is the intersection of a device characteristic curve with a load line.

Radio frequency (RF) Any frequency greater than 100 Hz.

Radio frequency interference (RFI) High frequencies produced when high values of current and voltage are rapidly switched on and off.

Range The set of values a transducer is designed to measure.

Recombination The process of a free electron in the conduction band falling into a hole in the valence band of an atom.

Rectifier An electronic-circuit that converts ac into pulsating dc.

Regulator An electronic circuit that is connected to the output of a rectifier and maintains an essentially constant output voltage despite changes in the input, the load current, or the temperature.

Relaxation oscillator A type of oscillator that uses an *RC* timing circuit to generate a nonsinusoidal waveform.

Repeatability A measure of how well a transducer responds to the same input multiple times.

Resistance temperature detector (RTD) A type of temperature transducer in which resistance is directly proportional to temperature.

Resolution The magnitude of the smallest detectable change in a quantity that is being measured.

Reverse bias The condition in which a *pn* junction blocks current.

Ripple voltage The variation in the dc voltage on the output of a filtered rectifier caused by the slight charging and discharging action of the filter capacitor.

Roll-off The rate of decrease in gain, below or above the critical frequencies of a filter.

Root mean square (RMS) The value of an ac voltage that corresponds to a dc voltage that produces the same heating effect in a resistance.

Saturation The state of a BJT in which the collector current has reached a maximum and is independent of the base current.

Schmitt trigger A comparator with hysteresis.

SCR Silicon-controlled rectifier; a type of three-terminal thyristor.

Semiconductor A material that has a conductance value between that of a conductor and that of an insulator. Silicon and germanium are examples.

Shell An energy level in which electrons orbit the nucleus of an atom.

Signal In electronics, an electrical voltage or current that contains information.

Silicon A semiconductive material used in diodes and transistors.

Single-ended mode The input condition of an op-amp in which one input is grounded and the signal voltage is applied only to the other input.

Sinusoidal wave A voltage or current waveform that has the same shape as the mathematical trig function called the sine wave.

Slew rate The rate of change of the output voltage of an op-amp in response to a step input.

Source One of the three terminals of a field-effect transistor; it is one end of the channel.

Spectrum A plot of amplitude versus frequency for a signal.

Stage Each functional part in a multistage amplifier that amplifies a signal.

Strain The elastic deformation of a body due to an applied force; it is the ratio of a change in length divided by the length of an object.

Strain gauge A thin conductor made in a back and forth pattern. It responds to force by changing its resistance as it is stretched or compressed.

Stress The force per area in a solid body; it can be either due to a tension force or a compression force.

Summing amplifier A variation of a basic comparator circuit that is characterized by two or more inputs and an output voltage that is proportional to the magnitude of the algebraic sum of the input voltages.

Switch An electrical or electronic device for opening and closing a current path.

Switching regulator A voltage regulator in which the control element is a switching device.

Temperature A measure of how vigorously the atoms of a substance are moving and colliding with other atoms.

Terminal An external contact point on an electronic device.

Thermistor A sensitive-resistive temperature sensor that is made of a sintered mixture of transition metal oxides; the resistance is inversely proportional to the temperature.

Thermocouple A temperature transducer that produces a current proportional to the difference in temperature between a measuring junction and a reference junction.

Thermocouple junction The junction between two metals.

Thevinen's theorem An equivalent circuit that replaces a complicated two-terminal linear network with a single voltage source and a series resistance.

Threshold The smallest detectable value of the measured quantity starting near the zero value.

Thyristor A class of four-layer (*pnpn*) semiconductor devices, such as the silicon-controlled rectifier.

Transconductance The ratio of output current to input voltage; the gain of a FET; it is determined by a small change in drain current divided by a corresponding change in gate-to-source voltage. Is is measured in siemens.

Transducer A device that converts energy from one form to another. For electronic systems, the output is an electrical parameter.

Transfer curve A plot of the output of a circuit or system for a given input.

Transistor A semiconductor device used for amplification and switching applications in electronic circuits.

Triac A three-terminal thyristor that can conduct current in either direction when properly activated.

Valence electron An electron in the outermost shell or orbit of an atom.

Vector Any quantity that has both magnitude and direction.

Voltage-controlled oscillator A type of relaxation oscillator whose frequency can be changed by a variable dc voltage; also known as a VCO.

Voltage-divider bias A very stable form of bias in which a voltage divider is connected between V_{CC} and ground; the output of the divider supplies bias current to the base of a BJT.

Voltage-follower A closed-loop, noninverting op-amp circuit with a voltage gain of one.

Voltage gain The ratio of the output voltage to input voltage.

Voltage regulation The process of maintaining an essentially constant output voltage over variations in input voltage or load.

Voltage-to-current converter A circuit that converts a variable input voltage to a proportional output current.

Wien-bridge oscillator A type of sinusoidal feedback oscillator that uses an *RC* lead-lag network in the feedback loop.

Zener diode A type of diode that operates in reverse breakdown (called zener breakdown) to provide voltage regulation.

Zero-voltage switching The process of switching power to a load at the zero crossings of an ac voltage to minimize RF noise generation.

Young's modulus A constant for an elastic material that depends on the geometry and material. It is the ratio of stress divided by strain; it is specified differently for stretching and bending.

INDEX

INDEX